TM 508 JAN

MAIN LIBRARY
QUEEN MARY, UNIVERSITY OF LONDON
Mile End Road, London E1 4NS
DATE DUE FOR RETURN.

TM 508 JAN

Advanced Polymer Composites

Bor Z. Jang

Acquisitions Editor

Veronica Flint

Production Coordinator

Randall L. Boring

**The Materials
Information Society**

First printing, August 1994

This book is a collective effort involving hundreds of technical specialists. It brings together a wealth of information from world-wide sources to help scientists, engineers, and technicians solve current and long-range problems.

Great care is taken in the compilation and production of this Volume, but it should be made clear that NO WARRANTIES, EXPRESS OR IMPLIED, INCLUDING, WITHOUT LIMITATION, WARRANTIES OF MER-CHANTABILITY OR FITNESS FOR A PARTICULAR PURPOSE, ARE GIVEN IN CONNECTION WITH THIS PUBLICATION. Although this information is believed to be accurate by ASM, ASM cannot guarantee that favorable results will be obtained from the use of this publication alone. This publication is intended for use by persons having technical skill, at their sole discretion and risk. Since the conditions of product or material use are outside of ASM's control, ASM assumes no liability or obligation in connection with any use of this information. No claim of any kind, whether as to products or information in this publication, and whether or not based on negligence, shall be greater in amount than the purchase price of this product or publication in respect of which damages are claimed. THE REMEDY HEREBY PROVIDED SHALL BE THE EXCLUSIVE AND SOLE REMEDY OF BUYER, AND IN NO EVENT SHALL EITHER PARTY BE LIABLE FOR SPECIAL, INDIRECT OR CONSEQUENTIAL DAMAGES WHETHER OR NOT CAUSED BY OR RESULTING FROM THE NEGLIGENCE OF SUCH PARTY. As with any material, evaluation of the material under enduse conditions prior to specification is essential. Therefore, specific testing under actual conditions is recommended.

Nothing contained in this book shall be construed as a grant of any right of manufacture, sale, use, or reproduction, in connection with any method, process, apparatus, product, composition, or system, whether or not covered by letters patent, copyright, or trademark, and nothing contained in this book shall be construed as a defense against any alleged infringement of letters patent, copyright, or trademark, or as a defense against liability for such infringement.

Comments, criticisms, and suggestions are invited, and should be forwarded to ASM International.

Library of Congress Cataloging-in-Publication Data

Jang, Bor Z.
Advanced polymer composites / Bor Z. Jang
  p.   cm.
Includes index.
ISBN 0-87170-491-9
1. Polymeric composites.
I. ASM International.
II. Title
TA418.9.C6J35   1994
620.1'92—dc20   94-18572
                                     CIP

SAN: 204-7586

ASM International®
Materials Park, OH 44073-0002

Printed in the United States of America

Dedication

To

Mei Shio
Justin
Susan
Kevin

for their understanding and love

Acknowledgments

This book was written while I was a visiting scholar with the Department of Materials Science and Metallurgy at the University of Cambridge. Dr. Julia King was my host during this wonderful time. This trip was made possible by grants from the U.K. Fulbright Commissions, Exxon Educational Foundation, and Churchill College. It gave me great pleasure to be an Overseas Fellow Commoner with Churchill College of Cambridge, where I had the opportunity to interact with many distinguished scholars. Thanks also go to my sister Michelle and my wife MeiShio for typing the manuscript. Dr. S.R. Krishna of Cambridge University, Dr. S.F. Yang of Auburn University, Dr. Doug Hirt of Clemson University, and Mr. John Dispennette patiently read the manuscript and made many constructive suggestions. My former students Dr. Z.W. Chiou and Dr. J.J. Tai put forward tremendous effort in helping me draw most of the figures used in this book.

Table of Contents

Part I

Materials-Dominated Issues for Composites

Chapter 1

Introduction

A fiber-reinforced composite is not simply a mass of fibers dispersed within a polymer, metal, or ceramic matrix. A composite consists of fibers embedded in or bonded to a matrix with distinct interfaces (or interphases) between the two constituent phases. The fibers are usually of high strength and modulus and serve as the principal load-carrying members. The matrix must keep the fibers in a desired location and orientation, separating fibers from each other to avoid mutual abrasion during periodic straining of the composite. The matrix acts as a load transfer medium between fibers, and in less ideal cases where loads are complex, the matrix may even have to bear loads transverse to the fiber axis. Since the matrix is generally more ductile than the fibers, it is the source of composite toughness. The matrix also serves to protect the fibers from environmental damage before, during, and after composite processing. In a composite, both fibers and the matrix largely retain their identities and yet result in many properties that cannot be achieved with either of the constituents acting alone.

A wide variety of fibers are available for use in composites. The most commonly used fibers in polymer matrices are various types of carbon, glass, and aramid (e.g., Kevlar®) fibers. Boron fibers are expensive and are used currently in military and aerospace applications only. Also still in limited use are alumina, silicon carbide, mullite, silicon nitride and other ceramic fibers and metal wires. Many of these fibers can be fabricated with a wide range of properties. The unique combinations of properties available in these fibers provide the outstanding structural characteristics of fiber-reinforced composites. All these fibers can be incorporated into a matrix either in continuous lengths or in discontinuous lengths (chopped fibers or whiskers). The key features of low fiber density and high strengths and moduli give rise to high specific strength (strength/density) and specific stiffness (stiffness/density) properties of composites. The matrix material may be a thermoplastic or thermoset polymer, a metal (for example, aluminum, titanium, and superalloys), or a ceramic (including carbon, glass, and intermetallics). A wide spectrum of chemical compositions and microstructural arrangements are possible in each matrix category.

Types of Fiber Composites

A key feature of fiber composites that makes them so promising as engineering materials is the opportunity to tailor the materials properties through the control of fiber and matrix combinations and the selection of processing techniques. Matrix materials and fabrication processes are available that do not significantly degrade the intrinsic properties of the fiber. In principle, an infinite range of composite types exists, from randomly oriented chopped fiber based materials at the low-property end to continuous, unidirectional fiber composites at the high-performance end. Figure 1 gives several examples of the possible composite reinforcement forms. Composites can differ in the amount of fiber, fiber type, fiber length, fiber orientation, and possibly fiber hybridization. In general, short-fiber composites are used in lightly loaded or secondary structural applications, while continuous fiber-reinforced composites are utilized in primary structural applications and are considered high-performance structural materials.

By nature, continuous-fiber composites are highly anisotropic. Maximum properties can be achieved if all the fibers are aligned in the fiber-axis direction. The properties, such as modulus and strength, decrease rapidly in directions away from the fiber direction. To obtain more orthotropic properties, alternate layers of fibers may vary between 0 and 90°, resulting in less directionality, but at the expense of absolute properties in the fiber direction. A laminate is fabricated by stacking a number of thin layers of fibers and matrix, consolidating them into the desired thickness. A laminate is the most common form of composites for structural applications. The fiber

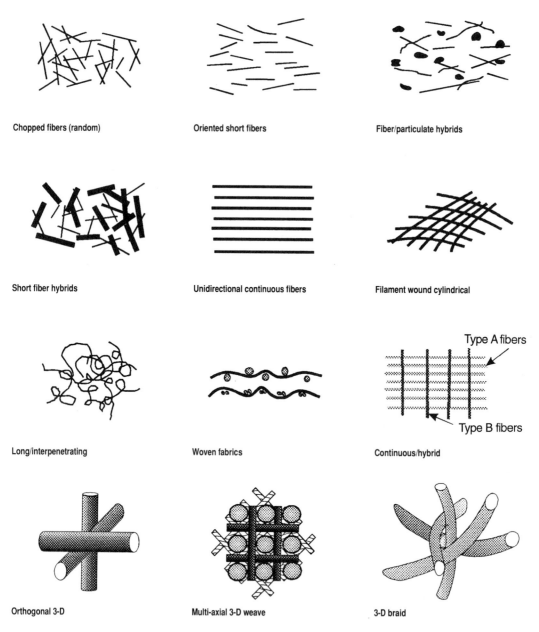

Chopped fibers (random)

Oriented short fibers

Fiber/particulate hybrids

Short fiber hybrids

Unidirectional continuous fibers

Filament wound cylindrical

Long/interpenetrating

Woven fabrics

Continuous/hybrid

Type A fibers

Type B fibers

Orthogonal 3-D

Multi-axial 3-D weave

3-D braid

Fig. 1 Examples of reinforcement styles, combinations, orientations, and configurations (composite types)

orientation in each layer as well as the stacking sequence of various layers can be manipulated to produce a wide scope of physical and mechanical properties.

For cylindrical structures such as pressure vessels, filament winding is the best process. Pultrusion is particularly suitable for producing long structural elements with constant cross-sections from composites. Laminated, filament-wound, and pultruded parts represent the three most common forms in which continuous fiber-reinforced polymer composites are fabricated for structural usage. Nonetheless, textile structural composites produced by 3-D weaving, braiding, knitting, etc. are under consideration for structural composites, particularly when interlaminar shear strength, damage tolerance, and thickness-direction properties of composites are important design considerations. Stiffness, strength, impact resistance, and damage tolerance of this new class of composites are discussed in various chapters of this book.

Table 1   Properties of selected conventional materials and composites

Material	Density (ρ), g/cm³	Tensile modulus (E), GPa	Tensile strength, σ_t MPa	E/ρ, 10^6 N · m/kg	σ_t/ρ, 10^3 N · m/kg
Al 6061-T6	2.70	68.9	310	25.7	115
SAE 1010 steel (cold-worked)	7.87	207	365	26.3	46.4
Ti-6Al-4V	4.43	110	1171	25.3	26.4
Nylon 6/6	1.14	2	7.0	1.75	61.4
Unidirectional high-strength carbon fiber/epoxy	1.55	137.8	1550	88.9	1000
Unidirectional E-glass fiber/epoxy	1.85	39.3	965	21.20	522
Unidirectional Kevlar®-49 fiber/epoxy	1.38	75.8	1378	54.9	999
Quasi-isotropic carbon fiber/epoxy	1.55	45.5	579	29.3	374
Random glass fiber/epoxy	1.55	8.5	110	5.48	71

Composites Versus Traditional Materials

A judicious selection of fiber, matrix, and interface conditions can lead to a composite with a combination of strength and modulus comparable to or better than those of many conventional metallic materials. Composites are superior to metals in specific strength and specific stiffness (Table 1). For a given design load, the weight of the component is lower if manufactured with a composite than with a metal. Weight reduction is a key consideration in many industries, notably automotive and aerospace. A lighter vehicle because of higher specific properties could mean better fuel efficiency.

Fatigue strength as well as fatigue damage tolerance of many composites are quite remarkable. Polymer-matrix composites normally exhibit higher corrosion resistance than metals. However, many polymer-matrix composites tend to absorb moisture from the surrounding environment, resulting in dimensional changes as well as adverse internal stresses within the material. Thermo-oxidative degradation of the polymer matrix can occur at elevated temperatures. In metal-matrix composites, oxidation of the matrix as well as undesirable reaction between fibers and matrix may present a problem for high-temperature applications. The anisotropic nature of a fiber composite adds a considerable degree of complexity to the design of a component. However, this same nature provides the unique opportunity of tailoring the properties according to the design requirements. One can reinforce a structure selectively in the directions of major stresses, for example. Coefficients of thermal expansion (CTE) for many fiber composites are much lower than those for metals. As a consequence, composite structures may exhibit better dimensional stability if a variation in service temperature is involved. The anisotropic nature of fiber composites may allow the production of structures with CTEs of practically zero.

Although yielding and plastic deformation are quite common to structural metals, most fiber composites are practically elastic in their tensile response. Nevertheless, the heterogeneous nature of these composites provides several energy-dissipation mechanisms on a microscopic scale comparable to the yielding process. These include matrix deformation and microcracking, fiber deformation and rupture, interfacial debonding, and fiber pull-out. These microfailure processes allow a composite laminate to exhibit a more gradual deterioration rather than a catastrophic failure. High internal damping in composites also makes these materials viable candidates for structural elements that require reduced noise and vibration.

Metal-matrix composites (MMCs) have additional advantages over monolithic metals including better wear resistance, better elevated-temperature properties, and better creep resistance. MMCs also have advantages over polymer-matrix composites, such as higher functional temperatures, higher transverse strength and stiffness, no moisture absorption, better conductivities, better electromagnetic interference (EMI) resistance, and better radiation resistance. However, MMCs suffer from high costs, new technologies, complex fabrication methods, and limited service experience. Polymer-matrix composites (PMCs) are more advanced in the fabrication technology and are lower in raw material cost and fabrication cost. When compared to MMCs, PMCs are lower in density.

Composite Applications

The types of composites and composite design technologies adopted by different sectors of industry can be quite specific to the particular requirements and practice of that particular sector. Since weight reduction in a structural design is critical to the aerospace industry and usually low-volume production is involved, more expensive fibers and resins, long fabrication time, and less automated processing techniques (e.g., hand lay-up) can be tolerated. However, in consumer-oriented industries (automotive and sporting goods, for example), high volume and high production rates are normally required. Automated fabrication, short processing time, and minimization of cost are vital to the success of these industries.

Potential advantages that can be gained by replacing metals with composites in a structural design include:

- Fewer components involved because complex shapes can be manufactured in single molding operations
- Freedom of shape because of the manufacturing processes available

Table 2 Applications of fiber composites

Industry	Examples	Comments
Aircraft	Doors, skin on the stabilizer box fin, elevators, rudders, landing gear, fuselage, tail spoiler, flap, body, etc.	• Usually result in 20 to 35% weight-savings over metal parts
Aerospace	Space shuttle, space station	• Great weight savings • Dimensional stability and low CTE. Carbon/carbon composites for thermal stability
Automotive	Body frame, chassis components, engine components, drive shaft, exterior body components, leaf springs, etc.	• High stiffness and damage tolerance • Good surface finish for appearance • Lower weight and higher fuel efficiency
Marine	Hulls and masts for recreational boats, submersibles, spars, decks, and bulkheads, etc.	• Weight reduction results in higher boat cruising speed and distance, fast acceleration, maneuverability, and fuel economy
Sporting goods	Tennis and racquetball racquets, golf club shafts and heads, bicycle frames, skis, canoes, helmets, fishing poles, tent poles, bobsleds and bobsled track, race cars, pole vaulting poles, etc.	• Weight reduction • Vibration damping • Design flexibility
Chemical	Pipes, tanks, pressure vessels	• Weight savings • Corrosion resistance
Construction	Structural and decorative panels, fuel tanks, portable bridges, housings, furniture, etc.	• Weight savings • Portable
Electrical	Printed circuit boards, computer housing, insulators, radomes, battery plates, smart structures	• High specific properties leading to weight savings

• Inexpensive prototypes because trial fabrication can be done with cheap machine tools.

Composite materials are used when the total product cost for a given performance is favorable. A need will always exist to develop new cost-effective manufacturing processes closely linked to material design. Although ample evidence can be found of the growing confidence in composite materials and of increasing penetration in many technologies, the application base of composites cannot be broadened without a significant improvement in manufacturing reliability and reproducibility. This issue is addressed later in this volume.

Selected examples for the applications of fiber composites in various industries are given in Table 2, illustrating the aggressive penetration of composites into various segments of technology. For further information on the definition, classification, applications and select engineering properties of composites, the readers may consult the handbooks and textbooks in the "Selected References" list.

Selected References

• D.J. DeRenzo, *Advanced Composite Materials: Products and Manufacturers*, Noyes Data Corp., Park Ridge, NJ, 1988

• I. K. Partridge, Ed., *Advanced Composites*, Elsevier, London, 1989

• J.A. Lee and D.L. Mykkanen, *Advanced Materials Research Status and Requirements*, Vol 1, *Technical Summary*, BDM/H-86-0115-TR, BDM Corp., 1986

• J.A. Lee and D.L. Mykkanen, *Advanced Materials Research Status and Requirements*, Vol 2:, Materials Properties Data Review, BDM/H-86-0115-TR, BDM Corp., 1986

• B. Agarwal and L. Broutman, *Analysis and Performance of Fiber Composites*, John Wiley and Sons, 1980

• M.U. Islam, W. Wallace, and A.Y. Kandeil, *Artificial Composites for High-Temperature Applications*, Noyes Data Corp., Park Ridge, NJ, 1988

• *ASM Engineered Materials Reference Book*, ASM International, 1989

• B.C. Hoskin and A.A. Baker, Ed., *Composite Materials for Aircraft Structure*, AIAA Education Series, AIAA, New York, 1986

• M.M. Schwartz, *Composite Materials Handbook*, McGraw-Hill, 1984

• *Composites and Laminates*, D.A.T.A., Inc., San Diego, CA

• *Engineered Materials Handbook, Vol 1, Composites*, ASM International, 1987

• *Engineered Materials Handbook, Vol 2, Engineering Plastics*, ASM International, 1988

• J.W. Weeton, D.M. Peters, and K.L. Thomas, Ed., *Engineers' Guide to Composite Materials*, ASM International, 1987

• P. K. Mallick, *Fiber Reinforced Composites*, Marcel Dekker, 1988

• K. Ashbee, *Fundamental Principles of Fiber Reinforced Composites*, Technomic Publishing, Lancaster, PA, 1989

• A.B. Strong, *Fundamentals of Composites Manufacturing*, SME, Dearborn, MI, 1989

• G. Lubin, Ed., *Handbook of Fiberglass and Advanced Plastics Composites*, Van Nostrand Reinhold, 1982

• D. Hall, *Introduction to Composite Materials*, Cambridge University Press, 1981

• *Reinforced Plastics for Commercial Composites: Source Book*, American Society for Metals, 1986

• J. Delmonte, *Technology of Carbon and Graphite Fiber Composites*, Van Nostrand Reinhold, 1981

Chapter 2

Fibers and Matrix Resins

Advanced polymer composites are composed of polymer matrix materials reinforced with high-strength, high-modulus fibers. This reinforcement may be continuous fibers or filaments, or chopped fibers or whiskers. Advanced polymer composites are being used increasingly in aerospace, automotive, and other industries. However, today's polymer composites have not yet reached their full potential as engineering materials. Researchers need to understand the various structural properties required and whether these properties depend on the fiber, matrix, or interface characteristics. Improvements in fibers and matrices are needed to obtain adequate fiber-matrix interface properties. While composite strength and modulus are primarily a function of the reinforcing fibers, the ability of the matrix to support these fibers is also important. Many properties, including impact resistance and damage tolerance, are critically dependent on the fiber-matrix interfaces. In this chapter, more commonly used reinforcement and matrix materials are introduced, and recent developments are discussed.

Fibers

Advanced reinforcement materials include, but are not limited to, high-strength, high-modulus carbon, boron, glass, silicon carbide, alumina, mullite, aramid, and polyethylene fibers. Fiber reinforcement is available in a wide range of sizes. Diameters of commonly used fibers vary from a few microns to more than one hundred microns, while whiskers can be in the nanometer range. Fiber lengths range from long continuous fibers through chopped short fibers (e.g., a few centimeters long) to sub-micrometer whiskers.

The strength properties of composites, except those that are interlaminar dependent, are mainly determined by the fiber strength. Composite stiffness is also dictated by the fiber stiffness. For many advanced fibers, attempts to improve fiber strength often led to reduced stiffness and vice versa. Optimization of fiber strength and stiffness has been a basic objective of fiber manufacturers. Fiber properties have improved steadily since the first

synthetic organic (nylon) fibers were produced with an elastic modulus around 5 GPa (725 ksi). Nylon and polyester fibers are produced by melt spinning and are drawn to increase their modulus and strength. These fibers alone are usually not utilized in high-performance composites and will not be further discussed in this chapter.

The schematic stress-strain curves of glass, carbon, and aramid fibers are compared in Fig. 1. Additional data comparing the properties of these and other fibers are given in Table 1 [1]. In this table, both high-modulus and high-strength carbon fibers are derived from polyacrylonitrile fibers. Kevlar®-49 is Du Pont's high modulus aramid fiber, and E-glass is the electrical grade glass fiber, the most commonly used grade of glass fibers used in composites. Spectra® fibers are ultrahigh-molecular-weight polyethylene (UHMWPE) fibers produced by Allied Signal using gel spinning. For comparison, data on selected metallic and ceramic fibers, and monolithic ma-

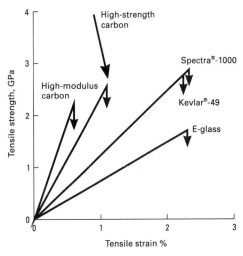

Fig. 1 Representative stress-strain curves of select advanced reinforcement fibers. Source: Ref 3

Table 1 Properties of fibers and conventional bulk materials(a)

Material	Tensile modulus (E), GPa(b)	Tensile strength (σ_u), GPa(b)	Density(c) (ρ), g/cm^3	Specific modulus(d) (E/ρ), $(10^9$ N · m)/kg	Specific strength(d) (σ_u/ρ), $(10^9$ N · m)/kg
Fibers					
E-glass	72.4	3.5(e)	2.54	28.5	1.38
S-glass	85.5	4.6(e)	2.48	34.5	1.85
Type I carbon (high modulus)	390.0	2.1	1.90	205.0	1.1
Type II carbon (high tensile strength)	240.0	2.9	1.77	135.6	1.64
Type III carbon	190.0	2.6	1.76	108.0	1.48
Boron	385.0	2.8	2.63	146.0	1.1
Silicon carbide (on tungsten)	400	3.5	3.50	114.0	1.0
Silica	72.4	5.8	2.19	33.0	2.65
Tungsten	414.0	4.2	19.30	21.0	0.22
Beryllium	240.0	1.3	1.83	131.0	0.71
PBT (heat treated)	331.0	4.2	1.58	209.0	2.65
Kevlar®-49	130.0	2.8	1.45	87.0	1.87
Kevlar®-29	60.0	2.8	1.44	42.0	1.80
Spectra®-900 (PE)	117.0	2.6	0.97	120.0	2.8
Spectra®-1000	172.0	3.0	0.97	180.0	3.2
Conventional materials					
Steel	210.0	0.34 to 2.1	7.8	26.9	0.043 to 0.27
Aluminum alloys	70.0	0.14 to 0.62	2.7	25.9	0.052 to 0.23
Glass	70.0	0.7 to 2.1	2.5	28.0	0.28 to 0.84
Tungsten	350.0	1.1 to 4.1	19.30	18.1	0.057 to 0.21
Beryllium	300.0	0.7	1.38	164.0	0.38

(a) In part from Ref 1. (b) To convert from Pa to psi, multiply by 1.45×10^{-4}. (c) To convert from g/cm^3 to lb/in.^{3}, multiply by 3.61×10^{-2}. (d) To convert from N · m/kg to (lbf · in.)/lb, multiply by 4.015×10^{-3}. (e) Virgin strength values. Actual strength values prior to incorporation into composite are approximately 2.1 GPa (304.6 ksi).

terials, are also given. All the fibers have a high elastic modulus, the modulus of the aramid being intermediate between those of the glass and carbon fibers. Although all the fibers shown are brittle (have low elongation), the carbon fibers are more brittle than the aramid and glass. The densities of the fibers vary from PE, which is the lowest, through Kevlar®, carbon, glass and silicon carbide fiber.

Glass Fibers

The most common reinforcement for polymer matrix composites is glass fiber. The advantages of glass fibers include low cost, high tensile and impact strengths, and high chemical resistance. The disadvantages include relatively low modulus, self-abrasiveness, low fatigue resistance, and poor adhesion to matrix resins. Typical compositions of three glasses used for fiber manufacture, E (electrical), C (chemical), and S (high tensile strength), are given in Table 2 [2]. Glass fibers can be produced in either continuous filament or staple form. The glass is melted and fibers formed by passing the melted glass through small orifices.

The polyhedron network structure of sodium silicate glass is shown in Fig. 2 [3]. Each polyhedron can be seen to be a combination of oxygen atoms around a silicon atom bonded together by covalent bonds. The sodium ions form ionic bonds with charged oxygen atoms and are

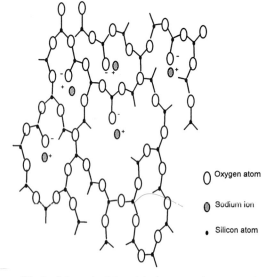

Fig. 2 Schematic of the polyhedron network structure of sodium silicate glass. Source: Ref 3

- ◯ Oxygen atom
- ◉ Sodium ion
- • Silicon atom

not linked directly to the network. The three-dimensional network structure of glass results in isotropic properties

Table 2   Compositions of glass used for fiber manufacture and their basic properties in fiber form

Constituent(a)	E-glass	C-glass	S-glass
SiO_2	55.2	65	65.0
Al_2O_3	14.8	4	25.0
B_2O_3	7.3	5	...
MgO	3.3	3	10.0
CaO	18.7	14	...
Na_2O	0.3	8.5	...
K_2O	0.2	...	...
Fe_2O_3	0.3	0.5	...
F_2	0.3	...	...
Liquidus temperature, °C	1140	...	...
Tensile strength of single fiber at 25 °C, kg/mm^2	370	310(b)	430
Tensile strength of strand, kg/mm^2	175 to 275	160 to 235(b)	210 to 320(b)
Elastic modulus at 25 °C, kg/mm^2	7700	7400(b)	8800
Density, g/cm^3	2.53	2.46(b)	2.45
Refractive index, n_D	1.550	...	...
Coefficient of linear thermal expansion, 10^{-6} °C	5	8(b)	5(b)
Dielectric constant at 25 °C and 10^{10} Hz	6.11	...	...
Loss tangent at 25 °C and 10^{10} Hz	0.006	...	...
Volume resistivity in ohm · cm	10^{15}	...	...

(a) Compositions in wt%. (b) Indicates estimated or extrapolated value

of glass fibers, in contrast to those of carbon and Kevlar® aramid fibers which are anisotropic. The elastic modulus of glass fibers measured along the fiber axis is the same as that measured in the transverse direction, a characteristic unique to glass fibers.

Sizing materials are normally coated on the surface of glass fibers immediately after forming as protection from mechanical damage. For glass fibers intended for weaving, braiding or other textile operations, the sizing usually consists of a mixture of starch and a lubricant, which can be removed from the fiber by burning after the fibers have been processed into a textile structure. For glass reinforcement used in composites, the sizing usually contains a coupling agent to bridge the fiber surface with the resin matrix used in the composite. These coupling agents are usually organosilanes with the structure X_3SiR, although sometimes titanate and other chemical structures are used. The R group may be able to react with a group in the polymer of the matrix; the X groups can hydrolyze in the presence of water to form silanol groups which can condense with the silanol groups on the surface of the glass fibers to form siloxanes. The organosilane coupling agents may greatly increase the bond between the polymer matrix and the glass fiber and are especially effective in protecting glass fiber composites from the attack of water. A more detailed discussion of the adsorption of organosilanes can be found in [4].

Carbon or Graphite Fibers

High-strength, high-modulus carbon fibers are manufactured by treating organic fibers (precursors) with heat and tension, leading to a highly ordered carbon structure. The most commonly used precursors include rayon-base fibers, polyacrylonitrile (PAN), and pitch. The various processes involved with each precursor are

Table 3   Comparative graphite fiber properties

Property	Pitch	Rayon	PAN
Tensile strength, GPa(a)	1.55	2.06 to 2.75	2.5 to 3.2
Tensile modulus, GPa(a)	370	380 to 550	210 to 400
Short beam shear, MPa(a)			
untreated	41	28	28 to 68
treated	68	56	56 to 120
Specific gravity	2.0	1.7	1.8
Elongation, %	1	...	1.2 to 0.6
Fiber diameter, μm	...	6.5	7.5

(a) To convert from Pa to psi, multiply by 1.45×10^{-4}. Source: Ref 4

briefly described in this section. Typical properties of the representative classes of carbon fibers are listed in Table 3. Other unique properties of carbon fiber materials can be found in Table 1.

Rayon carbon filaments are stretched in a series of steps in an inert atmosphere at temperatures of around 2700 °C (4900 °F). The tension, along with the high temperatures, causes the graphite layer planes to align with the fiber axis, imparting high strength and high modulus to the fibers. Coal tar pitch is heated for up to 40 hours at approximately 450 °C (800 °F), forming a viscous liquid with a high degree of molecular order known as mesophase. The mesophase is then spun through a small orifice, aligning the mesophase molecules along the fiber axis. The fibers are thermoset at relatively low temperatures and then heated at 1700 to 3000 °C (3100 to 5400 °F).

Polyacrylonitrile (PAN) is a long-chain linear polymer composed of a carbon backbone with attached carbonitrile groups. The conversion of PAN fibers into car-

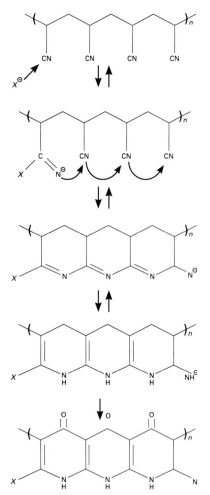

Fig. 3 Possible reactions during oxidation or "stabilization" of PAN fibers at 200 °C (392 °F). Source: Ref 20

bon fibers involves treating the fibers in an oxidizing atmosphere at temperatures typically in the range of 200 to 300 °C (392 to 572 °F) while under tension to avoid shrinkage. This treatment converts thermoplastic PAN into a non-plastic cyclic or ladder structure which is capable of withstanding the high temperatures present in the subsequent carbonization treatment.

During the oxidizing treatment, several thermally activated processes can occur [5,6], causing considerable reorganization of the polymer chains and three-dimensional linking of the parallel molecular chains by oxygen bonds. Oxidation mechanisms of PAN fibers have been studied quite extensively and consequently several different structures have been proposed for the cyclized PAN material [7]. FTIR studies of the PAN copolymers have been carried out at temperatures below the onset of the major exothermic reaction and under reduced pressure of 5×10^{-2} torr [8-10]. A possible sequence of reactions is given in Fig. 3 [7], although great uncertainty still exists

regarding the reaction mechanisms during the oxidation treatment of PAN in air [7].

According to Coleman and Sivy [10], the chemical reactions occurring during the oxidation step begin with the degradation of PAN, which involves initiation by an anion X, presumably derived from an impurity or degradation product. The initiation was followed by a cyclization to yield an imine structure, which tautomerized into an enamine structure, followed by oxidation to give the final pyridone structure. The sequences of cyclized groups in the ladder structure impart rigidity along the chain. In addition, the structure of the degraded material during the final pyrolysis to carbon fiber is maintained by extensive inter-chain bonding between the polymer chains through $C = O$ and N–H groups.

The stabilized PAN fibers can be further treated at higher temperatures in an inert atmosphere. The non-carbon elements are gradually removed as volatile gases such as H_2O, HCN, NH_3, CO, CO_2, N_2, etc. During the early stages of carbonization (e.g. 400 to 500 °C, or 752 to 932 °F, when heated at a few °C/min.), the hydroxyl group present in the oxidized PAN fibers start cross-linking condensation reactions which help in reorganization and coalescence of the cyclized sections (Fig. 4) [11]. The structure of the polymer is largely "fixed" by this cross-linking process, while the remaining linear segments either become cyclized or undergo chain scission evolving the gaseous products [7]. The cyclized structures undergo dehydrogenation [12] and begin to link up in the lateral direction [11].

The elastic modulus of a fully carbonized fiber is governed by the orientation of the graphite crystallites relative to its axis. The relationship between the preferred orientation and elastic modulus was explained on the basis of the wrinkled-ribbon model [13]. When compared to the iso-strain or iso-stress models, this model was found to fit the experimental values better for a wide range of fiber moduli, heat treatment temperatures and precursors [14]. The gradient in preferred orientation through the cross section of the fiber was also found to be an important factor [15].

The preferred orientation of a carbon fiber can be enhanced by stretching the fiber during one of the processing stages. This stretch orientation is usually carried out during the low-temperature heat treatment for carbon fibers derived from PAN [7]. Other factors that might influence the elastic modulus of carbon fiber include heat treatment temperature (HTT) [13,16,17], boron doping [18], and neutron irradiation [19].

The strength of a carbon fiber depends on the type of the precursor fiber, processing conditions such as HTT, preoxidation conditions, and the presence of flaws and defects [7]. The strength of PAN-based carbon fibers was observed [11,20,21] to increase with an increasing HTT and to reach a maximum value at about 1500 °C (2732 °F) before starting to decrease at higher HTTs. Various possible mechanisms have been given to explain this observation [7,17,22-26]. Just like the modulus, the strength of a carbon fiber is greatly influenced by the preoxidation conditions of PAN fibers [7,20]. However,

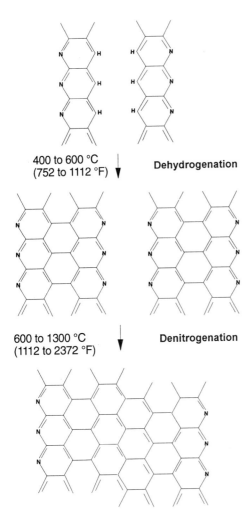

400 to 600 °C
(752 to 1112 °F) ↓   **Dehydrogenation**

600 to 1300 °C
(1112 to 2372 °F) ↓   **Denitrogenation**

Fig. 4 Carbonization reactions of oxidized PAN fibers to produce carbon fibers. Source: Ref 6

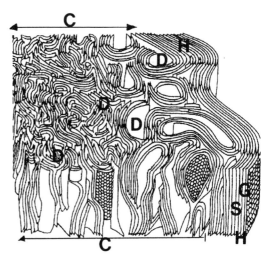

Fig. 5 A 3-D model representing the internal structure of a carbon fiber. Various types of defects are noted. C, carbon fiber core structure, oriented to a lesser degree; D, internal defects (e.g. voids, edges of graphitic planes, irregular stacking of graphitic planes); G, graphitic plane composed of fused benzene rings (highly aromatic); H, highly oriented skin layers; S, stacked graphitic planes. Source: Ref 28

defects in carbon fibers appear to be the most critical strength-limiting factor in this class of materials [7].

Many models have been proposed to illustrate the possible structure of carbon fibers [27]. For example, Wick's model [19] depicts continuous graphite layers having a random stacking sequence in a c direction and a highly preferred orientation parallel to the fiber axis. Regions of crystallinity are separated, in a longitudinal direction, by zones of extensive bending and twisting of the basal layers at tilt, twist and bend plane boundaries which consist of arrays of basal dislocations. Sharp-edged microvoids separate adjacent crystalline regions in the transverse direction. The effect of graphitizing treatments at elevated temperatures (>1000 °C, or 1832 °F) is viewed as a progressive elimination of these boundaries by migration and annihilation resulting in increased crystalline order. A three-dimensional model, shown in Fig. 5, was proposed by Bennett and Johnson [28], which indicates a difference in ordering and density of the carbon ribbons or micrcrystallites between the peripheral zone and the core of the fiber.

After the carbon fiber is formed, the surface of the fiber is usually oxidized to increase the adhesion of the resin matrix to the fiber. These oxidizing treatments include wet chemical treatments with sodium hypochlorite or nitric acid. Drzal *et al.* [29] proposed that these treatments produced two effects with epoxy resins: first, an outer, weak, defect-laden fiber surface layer is removed; second, surface oxygen groups are added which can interact with the resin matrix. In addition to oxidative surface treatments of carbon fibers, there may be additional surface treatment with finishes or sizings to enhance compatibility with the polymer matrix [30]. A more detailed discussion of surface treatment of carbon fibers is given in Chapter 7.

Carbon fibers are now available in a wide range of modulus and strength. Experimental carbon fibers with moduli up to 965 GPa (140 Msi) has been reported [31], while the theoretical modulus is in the vicinity of 1,035 GPa (150 Msi) [32]. A tensile strength of 20.7 GPa (3 Msi) and 6.9 GPa (1 Msi) has been reported with carbon whiskers [33] and continuous carbon fibers [34], respectively. Carbon fibers can be conveniently classified into five distinct performance groups in terms of their tensile modulus (E) and strength (σ_u):

- Ultra-high modulus ($E > 394$ GPa, or 57 Msi)
- High modulus ($317 < E < 394$ GPa, or $46 < E < 57$ Msi)
- Intermediate modulus ($255 < E < 317$ GPa, or $37 < E < 46$ Msi)

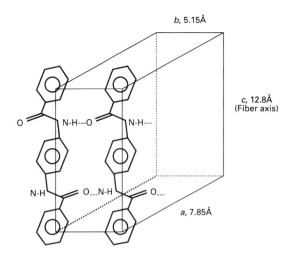

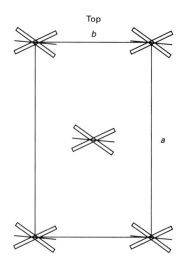

Fig. 6   The unit cell of PPD-T crystals. Source: Ref 38

- High strength ($\sigma_u > 3.8$ GPa, or 551 ksi)
- High strain carbon fibers ($\varepsilon_f > 1.8\%$)

Fiber producers are constantly improving mechanical properties of their carbon fibers, particularly in the areas of strength and modulus, higher strain-to-failure and reduced coefficients of variation in mechanical properties. Future carbon fiber development for primary structural composite applications will be directed toward concurrent improvement on the strength and strain-to-failure of carbon fibers.

Aramid Fibers

Aramid fibers is a generic term for aromatic polyamide fibers. As an example, Kevlar® fibers, developed by Du Pont, are composed of poly (1,4-phenyleneterephthalamide) [35]. The polymer can be prepared by solution polycondensation of p-phenylene diamine and terephthaloyl chloride at low temperatures [36]. The fiber can be spun by extrusion of a solution of the polymer, which is followed by a stretching and drawing treatment. The fibers are produced as Kevlar®-29 and Kevlar®-49, the latter having the higher modulus and being the one more commonly used in composite structures. The latest developments include Kevlar®-149 fibers, which possess a modulus of 186 GPa (27 Msi) being 74% of the theoretical modulus (248 GPa or 36 Msi). The tensile strength of Kevlar®-149 fibers is 3.47 GPa (503 ksi), comparable to 3.62 GPa (525 ksi) of Kevlar®-49 fibers. The structure of Kevlar®-49 as shown by X-ray diffraction studies of Northolt [37] is given in Fig. 6. The rigid linear molecular chains are highly oriented in the fiber axis direction, with the chains held together in the transverse direction by hydrogen bonds. The strong covalent bonds in the fiber axis direction provide high longitudinal strength, whereas the weak hydrogen bonds in the transverse direction result in low transverse strength.

The morphology of three types of polyamide fiber, PRD-49 [poly(p-benzamide)], Kevlar®-29, and Kevlar®-49 was studied by Li et al. [39]. These fibers exhibit skin and core structures. The core is a layered structure with layers stacked perpendicular to the fiber axis that are composed of rod-shaped crystallites with an average diameter of 50 nm. These crystallites are closely packed and held together with hydrogen bonds nearly in the radial direction of the fiber.

While the theoretical modulus of the poly (p-phenylene terephthalamide) molecule has been calculated to be greater than 1500 dN/tex in textile units [40 to 42], the modulus of commercially available p-aramid fibers ranges from 440 dN/tex to 900 dN/tex [38]. The elastic modulus of the fiber is closely related to the crystalline orientation angle, which can be determined by wide-angle X-ray diffraction [43]. The strength per unit weight (tenacity) of p-aramid fibers has been estimated to be above 120 dN/tex [44]. This prediction was based on the premise that a fiber is composed of a perfectly oriented single crystal consisting of infinitely long polymer molecules. The current commercially available p-aramid yarns possess a strength of up to 25 dN/tex [38].

The aramid fibers have an elastic modulus over twenty times greater than that of conventional polyamide (nylon) fibers (Table 4). The more rigid and complex molecules impart higher mechanical properties and improved thermal resistance to aramid fibers. These fibers with high specific strength have great cohesiveness and a tendency to form fibrils; they absorb much more energy than brittle fibers and are widely used in lightweight armor and other impact-resistant structures. However, they have a highly anisotropic structure that leads to low longitudinal shear moduli, poor transverse properties, and low axial compressive strengths. This directional tradeoff arises because of the much weaker lateral bonding mechanisms [46]. While glass and carbon fiber composites show similar tensile and compressive strengths, the compressive strength of unidirectional composites con-

Table 4   Kevlar® aramid fiber properties

	Akzo Twaron	Twaron® HM	Kevlar®-29	Kevlar®-49	Teijin HM-50
Density, g/cm^3	1.44	1.45	1.44	1.44	1.39
Tensile strength, MPa(a)	2800	2800	2758	2758	3030
Tensile modulus, GPa(a)	80	125	62	124	74
Tensile strain, %	3.3	2.0	3.5	2.5	4.2
Coefficient of thermal expansion, 10^{-6} °C					
Longitudinal:					
0 to 100 °C	...	...	−2.0	...	...
100 to 200 °C	...	...	− 4.1	...	...
200 to 260 °C	...	...	−5.3	...	...
Radial:					
0 to 100 °C	...	...	59	...	...

(a) To convert from Pa to psi, multiply by 1.45×10^{-4}. Source: Ref 45

taining Kevlar®-49 fibers could be as low as 25% of their tensile strength. The interfacial bond strength between aramid fibers and epoxy resins is normally lower than what is experienced with carbon fiber composites. Work has been carried out on various surface coating, sizing or plasma treatments to improve interfacial properties of aramid composites (Chapter 3).

The aramid family of fibers has been well accepted as reinforcements for composites. Still greater improvements in selected properties are possible as evidenced by the latest development of Kevlar®-149 from Du Pont. This new p-aramid fiber has a very high level of crystalline order, leading to a further improved elastic modulus. This value represents a 40% improvement over the previous aramid fibers and provides a specific modulus nearly equivalent to that of high-strength graphite fiber. Further improvement appears possible since calculations based on perfect alignment and crystalline order predict a theoretical modulus of about 220 to 245 GPa (32 to 35 Msi). The new Kevlar® fibers are slightly more compact with inherently low moisture absorption.

High-Strength Polyolefin Fibers

Another fairly recent advancement which shows considerable promise is the development of high-modulus polyolefins. The polymer chains are straightened during processing to take advantage of the strong covalent bonds in the backbone of the molecule.

Solid-state drawing can be used to produce high-modulus polyethylene fiber from high density PE. The linearity and crystallinity of this polymer impart a significant degree of order in the polymer structure prior to drawing, thus permitting a highly extended chain configuration to be attained through post-fiber-formation drawing. The draw ratio, or ratio of final fiber diameter to initial fiber diameter, is proportional to the final modulus of the fiber. Thus, high draw ratios are essential for producing high-strength fibers [46].

For an ultrahigh molecular weight PE (UHMWPE, MW $\approx 10^6$ g/mole), the natural draw ratio is around 5:1, where further orientation can no longer be imposed. However, draw ratios of 50:1 to 70:1 are necessary to obtain high-strength fibers. These draw ratios can be achieved by dissolving UHMWPE in a solvent (decalin, paraffin oil or paraffin wax) to reduce intermolecular entanglements, making it possible to achieve greater fiber orientation. Typically, a 5 to 10% solution of polymer in solvent is prepared at 150 °C (302 °F). The dilute solutions allow the polymer molecules to detangle. The solution is then cooled to ~135 °C (275 °F) and forms a "gel," a porous, crystallized solid with minimal entanglement density. This gel is spun into fibers and then ultradrawn at 120 °C (248 °F) to reach the final fiber diameters and strength [46].

Like other polymer-based high-performance fibers, the transverse and longitudinal compressive strengths of PE fibers are far from being satisfactory. The epoxy composites containing PE fibers show poor compressive response. Their interlaminar shear strength is lower than that achievable with carbon fiber composites, mainly due to the usually poor bonding between PE fibers and the epoxy matrix. Cold plasma treatments of PE fiber surface appear to be an effective technique for improving interfacial bonding in these composites [47].

The physical and chemical properties of the gel spun PE fibers are similar to other PE fibers (e.g., solid-state drawn) except that they are of extremely high strength and modulus. These fibers, similar to the aramid fibers, produce high strength and high modulus materials at a significant weight savings over conventional fibers. In fact, UHMWPE fibers, with a density of 0.97 g/cm^3, are even lighter than aramid fibers. This low density feature, coupled with high strength and modulus values, has made UHMWPE fibers the most outstanding among the currently commercial reinforcing fibers in terms of specific strength and specific modulus (Fig. 7 and Table 1).

Because of their low softening point, PE fibers do not retain their strength at temperatures above 145 °C (293 °F). Therefore, these fibers should not be used as reinforcements in many injection molding or sheet molding processes. The creep resistance of these fibers under high load is rather poor. However, their good resistance to aqueous and low temperature organic environments have made them attractive materials for marine tow ropes,

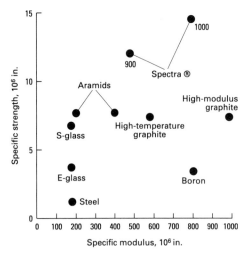

Fig. 7 Specific strength and specific modulus data of several reinforcement fibers. Source: Allied Fibers

Table 5 Room temperature mechanical properties of boron-on-tungsten filament

Ultimate tensile, GPa(a)	3.4+	3.4+	3.4+
Tensile modulus, GPa(a)	400	400	400
Shear modulus, GPa(a)	165	165	...
Coefficient of thermal expansion, 10^{-6}/°C	4.86	4.86	4.86
Density, g/cm^3	2.6	2.47	2.39
Length/weight, km/kg(b)	47	25.3	12.7

(a) To convert from Pa to psi, multiply by 1.45×10^{-4}. (b) To convert from km/kg to ft/lb, multiply by 1488.2. Source: Ref 4

mooring cables, fishing nets and sail cloth. Because of their high impact resistance, UHMWPE fibers have found their way into the markets of both soft and rigid (composite) armors. The low dielectric and low loss tangent properties, and hence good microwave transparency, of PE fibers have made them viable candidate reinforcements for radome systems. Future applications should include sporting goods, low-temperature pressure tanks and medical devices.

Boron Fibers

Boron filaments have been used for many years as a reinforcement for both polymers and metal matrices. Boron fiber reinforced composites offer a unique combination of mechanical and physical properties. The elastic modulus of boron fiber is extremely high (Table 5), providing good stiffness to the reinforced composites. Potential applications for boron-based materials include neutron shielding, cutting tools, heat-dissipator plates, armor penetrators, and antenna design. Boron fiber reinforced aluminum was employed for tube-shaped truss members to reinforce the space shuttle orbiter structure, and has been tested as a potential fan-blade material in turbofan jet engines [48]. However, the rapid reaction of boron fiber with molten aluminum and long-term degradation of diffusion-bonded B/Al at temperatures greater than 480 °C (896 °F) may preclude its use for high-temperature applications. These drawbacks will also be a real concern in using potentially more economical fabrication methods, such as squeeze casting or low-pressure high-temperature pressing for metal matrix composites.

Boron filament is manufactured by chemical vapor deposition (CVD), which involves the formation of solid by decomposition or reduction of gaseous molecules upon a heated substrate (tungsten or carbon filaments). The filaments are heated to approximately 1260 °C (2300 °F) and then pulled through a boron trichloride/hydrogen environment. The linear deposition rate is about 50

mm/h. (2 in./h.), usually requiring many reactors to be run in parallel to achieve an appreciable production rate.

The majority of the boron filaments are used in continuous fiber-epoxy composites. Boron-epoxy composites offer high strengths and stiffness in tension, compression, and bending. The high mechanical properties of boron-reinforced composites (Table 6) have led boron fibers to be applied for many high stress applications. Boron composites have been utilized in numerous aircraft and helicopter designs, e.g., stabilizer and rudder in the F-15 fighter plane. In sporting goods, a small quantity of boron-epoxy composites have been used to improve strengths, stiffness, and sensitivity of rackets, fishing rods, and golf-club shafts [48].

Boron fiber reinforced metal matrix composites typically contain approximately 50% filament by volume. Major applications for boron/aluminum include tubular struts used in the mid-fuselage section of the space shuttle and in heat dissipator plates to control the temperature of electronic modules. This is because boron/aluminum possesses high strength and stiffness, light weight, high thermal conductivity, and low thermal expansion. Boron reinforced metal composites are also being considered for uses in jet-engine fan blades, aircraft wing skins, structural support members and landing-gear components.

Silicon Carbide Fibers

Silicon-carbide filaments are also produced by the CVD process. The β-SiC is obtained by the reaction of silane and hydrogen gases with the carbon filament being the substrate for deposition. The SiC fibers have mechanical and physical properties equal to those of boron, and can be used at higher temperatures than are possible with boron. SiC fiber will be less expensive than the present boron fiber when available in production quantities. CVD SiC fibers are primarily used for reinforcing metal and ceramic matrices.

The SiC fiber is qualified for use in aluminum, magnesium, and titanium. An extensive design data base for hot-molded SiC reinforced 6061 aluminum is available. Property data on investment-cast SiC/aluminum are more limited. SiC fiber reinforced titanium parts typically are fabricated by press bonding or hot isostatic pressing. Composites of SiC/Ti have been shown to successfully withstand extended exposure at high temperature. SiC fibers retain usable strengths at up to 1000 °C (1832 °F),

Table 6 Properties of select boron and SiC-reinforced composites

Property	47 vol% SiC/Al 6061	50 vol% B/Epoxy R.T.	50 vol% B/Epoxy 177 °C (350°F)	Unidirectional B reinforced Al	Cast SiC/34 vol% Al
Longitudinal tensile strength, MPa(a)	1,460	1,700	1,450	1,100	1,030
Longitudinal tensile modulus, GPa(a)	204	210	210	235	172
Transverse tensile strength, MPa(a)	86.2	90	37	110	...
Transverse tensile modulus, GPa(a)	118	25	9.7	138	...
Longitudinal compressive strength, MPa(a)	2,990	2,600	950	1,215	1,900
Longitudinal compressive modulus, GPa(a)	218	210	210	207	186
Transverse compressive strength, MPa(a)	286	310	90	159	...
Transverse compressive modulus, GPa(a)	115	24	9.7	131	...
Shear strength (15% off-axis), MPa(a)	113	110	48	126	...
Shear modulus (15% off-axis), MPa(a)	40.5	...	...	...	...
Longitudinal Poisson's ratio (tension)	0.268	0.21	0.21	0.23	...
Transverse Poisson's ratio (tension)	0.124	0.019	0.008	0.17	...
Longitudinal Poisson's ratio (compression)	0.241	...	...	...	...
Transverse Poisson's ratio (compression)	0.174	...	...	...	...
Longitudinal coefficient of thermal expansion, 10^{-6}/°C	6.6	4.5	4.5	5.8	...
Transverse coefficient of thermal expansion, 10^{-6}/°C	21.3	23.6	36	19.1	...
Density, g/cm^3	...	2.00	2.00	2.71	...

(a) To convert from Pa to psi, multiply by 1.45×10^{-4}. Source: Ref 48

and have the unique characteristic of retaining basic fiber strength in molten aluminum for up to half an hour. This characteristic should greatly reduce the cost and difficulty of producing SiC based composites.

The use of advanced ceramics in applications such as gas-turbine engines has been limited partially because of their brittle nature, which often leads to catastrophic failure. However, the fracture toughness of a ceramic can be significantly improved by the addition of SiC fiber. The high mechanical properties of SiC fibers (Table 7) result in higher composite strength, stiffness, and toughness. The fibers have been evaluated in various glass, ceramic, and glass-ceramic matrices. The use of these high-density fibers in polymer matrix composites has yet to be explored.

The Quest for High-Strength and High-Modulus Fibers

The ever widening application scope of composites demands the availability of higher performance reinforcing fibers. In this section, polymer fibers will be used as an example to illustrate the concepts of high strength and high stiffness in fibers. Recent developmental efforts on various reinforcing fibers for polymer matrix composites are then reviewed. Directions of future research are then discussed.

Stiffness and Strength of Polymer Fibers

Linear organic polymers consist of carbon and other atoms joined by strong covalent bonds to form a long chain; but the bonding between the chains in a solid is accomplished by van der Waals' forces. Deformation of a crystal parallel to the chain direction involves stretching of the strong covalent bonds and changing the bond

Table 7 Silicon carbide filament room temperature properties

Diameter, μm (in.)	142 (0.0056)
Ultimate tensile strength, GPa (ksi)	3.34+ (485+)
Tensile modulus, GPa (Msi)	400 (58)
Shear modulus, GPa (Msi)	165 (24)
Coefficient of thermal expansion, 10^{-6}/°C (10^{-6}/°F)	4.86 (2.7)
Density, g/cm^3 (lb/in.^{3})	3.08 (0.111)
Length/weight, km/kg (lb/ft)	20.3 (30 500)

Source: Ref 45

angles or, if the molecule is in a helical conformation, distorting the molecular helix [49]. Deformation in the transverse direction is opposed only by relatively weak secondary van der Waals forces (possibly including hydrogen bonding). With knowledge of the crystal structure and the force constants of the chemical bonds, one can calculate the theoretical moduli of polymer crystals.

The chain-direction modulus of a polymer can be estimated by a method suggested by Treloar [50]. In this model (Fig. 8), a polymer is considered to crystallize with its backbone in the form of a planar zigzag. Deformation of this crystal is assumed to involve only the bending and stretching of the bonds along the molecular backbone. The chain is treated as consisting of n rods of length l which can be stretched along its length but can not be bent. The rods are joined together by torsional springs. The bond angles are taken to be θ initially, while the angle between the applied force f and the individual bonds is initially α. Then from the basic strength of material approach one obtains [49,50]:

$$E = [l \sin (\theta/2)/A] [\sin^{-2} (\theta/2)/K_l + l^2 \cos^2 (\theta/2)/(4K_\theta)]^{-1} \qquad \text{(Eq 1)}$$

where A is the cross-sectional area supported by each chain, K_l and K_θ are the force constants which can be determined from spectroscopic methods. Based on this model, a value of E for polyethylene is predicted to be about 180 GPa (26 Msi), which is substantially lower than values obtained from more sophisticated calculations [49,51].

Chain direction moduli can be estimated experimentally using an X-ray diffraction method, which involves measuring the relative change under an applied stress in the position of any (*hkl*) Bragg reflection which has a component of the chain-direction repeat. Presented in Table 8 are some calculated and experimentally measured values of chain-direction crystal moduli of several different polymer crystals. The current state-of-the-art gel-spun PE fibers (e.g., Spectra®-1000) have a modulus of 172 GPa (24.9 Msi), only about 50% of that achievable. The presence of an amorphous phase, some degree of chain misalignment and material defects such as chain ends are likely responsible for this discrepancy.

The theoretical tensile strength of a polymer can be estimated from the breaking load of a carbon-carbon bond per molecular cross-sectional area (*MA*). Using a more commonly used textile tenacity unit (*g/d*), where 1 [GPa] = 11.3 · [*g/d*]/material density (g/cm³) and taking the carbon-carbon bond force to be 6.1×10^{-4} dyne [52], Ohta [53] suggested that

Theoretical strength (*g/d*) =

$$\frac{69.0}{\text{Density (g/cm}^3) \times MA(\text{nm}^2)} \qquad \text{(Eq 2)}$$

Ohta [53] listed calculated tensile strengths for polymers in contrast to the strengths of commercial fibers

Fig. 8 Model of a polymer chain undergoing deformation. Source: Ref 49

(Table 9). These data clearly indicate that many highly oriented commercial polymers are still two orders of magnitude short of their theoretical limits. Considerable research effort has been made to determine why only a small portion of ultimate polymer strength has been attained and how processing can be developed to more closely approach those theoretical values. A comprehensive list of strength and modulus data on various high-tenacity and high-stiffness fibers derived from flexible polymers by the processing techniques of hot drawing, zone drawing, solid state extrusion, gel spinning, and fibrillar crystal growing can be found in [53].

Ultrahigh tenacity fibers can only be attained by processing techniques which meet the following structural requirements:

- A draw ratio of at least 20 is required for maintaining a high number of tie molecules and high degree of molecular orientation
- An increase in molecular chain length for reducing the intermolecular chain slip at polymer chain ends
- A reduced concentration of defects and flaws in microfibrils of the fiber structure

Only the two processing techniques of fibrillar crystal growing and gel spinning appear to satisfy these requirements simultaneously. The production of proper ultrahigh molecular weight polymers and the preparation of optimum gel are the two most critical factors to consider when attempting to produce high strength fibers from flexible chain polymers.

Developments of Rigid Rod Reinforcing Fibers

Poly (p-phenylene benzobisthiazole), or PBT, is a stiff chain polymer which can be synthesized [54,55] and spun [56] into the fiber form by a dry jet-wet spinning technique using a large air gap. A high draw ratio is imposed on the fiber following spinning to create a high level of orientation. Fibers of high strength and high modulus are obtained with a further heat treatment of drawn fibers in a nitrogen environment [46].

Table 8   Calculated and measured values of chain-direction methods for several different polymer crystals

| Polymer | Calculated E | | Measured E | |
	GPa(a)	Method(b)	GPa(a)	Method
Polyethylene	182	VFF	240	x-ray
	340	UBFF	358	Raman
	257	UBFF	329	Neutron
Polypropylene	49	UBFF	42	x-ray
Polyoxymethylene	150	UBFF	54	x-ray
Polytetrafluoroethylene	160	UBFF	156	x-ray
			222	Neutron
Substituted Polydiacetylene				
(I) Phenyl urethane derivative	49	VFF	45	Mechanical
(II) Ethyl urethane derivative	65	VFF	61	Mechanical

(a) To convert from Pa to psi, multiply by 1.45×10^{-4}. (b) VFF, Valence force field (Treloar's method[50]); UBFF, Urey-Bradley force field. Source: Ref 49

Table 9 Comparison between the theoretical strength of polymers and the strength of commercial fibers

Polymer	Density, g/cm³	Theoretical ultimate strength (g/d)	Strength of commercial fibers (g/d)
PE	0.96	372	9.0 (not gel spun)
Nylon 6	1.14	316	9.5
POM	1.41	264	...
PVA	1.28	236	9.5
Kevlar®	1.43	235	25.0
PET	1.37	232	9.5
PP	0.91	219	9.0
PVC	1.39	169	4.0
Rayon	1.50	133	5.2
PMMA	1.19	87	...

Source: Ref 53

PBT, in addition to possessing extremely high specific modulus and strength (Table 1), also has very good chemical resistance to all solvents but strong acids. Good thermal stability is another advantage associated with this fiber over all the currently available commercial polymer fibers. Physical properties are said to be retained at up to 700 °C (1292°F) for short term exposure and at 350 °C (662 °F) for an extended period of time [46].

Rigid rod polymers such as polybenzobisoxazoles (PBO) and self-reinforcing liquid crystalline polyesters can also be fabricated into a fiber form. Thermotropic liquid crystalline polymers can orient themselves into highly ordered states and can be thermally processed, as opposed to the lyotropic aramids which must be solution processed into reinforcing fibers [46].

Liquid crystals may be formed *in situ* in a polymer by taking advantage of the self-orienting characteristics of anisotropic melts and solutions of polymers. In this way, the liquid crystals, because of their high level of orientation, take the place of fibers in a composite, and the self-reinforced resin is its own composite [16-18]. Although most of the work in this area involves aromatic polyesters and copolyesters, recently published work gives a promising application of blending liquid crystal polymers with more conventional polymers in which the reinforcing component is dispersed microscopically or molecularly [19,20]. The concept of developing polymers that can form liquid crystals during fabrication into structures (for example by injection molding) is receiving considerable attention, for in this way it may be possible to produce composites with high performance economically [13].

Compressive Strength of Fibers

Perhaps the most serious drawback associated with polymer based advanced fibers for structural composite applications is their weak compressive strength. Many aerospace applications require that composites possess good compressive strength, which is not provided for by the state-of-the-art polymer fibers and high-modulus carbon fibers. Composites containing intermediate-modulus carbon fibers are known to have better compres-

sive strength. The poor composite strength is thought to be a result of the weak axial compressive strengths of the fibers themselves [57,58]. The axial compressive strength of most polymeric fibers is significantly lower than their tensile strength.

As opposed to the case of carbon fibers, the compressive strength of polymer fibers is relatively insensitive to improvement in modulus [59]. For example, the compressive strength of Kevlar®-29, -49, and -149 is about the same while the compression moduli for these Kevlar® fibers are quite different: 75.8 GPa (11 Msi) for Kevlar®-29 and 110 GPa (16 Msi) for both Kevlar®-49 and -149 [60]. This is a puzzling observation because the buckling of fibrils or microfibrils leading to kinking is a well-known compressive failure mechanism with most of the high performance polymer fibers [61,62] and the compressive stress for buckling of a "structural" member (such as a fibril) is closely related to its modulus [59]:

$$\sigma_c = C\pi E(r/l)^2 \qquad \text{(Eq 3)}$$

where the fibril is assumed to have a cylindrical shape with length l, radius r and axial modulus E, and C is a constant. It is possible that changing the polymer fiber processing conditions to vary E would at the same time alter the (r/l) ratio of the fiber microstructure in such a way that σ_c remains unchanged. More work is required to resolve this and other issues. For example, what is the maximum theoretical compressive strength for a given polymer chain and how would this strength be achieved? How can the modulus of carbon fibers be increased without compromising their tensile and compressive strength? These areas are of academic and technological significance.

Matrix Materials

The matrix materials in fiber composites are extensive and varied in properties. They may be classified as polymeric, metallic, or ceramic. The chief polymers used in high performance fiber composites are thermosetting resins converted from low molecular weight species, with low viscosity to three-dimensional cross-linked networks that are infusible and insoluble. Cross-linking can be accomplished with the application of heat and/or by chemical reaction. Thermoplastic polymers, although used to a lesser extent, are gaining increasing acceptance in composite applications. Metal matrix composites have been tested with many metals, but the most important are aluminum, titanium, magnesium, and copper alloys. Ceramic-based composites are particularly suitable for applications where elevated temperature or chemical stability is a concern.

The property requirements for a matrix material are different than those for a reinforcement. Since the fibers must serve as the principal load-bearing members in a composite, they must be of high strength and stiffness. With some exceptions, reinforcement fibers are usually of low ductility. In contrast, matrix materials usually have relatively low modulus and strength values and comparable or higher ductility values. The matrix serves to bind the fibers together, and therefore the thermal stability of the composite is determined by that of the matrix. The matrix protects the typically rigid and brittle fibers from abrasion and corrosion. The matrix transmits load in and out of composites, and in some cases, carries some transverse load. When the composite is under a compressive load, the matrix also plays a critical role in avoiding micro-buckling of fibers, a major compressive failure mechanism in continuous fiber composites. The matrix provides the composite with interlaminar fracture toughness, damage tolerance, and impact resistance.

Composite performance is influenced by the following matrix properties:

- Elastic constants
- Yield and ultimate strengths under tension, compression, or shear
- Failure strain or ductility
- Fracture toughness
- Resistance to aggressive organic liquids and moisture
- Thermal and oxidative stability

When selecting a particular resin for a specific composite application, service environment parameters such as temperature, stresses, moisture, chemical effects, and possibly radiation dosage must be considered. The processability and the processing history of the matrix must be taken into account, since it directly influences the formation of flaws and microstructure in the matrix, residual stresses, and the fiber-matrix interface. These factors ultimately determine the properties of a composite.

Thermosetting Resins

Several important general classes of thermosetting resins are used as composite matrices. These include the unsaturated polyesters, vinyl esters, epoxides, phenolics, polyimides (and bismaleimides), and the modified versions of these resins such as interpenetrating networks. A thermosetting resin begins with polyfunctional monomers or oligomers, which are then cured to produce a three-dimensional network of covalently bonded chains, which are insoluble and infusible. The initial uncured state, referred to as the A-stage, is characterized by low viscosity, allowing good flow and impregnation of the resin into the reinforcing fiber bundles. In processes such as resin transfer molding (RTM) or structural reaction injection molding (SRIM) the resin, once fully permeated into the fiber preform, will be cured within a time frame ranging from less than a minute to an hour. An alternative process is to combine the reinforcing fibers with a resin to form a pre-impregnated sheet or tape, called a prepreg. In a prepreg, the resin is generally advanced to the B-stage, which represents a partially cured and usually vitrified system prior to the gel point. Because an infinitely large three-dimensional network of chains (gela-

tion) has not been formed in the B-stage, prepreg layers are readily formed into a variety of shapes. Once cut and laid up in a proper orientation sequence, the prepreg layers are fully cured to form a consolidated composite. The resin now is at the C-stage, characterized by insolubility and infusibility.

Curing of a thermosetting resin involves chain extension, branching, and cross-linking. Cross-linking imparts to the resin rigidity, high strength, solvent resistance, and good thermal and oxidative stability. These properties are directly related to the cross-link density (ν_c) and the chain molecular weight between cross-links (M_c). A higher ν_c results in a greater matrix rigidity or higher elastic modulus. However, a highly cross-linked resin with a reduced M_c has limited molecular mobility, leading to brittleness, low failure strain, poor impact resistance, and inferior fracture toughness. Conversely, a reduced chain length between cross-links, although exhibiting increased ductility and toughness, tends to compromise the solvent stability. Toughening of thermoset resins by incorporating a second phase such as an elastomer, thermoplastic, rigid inclusions, and glass beads as well as by other means is discussed in later in this volume.

Catalysts and heat are typically used to effect the curing of thermosets. The molecular weight of the growing chains increases through chain extension and branching in the first stage of curing. Rheological properties such as viscosity and dynamic modulus (E') will increase rapidly. At the gel point, a three-dimensional network of chains with a practically infinite molecular weight (gel) is formed. Viscosity and dynamic modulus increase dramatically, while the loss tangent (tan δ) exhibits a peak when plotted against the extent of reaction. After gelation, more molecular chains will be incorporated into the cross-linked network at the expense of the soluble portion (sol). The viscosity or dynamic modulus will continue to increase, gradually reaching a plateau.

A second peak in the tan δ plot can sometimes be observed as a result of vitrification [63]. As curing proceeds, the glass transition temperature (T_g) of the growing chains increases as the network molecular weight increases. Should the cure temperature at any point become lower than the T_g of the growing polymer chains, chain mobility becomes severely restricted and further conversion to the network structure is limited. The resin is said to be vitrified and its T_g coincides with the maximum cure temperature achieved. Reapplication of heat to a higher temperature is required to complete the curing process and a post-curing procedure is performed to ensure that the ultimate glass transition be attained. The curing process of a resin under various curing conditions may be conceptually represented in a time-temperature-transformation (TTT) diagram [63]. Techniques that can be used to monitor and study the cure behavior of a thermosetting resin are discussed in Chapter 10.

Thermoset matrix materials result in composites that have higher specific tensile strength and stiffness properties than metal matrix composites. Thermosetting resin composites are also more advanced in fabrication technology and are lower in raw material cost and fabrication cost.

Unsaturated polyesters are very versatile polymers that impart to a fiber composite a variety of properties at moderate cost. They are used mainly where a good balance is required between mechanical properties and chemical resistance at moderate or ambient temperatures. Their main weaknesses are:

- Relatively high shrinkage on curing
- Sensitivity to some solvents and chemicals especially under alkaline conditions
- Appreciable water absorption under certain conditions

For about two decades, unsaturated polyesters have dominated the markets for reinforced glass fiber composite products. Major areas of application for reinforced polyesters include appliances (air conditioner ducting, shower enclosures and tubs), business equipment (computer housing), consumer goods, corrosion resistant products, sporting goods (boat hulls), transportation (automotive ducting, car bodies, and railroad cars) and construction (building wall and roof panels). Their poor impact resistance, inferior hot/wet mechanical properties, limited shelf life and high curing shrinkage preclude their utilization for high performance applications.

Unsaturated polyesters are made by reacting a glycol such as propylene glycol with an unsaturated acid (maleic anhydride) and cross-linking the resulting polymer with an unsaturated monomer (styrene, vinyl toluene, and chlorostyrene). The first reaction is a condensation-type reaction involving the formation of by-products such as water, which must be removed continuously. The cross-linking reaction occurs by addition-type polymerization reaction. A general purpose polyester is based on a blend of phthalic anhydride and maleic anhydride esterified with propylene glycol at a phthalic/maleic ratio ranging from 2:1 to 1:2 [64]. This polyester alkyd is then blended with styrene monomer in a ratio of about 2 parts alkyd to 1 part styrene [64].

A large variety of starting materials can be formulated to produce polyesters with desired properties. Flexibility can be imparted to the polyester chain by replacing the aromatic acid (phthalic anhydride) with straight chain dibasic acids such as adipic or sebacic. Toughness can be increased by using isophthalic acid in place of the phthalic anhydride. The resistance to alkaline hydrolysis can be improved by increasing the size of the glycol by substituting hydrogenated bisphenol A or the reaction product of bisphenol A and propylene oxide for the propylene glycol, for example. Improved fire resistance can be obtained by using halogenated intermediates such as tetrachlorophthalic or tetrabromophthalic acids and/or dibromopentyl glycols [64]. Although styrene is the most commonly used cross-linker, other monomers can be substituted, such as methyl methacrylate for improved clarity and weather resistance, or a multi-functional monomer such as diallyl phthalate for improved electrical and stability properties.

The unsaturated polyesters are cured by adding free radical catalysts or initiators such as methyl ethyl ketone peroxide, plus an accelerator such as cobalt naphthenate, diethyl aniline and dimethyl aniline [64]. Anhydrides can be used to avoid the problem of by-product formation, thereby reducing the processing difficulty. Long chain reactants may be used to improve chemical resistance, thermal stability and shrinkage.

Vinyl ester resins are made by the addition of unsaturated carboxylic acids (methacrylic or acrylic acid) to an epoxide resin (usually of the bisphenol A-epichlorohydrin type), as shown in Fig. 9 [65]. The unsaturation (C = C double bonds) occur only at the ends of a vinyl ester molecule. Therefore, cross-linking can take place at the ends. The vinyl ester resins are usually cross-linked with styrene or other vinyl-type reactive monomer when used in composites. As with the unsaturated polyester resins, cross-linking proceeds through a free radical

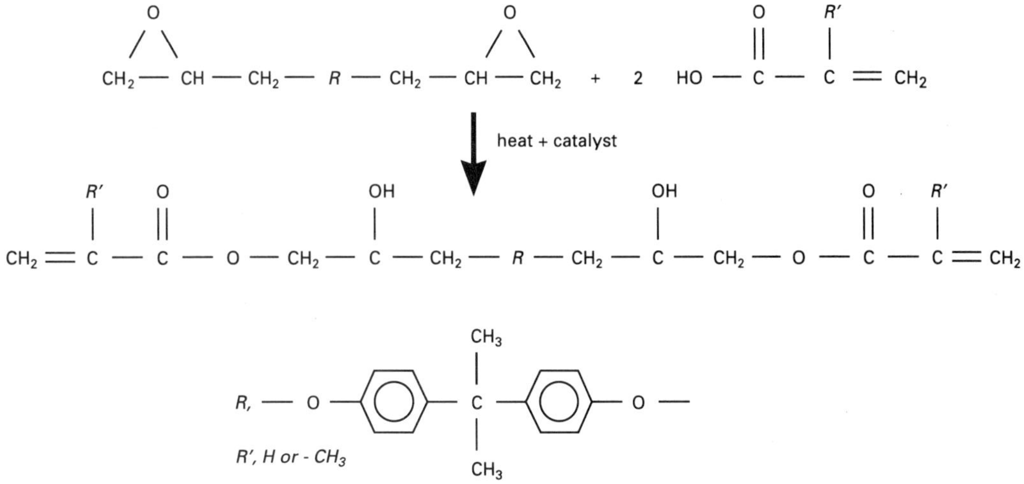

Fig. 9   Preparation of vinyl ester resin

mechanism (addition-type reaction). Due to fewer cross-links, cured vinyl ester resins typically are more flexible and have higher fracture toughness than polyester resins. A number of hydroxyl (OH) groups exist along the vinyl ester chains, allowing hydrogen bonds to form with similar groups on the glass fiber surface. This results in better fiber wet-out and improved adhesion with glass fibers, as compared with polyester resins. However, they only exhibit moderate adhesive strengths compared with epoxy resins.

Although vinyl ester resins are sometimes classified as polyesters, they are typically di-esters that contain recurring ether linkages. Unlike polyesters, vinyl ester resins possess a low ester content and a low vinyl functionality, which result in greater resistance to hydrolysis with less susceptibility to hydrolysis than the acrylate vinyl ester resins and are therefore preferred in applications requiring chemical resistance. The heat deflection temperature and thermal stability can be improved by using more heat-resistant parent epoxy resins such as phenolic-novolac types. The volume shrinkage (5 to 10%) of vinyl ester resins, although lower than that of polyester resins, is higher than that of the parent epoxy resins. The largest use of the vinyl ester resins in composites is in chemically resistant equipment such as pipe, ducts, scrubbers, and storage tanks [65].

Epoxy resins were first used for composite applications in the early 1950s. This family of oxirane-containing polymers can be made from a wide range of starting components and provide a broad spectrum of properties. Their good adhesion characteristics with glass, aramid and carbon fibers have resulted in remarkable success as matrices for fiber composites. They also have a good balance of physical, mechanical, and electrical properties and have a lower degree of cure shrinkage than other thermosetting resins such as polyester and vinyl ester resins. Other attractive features for composite applications are relatively good hot/wet strength, chemical resistance, dimensional stability, ease of processing, and low material costs. Epoxies may be more expensive than polyester resins and may not perform as well at elevated temperatures as polyimide resins, but overall their properties are excellent. Over the past two decades the heat stability and resistance of this family of resins have been improved considerably to meet the requirements of aerospace industry. Epoxies for aerospace composites are generally cured at 177 °C (351 °F). Problems with moisture pickup and loss of some properties may occur at 120 to 160 °C (248 to 320 °F) with this class of materials.

Epoxy resins are characterized by the existence of the epoxy group, which is a three-membered ring with two carbons and an oxygen (Fig. 10). This epoxy group is the site of cross-linking and provides for good adhesion with solid substrate like a reinforcement surface. Many epoxies use the slightly modified epoxy group, called glycidyl, containing one additional carbon. Aromatic groups are often chosen for improved stiffness, thermal stability, and higher glass transition temperature. Figure 10b and c shows an example of epoxide reacting with a curing agent (diethylene triamine, DETA) to form cross-links, eventually leading to the formation of a three-dimensional network.

Four more commonly used epoxy resin systems are shown in Fig. 11. Diglycidyl ethers of bisphenol are the most widely used epoxy resin. They have found applications in protective coatings, adhesives, sealants, impregnants, bonding and laminating materials. Low molecular weight resins are liquid, while high molecular weight versions are in solid powder form. Epoxidized phenolic novolacs resin is available in solid or liquid form, depending on the molecular weight. This resin generally reacts more quickly than standard epoxies, and it is considerably more thermally stable. Tetraglycidyl ether of tetrakis (hydroxyphenol) affords greater cross-linking density and hence better heat stability. An example is the Epon®-1031 from Shell Chemical Co., which in solid form is usually used in conjunction with liquid nadic methyl anhydride cure agents to achieve greater fluidity. Tetraglycidyl ether of 4,4-diaminodiphenyl methane resin is a viscous liquid at room temperature. This tetra-functional epoxy resin (Ciba-Geigy's MY-720) has been widely used as the matrix for graphite fiber composites. Both this resin and its curing agent (anhydrides) are hygroscopic and care must be taken to avoid moisture pickup during the application and lay-up procedures.

Usually epoxy resins are used in conjunction with a curing agent to reduce curing time and to achieve desirable properties. Properties such as chemical resistance, thermal stability, and glass transition temperatures are controlled by curing agents. Anhydrides provide good electrical insulating properties, thermal resistance, and environmental stability. Aromatic amines give higher thermal resistance, but require a higher cure temperature. Aliphatic amines lead to fast cures and are suited to room temperature curing of epoxy. The properties of different types of cast epoxy resins all cured with aromatic diamines are compared in Table 10 [66].

In addition to hardeners, catalytic curing agents (Lewis acids) or accelerators may be used to aid in the curing of epoxide resins. These agents include boron trifluoride, complexes of boron trifluoride and monoethyleneamine, and stannous octoate. Diluents are sometimes added to reduce the resin viscosity, improve shelf and pot life, lower the exotherm, reduce shrinkage, and possibly lower the material cost. Examples of diluents include butyl glycidyl ether, cresyl glycidyl ether, and phenyl glycidyl ether.

Table 10 compares the properties of typical epoxy and polyester resins. These properties, obtained from specimens prepared by casting the uncured resins into molds followed by curing, are subject to wide variations depending on the materials used and the curing conditions. This table indicates that the epoxy resins have lower shrinkage on curing and a lower coefficient of thermal expansion, which are very important properties in the performance of composites. Aerospace applications require that the toughness of the first-generation epoxy resins be improved considerably to impart necessary interlaminar fracture toughness and damage tolerance to carbon fiber composites. Approaches used to improve the

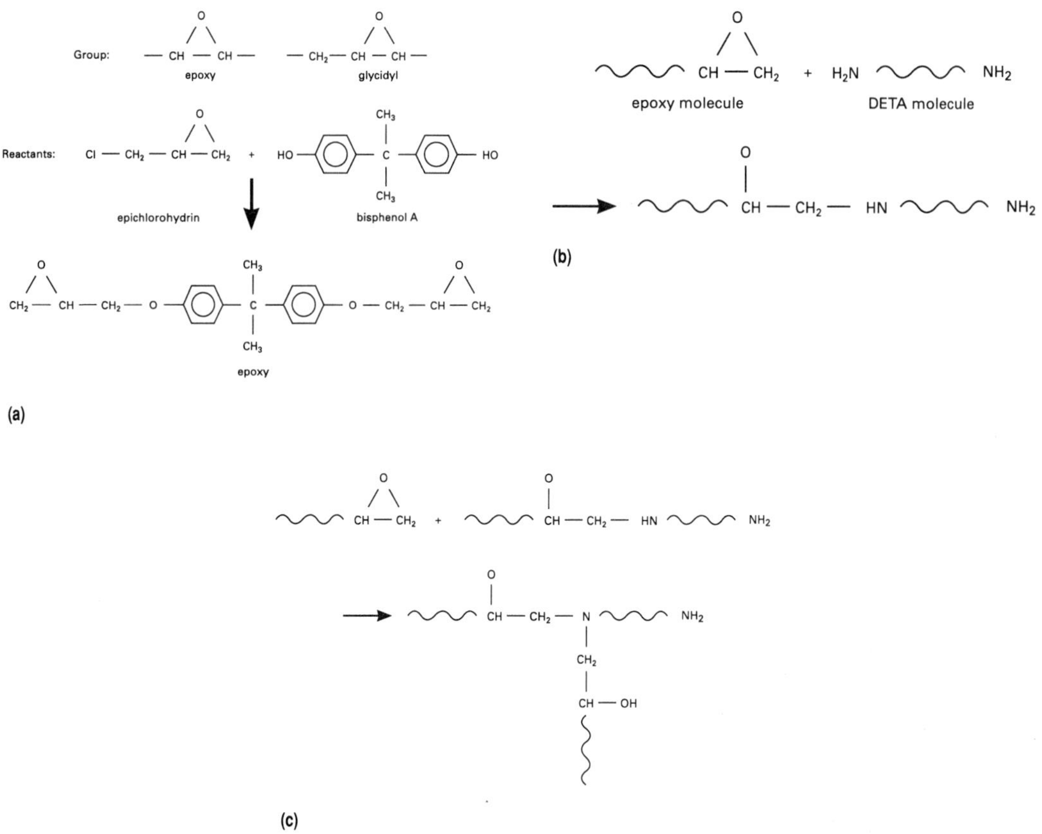

Fig. 10 Preparation of epoxy resins. (a) Epoxide groups and related reactants. (b) Reaction of an epoxide group with a DETA molecule. (c) Formation of cross-links

epoxy resin toughness have included rubber-toughening, forming a lightly cross-linked network, thermoplastic toughening, and forming an interpenetrating network (IPN) or semi-IPN.

Phenolic resins are produced by condensation or stepwise growth polymerization from the reactions of phenols with aldehydes, in which water is formed as a byproduct. In the initial stage (A-stage), the polymer is of low molecular weight, soluble and fusible. As the condensation reaction continues, more molecules are involved and the resin becomes a rubbery, thermoplastic phase which is only partially soluble (B-stage). The resin is then cured to a fully cross-linked, intractable material (C-stage). The low molecular weight prepolymers can be classified into either resoles or novolacs, depending upon whether the phenol or aldehyde is used in excess during A-stage. Resoles are obtained when an excess of formaldehyde is allowed to react with phenol under basic conditions, whereas novolacs are produced with an excess of phenol under acidic conditions. While resoles are cured by simple heating that allows condensation of the methylol groups to result in gelation (one-step systems), a cross-linking agent is necessary for network formation with the novolacs (two-step systems).

Lower in cost than epoxies or polyimides, phenolic resins still have their share in important markets such as binders, adhesives, molding, coatings, and laminating compounds. In the areas of advanced composites, phenolic resins have gradually given way to the epoxy resins because of the better adhesion, lower cure shrinkage, and the absence of volatiles formed during curing of epoxy resins. The reaction by-products that could lead to void formation include water, ammonia, formaldehyde, and possibly decomposition by-products of the cross-linking agent when used.

Phenolic resins are the precursor to carbon matrix systems, which are of increasing importance in the field of carbon-carbon composites. This is largely due to the high char yield and strength of phenolic resins. Phenolic resins also function as cure agents for some of the epoxies, through their hydroxyl group, as well as serving as raw materials for epoxidation. The epoxidized phenolic-novolacs are available as candidate materials for advanced composites. The phenol or thiophenol group in epoxy resin polymerization serves to accelerate the rate of reaction. The low level of flammability and minimum smoke evolution associated with select phenolic resins are desirable properties in many industrial applications.

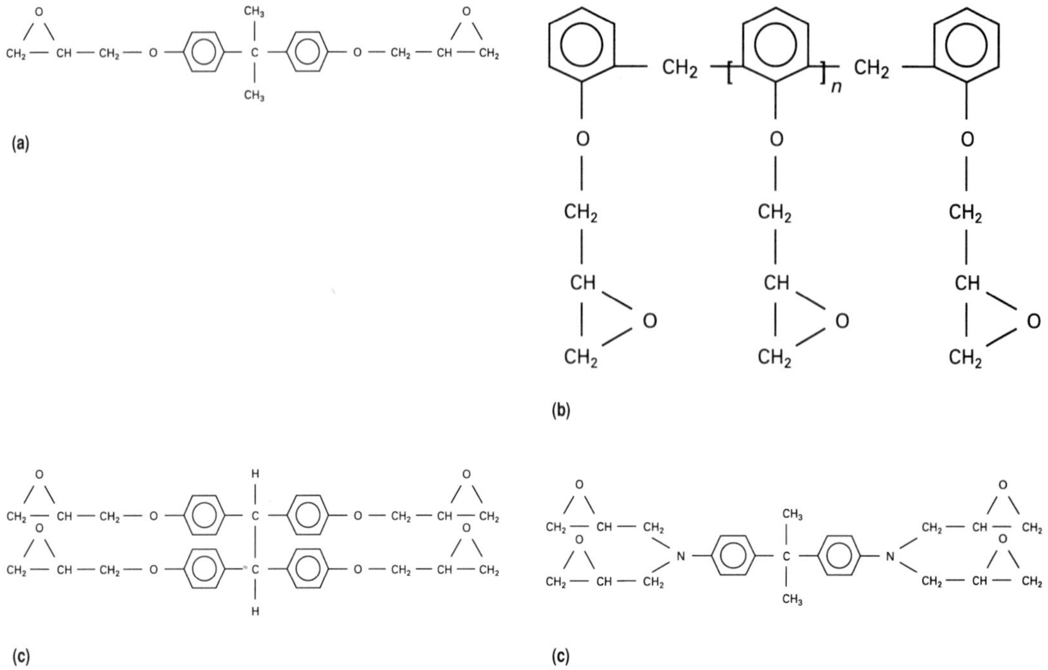

Fig. 11 Four basic epoxy resins used in advanced polymer composites. (a) Kiglycidyl ether of bisphenol A (DGEBA); (b) Epoxy novalac (e.g., Dow DEN 438); (c) Tetraglycidyl ether of tetrakis (e.g., Shell EPON 1031); and (d) Tetraglycidyl ether of 4,4-diamino diphenyl methane (e.g., Ciba Geigy MY-720).

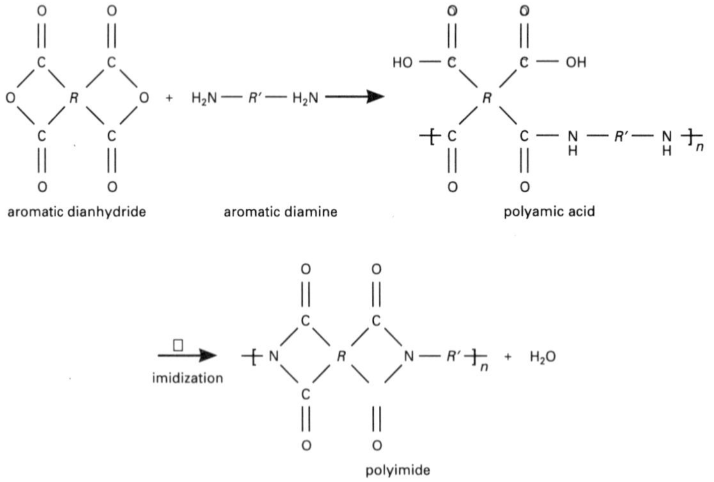

Fig. 12 Formation of a condensation polyimide. Source: Ref 33

Conventional phenolics are mainly reinforced with glass fibers, except as a precursor to carbon-carbon composites. Glass-reinforced phenolics exhibit excellent initial retention of strength but a considerable drop in properties is observed after long-term heat aging [67]. A much more recent class of phenol-based thermosets is phenol-aralkyl resin, which is obtained by the condensation of a dimethoxyl compound with phenols [68,69]. Cross-link-ing is achieved by heating with hexamine or a di- or poly-epoxide. Hexamine cure leads to the evolution of ammonia during the reaction, making it difficult to fabricate thick sections. Similar to the conventional phenolic resins, phenol-aralkyls possess very good flame resistance and low smoke generating characteristics.

Polyimides. For high temperature composites applications, epoxy resins appear inadequate while polyimide

Table 10 Comparison of properties of cast epoxy resins

Property	Bisphenol-A epoxy DDM cure	Bisphenol-A epoxy DDS cure	Cyclo-aliphatic epoxy MPD cure	Tetrafunctional epoxy DDS cure	Polyester resins
Tensile strength, MPa					
at 20 °C	53	59	89	41	40 to 90
at 150 °C	19	37	78	33	...
Tensile modulus, MPa					
at 20 °C	2750	3070	6280	4100	2000 to 4500
at 150 °C	1540	1470	432	3230	...
Compressive strength, MPa					
at 20 °C	>111	107	227	178	90 to 250
at 150 °C	>29	63	65	115	...
Compressive modulus, MPa					
at 20 °C	2670	2000	4130	2930	2000 to 4000
at 150 °C	721	1280	1290	2200	...
Flexural strength, MPa					
at 20 °C	116	102	159	86	...
at 150 °C	41	49	84	86	...
Flexural modulus, MPa					
at 20 °C	2730	2790	6450	3920	...
at 150 °C	1680	1160	2640	3200	...
Impact strength, MN/mm					
at 20 °C	0.21	0.17	0.21	0.083	...
at 150 °C	0.19	0.21	0.15	0.073	...
Elongation at break, %					
at 20 °C	4.9	3.3	2.1	1.1	2
at 150 °C	2.7	8.0	12.3	1.0	...

DDM, diaminodiphenylmethane; DDS, diaminodiphenylsulfone; MPD, m-phenylene diamine Source: Ref 66

and bismaleimide resins are good candidate materials. However, these two materials are more expensive than epoxies. Some polyimide resins can be used at temperatures up to 316 °C (600 °F), or even 370 °C (698 °F). Polyimides have predominantly found markets in high-quality and performance applications such as aerospace and electronics, where the increased cost can be justified. When used in conjunction with graphite fibers, polyimide composites are used in high-speed missiles and space shuttle type transportation systems where, depending on configuration, location, and thermal protection system, temperatures over 260 °C (500 °F) could be experienced. The aromatic-heterocyclic structure of the polymer backbone imparts the great elevated temperature stability to this class of material. The chemical structure of polyimide is schematically shown in Fig. 12, where R and R' can be varied. This type of structure is thermally and thermo-oxidatively stable, with a high glass transition temperature. Three important classes of polyimide resins have been used as composite matrices:

- Condensation polyimides
- Polymerization of monomeric reactants (PMR) polyimides
- Bismaleimides (BMI)

Condensation Polyimides. Figure 12 shows the two reactions involved in the two-step process of preparing a typical condensation polyimide. In the first step, an aromatic tetracarboxylic dianhydride (benzophenone tetra-carboxylic dianhydride, BTDA) and an aromatic diamine (methylene dianiline, MDA) first combined to form a polyamic acid, which is soluble in aprotic solvents such as N-methylpyrolidonc (NMP), dimethylformamide (DMF), dimethylacetamide (DMAc), and dimethylsulfoxide (DMSO). This solution can be used to impregnate reinforcements to produce prepregs. Because of the high viscosity associated with polyamic acids, prepregs are preferably made by impregnating the fibers with monomer solutions.

The next reaction step then involves cyclization of the amic acid structure to the final imide ring structure [33,70], chain extension of the polymer, and cross-linking (if desired). The imidization, however, involves the release of condensation by-products such as water and the loss of residual solvent or volatiles. These volatiles have made it difficult to fabricate void-free composites. One way to minimize porosity in the cured composite is to carefully design a cure cycle, which may contain long hold times at various temperatures to permit the gradual evolution of volatiles and may also require the application of pressure at precise moments in the cycle.

Major condensation polyimides include Avimide K (T_g 254 °C, or 489 °F) and Avimide N (T_g 349 to 371 °C, or 660 to 700 °F) from Du Pont. In both materials, essentially linear polymers are produced which have thermoplastic behavior above their T_gs. They can be processed like thermoplastics for only a few cycles. Subsequent processing, if possible at all, requires higher

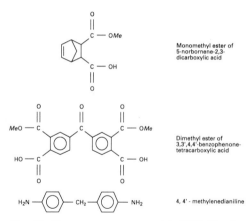

Fig. 13 Component reactants used in PMR-15 polyimide resin. Source: Ref 71

temperatures and pressures. In Monsanto's Skybond polyimide, some cross-linking appears to occur during the cure of the polyimide precursor solutions by the reaction of the amine end-groups with the carbonyl bridge.

Some performance characteristics for an E-glass reinforced condensation polyimide are shown in Table 10. These composites appear to possess outstanding properties when exposed to thermooxidative environments. The properties of the final polyimide product can be tailored by the choice of R' and R. Polyimide composites have been used in aerospace structures such as radomes and engine components.

PMR Polyimides. Originally developed at the NASA Lewis Research Center [71], PMR resins contain monomeric reactants which are soluble in low boiling, easily removed solvents such as methanol. Removal of these solvents is relatively easy as compared to the solvents used in processing condensation polyimide composites. Starting materials (Fig. 13) dissolved in methanol or ethanol are used to impregnate a variety of reinforcements. The processing of PMR prepregs consists of a series of steps. In the first step, the solvent is removed at a relatively low temperature (50 to 120 °C, or 122 to 248 °F). This is followed by the early stage of curing, which involves reaction of the monomeric reactants and initiation of imidization at about 205 °C (401 °F). This imidized structure is end-capped with the norbornene functionality. At higher temperatures (275 to 350 °C, or 527 to 662 °F), the resin is thought to undergo the final cross-linking reactions by the reverse Diels-Alder reaction which does not involve volatile evolution and therefore makes for easier composite processing. Normally the composite is then subjected to a postcure treatment to ensure completion of the reactions.

Early research efforts on PMR systems showed that the molecular ratio between the monomers affect the thermal stability and other physical properties of the polymer. The best overall balance of properties was obtained with a molecular ratio corresponding to a formulated molecular weight of 1500, hence the name PMR-15.

PMR/carbon composites exhibit good flexural and short beam shear properties and good retention of properties up to 316 °C (601 °F). One of the most serious shortcomings of this class of materials is the brittleness and low fracture toughness of the matrix resins and hence the poor damage tolerance of composites. This problem can be alleviated to some extent by the various toughening mechanisms.

Another type of end-group cross-linking polyimide is the acetylene-terminated polymer (Thermid 600). Chain extension occurs by joining two polyimide oligomers (with triple bonds near chain ends) to form a butadiene linkage, which is composed of two adjacent double bond segments. Each of these double bonds can then react with double or triple bonds to form a highly cross-linked network. The fully cured resin is typically hard and brittle.

BMI Polyimides. The baseline bismaleimide (BMI) is formed by the reaction of a diamine with maleic anhydride:

$$R' \diagdown \overset{\overset{O}{\parallel}}{\underset{\underset{O}{\parallel}}{\diagup}} O + H_2N - R - N_2H \rightarrow R' \diagdown \overset{\overset{O}{\parallel}}{\underset{\underset{O}{\parallel}}{\diagup}} N - R - N \overset{\overset{O}{\parallel}}{\underset{\underset{O}{\parallel}}{\diagdown}} R'$$

The group R, either aromatic or aliphatic, can be varied to attain a variety of properties. This BMI can be used by itself, with other BMIs as a comonomer, or with diamines to form a final resin system. If diamines are used, two steps are involved in the cure of the final resin: a Michael addition reaction of the diamine across the double bond and a free radical polymerization of the double bonds. The cure of a BMI without diamines proceeds via the free radical reaction only [70]. The BMI monomers and possibly other comonomers can be dissolved in solvents to form varnishes for fiber impregnation. They can also be formulated with reactive diluents to allow for solventless, hot-melt processing of composites. Various prepreg products can be achieved with BMIs, ranging from dry and boardy to tacky and drapable epoxy-like prepregs with a variety of fiber reinforcements.

Cure conditions of BMI-based materials vary with the product form. Dry, boardy prepregs and molding compounds are typically fabricated with temperatures of 230 to 280 °C (446 to 536 °F) under high pressure using a compression press or an injection molding machine, respectively. Wet, solvent-based or hot-melt solventless prepregs are processable with epoxy-like cure conditions [70]. They can be vacuum-bag, autoclave cured at temperatures of 177 °C (351 °F) and pressures of 690 kPa (100 psi) and then postcured at temperatures of 210 to 260 °C (410 to 500 °F).

BMIs in general are not as thermally stable as condensation or PMR polyimides, but are much better than epoxies in thermo-oxidative stability. Their ability to provide higher service temperatures than epoxies while maintaining epoxy-like processing have made them good candidate materials for aircraft applications such as wings or fuselage [70].

In common with other polyimides, BMIs suffer from extreme brittleness, which can be improved by a number of approaches. One approach involves increasing the chain length between the terminal maleimide groups to decrease the cross-link density, hence improving the material's ability to undergo plastic deformation. Other techniques include judicious choice of co-reactants, addition of a toughening phase, or Michael addition reactions with aromatic diamines. The thermoplastics which have been successfully used to toughen BMI include polysulfone, polyetherimide, polyhydantoin and poly(arylene ether ketone) [72,73].

Thermoplastic Resins

Although thermosetting resins have played a dominant role in the composite industry over the past three decades, these resins typically exhibit low ductility, fracture toughness and impact resistance, leading to a composite with poor damage tolerance. Even though thermoset resins can be toughened by several techniques, one alternative solution to this brittleness problem is to specify a high performance thermoplastic resin as the composite matrix. Thermoplastics typically consist of high molecular weight, linear or slightly branched chains. In thermosets the chains are covalent-bonded together to form cross-links, severely limiting the resin's ability to undergo plastic deformation. In contrast, the physical entanglements between chains impart strength to a thermoplastic while still permitting a high level of plastic flow mainly through chain stretching and pulling. The effect of entanglements become significant when the chains exceed a critical molecular weight (M_c). With an increase in molecular weight, the polymer will exhibit an improved performance at the cost of increased viscosity and therefore difficulty in processing.

Most thermoplastic resins for high performance composite applications are synthesized by condensation routes. They must meet certain requirements to attain high molecular weights, including difunctional monomers of extremely high purity, no side reactions, a tight control on the reactant stoichiometry and accessibility of reactive end groups [74]. Molecular weights may be controlled by either incorporating a monofunctional reactant or by slightly offsetting the reaction stoichiometry. These methods are also useful in controlling the end groups, yielding either potentially reactive or non-reactive chain ends. An important function of non-reactive end-caps is to provide well defined and stable melt flow. Most advanced thermoplastics to be processed by compression, transfer, or injection molding have controlled molecular weights and end caps [74]. The properties of a given thermoplastic resin are much influenced by the degree of crystallinity, crystalline morphology and orientation, which are in turn governed by the processing conditions. The presence of reinforcement fibers may also affect the crystallization behavior of a semi-crystalline matrix resin, forming a "transcrystalline" zone at the fiber-matrix interface.

The chemical structures of some thermoplastic polymer composite matrices are shown in Fig. 14. These polymers are heavily aromatic in nature, usually with flexibilizing groups in the backbone to reduce the processing difficulty. Those polymer chains which do not possess bulk side groups tend to exhibit some crystallinity because of better structural symmetry. Examples of these semi-crystalline polymers include poly (phenylene sulfide) (PPS), poly (ether ether ketone) (PEEK), and poly (ether ketone) (PEK). In general, the glass transition temperature (T_g) of a thermoplastic, either semi-crystalline or amorphous, determines the upper service temperature of composites. However, the semicrystalline polymers retain some stiffness and strength above T_g, and the retention depends on the degree of crystallinity. Under normal processing conditions, PPS can develop up to 50% crystallinity, and PEEK and PEK up to 40%. The upper service temperature requirements are approximately 100 °C (212 °F) for civil transport aircraft and helicopters, up to 175 °C (347°F) for fighter aircraft [75] and much higher temperatures for applications such as the National Aerospace Plane (NASP). Selected physical and mechanical properties of leading thermoplastic matrices are presented in Table 11. Commercial sources, chemical names and crystallinity of these resins are given in Table 12 [75].

Polyarylene Ethers. This family of polymers are prepared by either a nucleophilic or electrophilic reaction [76-78]. Amorphous polymers of this family, (for example, poly (ether sulfone) (ICI Victrix®), polysulfone (AMOCO UDEL®) and poly (phenyl sulfone) (AMOCO Radel®)), are excellent engineering thermoplastics, but do not provide adequate chemical resistance for high performance applications. Among the most advanced poly (arylene ethers) for composite applications is PEEK, a semi-crystalline, aromatic polymer. The semi-crystalline polymers possess higher moduli and yield strengths compared to the amorphous versions. Toughness values of these semi-crystalline polymers vary considerably, partially due to the different degrees of crystallinity and different crystal morphologies. The heat distortion temperatures of the semi-crystalline polymers are similar to or greater than T_g (Table 11), suggesting that mechanical properties are retained at temperatures up to or above the T_g. However, there are concerns in using polymers above their T_g due to potentially high creep effects. The semi-crystalline polymers, in general, possess high resistance to most environments due to the crystalline order that makes permeation of foreign molecules difficult. The amounts of the absorbed moisture and their effects on properties are generally negligible. At the present time, PEEK is available from ICI in either natural grades (Victrex), or as continuous fiber-reinforced prepregs (APC-2).

Thermoplastic Polyimide. Linear high molecular weight polyimides are typically synthesized by a solution reaction of an aromatic dianhydride and an aromatic diamine. An intermediate poly (amic acid) is obtained, which is then cyclodehydrated by chemical or thermal means to form the polyimide [79,80]. Several other stepwise or condensation reactions can also be used to yield high molecular weight polyimides [74]. A fully cyclized polyimide typically is intractable and difficult to process.

	Tg	
	°C	°F
Poly(phenylene sulfide)	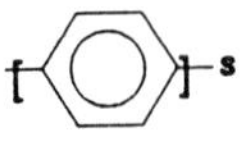 85	185
Poly(ether ether ketone)	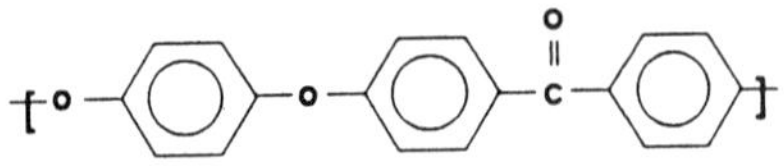 143	289
Poly(ether ketone ketone)	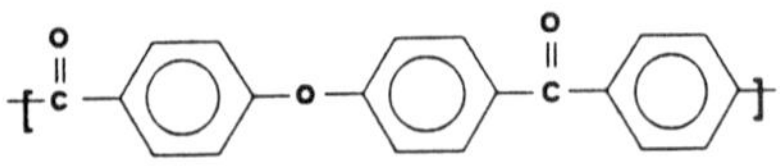 156	313
Poly(ether ketone)	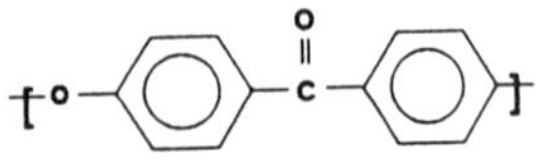 165	329
Poly(sulfone)	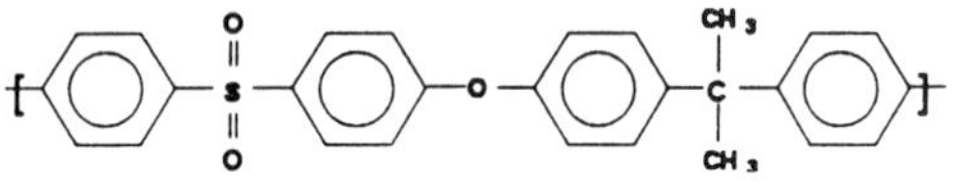 190	374
Poly(phenylene sulfide sulfone)	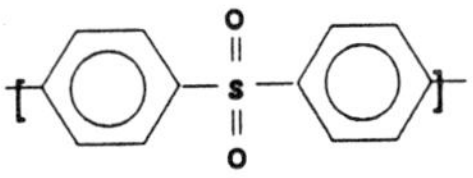 213	415
Poly(ether imide)	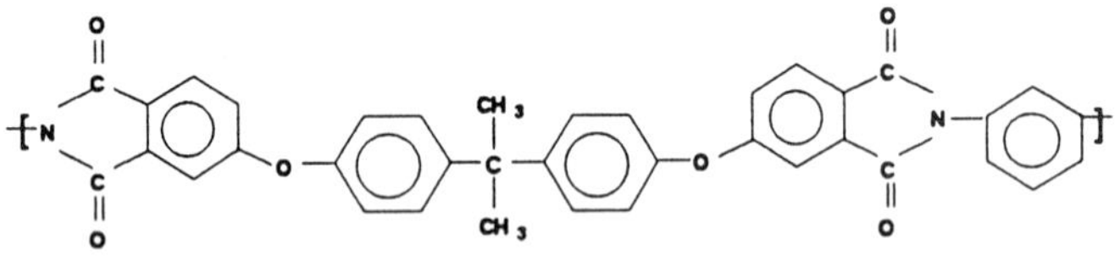 216	421
Poly(phenyl sulfone)	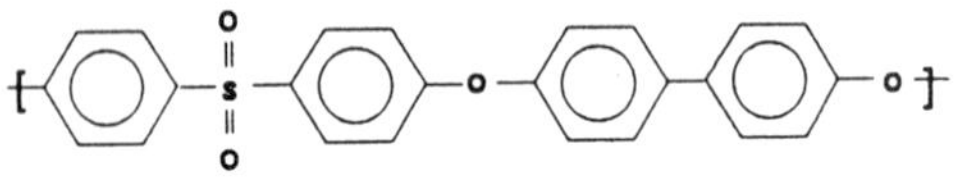 220	428
Poly(ether sulfone)	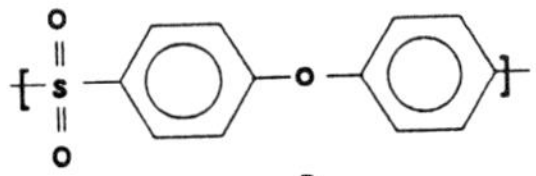 230	446
Poly(amide imide)	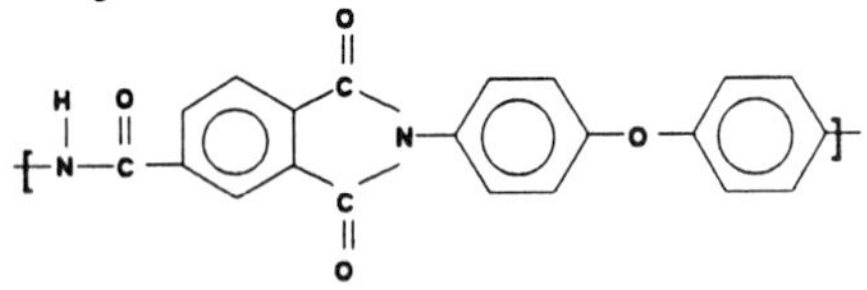 249 to 288	480 to 550
Poly(imide)	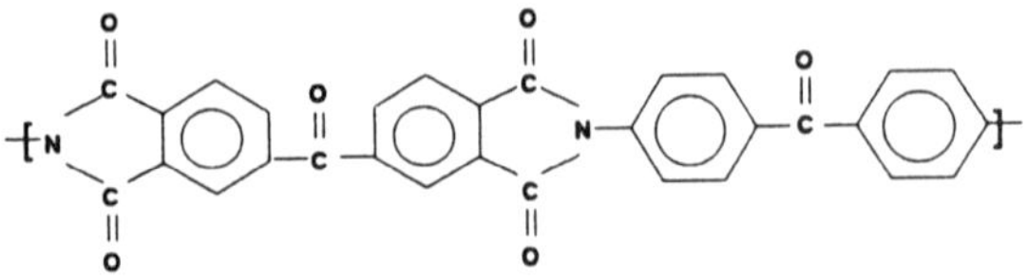 256	493
Poly(imide sulfone)	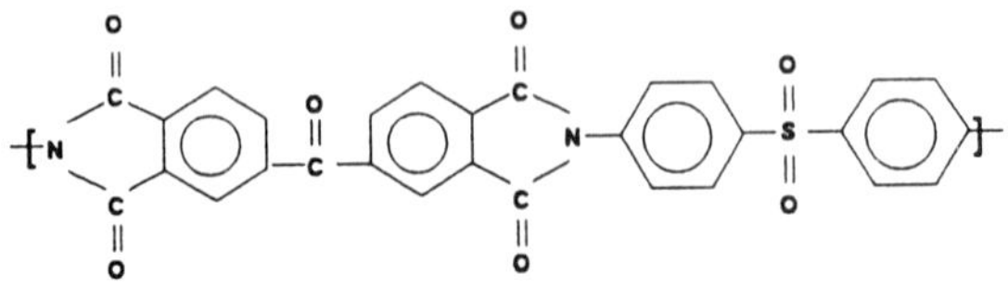 273	523

Fig. 14 Chemical structures of thermoplastic resins for advanced composites

However, processability can be improved by incorporating flexible linking units or bulky side groups into the polyimide backbone, or by co-polymerizing with more flexible components [81,82]. Alternatively, poly (amic acid) solution can be processed directly, followed by a thermally-induced cyclodehydration and removal of the solvent and by-product (water).

Table 11 Properties of select thermoplastics used as composite matrices

	Semi-crystalline polymers				Amorphous polymers				Typical epoxy
		VICTREX							3501-6
Property	RYTON PPS	PEEK 450G	PEK	LARC-TPI	VICTREX PES 4100G	UDEL P1700	ULTEM 6000	AVIMID K-III	Hercules
Specific gravity	1.36	1.30	1.30	1.40	1.37	1.27	1.29	1.31	...
Tensile modulus, GPa	3.4	3.7	4.0	3.5	2.6	2.6	3.0	3.7	3.8
Tensile yield strength, MPa	79	92	105	166	84	70	103	72	46
Elongation, %	2 to 20	50	5	9	40 to 80	50 to 100	30	14	1
Fracture toughness, kJ/m^2	...	4.8	...	...	3.0	2.5	...	1.5	0.08
Notched Izod impact strength, J/m	21	83	...	21	84	69	40	...	...

Source: Ref 75

Table 12 Select thermoplastic polymers: manufacturers, chemical names, and morphology

Manufacturer/ Trade name	Polymer	T_g °C	T_m °C	T_p °C	Morphology (a)(b)
Amoco					
UDEL	Polysulfone, PS	190	...	300	A
RADEL	Poly(phenyl sulfone), PPYS	220	...	330	A
PXM	Poly(arylene ketone), PAK	265	...	...	A
Torlon	Poly(amide imide), PAI	245 to 280	...	400	A(P)
XYDAR	Aromatic polyester, APE	350	421	400	LC
BASF					
ULTRAPEK	Poly(ether ketone ether ketone ketone), PEKEKK	175	375	400	SC
Du Pont					
PEKK	Poly(ether ketone ketone), PEKK	156	180	338	SC
J-2	Polyamide, PA	160	...	300	A
AVIMID K-III	Polyimide, PI	250	...	350	A(P)
NR150B2	Polyimide, PI	350	...	350 to 400	A(P)
General Electric					
ULTEM	Poly(ether imide), PEI	216	...	360	A
Hoechst-Celanese					
FORTRON	Poly(phenylene sulfide), PPS	85	285	340	SC
VECTRA	Aromatic polyester, APE	70	280	285	LC
HOSTATEC	Poly(ether ketone), PEK	167	360	400	SC
PISO$_2$	Polyimide sulfone, PISO$_2$	273	...	350	A(P)
VICTREX PEEK	Poly(ether ether ketone, PEEK	143	334	380	SC
VICTREX PEK	Poly(ether ketone), PEK	162	365	400	SC
VICTREX PES	Poly(ether sulfone), PES	230	...	350	A
Mitsui-Toatsu					
LARC-TPI	Polyimide, PI	264	275 to 305	350	SC(P)
"New TPI"	Polyimide, PI	250	388	400	SC (P)
Phillips					
RYTON	Poly(phenylene sulfide), PPS	90	290	340	SC
PAS-2	Poly(arylene sulfide), PAS	215	...	329	A
RYTONS	Poly(phenylene sulfide sulfone), PPSS	213	...	300 to 340	A
DURIMID	Polyimide, PI	256	305	340	SC(P)

(a) A, amorphous; LC, liquid crystal; SC, semi-crystalline. (b) Chemical nature conventional unless noted with a (P) for pseudo-.
Source: Ref 75

As discussed earlier, thermosetting polyimides (addition types) are easier to process with little by-products evolved, but exhibit higher cure shrinkage and extreme brittleness. In contrast, the condensation-type polyimides are more difficult to process; the amic acids tend to have a poor shelf life. However, thermoplastic polyimides exhibit good toughness, fatigue resistance, thermo-oxidative stability and a good retention of properties at elevated temperatures. New commercially available thermoplastic polyimides for potential composite applications include Mitsui Toatsu's LARC-TPI and "New TPI" [83], Hoechst-Celanese's fluorinated polyimides [84] and Ethyl's Eymyd series [85].

Other imide-containing thermoplastic matrix resins include poly(ether imide) (PEI), polyamide imide (PAI) and poly(imide sulfone) (PISO$_2$). The imide link in PEI is very stiff and is very thermally stable. However, in the case of PEI the imide is present in relatively low proportions and therefore the polymer has a relatively low T_g (216 °C, or 421 °F) but is easy to process [75]. The polymers containing a high imide content may have very high T_g values but can also be difficult to process due to the higher viscosities imparted to them by the rigid imide group. This is exemplified by poly (amide imide) which has a very high viscosity and requires annealing to achieve the higher T_g and improved properties.

Poly(phenylene Sulfide) (PPS) is a semi-crystalline, aromatic thermoplastic which exhibits excellent chemical and thermal stability, and resistance to prolonged combustion. The tensile properties of PPS at room temperature are comparable to or slightly lower than those of other high performance thermoplastics discussed so far. However, their impact properties are typically on the lower end of the spectrum. Perhaps the most serious problem of this resin is the low T_g (85 °C, 185 °F) which falls short of aerospace requirements, even though the melting point is as high as 285 °C (545 °F).

At high temperatures in the presence of oxygen, PPS resins tend to undergo cross-linking and/or chain extension, resulting in significant property variations. These variations include increased molecular weight, enhanced toughness, higher melt viscosity, reduced solubility and lower crystallinity. A recent version of PPS developed by Phillips was reported to have improved adhesion property with carbon fibers and provide a better impact resistance to carbon fiber composites.

Polybenzimidazole (PBI) resins are less prevalent than polyimides, but have equivalent and sometimes superior physical and thermal properties. PBI may be produced by reacting aromatic diamines with aromatic diacids to form relatively low molecular weight prepolymers. Polymers prepared from 3,3'-diaminobenzidine and diphenyl isophthalate have shown good stability. But the reactions could yield significant amounts of phenols and water as by-products, causing processing difficulties. This problem can be reduced by replacing the diphenylester with isophthalamide in the reaction. The processing cycle of PBI is fairly long and complex, and the development of the optimum properties demands the utilization of high temperatures and an inert atmosphere.

The T_g of PBI can approach 420 °C (788 °F) and are potentially useful in high temperature laminate and adhesive formulations [86,87]. PBI provides composites with good mechanical properties to about 250 °C (482 °F), and offer good stability after aging [74].

Aromatic liquid crystalline polymers (LCPs) are characterized by a train of para-oriented ring structures to give a stiff chain with a high molecular length-to-width ratio. LCPs have superior tensile properties along the axis of orientation and yet are easy to process. Melt viscosity can be low and relatively insensitive to temperatures and shear rates. However, good mechanical properties in a typical composite may be difficult to achieve. The liquid crystalline morphology may not be possible in a densely packed reinforcement network. The LCPs would have a major property advantage over the other groups of polymers, provided the polymers could be oriented to provide the maximum properties in the desired directions [88,89].

Comparison Between Thermoplastic and Thermoset Matrices

To specify a composite for a given application, difficulty may arise in choosing between thermoset and thermoplastic matrices. A proper decision may not be made without a clear understanding of the relative advantages and disadvantages associated with each class of materials. In this section, we discuss the processing and performance aspects of both matrix types with emphasis on current needs and issues.

The composites industry is currently dominated by thermosetting resins because of their historical availability, relative ease of processing, lower cost of capital equipment for processing, the existence of a large database for commercially available resins, and low material cost [74]. Thermosetting resins such as epoxies are available in oligomeric or monomeric low-viscosity liquid forms that have excellent flow properties to facilitate resin penetration of fiber bundles and proper wetting of fiber surface by the resin. The quality and reproducibility of epoxy prepreg have improved greatly over the last decade. When handled properly, the prepreg can be used to yield high-quality composite parts with low scrappage. In contrast, thermoplastics have not reached this level of sophistication and problems still exist with poor resin uniformly on the fiber, poor fiber wetting and excess void formation. Nevertheless, thermoplastics do offer other significant advantages, among them cost, quality control, and toughness and damage tolerance.

Manufacturing Cost. Although currently the cost for manufacturing thermoplastic composites is rather high, it can be significantly curtailed in the future in view of the following observations. Most epoxy prepregs are formulated to last for approximately two weeks on the shelf. A large amount of epoxy prepreg is discarded annually because it has exceeded the recommended shelf life. In contrast, high molecular weight thermoplastic prepreg has practically an unlimited out life; no refrigerated storage is required and material is less hygroscopic. In most cases, no solvent is involved and therefore is environmentally safer. Thermoforming of flat sheet stock

Table 13 Toughness of carbon fiber/thermoplastic composites

Composite	Polymer/ Fiber	Interlaminar fracture toughness, G_{IC}, kJ/M^2(a)	Post-impact compressive strength, MPa(b)	Short-beam shear strength, MPa(b)
Unidirectional	**Carbon fibers**			
APC-/AS4	PEEK/AS4	2.1 to 2.7	338	105
RYTON	PPS/AS4	0.8	...	69
J-Polymer	PA/AS4	2.0	283 to 400	86
TORLON	PAI/...	1.2	327	110
Avimid K-III	PI/...	1.4	274	99
LARC-TPI	PI/HTA-7	...	...	80
Woven fabric				
TORLON	PAI/plain weave	1.75	296	90 (Celion 3000/4HS)
Cypac	PEI/plain weave	4.0	320	66
Typical epoxy				
3501-6	Epoxy/AS4	0.22	165	110
	Aramid fibers			
VICTREX	PEEK/Aramid	...	...	44
RYTON	PPS/Kevlar®-49	0.7	...	...
J-Polymer	PA/Kevlar®-49	2.0	...	55 to 62
	Glass fibers			
VICTREX	PEEK/E-glass	3.1	...	95
	PEEK/S-2	2.5	...	70
RYTON	PPS/E-glass	1.3	...	43
	PPS/S-2	1.3	...	32
ULTEM	PEI/E-glass	...	...	70

(a) To convert from J/m^2 to ft · lbf/in.^{2}, multiply by 4.16×10^{-4}. (b) To convert from Pa to psi, multiply by 1.45×10^{-4}. Source: Ref 75

may be possible with thermoplastic composites. Manufacturing flaws can possibly be corrected with re-processing and processing cycles are much faster.

Quality Control. Epoxy formulations are generally more complex because of the large number of ingredients involved: base epoxies, curing agents, catalyst, flow control agent, and property modifiers. Thermoplastic formulations are much simpler, possibly having only the basic thermoplastic chains or monomers in solution [90].

Toughness and Damage Tolerance. The toughness of thermoplastic composites is higher than that of thermosets and this has provided an impetus for the development of thermoplastic matrices for advanced composite applications. The most commonly used technique to measure the interlaminar fracture toughness (G_{IC}) of composites is the double cantilever beam (DCB) test. The interlaminar fracture toughness values of various thermoplastic composites and a typical epoxy matrix composite is given in Table 13. Although the toughness of Hercules 3501-6/ASA is 0.2 kN/m (13.7 lbf/ft), that of a thermoplastic typically lies in the range of 0.8 to 4.4 kN/m (54.8 to 301 lbf/ft). Another frequently used test to measure the toughness of composites is a compression-after-impact (CAI) test in which a laminate is subjected to a low energy impact followed by a compression test to fracture. This test is used to simulate the in-service requirement to endure low energy impact, with a typical incident energy of 6.7 J/mm (1500 in. · lbf/in.). Table 13 also shows the CAI strength values of various composites. The high interlaminar fracture toughness of thermoplastics imparts a much greater resistance to delamination, which is one of the most common failure modes of laminated composites. This high toughness leads to a much reduced area of delamination during impact, and improved resistance to the propagation of the delamination during subsequent compressive loading [91].

The high interlaminar fracture toughness of thermoplastic matrix composites is normally attributed to the high toughness of the matrix resin (Table 11). However, the fiber-matrix interface also plays a significant role and a strong interfacial adhesion may be needed in order to fully realize the potential toughness of the thermoplastic matrix [92]. In general, the short beam shear strengths of thermoplastic composites are found to be lower than that of a representative epoxy, indicating that the interfacial adhesion in these thermoplastic composites has yet to be optimized.

Despite the above major advantages associated with the thermoplastic composites, many major difficulties have yet to be overcome before these composites can be more widely used. The current problems include prepreg preparation difficulty, composite processing problems, and property variations.

Prepreg Preparation Difficulty. High molecular weight thermoplastics may be prepregged either at high temperatures to achieve the required low viscosity conditions or in polar solvents that must later be removed. Higher molecular weights tend to result in poor fiber wettability and non-uniformity in fiber-resin dispersion. Poor wetting in hot-melt prepregging and the presence of solvent in solution prepregging could produce a high void content. The use of hybridization by commingling, co-weaving or co-winding of reinforcement and matrix filaments to yield thermoplastic prepreg has not reached a high level of sophistication. Also at the developmental stages are the techniques of powder impregnation, extrusion coating and slurry coating. The quality consistency of the prepregs prepared by these new techniques has yet to be demonstrated. Thermoplastic prepreg is generally boardy (except for powder-, slurry-impregnated or commingled ones) and tack-free, unless heat is applied or solvent is added that later chemically react or become volatile.

Composite Processing Problems. Processing of thermoplastic normally involves the use of higher temperatures and higher pressures. Tooling costs are rather high.

Property Variations. Thermoplastic resins are subject to environmental stress cracking and possibly solvent susceptibility. They generally show a higher coefficient of thermal expansion at equivalent fiber volume fractions. Crystalline morphologies in semi-crystalline thermoplastics are influenced by process conditions (cooling rates, post annealing) and the chemical and physical state of fiber surface. Variations in properties can arise as a result of changing processing conditions. Limited database is available for long-term performance (creep and fatigue) of thermoplastic composites.

Material Design of High-Temperature Matrix Resins

High-Temperature Matrix Resins

The previous section has been devoted to the description of most commonly used resins as composite matrices. Many of these polymers are considered to be high temperature polymers. However, the term "high temperature" requires some qualification. The temperature of 93 °C (199 °F) is frequently cited as a lower limit for continuous use temperature for civil aircraft. About 180 °C (356 °F) is required for skins of supersonic jets and military fighter aircraft. Thermally critical areas of aircraft such as engines have to endure a continuous exposure to a temperature of higher than 300 °C (572 °F). Missile applications demand short-term stability at 500 °C (932 °F) or above [93].

Demand for the low density, high specific moduli polymer-based composites for high-temperature applications is increasing, partly due to the trend in pushing commercial and military aircraft toward a status of higher speed and higher performance. The basic requirement for these materials is that they are capable of withstanding high temperature throughout the intended service life designated for the structure. For polymers to endure higher temperature (> 230 °C, or 446 °F) for a long period, the polymers must have high glass transition temperature (T_g) or melting temperature (T_m) and good thermostability, in addition to the other properties required for a specific application. The mechanical properties of polymers normally change dramatically and reversibly near T_g or T_m (depending on the crystallinity). Even when the T_g or T_m is high, degradations such as thermal oxidation and chemical attack can change the properties of a polymer irreversibly.

Common high-temperature polymers being considered for composite matrix use include polyimides, polyamide-imides, polybenzimidazole, polyetherimides, polyphenylene sulfide, polyimidesulfone, polyethersulfone, and polyetherketone. Representative chemical structures of these polymers are shown in Fig. 13. These structures appear to show a dominant aromatic character. The presence of aromatic groups in the main chain contributes to high T_g value and also contributes to high tensile and shear properties.

As discussed earlier, the matrix in a fiber-reinforced composite serves to transfer the load between fibers and to integrate the whole structure to form a useful shape. Other functions of the matrix include providing a barrier against an adverse environment and protecting the surface of the fibers from mechanical damages. The modulus of a matrix is not particularly important in terms of tensile load-carrying capacity of a composite. However, it exerts important influence on the interlaminar shear and in-plane shear properties of a composite material. The interlaminar shear strength is an important design consideration for structure under bending loads whereas the in-plane shear strength is important under torsional loads. The matrix provides lateral support to keep the fibers from buckling under compression, so to some extent it dictates the 0° compressive strength of a composite.

Factors That Influence Heat Resistance

Important factors to consider include those that will influence T_g/T_m and those that will influence thermal stability. T_g/T_m can be raised by increasing the intermolecular forces between chains. This can be done by incorporating polar side groups, by increasing the opportunity of hydrogen bonding, by actual chemical cross-linking of the chains and by incorporating the bulky cyclic groups (especially para-linked aromatic ring) in the main chain. The use temperature can also be raised by stereospecific polymerization to increase the regularity of the chain, with possible consequent increase in the degree of crystallinity or packing density [94,95].

Chemical factors which influence heat resistance include primary bond strength, secondary bond strength,

Table 14   Strength of select chemical bonds

Bond	Strength, kJ/mole	Bond	Strength, kJ/mole	Bond	Strength, kJ/mole	Bond	Strength, kJ/mole
C–H	416	C–Si	328	C=C	609	Si–O	445
C–B	374	C–S	273	C=N	617	P–O	528
C–C	349	C–Cl	340	Si–H	319	B–H	294
C–O	361	C–P	580	Si–N	437	B–N	386
C–N	307	C–F	428	B–O	777		

Adopted from Ref 94, 95

resonance stabilization, mechanism of bond cleavage, rigid intrachain structure and cross-linking and branching. Thermal stability is influenced primarily by the strength of the chemical bonds. Estimated bond strength values of selected bonds are listed in Table 14.

However, bond strength is not the only factor that controls thermal stability. Some high-strength bonds are more vulnerable to chemical attack leading to bond cleavages. As an example, the C–F bond is relatively high in strength. If a fluorine atom is next to a hydrogen atom, hydrogen fluoride is given off at elevated temperatures. As long as there is no alternative path of lower activation for rearrangement reactions, most bonds are strong enough to give adequate thermal stability [94].

The resonance stabilization is a well recognized mechanism for heat resistance and is known to enhance bond strength considerably. Aromatic and heterocyclic rings are widely used in thermally stable polymers. The N–N bond is relatively weak (about 160 kJ/mole) and should be expected to be absent in a thermally stable chain. Yet, most likely because of resonance stabilization, the N–N bond in heterocyclic rings exhibits reasonably high thermal stability [94].

Secondary bonding force also plays a key role in molecular stability and cohesion energy density, which affect chain stiffness, T_g, and the solubility. Thus, thermally stable polymers often contain –CO– and –(SO$_2$)– groups which participate in strong intermolecular association. Polymers containing electron-withdrawing groups, like –CO–, as the connecting group are generally more thermally stable than those containing electron-donating groups, like –O–.

Chemical cross-linking promotes heat resistance simply because more bonds must be cleaved in the same vicinity for the polymer to exhibit the weight loss or reduction in mechanical properties. For linear polymers, physical factors to consider include molecular weight and molecular weight distribution, crystallinity. However, these are generally considered to be of secondary importance. Higher molecular weight polymers are more thermally stable because of more entanglements and the ability to accommodate more chain cleavage without significant property reduction. A polymer with higher crystallinity will have better thermal resistance.

In summary, the following points serve as general guidelines for designing thermally stable polymers [94-96]:

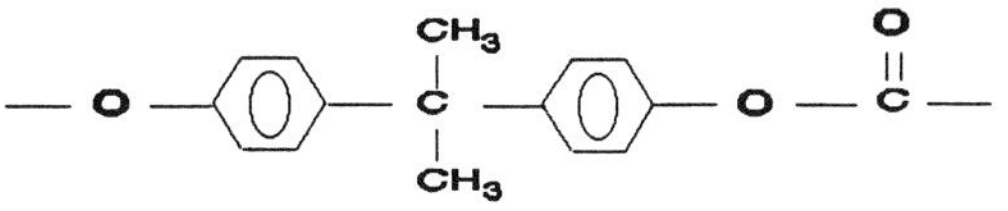

Fig. 15   Chemical structure of polycarbonate

- Chains containing para-linked rings have the highest thermal stability. They will also have highest softening points and lowest solubility. To compromise between processability and thermal stability, meta-linked units have to be introduced in lieu of some or all the para-linked rings. Ortho catenation was not found to be effective in lowering T_g.

- Rings containing only hydrogen substituents give optimum heat resistance. However, at higher temperature in an oxidizing atmosphere, hydrogen substituents themselves become reactive.

- To improve the processability, some flexible linking groups often have to be introduced between rings. These cause a reduction in stability. Those having the least effect are : –CO–, –COO–, –CONH–, –S–, –SO$_2$, –O–, –CF$_2$–, –C(CF$_3$)$_2$–.

The Group Additivity Approach

The Estimation of T_g. The method of theoretical estimation of the glass transition, the *group additivity approach* is based on the assumption that the structural groups in the repeating unit provide weighted additive contributions to the T_g. It is further assumed that the contribution of a given group is independent of the nature of its adjacent groups. Van Krevelen and Hoftyzer propose the following additive quantities [97]:

$$Y_g = \sum_i Y_g i = T_g \cdot M \qquad \text{(Eq 4)}$$

so that

$$T_g = \frac{Y_g}{M} = \frac{\sum_i Y_g i}{M} \qquad \text{(Eq 5)}$$

There is a marked deviation when a polar group exists in the repeating unit. The interaction of polar groups can be

Table 15 Group contributions to Y_g of commonly used chemical groups

Group	Y_{gi}, kg/mol	Group	Y_{gi}, kg/mol	$Y_g(I_x)$
$-CH_2-$	2,700	$-(CH_3)(C_2H_5)-$	17,700	...
$-CH(CH_3)-$	8,000	$-(CH_3)(C_6H_5)-$	50,000	...
$-CH(C_2H_5)-$	10,500	$-(CH_3)(COOCH_3)-$	13,000	...
$-CH(C_3H_7)-$	13,100	$-CH(OH)-$	13,000	...
$-CH(C_6H_5)-$	35,000	$-CHF-$	11,000	...
$-CH(C_6H_4CH_3)-$	42,000	$-CHCl-$	20,000	...
$-CH(OCH_3)-$	11,900	$-CF_2-$	13,000	...
$-CH(COOCH_3)-$	21,300	$-CCl_2-$	25,000	...
$-C(CH_3)_2-$	15,000	$-CFCl-$	23,000	...
	8,400(a)		...	

Group	Y_{gi}, kg/mol	Group	Y_{gi}, kg/mol	$Y_g(I_x)$
(para-phenylene ring)	32,000	(meta-phenylene ring)	4,000	...
(chloro-methyl substituted phenylene ring, Cl)	51,000	$-C(=O)-$	27,000	...
(methyl substituted phenylene ring, CH$_3$)	35,000	$-C(=O)-O-$	8,000	$12,000I$
(trimethyl substituted phenylene ring, CH$_3$, CH$_3$)	55,000	$-O-C(=O)-O-$	16,000	$10,000I$
(trans-cyclohexylene ring) (trans)	31,000	$-C(=O)-O-C(=O)-$	20,000	...
(naphthylene ring)	58,000	$-S-$	7,500	...
(fused cyclopentane aromatic ring)	7,000	$-S(=O)_2-$	58,000	...
(indane-type fused ring)	28,000	$-S(=O)_2-O-$	31,000	...
(methyl substituted indane-type ring, CH$_3$)	30,000	$O-C(=O)-NH-$	12,000	$1,800\,I^1 + 2 \times 10^6$ N/m
		$O-C-NH-$	25,000	...
		$-NH-C(=O)-NH-$	20,000	$2,100I^{-1}$
		$-S(CH_3)_2-$	8,000	...

(a) For polyisobutylene only. Source: Ref 97

formulated by introducing an interaction factor I_x. For a polymer of this configuration $-(CH_2)_n 1-X-(CH_2)_n 2-$ where X contains n_x chain atoms.

$$I_x = \frac{n_x}{n_x + n_1 + n_2} \qquad \text{(Eq 6)}$$

and Eq 4 must be modified as:

$$Y_g = \sum_i Y_g i + \sum Y_g(I_x i) \qquad \text{(Eq 7)}$$

The Y_{gi} and $Y_g(I_{xi})$ values of several more commonly used chemical groups are listed in Table 15. For certain classes of polymers, a number of special rules must be followed. For instance, if two phenylene groups are directly connected, this causes a structural Y_g correction factor (+13,000 for p-phenylene groups and +47,000 for o-phenylene groups). If two single p-phenylene groups are connected by $-O-$, this causes a depression of Y_g by 5,000. These corrections were derived empirically. The melting temperature of a semicrystalline polymer can be estimated in a similar manner [97].

The Estimation of Elastic Moduli. In this section, elastic modulus is denoted as E, bulk modulus as B, shear modulus as G and Poisson ratio v. The relationships between these quantities are:

$$E = 2G(1 + v) = 3B(1 - 2v) \qquad \text{(Eq 8)}$$

These quantities can be estimated by means of additive quantities. The bulk modulus can be estimated by:

$$\frac{B}{\rho} = \left(\frac{U}{V}\right)^6 \qquad \text{(Eq 9)}$$

where U is the Rao function, V is the molar volume, and ρ is the density. Both U and V are additive quantities (see [97]. Poisson's ratio, if not known, can be approximated by:

$$\log\left(\frac{B}{\rho}\right) = 8.3 - 4v \qquad \text{(Eq 10)}$$

Example 1. Calculate the glass transition temperature and elastic modulus of polycarbonate.
Solution: Consider the chemical formula of polycarbonate (Fig. 15).

Molar weight of repeating unit = 254

Y_{gi}:		
	Aromatic ring	32,000
	$-C(CH_3)_2-$	15,000
	$-O(CO)O-$	$16,000 + 10,000 \times I$

$I = 2/2 = 1$
$Y_g = 32,000 \times 2 + 15,000 + 16,000 + 10,000 \times 1$
$\quad = 105,000$

$T_g = 105,000/254 = 413.4$ K

The experimental glass transition temperature of this polycarbonate is about 414 K (141 °C, or 286 °F). The Rao function U_is and molar volume V_is for each group:

	U_i(cm$^{10/3}$/ s$^{1/3}$ · mol)	V_i (cm^3/mol)
Aromatic ring	4100	65.5
>C<	50	4.6
$-(CH_3)$	1400	23.9
$-O(CO)O-$	1600	31.4

$U = 4100 \times 2 + 50 + 1400 \times 2 + 1600 = 12,650$
$V = 65.5 \times 2 + 4.6 + 23.9 \times 2 + 31.4 = 214.8$
$\rho = 254/214.8 = 1.18$ g/cm^3
$B = 4.92$ GPa
$v = 0.42$ (from Eq 7)
So $E = 2.35$ GPa from Eq 5.

The experimental values of the bulk and elastic modulus of polycarbonate are 5 GPa and 2.47 GPa, respectively.

Example 2. Estimate the glass transition temperature of poly(ether ether ketone), PEEK.
Solution: The structural formula for PEEK is:

$$-[-O-\varphi-O-\varphi-CO-\varphi-]-$$

The group contributions are:

Group	No.	Y_{gi}	Contributions	M_i
$-O-$	2	4,000	+8,000	16
$-CO-$	1	27,000	+27,000	28
$-\varphi-$	3	32,000	+96,000	76
			$-10,000*$	

* p-phenylene groups connected to two ether linkages

Therefore,

$$T_g = \frac{2(4,000) + 3(32,000) + 1(27,000) - 2(5,000)}{2 \times 16 + 3 \times 76 + 1 \times 28}$$
$$= 420\text{K} \ (147 \ °\text{C, or } 297 \ °\text{F})$$

The actual T_g value of PEEK was found to lie between 140 and 170 °C (284 and 338 °F) (143 °C, or 289 °F, from Fig. 13).

References

1. B.D. Agarwal and L.J. Broutman, *Analysis and Performance of Fiber Composites*, 2nd ed., Wiley Interscience, New York, 1990
2. K.L. Loewenstein, *The Manufacturing Technology of Continuous Glass Fibers*, Elsevier Science Publishing Co., Amsterdam, 1973, p 29

3. D. Hull, *An Introduction to Composite Materials*, Cambridge University Press, Cambridge, England, 1981
4. A. Watts, Ed., *Commercial Opportunities for Advanced Composites*, STP 704, ASTM, Philadelphia, 1980
5. V. Raskovic and S. Marinkovic, *Carbon*, Vol 13, 1975, p 535
6. J.B. Donnet and P. Ehrburger, *Carbon*, Vol 15, 1977, p 143
7. J.B. Donnet and R.C. Bansal, *Carbon Fibers*, Marcel Dekker, Inc., New York and Basel, 1984
8. M.M. Coleman and R.J. Petcavich, *J. Polymer Sci., Polymer Phys. Ed.*, Vol 16, 1978, p 821
9. R.J. Petcavich, P.C. Painter, and M.M. Coleman, *J. Polymer Sci., Polymer Letters*, Vol 17, 1979, p 165
10. M.M. Coleman and G.T. Sivy, *Carbon*, Vol 19, 1981, p 123, 127, 133, and 137
11. P.J. Goddhew, A.J. Clarke, and J.E. Bailey, *Materials Sci. Eng.*, Vol 17, 1975, p 3
12. V.I. Kasato Chikin and V.A. Kargin, *Doklady Phys. Chem.*, Vol 191, 1970, p 303
13. A. Fourdeux, R. Perret, and W. Ruland, *Carbon Fibers—Their Composites and Applications*, The Plastics Institute, London, 1971, p 57
14. W. Ruland, *Appl. Polymer Symposium*, Vol 9, 1969, p 293
15. E. Tokarsky and R.J. Diefendorf, *12th Bienn. Conf. on Carbon*, Pittsburgh, PA, 1975, Extended Abstracts, p 301
16. D.J. Johnson, *Carbon Fibers—Their Composites and Applications*, The Plastics Institute, London, 1971, p 52
17. A.A. Bright and L.S. Singer, *Carbon*, Vol 17, 1979, p 59
18. S. Allen, G.A. Cooper, D.J. Johnson, and R.M. Mayer, *Proc. 3rd Conf. Industrial Carbon and Graphite*, Soc. Chem. Ind., London, 1971, p 456
19. B.J. Wicks, *J. Nuclear Mater.*, Vol 56, 1975, p 287
20. W. Watt and W. Johnson, *Proc. 3rd Conf. Industrial Carbon and Graphite*, Soc. Chem. Ind., London, 1973, p 417
21. R. Moreton, W. Watt, and W. Johnson, *Nature*, Vol 222, 1967, p 690
22. B.F. Jones and R.G. Duncan, *J. Mater. Sci.*, Vol 6, 1971, p 289
23. D.J. Johnson and C.N. Tyson, *J. Phys. D: Appl. Phys.*, Vol 3, 1970, p 526
24. J.V. Sharp and S.G. Burnay, *Carbon Fibers—Their Composites and Applications*, The Plastics Institute, London, 1971, p 68
25. J.W. Johnson and D.J. Thorne, *Carbon*, Vol 7, 1989, p 659
26. D.J. Thorne, *Nature*, Vol 225, 1970, p 1039
27. O.P. Bahl and L.M. Manocha, *Carbon*, Vol 13, 1975, p 297
28. S.C. Bennett and D.J. Johnson, *Proc. Fifth London Carbon and Graphite Conf.*, Soc. Chem. Ind., Vol. 1, 1978, p 377
29. L.T. Drzal, M.J. Rich, and P.F. Lloyd, *J. Adhesion*, Vol 16, 1983, p 1
30. L.T. Drzal, M.J. Rich, M.F. Koenig, and P.F. Lloyd, *J. Adhesion*, Vol 16, 1983, p 133
31. E.G. Wolff, *J. Compos. Mater.*, Vol 21, 1987, p 81
32. S. Chwastiak and R. Bacon, *ACS Polymer* Preprint, Vol 22, No. 2, 1981, p 222
33. J. Delmonte, *Technology of Carbon and Graphite Fiber Composites*, Van Nostrand Reinhold Co., New York, 1981
34. S. Yamane, T. Hiramatsu, and T. Higuchi, *32nd Intl. SAMPE Symp.*, April 1987, p 928
35. M. Takayanagi, T. Kajiyoma, and T. Katayose, *J. Appl. Polym. Sci.*, Vol 27, 1982, p 3903
36. H. Blades, "Dry-Jet Wet Spinning Process," U.S. Patent No. 3,767,756, Oct. 23, 1973
37. M.G. Northolt, X-Ray Diffraction Study of Poly(p-phenylene Terephthalamide), Fibers, *European Polym. J.*, Vol 10, 1974, p 799
38. D. Tanner, J.A. Fitzgerald, P.G. Riewald, and W.F. Knoff, in *High Technology Fibers*, Vol III, Part B, M. Lewin and J. Preston, Ed., Marcel Dekker, Inc., New York and Basel, 1989, p 35
39. L.S. Li, L.F. Allard, and W.C. Bigelow, *J. Macromol. Sci.*, Vol B22, No. 2, 1983, p 269
40. G.S. Fielding-Russell, *Text. Res. J.*, Vol 41, 1971, p 861
41. K.E. Perepelkin and Z.U. Chereiskii, *Mekhanika Polimerov*, Vol 6, 1977, p 1002
42. T. Ito, *Sen i Gakkaishi*, Vol 38, 1982, p 54
43. M.G. Northolt and R.V.D. Hout, *Polymer*, Vol 26, 1985, p 310
44. W.B. Black and J. Preston, *High Modulus Wholly Aromatic Fibers*, Marcel Dekker, Inc., New York, 1973
45. Ref 4, p 23
46. G.R. Hattery and M.E.D. Hillman, in *High Performance Polymers*, E. Baer and A. Moet, Ed., Hanser Publishing, New York, 1991, p 255
47. H.X. Nguyen, G. Riahi, G. Wood, and A. Poursartip, in *33rd Intl. SAMPE Symp.*, Anaheim, CA, 1988
48. M.E. Buck, *Adv. Mater. Proc.*, Vol 9, 1987, p 61
49. R.J. Young, *Introduction to Polymers*, Chapman and Hall, 1983
50. L.R.G. Treloar, *Polymer*, Vol 1, 1960, p 95, 279, and 290
51. Odajima and Maeda, *J. Polym. Sci.*, Vol C15, 1966, p 55
52. A. Kelly, *Strong Solids*, Clarendon Press, Oxford, 1973
53. T. Ohta, *Polym. Eng. Sci.*, Vol 23, 1983, p 697
54. J.F. Wolfe, B.H. Loo, and E.R. Seville, *Polymer* Preprint, ACS Vol 22 No. 1 1981, p 60
55. J.F. Wolfe, B.H. Loo, and F.E. Arnold, *Macromolecules*, Vol 14, 1981, p 915
56. J.F. Wolfe, P. Sybert, and J. Sybert, U.S. Patent 4,533,693, Aug 1985
57. J.H. Greenwood and P.G. Rose, *J. Mater. Sci.*, Vol 9, 1974, p 1809
58. H.M. Hawthorne and E. Teghtsoonian, *J. Mater. Sci.*, Vol 10, 1975, p 41
59. S. Kumar, *SAMPE Q.*, Jan 1989, p 3-8
60. Du Pont technical literature E-95614 and E-95612, June 1987
61. S.J. Deteresa, US AFWAL-TR-85-4013, 1985
62. M.G. Dobb, D.J. Johnson, and B.P. Saville, *Polymer*, Vol 22, 1981, p 960
63. J.K. Gillham, in *Encyclopedia of Polymer Science and Engineering*, Vol 4, 2nd ed., John Wiley and Sons, New York, 1986, p 519
64. I.H. Updegraff, Unsaturated Polyester Resins, in *Handbook of Composites*, Section 2, G. Lubin, Ed., Van Nostrand Reinhold Co., New York, 1982
65. M.B. Launikitis, Vinyl Ester Resins, in *Handbook of Composites*, Section 3, G. Lubin, Ed., Van Nostrand Reinhold Co., New York, 1982
66. D. Betty, B. Buck, and A. Cornwell, *Handbook of Resin Properties, Part A: Cast Resins,* Yarsley Testing Lab., Ashstead, 1975
67. J.P. Critchley, G.J. Knight, and W.W. Wright, *Heat-Resistant Polymers—Technologically Useful Materials*, Plenum Press, New York and London, 1983
68. G.I. Harris and F. Coxon, British Patent 1150203, 1969
69. G.I. Harris, Chapter 4, in *Developments in Reinforced Plastics*, Vol 1, G. Pritchard, Ed., Applied Science Publishing, London, 1980

70. J.M. Margolis, Ed., *Advanced Thermoset Composites: Industrial and Commercial Applications*, Van Nostrand Reinhold Co., New York, 1986

71. T.T. Serafini, *et al.*, U.S. Patent 3,745,149, July 1973

72. H.D. Stenzenberger, W. Romer, P.M. Hergenrother, and P. Konig, *Proc. SAMPE Intl. Symp.*, Vol 33, 1988, p 1546

73. H.D. Stenzenberger, W. Romer, P.M. Hergenrother, and B.J. Jensen, *Proc. SAMPE Intl. Symp.*, Vol 34, 1989

74. C.A. Arnold, P.M. Hergenrother, and J.E. McGrath, An Overview of Organic Polymeric Matrix Resins for Composites, *ACS Symposium*, T. Vigo and B. Kinzig, Ed., 1990

75. D.C. Leach, Chapter 2, in *Advanced Composites*, I.K. Partridge, Ed., Elsevier Applied Science Publishing, London and New York, 1989

76. R.N. Johnson, A.G. Farnham, R.A. Clendinning, W.F. Hale, and C.N. Merriam, *J. Polym. Sci.*, A-1, Vol 5, 1967, p 2375

77. T.E. Attwood, P.C. Dawson, J.L. Freeman, L.R.J. Hoy, J.B. Rose, and P.A. Staniland, *Polymer*, Vol 22, 1981, p 1096

78. R. Viswanathan, B.C. Johnson, and J.E. McGrath, *Polymer*, Vol 25, 1984, p 1827

79. G.M. Bower and L.W. Frost, *J. Polym. Sci., A-1*, Vol 1, 1963, p 3135

80. J.A. Kreuz, A.L. Endrey, F.P. Gay, and C.E. Sroog, *J. Polymer Sci., A-1*, Vol 4, 1966, p 2607

81. J.G. Wirth and D.R. Heath, U.S. Patent 3,730,946 (to General Electric), 1973; I.W. Serfaty, in *Engineering Thermoplastics*, J.M. Margolis, Ed., Marcel Dekker, New York, 1983, p 283

82. C.J. Billerbeck and S.J. Henke, in *Engineering Thermoplastics*, J.M. Margolis, Ed., Marcel Dekker, New York, 1983, p 373

83. Mitsui Toatsu Chemicals, Inc., 140 E. 45th Street, New York, NY 10017

84. Hoechst-Celanese, Specialty Products Group, 500 Washington Street, Conventry, RI 02816

85. Ethyl Corporation, Baton Rouge, LA 70801

86. K.C. Brinker and I.M. Robinson, U.S. Patent 2,895,948 (to Du Pont), 1959

87. G.M. Moelter, R.F. Tetreault, and M.J. Hefferson, *Polymer News*, Vol 9 No. 5, 1983, p 134

88. M. Gordon, Ed., *Liquid Crystal Polymers*, Vol. 59-61 of Advances in Polymer Science Series, Springer-Verlag, New York, 1984

89. J.F. Wolfe, in *Recent Advances in Polyimides and Other High Performance Polymers*, workshop sponsored by the Am. Chem. Soc., P.M. Hergenrother, Chairman, 1987

90. N.J. Johnston, T.W. Towell, and P.M. Hergenrother, Physical and Mechanical Properties of High Performance Thermoplastic Polymers and Their Composites, in *Thermoplastic Composite Materials*, Vol 7, L.A. Carlsson, Ed., Elsevier Science Publishing, Amsterdam, 1991

91. D.C. Leach, D.C. Curtis, and D.R. Tamblin, in *Toughened Composites*, STP 937, N.J. Johnston, Ed., ASTM, Philadelphia, 1987, p 358

92. B. Fife, J.A. Peacock, and C.Y. Barlow, in *Proc. 6th Intl./2nd European Conf. Composite Materials*, Vol 5, Imperial College, July 1987, Elsevier, London, 1987, p 439

93. I.K. Partridge, Chapter 2, in *Advanced Composites*, I.K. Partridge, Ed., Elsevier Applied Science Publishing, London and New York, 1989

94. P.M. Hergenrother, Heat-Resistant Polymers, in *Encyclopedia of Polymer Science and Eng.*, Vol 7, 2nd ed., John Wiley and Sons, New York, 1987

95. J.P. Critchley, G.J. Knight, and W.W. Wright, *Heat-Resistant Polymers*, Plenum Press, New York, 1983

96. T.L. St. Clair, Structure-Property Relationships in Linear Aromatic Polyimides, in *Polyimides—Chemistry and Applications*, Blackie and Sons, Glasgow, 1991

97. D.W. Van Krevelen, *Properties of Polymers*, Elsevier Scientific Publishing Co., New York, 1976

Chapter 3

Interfaces and Interphases in Composites

Both short-term and long-term properties of a composite depend critically on the microstructure and properties of the interface or interphase between the fiber and the matrix. The word "interphase" refers to a region (see Fig. 1) where the fiber and matrix phases are chemically and/or mechanically combined or otherwise indistinct [1]. The interphase may be a diffusion zone, a nucleation zone, a chemical reaction zone, a thin layer of fiber coating, or any combination of the above. An "interface" is a boundary demarcating distinct phases such as fiber, matrix, coating layer, or interphase.

The formation of the interphase during processing and the resulting properties are complex and poorly understood. Partially because of this incomplete understanding and a lack of effective control of the interphase, the current use of advanced thermoplastic matrix composites and high-temperature metal matrix and ceramic matrix composites remains limited. Although this book concerns itself primarily with polymer matrix composites, related interface subjects in metal matrix and ceramic matrix composites are also discussed here since the interfaces in these three materials exhibit many common features.

Theories and Significance of Interfaces

Interfaces in Metal Matrix and Ceramic Matrix Composites

Several theories have been developed regarding interface in fiber-reinforced composites. The most common description of the interface structure is called the coincidence site lattice (CSL) model [2-4]. This model is similar to the Read-Shockley description for low-angle boundaries. It predicts a dependence of the surface energy with misorientation. The mechanical properties of the composite have been predicted in terms of properties of the interface by the Aveston, Cooper, and Kelly (ACK) and

Budiansky, Hutchinson, and Evans (BHE) models [5,6]. Application of these models is limited by an assumption that the cracking strength of the matrix is deterministic. Korchynski [7] proposed a model for the upper limit for fiber volume fraction. Faber [8] proposed that a small inter-fiber spacing in ceramic matrix composites should limit the size of Griffith flaws. Faber and Evans [8,9] have analyzed the crack deflection mechanism at the interface of ceramic matrix composites. Porter [10] showed that the application of the Griffith equation in determining the fracture toughness of continuous fiber-ceramic matrix composites is inappropriate due to the difficulty of determining the crack length.

Fredrick [11] proposed a model which predicts the strength of metal matrix composites as a function of the thickness of the brittle interphase. This model predicts an interphase thickness at which the composite reaches the

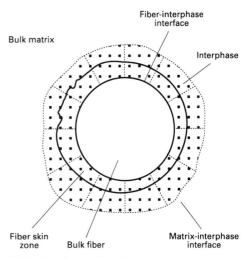

Fig. 1   Interface and interphase

maximum strength. Kiiko [12] analyzed the relative contribution of each mechanism of energy dissipation to composite fracture toughness. Cooper [13] proposed a model to predict fracture behavior of metal matrix composites, where large diameter fibers were said to increase the work to fracture.

Several types of interfaces can exist in composite materials, including mechanical, chemical, and physical interfaces. Mechanical interfaces are realized by the "key and lock" mechanisms. Primary factors controlling this type of interface are fiber surface roughness, porosity, surface morphology, and matrix pore-filling capabilities. This type of interface plays a very important role in ceramic matrix composites by providing a necessary degree of matrix debonding when the composite is under loading conditions, which can lead to the optimization of the composite mechanical properties. Prewo [14] postulated that carbon fiber-reinforced glass and glass-ceramic composites have predominantly mechanical interface bonding, as evidenced by a low transverse strength. Lowden [15] showed that a pyrolytic carbon coating on Nicalon SiC fibers results in primarily mechanical bonds between the fibers and the SiC matrix. Hill [16] evaluated the contribution of mechanical and chemical bonding for tungsten filaments in an aluminum matrix.

In SiC matrix composites, chemical bonding between the fiber and matrix may be formed as a result of the chemical interactions between the fiber, coating and the matrix, including the possible formation of SiO_2, Si, and SiO. The presence of silica at high temperature can lead to the formation of silicon resulting in strong chemical bonding. The wetting and viscous characteristics of silica can also result in the formation of strong physical and mechanical bonds between the silica layer present on the fiber surface and SiC matrix. Mah [17] showed that surface diffusion of silica can occur during the heat treatment of Nicalon fibers. This silica layer during composite consolidation and densification processes can form a chemical bond with the matrix. Similarly, Kowbel [18] used TEM to demonstrate the existence of a silica layer at the fiber-matrix interface in SiC/SiC composites. Lowden [19] showed that the application of a pyrolytic carbon to SiC fibers can significantly increase the strain to failure and the flexural strength of SiC/SiC composites. The thickness of the carbon layer had a strong effect on fracture behavior. An increase in the coating thickness was found to result in a significant improvement in the flexural strength and strain-to-failure. No detailed mechanism of the load transfer between the coated fiber and SiC matrix was proposed.

Fretty [20] showed the fracture behavior of SiC/SiC composites changed after long-term aging in air at 1400 °C (2550 °F). The uncoated fiber composites exhibited a large degree of fiber pull-out, while the fiber composites made with carbon-coated fibers behaved in a brittle manner. Thus, the carbon-rich layer promoted a high temperature oxidation embrittlement similar to that observed in SiC (Nicalon)-glass ceramic composites [21]. The key to controlling the fracture behavior of SiC matrix composites lies in the optimization of the interfacial properties,

primarily in the elimination or reduction of the chemical bonds. Bender [22] correlated interfacial characteristics with fracture toughness of ceramic matrix composites and found that the formation of chemical bonds was the most detrimental factor.

Naslain [23] studied the fiber-matrix interactions in titanium matrix composites and found that boron carbide was the best diffusion barrier. He also found that the fiber-matrix interactions in SiC/Ti composites were governed by diffusion and formation of carbides and silicides. The interfacial strength was found to be a function of the type of coating. Smith [24] found that the degradation of the fiber strength was the most critical factor in controlling the strength of SiC/Ti composites. Composites with AVCO SCS6 fibers possessed the highest strength values due to debonding at the buffer zone. Jones [25] carried out a thorough interfacial analysis in a SCS6/TiC composite and found extensive chemical interactions between the coated fiber and matrix resulted in the formation of TiC and Ti_xC_y. Results from *in situ* fracture experiments revealed that the fracture took place between the TiC and an amorphous carbon layer in the coating. The mechanisms for the degradation of composite mechanical properties due to interfacial reactions have been reviewed by Ochiai *et al.* [26]. Although much has been accomplished in the titanium-base metal matrix composite technology, additional work is required to identify unambiguously the fundamental factors that control the properties of the interphase.

Physical bonding (including van der Waals bonds) is particularly important in SiC/SiC and SiC/Ti composites if pyrolytic carbon coatings are used. The prismatic edges on the coating surface can result in a strong physical bond with the SiC matrix. Physical bonding will also be important if a liquid infiltration process is used. The subject of interface in composites has been the theme of several recent symposia [27-30].

Interfaces in Polymer Matrix Composites

The fiber-matrix interfacial adhesion plays an important role in determining the mechanical properties of a polymer composite. A better interfacial bond will impart to a composite better properties such as interlaminar shear strength, delamination resistance, fatigue and corrosion resistance.

Much work has been done in the characterization of carbon/graphite fiber surfaces, and a variety of surface coating and modification techniques have been developed to improve interfacial bonding between such fibers and epoxy matrix [31,32]. Oxidation etches the fiber surfaces and yields a substantial increase in composite shear strength with an accompanying loss in fiber tensile strength. Other techniques in the surface treatment of carbon fibers include vapor phase deposition and electrochemical polymerization [33]; both techniques show promise. The plasma polymerization technique was used by Sung *et al.* [31] to improve graphite-epoxy interfacial bonding. Acrylonitrile gas plasma treatment was reported to improve composite interlaminar shear strength greatly without loss of tensile properties.

Since the interface is the most highly stressed region of a composite material, several attempts have been made to lower these stress concentrations by placing either a material of intermediate modulus or an elastomer phase between the fiber and matrix [34-37]. The concept involves lowering the modulus ratio of any two neighboring components, thus lowering the stress concentrations caused by modulus mismatch. Local deformation capability also may be built into the interfacial region so that the modulus mismatch stress concentrations are damped out at least partially. The role of the interface in polymer composites has recently been critically reviewed by Kardos [38] and by Ehrburger and Donnet [39]. Finally, a comprehensive study was conducted by Theocaris [40] on the modeling of interphase and its role in the prediction of composite properties.

Thermoplastics are receiving ever increasing attention as matrix materials in structural composites. Thermoplastic composites can be fabricated by novel techniques less cumbersome and potentially faster than thermoset curing. Selected thermoplastic composites may also show superiority over epoxy composites in service temperatures [41,42]. Thermoplastic composites are also well known for their outstanding impact resistance and damage tolerance. However, interfacial adhesion between fiber and thermoplastic is difficult to achieve. The conventional coupling agents such as silanes appear to be much less successful when applied to reinforced thermoplastics [43]. Further, the proposed coupling agent theories do not adequately describe the behavior of thermoplastic systems [42]. A need exists for effective adhesion promotion agents and a new means of controlling adhesion in composites.

Plasma treatment was used with some success to improve the composite properties of aramid fiber-epoxy resin [44-47]. For instance, Allred, Merrill and Roylance [47] were successful in incorporating reactive amine groups into the aramid filament surface structure by exposing the fibers to amine-containing plasmas. Subsequent reaction of the amine groups with epoxides form strong bonds at the composite interface; the bonding may be covalent or hydrogen in nature. Surface chemical modification of aramid fibers has been studied by various investigators [48-50]. The goals of these studies were to improve wetting or to provide a fiber-matrix coupling. The results indicated that some gains may be made by improving wettability, if the chemical modification is covalently bonded to the substituent phases. However, some loss in filament strength often accompanies the adhesion improvements.

Research efforts on interfaces were usually aimed at a better understanding of the physical and chemical characteristics of the fiber-matrix interface in advanced composites. It is particularly important to clarify the adhesion mechanisms through which the surface treatments promote interfacial bonding in such materials. The following considerations are important in the adhesion promotion in such materials:

- The removal of a weak boundary layer or contaminants
- Improvement in wettability of fiber surface by thermoplastics
- Creation or addition of chemical groups
- Variation in surface topography (mechanical interlocking)

Surface Characterization

Experimental methods that can be used to characterize the fiber surface include:

- Electron spectroscopy for chemical analysis (ESCA) or X-ray photoelectron spectroscopy (XPS) can be employed to determine if any functional groups have been deposited or chemically active sites created.
- Fourier transform infrared analysis (FTIR) can be utilized to determine the chemical modification of fiber surfaces caused by surface treatment. Attenuated total reflectance (ATR) and diffuse reflectance attachments can be used in conjunction with the spectrometer.
- Dynamic contact angles can be calculated from wetting force measurements carried out on an electrobalance.
- Scanning tunneling microscopy (STM) and Scanning Electron Microscopy (SEM) can be applied to examine the fiber surface topography (shape, width, and depth of porosity). The fiber surface area is important in considering the lock and key configuration between fiber and matrix.
- Thermal desorption method can be conducted to measure the amount and composition of the species (H_2O, CO, CO_2) adsorbed that can serve as void generators within the composites.

Surface Spectroscopy

Fundamental information about the atomic and molecular composition of the fiber surface can be acquired by using surface spectroscopy. The techniques operate by bombarding the surface of interest with a probe ion (SIMS), electron (Auger), or photon (ESCA or XPS). These incident species either interact with the surface and are energy analyzed, or cause the emission of surface species (X-rays, electrons, or ions) characteristic of the surface environment. Each spectroscopic technique has its own advantages and limitations when applied to carbon fibers, but they provide complementary information.

Auger electron spectroscopy (AES) probes the surface with high energy electrons, which interact with atoms at and below the surface causing the emission of Auger electrons. These electrons can be emitted following the creation of a core hole in the electron shells and the occupation of this hole by another electron. The Auger electron spectrum typically contains peaks corresponding to the intensity of Auger electrons as a function of their kinetic energy, which is related to the energy released when the hole is filled and is independent of the energy of the incident beam. However, fine structure appears as a result of relaxation processes in the solid which contribute to the final energy state of the excited atom and its environment [51]. The intensity reflects the concentration of atoms or ions in the volume analyzed, while the fine structure relates to changes in chemical states. The surface atomic composition and concentration can be acquired quantitatively and in some cases molecular information is possible from the Auger data [52]. Auger spectroscopy can be useful in tracing the existence of any specific ions involved in degradation of a composite, but this technique alone will unlikely provide adequate information on the nature of interface bonding [53].

Electron spectroscopy for chemical analysis (ESCA), also known as X-ray photoelectron spectroscopy (XPS), uses X-ray photons as the probing species, which cause the ejection of core electrons from atoms in the material of interest. Due to the short mean free path for these core electrons, only electrons from the first few atomic layers are able to escape from the surface to the collector. The XPS analysis is therefore a surface sensitive technique. Furthermore, unique binding energies are associated with the molecular environment of each core electron, in principle the molecular environment where each electron came from can be determined. In XPS, counting techniques are quantitative and surface concentrations can be accurately and reproducibly determined [54]. However, the absolute values of the binding energies determined for each photoelectron are not unique. The photoelectrons collected can arise from the surface layer as well as from up to ten atomic layers below the surface. For carbon fibers, relative comparisons are straight forward and a finite number of surface species and molecular states exist for the atoms typically found on the fiber surfaces [55]. Data reduction therefore is not a particularly difficult task.

Secondary ion mass spectroscopy (SIMS) involves bombardment of the target material with an energetic beam of ions. This destructive process generates both monatomic ions and ion clusters which can be recognized by their mass and traced back to yield an indication of their chemical environment on the target surface. Cracking patterns having unique molecular characteristics can be analyzed to give quantitative information about the identification of parent surface molecules and their concentration. The species detected are only from the surface layer. SIMS techniques are subject to the limitations that the cracking patterns of the parent molecules must be known for the conditions of the experiment and the techniques are destructive in nature.

Microscopy

Scanning Electron Microscopy (SEM) and Scanning Tunneling Microscopy (STM). In SEM, electrons are used as the probing medium and therefore SEM has a much better resolution than does optical microscopy. SEM also provides a larger depth of field which facilitates the observation of three dimensional features on a surface. Under good operational conditions, SEM reveals surfaces down to 5 nm. STM is capable of giving quantitative information on the depth as well as the width/length of a surface feature such as a pore. Atomic level resolution can be achieved with STM and atomic force microscopy (AFM); this latter technique is useful even for nonconducting surfaces. Both STM and AFM are relatively new techniques and their potential as surface analysis tools has yet to be fully exploited.

Transmission electron microscopy (TEM) provides high resolution of a microstructural image. The requirement that the electrons be transmitted through the sample limits the TEM technique to either thin specimens (<100 nm in thickness) or to replicas of the fiber surface. A composite replica made from a thin polymer and surface coating (for example, sputter-coated carbon or gold) on the fiber surface can be removed intact from the fiber for observation in the TEM. Direct TEM observation of the fiber surface can be achieved through the use of ultramicrotomy. This technique involves encapsulating the fiber in a resin to immobilize the fiber during the cutting process. A sapphire or diamond knife edge cuts through the sample, which is controlled to move up and down and to advance a fixed distance in each cycle. Consecutive slices of material, typically about 50 nm thick, float on the surface of water and are collected on a copper grid for TEM observation. This technique permits observation of the fiber surface and adjacent regions provided the microtome knife edge cuts through these regions smoothly.

Wettability Measurements

A necessary condition for the formation of a proper fiber-resin interface is that the liquid resin "wet" or "spread" on the fiber surface. The contact angle which a drop of liquid forms when placed in contact with a surface is often taken as an indication of the compatibility between these two components. If the contact angle θ formed is less than 90° then the liquid is said to "wet" the solid. At the extreme case where θ is 0°, the liquid is said to "spread" on the solid substrate. Either wetting or spreading insures an acceptable interfacial free energy for adhesion in that the thermodynamic work of adhesion is positive.

Experimental measurements of contact angles on fibers of approximately 7 µm in diameter are difficult to perform optically. Contact angle values can be obtained indirectly by immersing the fiber into the liquid of interest and measuring the force of immersion or emersion. A simple force balance permits the calculation of the contact angle provided the fiber perimeter and surface free energy of the liquid are known.

The surface energy of a carbon fiber can be determined by measuring the contact angles of a variety of liquids according to the method proposed by Kaelble [56]. Using the Wihelmy plate technique [57], the contact force M (µg) between a single fiber of circumference C and a liquid of surface tension r_{lv} is described by the following equation:

$$M = \frac{C r_{lv} \cos\theta}{g} \qquad \text{(Eq 1)}$$

where θ is the advancing liquid-solid contact angle and g = 980.6 dyne/cm. Since M, r_{lv} and C can be evaluated independently, $\cos\theta$ can be calculated from Eq 1. In this study, M is measured by using a microbalance, and C from micrographs. In a typical Wilhemy setup (Fig. 2), a carbon filament was attached to a small plastic plate using adhesive tape and then suspended from the loop of a microbalance. A beaker containing a test fluid was raised by an elevator at a constant speed of 0.005 mm/s (0.0002 in./s), sufficiently slow so that the dynamic contact angle was independent of velocity and identical to the static angle [58].

The surface energies of solids and liquids are considered to be the sum of separate dispersive (London-d) and polar (Keesom-p) contributions. From such a two-component model the following relations can be derived for the polar and dispersive interaction between liquids and solids:

$$r_{lv} = r_{lv}^{\,d} + r_{lv}^{\,p} = a_l^2 + b_l^2 \qquad \text{(Eq 2)}$$

$$r_{sv} = r_{sv}^{\,d} + r_{sv}^{\,p} = a_s^2 + b_s^2 \qquad \text{(Eq 3)}$$

where r_{lv} is liquid-vapor surface tension, r_{sv} is solid-vapor surface tension, a_l and b_l are square roots of the respective dispersive $r_{lv}^{\,d}$ and polar $r_{lv}^{\,p}$ parts of r_{lv}, respectively, and a_s and b_s are square roots of respective dispersive $r_{sv}^{\,d}$ and polar $r_{sv}^{\,p}$ parts of r_{sv}, respectively.

The work of adhesion, W_a, is the decrease of Gibbs free energy per unit area when an interface is formed from two individual surfaces [58]. The greater the work of adhesion, the greater the interfacial attraction. The work of adhesion is defined as:

$$W_a = r_{lv}(1 + \cos\theta) \qquad \text{(Eq 4)}$$

From the relationship between W_a and the polar and dispersive components of the solid of interest, the work of adhesion can be expressed as

$$W_a = 2[a_l\, a_s + b_l\, b_s] \qquad \text{(Eq 5)}$$

which can be rearranged to give the following:

$$W_a\,/\,2a_l = a_s + b_s(b_l\,/\,a_l) \qquad \text{(Eq 6)}$$

The values of r_{lv}, $r_{lv}^{\,d}$, $r_{lv}^{\,p}$ are known for the liquids used and the contact angle θ can be measured. A plot of $(W_a\,/\,2a_l)$ versus $(b_l\,/\,a_l)$ will yield a straight line; its slope and intercept will give the value of $r_{sv}^{\,d}$ and $r_{sv}^{\,p}$, respectively, for the solid surface of interest. The surface free energies of liquid polymers can be measured directly through the determination of the shape of a molten drop of polymer either hanging or resting on a surface [59,60].

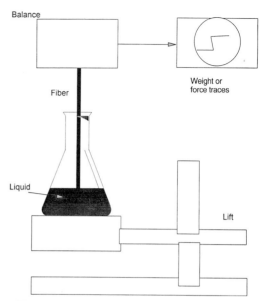

Fig. 2 A typical Wihelmy microbalance setup for determining the surface energy of a fiber

Chemical Analyses

Surface groups on carbon fibers can be analyzed by various chemical methods, including acid-base titration, specific chemical reactions such as diazomethane to detect hydroxyl groups, and the application of radioactive tracer labeled reactants [61]. The chemical reactivity of the fiber surface can also be determined by oxygen chemisorption measurements. These involve heating a fiber sample in a high vacuum to remove the surface species. Then oxygen is carefully introduced into the sample chamber with the amount of oxygen chemisorbed being measured through weight changes or volume changes, allowing quantification of the chemisorption sites. Relative site reactivities can be differentiated by varying the thermal conditions associated with the chemisorption. By using a mass spectrometer to monitor the desorption process, one can obtain information about the atomic and molecular character of the fiber surface [62,63]. Determination of the fiber surface area is possible through adsorption studies of inert gas molecules [62,63].

Specialized Techniques

Many other methods, although not commonly used, can provide specific desirable information on the surface state of fibers. Positron annihilation spectroscopy (PAS) utilizes positrons from a radioactive source (e.g., ^{22}Na) to react with bound electrons in the specimen and investigates the photons which are emitted as the positrons are annihilated. This technique can be used to measure moisture content and volume fraction of carbon fibers in a resin matrix and annihilation can take place at the interface [65,65]. Raman spectroscopy can be used to elucidate the surface structure of carbon fibers [66,67].

The solid-state nuclear magnetic resonance (NMR) techniques of cross polarization (CP), high power proton decoupling and magic angle sample spinning allow the study of the binding of coupling agents to reinforcement surfaces [68]. Although the transmission mode of FTIR does not provide spectra restricted to the surface material, the ability to obtain high sensitivity and to subtract spectra of the substrate from those of modified materials makes FTIR a useful tool [69]. FTIR can be used in conjunction with a microscope, an attenuated total reflection (ATR) attachment, or a diffuse surface reflectance unit for the investigation of fiber surface properties. Quantitative information about the degree of surface coverage or the compositions of mixtures of treated and untreated substrate can be obtained with ultraviolet-visual (UV-VIS) photoacoustic spectra [69,70].

SiC/SiC and SiC/Ti as Examples

Fiber Characterization. The interfacial strength in SiC/SiC and SiC/Ti composites is affected by the type of surface silica (amorphous or quartz), type of carbon (pyrolytic or amorphous), surface impurities, and surface topological ordering. A complete chemical characterization of the fiber surface can be carried out using ESCA, FTIR, SEM and STM. ESCA spectra will reveal information concerning the bonding state of the carbon and SiC in the surface layers [71]. The presence of silica, carbon, silicon, carbides, silicides, and surface impurities can be determined. Distinction between amorphous and pyrolytic carbon can be obtained.

FTIR may be used to analyze the presence of different layers on the fiber surface at different depths. Diffuse reflectance [72], photo-acoustic Spectroscopy, and ATR modes may be used in order to improve the sampling depth and resolution. Spectra obtained by FTIR techniques may be used to complement the ESCA analysis with respect to amorphous or crystalline silica. SEM and STM can be used to provide information on the morphology of the fiber surface, such as surface roughness, porosity, presence of pits, crenelation and slits.

Information provided by ESCA, FTIR, SEM and STM may be directly correlated with the mechanism of fiber-matrix interactions. The presence of surface layers, such as amorphous silica or silicon, will indicate the chemical nature of the interfacial bonding. The morphology of the fiber surface as studied by the SEM and STM will provide information on the contribution of the mechanical bonding to the total interfacial bond.

Surface structure of coated and modified fibers may be studied by the Reflection High Energy Electron Diffraction (RHEED), and TEM. In the case of CVD-coated fibers, a specimen preparation technique developed by Heuer [73] to study SiC fibers can be used. In the case of etched fibers and ion modified fibers, the RHEED technique is quite useful. This method was used by Kowbel *et al.* [74] to study the near-surface structure of amorphous Fe_4B. The application of these techniques allows determination of structural and compositional changes in the near-surface region.

Interfacial Characterization. Comprehensive characterization of the interfacial structure of composites such as SiC/SiC and SiC/Ti composites can be performed using a high-temperature X-ray diffractometer, a TEM/STEM (scanning transmission electron microscope) and a SEM. High-resolution lattice imaging and other TEM techniques can provide useful information regarding the nature of the fiber-matrix interface. Bright and dark field micrographs reveal information on grain size, relative grain orientation and possible porosity near the interphase zone. Convergent beam and STEM microdiffraction allows structural analysis and phase identification on the sub-micron level in the interfacial region [75]. Ultra-thin window energy-dispersive X-ray analysis (EDX) provides information on chemical composition. Utilization of the EDX in the STEM mode allows a compositional analysis below 500 Å.

The *in situ* fracture stage on the Scanning Auger Microprobe can be used to complement TEM observations [76]. The scan profile can be used with respect to O, C, Si, Ti, B, Al, and N in the cases of SiC/SiC and SiC/Ti. Such information aids in understanding the chemical composition changes and the type of the chemical bonding at the fiber-matrix interface. All these data complement one another and provide a comprehensive structural analysis of the degree of matrix cracking, porosity and fiber-matrix interface mechanical compatibility. All these characterization methods can be performed on the fibers before and after surface treatment. The information on microstructural variations will provide the needed fundamental understanding of the nature of the interfacial bonding, such as chemical or mechanical stress developments, cracks and porosity. These factors control the fracture properties of composites.

Diffusion couples can be utilized to simulate and study the phase reactions at the matrix/reinforcement interface. The diffusion couples can be fabricated by hot-pressing bulk pieces of the matrix alloy and reinforcing phases together; any coatings or other interface layers can be simulated by placing foils of that composition between the matrix and ceramic layers. The hot-pressing times and temperatures can be identical to those used for composite consolidation. These couples can then be heat-treated for various times and temperatures in order to determine the identity and growth rate of the interfacial phases. The increased physical dimensions of the diffusion couples (compared to the composite materials) will also reduce the experimental difficulties associated with characterizing very thin layers.

The diffusion couple method allows the thickness and properties, such as microhardness, of the interfacial phases to be determined as a function of time and temperature. With these data, the activation energies and rate constants for a variety of simulated interface modifications can be calculated. The understanding of interface phase formation and growth kinetics which results from these studies can be applied to the actual composite material by modifying the processing schedule and interface coatings so as to give the desired extent of reaction (if any) between the reinforcing phase and matrix. The

Table 1 Elements present on carbon fiber surfaces as revealed by XPS analysis

Fiber	Element, wt %						Ref
	C	O	N	Si	S	Na	
AU1	86	9	2	…	…	3	[82]
AS1	70	20	7	…	…	4	[82]
HMU	95	5	…	…	…	…	[82]
HMS	89	9	…	…	…	…	[82]
AU4	79	14	2	5	…	…	[82]
AS4	80	15	6	…	…	…	[83]
T300	96	2	3	…	…	…	[83]
C600	81	14	4	…	…	1.0	[83]
T500	82	16	3	…	…	…	[83]
AS1	84	11	4	…	0.2	1.0	[84]
AS4	83	12	4	…	0.2	0.7	[84]
AS6	85	9	4	…	0.4	1.7	[84]
IM6	87	9	3	…	0.3	0.6	[84]

Source: Ref 55

knowledge accumulated by these fundamental studies can also be used to fabricate non-optimum interfaces in order to test various microstructure/mechanical theories.

Surface Properties of Carbon Fibers

The fiber-matrix interfacial adhesion depends to a great extent on the physical and chemical state of the fiber surface. A complete characterization of this state is essential to the efforts of controlling the interfacial adhesion through the modification of fiber surfaces. This requires a detailed knowledge of the amount and nature of the surface atoms and functional groups, physi- or chemi-adsorbed species, and surface energetics. Also required is a clear understanding of the internal and surface structure, defects, pore size and structure, and surface area.

Surface Morphology of Carbon Fibers

As discussed in Chapter 2, various models have been proposed to describe the structure of carbon fibers. Although the detailed distribution of structural ordering and defects as conceived in each model might be different, essentially all models maintain that the basic building blocks of a carbon fiber are the graphitic crystallites, which are composed of layers of graphite basal planes arranged into a "turbostratical" layered structure. The size of these crystallites is controlled by the heat treatment time and temperature experienced by the fiber during the conversion process [77]. Higher heat treatment temperatures and longer exposure times tend to produce larger planar graphitic planes and larger layered crystallites. These crystallites are organized in a fibril or ribbon-like morphology roughly parallel to the fiber axis. Tension imposed upon the fiber during various pyrolysis and heat treatment steps causes the fibrils or ribbons to align better with respect to the fiber axis.

The high-modulus fibers, with better axial alignment of ribbons, tend to have surfaces composed predominantly of graphitic basal planes. A lower degree of preferred orientation of the crystallites is usually observed with the lower modulus fibers, which are produced at relatively low graphitization temperatures. These lower modulus fibers therefore have a greater percentage of the crystallites intersecting the fiber surface with edges and corners, which are preferential sites for surface modifications. In contrast, the graphitic basal planes found on the surface of the high-modulus fibers are very inert to chemical attack.

Surface area determines the interaction of a carbon fiber with the matrix material. An important part of the surface area is made up of the internal surfaces of micropores. The stacking of micro-fibrils or ribbons of graphite basal planes in carbon fibers is far from being perfect. This imperfect stacking gives rise to empty spaces between ribbons or microfibrils which form pores or voids [61]. A primary source of pores in carbon fibers is the volatile materials produced by PAN decomposition.

The surface area of carbon fibers may be obtained by the adsorption of nitrogen or krypton at 770 K using the Brunauer-Emmett-Teller (BET) equation [78]. This quantity is governed by the nature of the precursor material (pitch, rayon, or PAN), the temperature of heat treatment, and the subsequent fiber surface treatments. The carbon fibers prepared at lower temperatures have less well organized structures, are highly porous and hence have high surface areas.

Surface Chemistry of Carbon Fibers

Carbon represents the main atomic constituent of the carbon fiber surface. Oxygen is also present, possibly in four forms: carbonyl, phenolic, ether, or lactone structure [79]. Auger and XPS measurements have shown the presence of other elements including nitrogen, sulfur, silicon and a trace amount of metals [80,81]. Bascom and

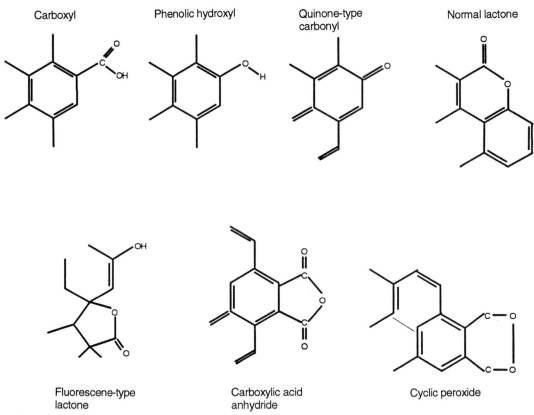

Fig. 3   Functional groups possibly present on the carbon fiber surface

Drzal [55] have compiled a list of the amounts of various elements present on the carbon fiber surface (Table 1). The total oxygen content typically lies between 10% and 20% of the fiber surface with nitrogen from 4 to 6%. The nitrogen has been assigned to an amine-like state and the oxygen to phenolic, carbonyl and other structures [82]. Figure 3 shows the functional groups thought to be present on oxidized carbon surfaces [85].

The surface composition is governed partly by the nature of the precursor fiber (rayon, PAN, or pitch) and partly by the reactivity of the fiber carbon with oxygen. The oxygen content may be derived from the starting material and becomes part of the chemical structure of the carbon fibers as a result of imperfect carbonization [86]. Oxygen may be part of chemical species such as carboxyl, carbonyl and hydroxyl groups that are bonded to the carbon atoms at the edges or corners of the graphitic crystallites or at defect positions [86].

It is also important to recognize the possible presence of physisorbed species on the fiber surfaces. Volatile species such as water, carbon monoxide and carbon dioxide have been detected desorbing from the fiber surface with mass spectroscopy at temperatures up to 150 °C (302 °F) [63]. At higher temperatures, desorption of chemisorbed species begins to occur. The weakly physisorbed species are always present on every fiber surface and, if not removed properly, can create interfacial voids during composite processing [87].

Surface Free Energy of Carbon Fibers

The surface free energies for the low-modulus carbon fibers produced at lower graphitization temperatures have been determined to be up to 50 mJ/m^2 (erg/cm^2). With the percentage of the inert graphitic basal plane increasing with the heat treatment temperatures, the presence of surface chemical groups diminishes. The higher modulus fibers therefore tend to have lower surface free energy, typically about 40 mJ/m^2. Since most polymers have surface free energies of 40 mJ/m^2 or less, the thermodynamic criterion for wetting of the carbon fiber surface is met and intimate contact between fiber and matrix is possible. However, the required equilibrium may not take place between a highly viscous resin and the fiber surface if low processing temperatures and short times are imposed during composite fabrication.

The dynamic wettability data obtained from the Wilhelmy technique permits the separation of the surface free energy into two components: polar and dispersive. The polar component is viewed to be related to the usually polar functional group attached to the edges and corners of the graphitic basal planes [88]. This component therefore declines in magnitude with increasing modulus of

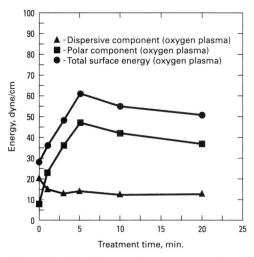

Fig. 4 The effect of oxygen plasma treatment on the polar and dispersive components of the carbon fiber surface free energy

carbon fibers. For the intermediate modulus fibers the polar component is only approximately 20% of the total surface free energy. The polar component of the surface free energy can amount to about 50% of the total value of the surface free energy of carbon fibers and is sensitive to the surface treatments. Figure 4 shows that both the polar component and the total value increase while the disperse component increases when the oxygen plasma treatment time increases [89]. This is related to the increased amount of oxygen-containing functional groups attached to the carbon fiber surface [89-91].

Fiber Surface Treatments and Sizing

Fiber surface modifications and coatings can be used to control the properties of the fiber-matrix interface; in particular, to prevent the formation of strong chemical bonds in ceramic matrix composites, to prevent the formation of a brittle interphase in metal matrix composites, and to form an adequate adhesion (preferentially a chemical bond) in polymer matrix composites. Poor bonding will lead to ineffective load transfer from the matrix to the fiber. A strong interfacial bond tends to reduce the fracture toughness of a composite, particularly a CMC. Surface treatments can be divided into chemical etching, plasma surface etching, Chemical vapor deposition (CVD), and ion implantation. In this section, SiC fiber-reinforced titanium and SiC matrix composites are used as examples for illustrating the issues of fiber surface treatments in MMC and CMC, respectively.

Metal and Ceramic Matrix Composites

Chemical etching can be performed using nitric and chromic acid solutions. The purpose of these treatments is to introduce controlled amounts of fiber surface roughness. Plasma etching offers a unique way to modify the

roughness of the surface on the atomic, microscopic and macroscopic levels [92]. These treatments can be performed using a 13.56 MHz radio frequency (RF) generator at a total pressure of 0.1 torr, using argon and oxygen. Treatment with argon should generate surface roughness on an atomic level in a controlled manner. The treatment can also involve a more reactive oxygen plasma, which can remove surface impurities and poorly bonded surface layers yielding a less irregular surface structure.

The types of surface roughness introduced by the plasma treatments should include atomic level structural changes as well as microscopic modifications like slits, crenelations, pits and porosity. These treatments will predominantly control the degree of mechanical bonding, and will also provide an answer to the question of whether "key and lock" bonding is facilitated by structural irregularities on the atomic, micro-, or macro-level.

Chemical vapor deposition (CVD) is widely used for the purpose of applying thin coatings on carbon and SiC fibers [19,21]. CVD provides a strong bond between the substrate and coating and allows proper control of the structure, composition, thickness and uniformity of the coating. The CVD deposition rate is governed by surface kinetics or diffusion. If the reaction is performed at low temperature, the kinetics of the surface reaction becomes the predominant factor, and a good mechanical bond between the coating and substrate can be achieved. The application of plasma-assisted CVD will produce an amorphous carbon. Varying both the deposition temperature and the supersaturation can result in the formation of highly oriented to fully isotropic coating structures. The degree of preferred orientation can influence the mechanical compatibility in SiC/C/SiC composites.

The coating morphology can be varied by changing the degree of the supersaturation. This may allow the study of the effect of coating morphology on interfacial bonding. A pyrolytic carbon coating can be used to eliminate chemical bonding in SiC/SiC composites so that the effect of mechanical bonding on the composite and interfacial properties can be examined. The boron nitride (BN) coating is an alternative material for fiber surface modification. Plasma-assisted CVD can be used to provide purely physical bonding in SiC/SiC by the application of a smooth stoichiometric SiC layer.

The composition control of the CVD process can also be used to produce graded SiC/C coatings on Nicalon SiC fibers, varying from stoichiometric SiC to pyrolytic carbon. The CVD phase diagram for the SiC/C system allows the prediction of the deposition conditions resulting in composition-graded structures. The coating thickness may vary from 0.1 to 1 μm in order to study the mechanical compatibility in the SiC/C/SiC system and the effectiveness of the pyrolytic carbon and h-BN as diffusion barriers.

In the case of SiC/Ti, the use of AVCO SCS-6 SiC offers an excellent example of the practical implementation of the concept of graded coatings [24]. The application of graded SiC-C coating can address two fundamental problems in metal matrix composites. First, the outer coating which is a buffer zone can form an interphase

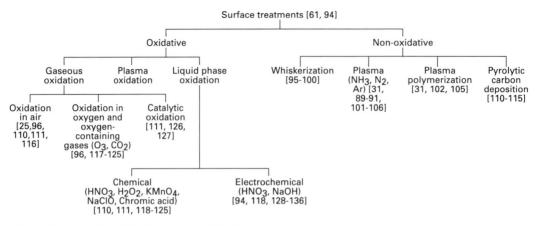

Fig. 5 Summary of available fiber surface modification techniques

layer with the matrix. Second, the inner zone should prevent the degradation of the fiber strength because of shear failure of the carbon-rich layer. The graded SiC/C/SiC coatings can be obtained by both thermal and plasma CVD processes. The composite SiC/C coatings can be obtained with the control of the morphology including layers, particulates, and flakes. Different morphologies result in different thermal and mechanical properties due to the high anisotropy of graphite. Other coatings that can be applied to AVCO SiC fibers include alumina, boron carbide, titanium diboride and boron.

Ion implantation introduces a new chemical species into a solid or a physical process that is not obtained by usual thermodynamic or kinetic considerations. This process may be viewed as atomic-level mechanical mixing [93]. Ion implantation with B may be used to alter the surface of SiC fibers. Several processes may be involved in the near surface modification. These include precipitate formation, ion induced amorphization, and residual surface stress.

The precipitate formation possibly involves B_4C. The structure and concentration are a function of the energy and flux of the ions. The ion implanted B_4C may be grossly different from the CVD deposited B_4C. Since ion implantation is a surface modification, no sharp coating-substrate interface is introduced. At high doses amorphous B_4C can be formed, which will vary the interface interactions in SiC/Ti matrix composites after the fiber is ion implanted.

Polymer Matrix Composites

Sizing on Carbon Fibers. All carbon fiber manufacturers apply sizing, similar to textile fiber sizings or finishes, to make the carbon fibers easier to handle and to protect them against mechanical damage during processing. The sizings that are applied to the carbon fiber surface are typically an epoxide pre-polymer or polyvinyl alcohol. The amounts of coating vary between 0.5 and 5%. The coating adheres to the untreated carbon fibers virtually by physical adsorption only and can be completely removed under mild chemical conditions, for example, by treatment with acetone, chloroform, water, etc. Sizings are not generally intended for improving the mechanical properties of the composite, although it is conceivable that specialty coatings can serve dual purposes.

Surface Treatment of Carbon Fibers. The adhesion between untreated carbon fibers and common aerospace-grade resins is usually poor. To achieve good interfacial adhesion between a matrix resin and carbon fibers, the possibility of forming a chemical bond between fiber surface groups and the functional groups in the resin matrix should be utilized. The creation of useful surface functional groups depends on the chemical nature of the fiber surface and the method of treatment. A variety of surface modification techniques developed to promote the interfacial bonding in carbon fiber composites are summarized in Fig. 5.

In one approach, fibers were oxidized with either liquid oxidizing agents [117], such as concentrated nitric acid, or gaseous media [137] such as air, oxygen, carbon dioxide and ozone. Generally, oxidation etches the fiber surface and possibly implants carbonyl and hydroxyl groups to the fiber surface, resulting in an increase in interfacial bond strength. Boehm et al. [138] have indicated that oxide functional groups can become attached to points on the carbon fiber surface where free valences of the graphite carbon atoms have not been saturated. Such free valences are formed by faults in the graphite layers at the surface (prismatic surfaces). In high-modulus graphite fibers where the graphite layers are highly oriented to the fiber axis, leaving behind a smaller concentration of surface defects and lesser extent of prismatic surfaces, the sensitivity toward oxidation would be relatively low. Contrarily, low-modulus carbon fibers can be more readily oxidized for improved interfacial adhesion.

Oxidation can also be carried out using an electrochemical technique, where the fibers are moved continually through an electrolyte bath, usually as an anode [128]. Possible electrolytes include caustic soda and potash, nitric, sulfuric and phosphoric acids, potassium dichromate and potassium permanganate [94]. Electrochemical techniques are fast, uniform and suited to mass production processes.

In a second approach, polymers such as vinyl, phenolic, and epoxy resins have been coated on the order of 1 to 2 wt. % onto fibers. A special example of improving adhesion through polymer coatings is electrochemical polymerization directly on the fiber surface [139,140]. These coatings resulted in a varying degree of improvement, but thickness control of the coating has been a problem.

Solutions of reducing agents, such as $FeCl_3$, have also been used to improve the carbon fiber-epoxy interfacial bond strength [141,142]. A substantial improvement in composite interlaminar shear strength has been observed without an appreciable loss in fiber strength. However, the iron compound residues could possibly catalyze oxidation reactions during the intended service period, leading to high-temperature instability.

Vapor phase deposition, in which the fiber surface is "whiskerized" by the deposited whiskers (such as silicon carbide) to provide mechanical bonding sites, led to a significant improvement in the interlaminar shear strength [95]. Nevertheless, the grown silicon carbide whiskers added additional weight to the composite. Fiber surfaces can also be coated with a thin layer of polymer through electrodeposition or electropolymerization. Moderate improvements in the interlaminar shear strength and the impact strength of composites were reported [33,143].

Gas plasma activated by microwave or radio frequency (RF) radiation has been extensively applied for processing semiconductor and other materials. In the cold plasma state, ionization, excitation, dissociation, recombination, and other reactions can occur due to the collision of electrons and other species in the plasma medium. Consequently, when the plasma contacts a solid material, a highly efficient energy exchange can occur. Plasma CVD, plasma etching, and plasma polymerization are all plasma processing methods that make full use of this energy exchange mechanism. If properly controlled, plasma can be used to modify the physical and chemical state of the material surface without significantly altering the bulk properties. Because of these attractive attributes, plasma treatments of fiber surfaces have been considered to be a prime candidate technique for the control of interfacial adhesion in composites.

Plasma treatments are known to significantly enhance the adhesion of polymer fibers to epoxy resins [44, 101,144,145]. Amine plasma was claimed to promote formation of covalent bonds with epoxy resins, leading to a considerable enhancement of the epoxy composite strength [101]. Carbon fibers and pyrolytic graphite blocks were treated with plasma polymer coatings from acrylonitrile and styrene monomers [31,107,108]. Additional work [102,109] using plasma polymerization for modifying fibers also showed encouraging results. Treated fibers in some cases even exhibited higher tensile strengths than the untreated counterparts, suggesting that plasma coatings possibly healed some of the surface flaws of the fiber [43,102,107-109]. However, more ef-

forts are needed to verify this speculation. The plasma-coated fibers also exhibited increased functionality and lower contact angles. Single-fiber interface testing results indicate a 100% increase in the interfacial shear strength. The interlaminar shear strength of epoxy composites improved approximately 30% with minimal degradation in the flexural properties [102,109].

When treated with oxygen plasma, the surfaces of PE fiber exhibited a high concentration of hydroxyl and carboxylic acid groups, which could permit chemical bonding between the fiber and the resin. An increase in interlaminar shear strength was obtained with epoxy resin matrices [146-148]. Similar functional groups were observed on the carbon fiber surfaces treated in an oxygen plasma [149,150]. The presence of a significant concentration of acid functionality on these fibers was confirmed by ESCA and titration techniques. The physical and chemical characteristics of carbon fiber surfaces were altered by the plasmas [89-91,103-105,150-155]. The level of surface roughness in the form of pits and crevices was increased [89-91,105,150,151,155]. Surface wetting properties of fibers were also improved, permitting a more intimate fiber-resin contact [89-91,105]. Different functional groups were observed on different types of carbon fibers; both carboxyl and hydroxyl groups were observed on PAN-based fibers while very little carboxyl was seen on pitch-based fibers [156]. Changes in the composite failure mode were observed in carbon-bis-maleimide composites when the fibers were modified with plasma [106].

Continuous plasma treatments of fibers have been proven feasible [109,148,150,151,157]. However, mechanisms of interfacial adhesion as promoted by plasma treatments are not yet well understood. The technique of plasma etching/polymerization appears to be a highly versatile and promising technique for the surface treatments of fibers. Such treatment offers many potential advantages over conventional surface treatments. The chemical and physical structure of the fiber surface can be systematically tailored for different purposes with relatively simple operations. Specialized functional groups may be deposited on the fiber surface to promote the subsequent formation of primary or hydrogen bonds with the matrix. Further, plasma polymerization of monomeric gas can generate a thin layer of coating on the surface with desired properties, and the coating may also be grafted to the substrate.

One example is given here to illustrate the effectiveness of plasma techniques in tailoring the interface in polymer composites. Figure 4 shows the surface energies of carbon fibers that were plotted versus oxygen plasma treatment time. Both the polar component and the total surface free energy are found to increase with the treatment time. A corresponding increase in the transverse tensile strength of composites and a minute reduction in fiber strength were observed [89-91,105]. SEM micrographs taken from the flexural test specimens indicated the reduced trend in fiber pull-out, indicating a better interfacial adhesion, as the treatment time increased

[89]. The following remarks can be drawn from this series of studies [89-91,105]:

- The effectiveness of plasma treatments in improving the interfacial adhesion in polymer composites has been demonstrated. For BMI matrix composites, ammonia/argon plasma appears to be the best plasma system in enhancing the interfacial adhesion without producing undesirable reduction in the fiber strength. In contrast, oxygen and argon plasmas show a greater etching effect on the carbon fiber surface and could degrade fiber integrity.

- Various adhesion mechanisms can be promoted by plasma treatments. These include: (a) possible removal of surface contaminants to provide a better fiber-resin contact; (b) the enhanced degree of mechanical keying between the fiber and the matrix because of the increased fiber surface roughness; (c) the increased surface energy which would promote wetting of the fiber by the matrix; and (d) deposited functional groups for possible chemical interactions between the fiber and the matrix resin. Mechanisms (a) and (d) could affect (c).

- Each mechanism works to a different extent in a different plasma. For oxygen, nitrogen and argon plasmas, mechanisms (a), (b) and (c) all work to some extent; however, oxygen plasma is the most effective in improving the wettability. This vastly improved wettability could be a result of mechanism (d), the plasma-deposited oxygen-containing groups on the carbon fiber surface. For ammonia and ammonia/argon plasmas, the chemical bonding between amine groups and bismaleimide (mechanism d) and the enhanced surface wettability (mechanism c) are the two important factors in increasing the interfacial adhesion in carbon fiber/BMI composites.

- Plasma polymerization can be used to promote interfacial adhesion between carbon fibers and thermoplastic resins. Plasma polymer is capable of adhering well to any substrate. The plasma polymer coating on the fiber surface can be designed to provide necessary chemical characteristics (functional groups, compatibility, and so on) to facilitate fiber-matrix adhesion.

- The operating parameters for oxygen and nitrogen plasma treatments of carbon fibers can be optimized for composite performance. Gas flow rate and treatment duration are two of the more important parameters to manipulate.

The surface of carbon fiber is a poorly defined carbonaceous layer that is partly aromatic and partly aliphatic with many adsorbed species and highly reactive sites [55]. How much of the surface treatment process is a mechanical cleaning of these "loose" species and how much is chemical alteration remains to be resolved. How the chemical constitution, scale of porosity, and hence the mechanical properties of the treated fiber surface vary with depth remains to be determined. Also unclear are the questions of whether a distinct layer or "interphase" always exists in a real composite, and how this interphase would be influenced by the various surface treatments [55]. The microstructure and micro-properties of such an "interphase" in advanced polymer composites need to be properly characterized before one can adequately predict the global properties of composites.

Interfacial Adhesion Between Carbon Fibers and the Matrix

Experimental determination of the micro-properties of the interphase in relation to the microstructure specifically at the interface is the first critical step to understanding the behavior of composite systems. Several test methods are available in the literature for measuring interfacial shear strength, tensile strength, debonding energy, and frictional forces. These techniques include fiber pull-out, fiber push-out, critical fiber length, short beam shear, fiber indentation, micro-debonding, and transverse tension [1,57]. Further, the degree of interfacial adhesion and the bond conditions have been investigated by polarized light microscopy, dynamic mechanical analysis [56,57], Rayleigh surface wave measurement, photoelasticity and others. These techniques have recently been reviewed by several workers [1,57,58]. Emphasized here will be the techniques that can be applied to characterize the micro-properties of the interface or interphase.

The Concept of Interfacial Strength

As pointed out by Bascom and Drzal [55], the term "adhesion" or "adhesion strength" is commonly used to describe the load or stress required to separate two dissimilar solids at or near their common boundary. However, the adhesion between two solids is thermodynamically defined as the "work of adhesion," which is based purely on the net work done for the generation of two new surfaces and the elimination of an interface between these two bodies. In general, this work of adhesion represents a small percentage of the total energy required to separate the two solids. Deformational energy that is dissipated before the creation of new surfaces is typically orders of magnitude greater than the work of adhesion. Bascom and Drzal [55] maintain that what is usually measured should be more correctly called a boundary strength, bond strength, or joint strength.

Evaluation of Interfacial Shear Strength

Measurements of interfacial properties must be conducted in order to evaluate mechanics models or to empirically correlate composite behavior with the interface properties [158]. Once the fundamentals of the mechanical behavior of composites are understood, such measurements can be used to identify the desirable combinations of fiber, matrix, surface treatment or coating, and processing conditions for a composite with the desired properties. Many techniques have been developed to meet this need and they depend on the validity of mathe-

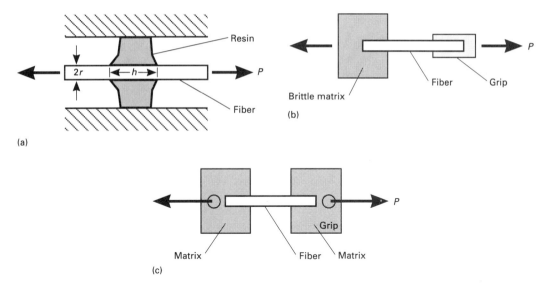

Fig. 6 Single fiber pull-out tests for evaluating the interfacial shear strength. (a) Single fiber pull-out test geometry for polymer composites. (b) One end of the fiber is embedded in a brittle matrix, such as glass or ceramic. (c) Both ends are embedded in the matrix

matical models of the failure process to obtain basic properties such as interfacial bond strength.

Interfacial shear strength can be measured in two ways. One technique is based on the fabrication of a model composite containing only a single fiber. This approach is subject to the limitation that the test is conducted on an isolated single fiber and it does not represent a true composite test. These tests make it easier to delineate the interfacial failure mechanisms in composites. Comparisons between fibers with different surface treatments can be made with this approach.

A second approach involves the fabrication of composite samples for standard short beam shear, four-point bending, or flexural strength tests. These tests would generate a value which is attributable to an interfacial shear strength. However, the state of stress in these specimens is in general not a pure interlaminar shear at the ply boundary nor a simple in-plane shear at the fiber-matrix boundary. Interpretation of these shear strength data is not without ambiguity. However, sample preparation and test execution of these techniques are rather easy to do and therefore they have been widely used despite this serious limitation.

The Fiber Pull-Out Test. In this test, the specimen contains a single fiber oriented through a thin disc of the resin of interest (Fig. 6a). The disc must be thin enough so that the fiber can be pulled out of the specimen without breaking. In a typical test, a peak load is reached when the interface is thought to "debond". The load then drops to a lower level corresponding to the presumed frictional resistance from the interface. This is followed by the pull-out of the fiber from the matrix. For polymer matrix composites, the experiments are conducted where the embedded length is increased up to the point of fiber

fracture and the measured force for debonding is plotted versus embedded length. The slope of this line is taken as the shear debonding strength [159,160]. A uniform shear stress distribution along the embedded fiber length is often assumed. This test tends to exhibit a great degree of data scattering. The polymer meniscus shape at the fiber exit point causes stress concentration which likely contributes to high data scatter.

Recently, this technique has been exploited to measure interface properties of brittle matrix composites [161,162]. In these applications, specimen preparation is different in that either part of the fiber sticks out of the matrix (Fig. 6b) or the two ends of a single fiber are embedded in the same matrix (Fig. 6c). The tensile stress imposed on the fiber will result in a Poisson radial tensile stress at the interface aiding debonding and pull-out. This effect, and that caused by the nonuniform shear stress along the fiber, complicates the analysis of the experimental data. Modeling of the fiber pull-out test [163-166] has led to the following useful expressions:

$$P_d = \frac{2\pi r \tau_d}{\alpha} \tanh{(\alpha L)} \qquad \text{(Eq 7)}$$

$$P_f = \frac{\pi r^2 \sigma_0}{K}\left[1 - \exp\left(-\frac{2\mu K L}{r}\right)\right] \qquad \text{(Eq 8)}$$

where P_d is the peak or debonding load, P_f is the frictional load, τ_d is the debonding stress, L is the embedded length, $2r$ is the fiber diameter, α is a shear-lag parameter, μ is the friction coefficient, σ_0 is the residual compressive stress on the interface, K is $E_m \nu_f / [E_f (1 + \nu_m)]$, and ν_f and ν_m

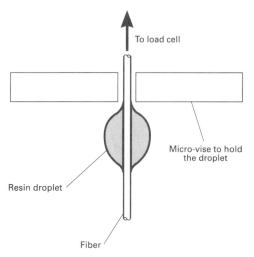

Fig. 7   A microbond test method

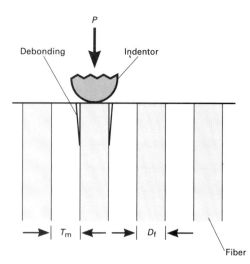

Fig. 8   Schematic of a micro-debond test method

are Poisson's ratios of the fiber and the matrix, respectively.

Here the parameter α depends on the elastic constants and its definition varies with the model details. In the case of resin composites where the embedded length is necessarily small relative to the fiber diameter (to ensure fiber pull-out rather than fiber fracture), Eq 7 and 8 reduce to:

$$P_d = 2\pi rL\tau_d \qquad\qquad\qquad \text{(Eq 9)}$$

$$P_f = 2\pi rL\tau_f \qquad\qquad\qquad \text{(Eq 10)}$$

where τ_f is the frictional stress.

The microbond method [167,168] is a variation of the fiber pull-out test for polymer composites. A droplet of polymer is formed on the fiber instead of the fiber passing through a supported disc (Fig. 7). The droplets normally form concentrically around the fiber in the shape of ellipsoids and retain their shape after curing. One end of each fiber specimen is glued to a metal tab which can be connected to a load cell. The droplet is gripped with a micro-vise which is mounted on the crosshead of the tester. The fiber specimens are then pulled out of the droplets just like the traditional pull-out test discussed earlier. The variability in the data is reduced with this microbond method. Average bond strengths between an epoxy resin (EPON 828) and a few commonly used fibers are 33 MPa (4.8 ksi) for E-glass, 49 MPa (7.1 ksi) for E-glass + silane, 38 MPa (5.5 ksi) for Kevlar®, and 57 MPa (8.3 ksi) for Celion®-3000 carbon fiber.

The Fiber Push-out Test. For large diameter fibers or bundles of thin filaments as in carbon-carbon composites, interfacial shear properties can be obtained by pushing the "fiber" out of a thin slice of the "matrix." Laughner *et al.* [169] proposed a fiber push-out test to measure interface properties of CMCs such as Nicalon® fibers in silicon nitride matrices. This technique is a variation of the indentation or micro-debond techniques. The speci-

men and indentation geometry was modified to measure both the debonding stress and the friction stress of the interface in shear. The load-displacement curve was found to consist of a peak load, which was assumed to correspond to interface debonding, followed by a lower frictional load. After debonding, the specimen was reversed and reloaded to check the frictional stress.

This test method is tailored for the determination of interface shear strength. Both chemical bonding and frictional resistance can contribute to this interfacial strength, though their contributions are often difficult to separate. A fiber push-out test has been developed to determine the relative contributions of debonding and frictional shear stresses [76]. The contributions of various treatments which introduce different degrees of chemical and mechanical bonding at the interface can be examined. By correlating results from this test and macrostructure analysis, the factors that control shear deformation and load transfer can be identified.

In the micro-debonding test, [170,171], a polished cross-section of a composite is prepared and placed under a specially configured optical microscope. A spherical indentor of about the same size as the fiber is positioned over a fiber end. The fiber is loaded in steps of increasing force, with the bond viewed optically between steps, until debonding is observed. The measured debonding force is divided by the fiber cross-sectional area to obtain the average applied compressive stress on the fiber end at debond initiation. The interfacial shear strength is then calculated from the applied stress using results from finite element analysis [171]. Failure is dominated by the interfacial shear stress component, which reaches a peak value at a distance of approximately one-half the fiber diameter below the contact stress. This technique assumes that debonding initiates at this point due to the maximum shear stress, and quickly spreads over an area of the circumference and up to the surface [171]. An improved version of this technique [172] uses a boundary element method to analyze the stress at the interface and allows for continuous loading of the fiber. Possible automatic determination of the debonding stress with the assistance of acoustic based methods was suggested [172].

The advantage of the method is that it is applied to an actual composite sample. However, several technical concerns have yet to be resolved, including: effects of the local fiber arrangement and fiber spacing; effects of the free-edge singularity; effects of specimen surface topography; probe alignment accuracy; and fiber diameter measurement uncertainty [76]. Uncertainty also exists about the deformation of fiber at the indentor contact area. The load measured has not been shown to be attributable entirely to the debonding process [171]. The micro-debonding test is shown in Fig. 8. The following equation may be used to calculate the interfacial shear strength τ:

$$\tau = \frac{\sigma}{2}\left(\frac{G_m}{E_f}\right)^{1/2}\left(\frac{D_f}{T_m}\right)^{1/2} \qquad \text{(Eq 11)}$$

where G_m is the matrix shear modulus and σ is the imposing compressive stress at which debonding occurs.

The Microindentation Test. A Vickers indentation instrument was used by Marshall [173,174] to exert an axial force on fibers in a polished section cut normal to the fiber alignment. If the fiber is weakly bonded to the matrix (frictional forces only) the exertion by the diamond tip will result in a compression of the fiber with concomitant axial displacement below the polished section surface over a debonded length. On removal of the load, the extent to which it recovers depends on the frictional forces exerted on the fiber by the matrix [58]. The relationship between interface friction stress τ, the force on the indentor P, and the distance μ that the fiber is depressed with respect to the matrix, is given by

$$\tau = \frac{P^2}{4\pi^2\mu r^3 E_f} \qquad \text{(Eq 12)}$$

where r is the fiber radius and E_f is the axial Young's modulus of the fiber. Marshall and Oliver [174] later relaxed the assumption that the indentor load was supported only by the frictional forces in the debonded region. The external load was assumed over a length l, part of which is debonded. An energy-balance approach was used to determine the debonded length u leading to the following equation:

$$u = \frac{P^2}{4\pi^2\tau r^3 E_f} - \frac{2\Gamma}{\tau} \qquad \text{(Eq 13)}$$

where 2Γ is the fracture-surface energy per unit area of the interface. The nanoindentor was used to measure the displacement as a function of the indentor load. A plot of P^2 versus u will provide an estimation of the friction stress (slope) and the fracture energy (intercept).

This technique is effectively a variation of the micro-debonding test, which was initially developed to measure the interface bond strength in polymer composites and

later applied, with a finite element analysis, to determine the debonding and the friction stress separately in glass, ceramic and metal matrix composites. This technique was used for assessing the interfacial adhesion in SiC matrix composites containing coated SiC fibers [175]. A simple tensile test proposed by ORNL [176] may also be used for determining interfacial frictional stresses.

The Embedded Fiber Fragmentation or Critical Length Method. Several versions of this method have been used to measure interfacial shear strength [177-181]. The method involves fabrication of a tensile coupon containing an embedded single fiber oriented axially along the tensile loading direction. Under a monotonically increasing tensile load, shear forces are transferred from matrix to fiber at the interface, which cause the build-up of tensile stresses in the fiber. If the tensile stress builds up to exceed the fiber tensile strength (σ_{fu}), the fiber fractures within the polymer. This process is repeated until the fragments remaining are no longer enough to transfer sufficient tensile stress to exceed the fiber tensile strength. At this point the fragment lengths remaining represent the critical transfer length (l_c) for reinforcement:

$$l_c = \frac{\tau\sigma_{fu}}{\tau_y} \qquad \text{(Eq 14)}$$

where τ_y is the shear yield stress of the matrix in the original derivation of this fragmentation critical length equation [182]. Researchers use Eq 14 to calculate τ_y from experimentally measured l_c values and consider this τ_y value the "interfacial shear strength" [179-181]. This is based on the assumption that there is a constant shear stress across the interface, ignoring the fact that τ_y is a matrix property [56].

The fracturing process actually generates a distribution of fragment lengths. A large number of fiber fragments are measured to form an adequate population to which a two-parameter Weibull distribution is fitted. Subsequently the Weibull mean and variance of the critical transfer length, or interfacial shear strength, are obtained for a given fiber-resin combination. The variance values are typically ±50% or higher.

The method provides qualitative indications for the fiber-matrix interfacial bonding. It actually prescribed that the upper limit to interfacial shear strength is the matrix shear strength. In this test, the state of stress more closely resembles what is expected in a real composite and the actual interfacial failure process can be observed with polarized transmitted light.

The short beam shear test (ASTM D2344-84) can also be conducted to assess the interlaminar shear strength of composites. The short beam shear test involves loading of unidirectional laminates in such a way that failure occurs in a shear mode parallel to the fibers. This is normally achieved by choosing a small specimen span-to-depth ratio.

This test does not measure interfacial bond strength directly and should be regarded as a simple engineering

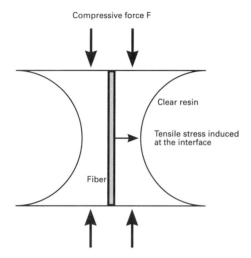

Compressive force F

Clear resin

Tensile stress induced
at the interface

Fiber

Fig. 9 Schematic of a curved neck specimen

test. On a relative basis, the test results may indicate tendencies of the bond strength in a given composite system where only the bonding level is a variable. This test becomes invalid when flexural failure mode (tensile or compressive) precedes shear failure, or mixed mode fracture occurs. The short beam shear strength represents the failure stress at the mid-plane of a laminate. This failure stress is a function of the fiber volume fraction which is not accounted for in the calculation of the short beam shear strength [56].

Dynamic Mechanical Analysis. Energy is expected to be dissipated at the fiber-matrix interface when the composite is subjected to continuous cyclic loading. The amount of internal energy dissipated at the interface depends upon the degree of adhesion. A weaker bond is expected to result in more energy loss which should be reflected by a higher damping coefficient. Thus, it is possible to determine the amount of energy dissipation due to poor interfacial adhesion by any vibration-based technique [57,183].

Interfacial Studies using Laser Raman Spectroscopy (LRS). The Raman spectra of polymer fibers, carbon fibers and ceramic fibers can be altered with the application of an external force, pressure, or heat. The interatomic force constants which determine the atomic vibrational frequencies are directly related to interatomic separation [184]. As the bond lengths increase the force constants and the vibrational frequencies decrease. This concept can be applied to monitor the stress or strain that a reinforcing fiber experiences in a composite. Such a "molecular" strain gage has been successfully applied to follow the micromechanics of fibers, the effect of resin shrinkage, the interfacial stress distributions and deformation properties of composites [184-194]. Some Raman-sensitive fibers can also be implanted between the plies of a laminate and laser light guided to the fiber using fiber optics. In this way, strain at any point within the laminate can be measured and first-ply failure and delamination detected [195]. By subjecting composites to various degrees of mechanical deformation a number of parameters, such as the transfer length, the stress transfer efficiency and the initiation of fiber debonding or matrix yielding can be assessed [194].

Evaluation of Interfacial Tensile Strength

The Curved Neck Specimen. Under transverse loading, the mechanical behavior of composites largely depends on the fiber-matrix interface tensile strength normal to the fiber surface. A curved neck single-fiber specimen is available for measuring that component of the interfacial strength (Fig. 9). A single fiber is placed centrally through the axis of the polymer specimen, which is loaded in compression. The hour-glass shape of the sample maintains a stable compressive load on the specimen. The Poisson's effect produces a normal tensile stress perpendicular to the fiber axis at the center of the specimen. The load at which the center of matrix debonds from the fiber may be detected by using reflected light, or possibly by acoustic emission. This value, in terms of stress σ, can be used to calculate an interfacial tensile strengths:

$$S = \frac{\sigma(\nu_f - \nu_m)E_f}{(1 + \nu_m)E_f + (1 - \nu_f - \nu_f^2)E_m} \qquad \text{(Eq 15)}$$

where the axial stress σ is calculated at minimum cross-section, ν_f and ν_m are the Poisson's ratios, while E_f and E_m are the Young's modulus of the fiber and the matrix, respectively. Because of the experimental difficulties, this method has not become popular.

The transverse tensile test involves testing unidirectional composites in the transverse (90°) direction. This test was found [176] to be one of the more sensitive techniques for assessing the relative interfacial adhesion strength in composites. Transverse toughness measurements can also be applied to assess the interfacial adhesion in composites [196].

Evaluation of Interfacial Toughness

Argon *et al.* [197,198] have developed a method to determine the intrinsic delamination toughness of interfaces between thin coatings of SiC and substrate of either silicon single crystals or carbon fibers from an analysis of the spontaneous delamination phenomenon of such coatings while under residual tension or compression. When the thickness of such stressed coatings reaches a critical threshold value, the elastic strain energy of material misfit stored in the coating becomes a driving force for delamination of the coatings. This method will be most suitable to the determination of interface toughness between SiC fiber and various coatings and that between a layer of SiC and Ti-based substrate. Interface toughness values can then be correlated with bulk toughness.

Quantitative correlation among the various measures of "interfacial adhesion" in composites has yet to be established. Which of these values best represents the interfacial bond and can be better correlated with the composite properties has yet to be determined.

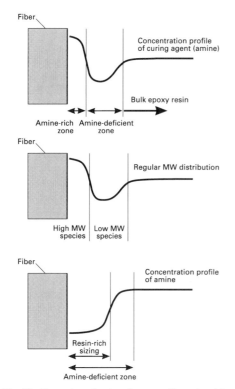

Fig. 10 Formation of an interface zone (interphase) in a thermoset matrix composite

Polymer Microstructure and Properties Near the Interface

The interactions between the fiber and the matrix resin (in melt or solution) before or during curing are rather complex but important phenomena. These interactions could lead to the formation of the interphase, which most likely has different composition, microstructure and properties than the bulk resin. Bascom and Drzal [55] have cited a number of possible polymer-fiber interactions: selective adsorption of matrix components, conformational effects, penetration of polymer molecules into the fiber surface, diffusion of low molecular weight (MW) components from the fiber, and catalytic effects of the fiber surface on polymers. Surface-induced or surface-modified crystallization phenomena can also occur at or near the fiber surface. In thermoset matrices, the presence or absence of a fiber surface sizing or coating layer can have a profound influence on the development of the interphase zone. These and other related issues are discussed in this section.

Thermosetting Resins

The possible presence of boundary layers in thermoset resins near or at a solid substrate was implicated in many early studies on thermoset morphology [199-204]. Kardos [199] gave some morphological evidence in

glass-epoxy that could be interpreted in the light of curing agent migration to the glass-epoxy interface. The existence of a distinct region approximately 100 nm thick around filler particles in epoxy resins was noted by Kumins *et al.* [201] and Kwei [202]. The material in this boundary layer was found to have a significantly lower molecular mobility than that in the bulk resin. Calculations of viscoelastic relaxation time spectra in filled epoxies led Lipatov *et al.* [200] to suggest a similar reduction in molecular mobility. Racich and Koutsky [203,205] have observed the preferential diffusion and chemisorption of the amine curing agent from the bulk resin to the polar adherend substrate such as copper wires. This selective adsorption and the concurrent reaction of the amine created an amine-rich region directly at the interface and depleted amine in the adjacent boundary layer. Similar results were obtained by Sukhareva *et al.* [204]. This situation is illustrated schematically in Fig. 10(a).

Bickerman [206] maintained that low molecular weight "impurities" could migrate from the bulk of adhesives to form a weak boundary layer (WBL) onto adherents. In thermoset systems where polar adherents (e.g., carbon fiber surface) strongly attract polar molecules (amine, or other catalysts), a boundary layer of high cross-link density can develop. This microstructural gradient can promote failure initiation or crack propagation through this boundary layer [207-209].

The presence of a surface sizing or coating also has a profound influence on the formation of a boundary layer. A polar coating would tend to attract polar species and vice versa. No systematic study was reported on the effect of fiber surface finish or coating polarity upon the interfacial behavior. Drzal *et al.* [210] conducted a study directed at understanding the role of a resin-rich finish on carbon fiber-matrix adhesion. Inadequate mixing of the fiber with the nearby resin/hardener mixture created a hardener-deficient layer around the fiber (Fig. 10c). Interface failure was noted with unsized fibers, but fracture propagated into the matrix when using sized fibers. The properties of this amine-deficient layer were believed [210] to be more brittle than the bulk matrix having the stoichiometric ratio between epoxy and curing agent.

Additional complications can come from the selective adsorption of polymer chains of differing lengths on the carbon fiber surface [211]. This can be understood from the difference rule [211]:

$$\gamma_{sp} = \gamma_s - \gamma_p \qquad (Eq\ 16)$$

where γ_s and γ_p are surface tension of solid substrate and polymer and γ_{sp} is interfacial tension (interface free energy). The dependence of γ_p on the chain molecular weight (MW) may be given as [212]:

$$\gamma_p = \gamma_{D,P} - K\ /\ MW^{2/3} \qquad (Eq\ 17)$$

where $\gamma_{D,P}$ and K are experimentally determined constants for a given type of polymer. A longer chain (larger MW) would mean a larger value of γ_p or a reduced γ_{sp}. This suggests that selective adsorption of longer chains to the fiber surface could lead to a reduction in overall surface free energy.

Lipatov [211] used narrow MW fractions of polystyrene and their mixtures to investigate the selective adsorption phenomenon on the glass surface. After a prolonged contact of the polymer melt with the glass surface, average molecular weight was found to be a strong function of the distance from the interface (Fig. 10b). A zone of about 20 microns was composed of preferentially longer chains, which was followed by a nearby region of predominantly shorter chains. This study demonstrated the existence of molecular mass segregation, although the results may not be extrapolated to give quantitative information on the oligomer dispersion near highly packed carbon fiber surfaces. It is conceivable that longer chains would lead to a tougher epoxy resin of lower degree of cross-linking. But the neighboring zone would be heavily cross-linked and such a structural gradient could have an important implication in the microfailure response of fiber-matrix interface. A systematic study directed at a better understanding of the property-structure relationship at the interface needs to be performed.

Tsai et al. [213] developed an elastic shear lag analysis and correlated the analysis results with the microdebonding test data [171] to determine the thickness and material properties of the interphase. This analysis assumed the existence of an interphase region of uniform modulus and fixed radial dimension. If the interphase thickness can be obtained from an independent measurement, then the interphase shear modulus can be computed from the equations derived by Tsai et al. [213]. Since this data was not readily available, the authors suggested utilization of an interaction procedure to estimate these properties. They concluded that an interphase exists and is softer than the bulk matrix. Whether this is due to the use of sizing or the selective adsorption of longer epoxy molecules, or possibly other unknown reasons, remains to be elucidated.

Thermoplastic Matrices

Semicrystalline thermoplastic matrices were shown to nucleate and crystallize on the surface of carbon fibers in a manner different from the bulk polymer [214-221]. The surface crystallization, typically resulting in the formation of transcrystallinity, was hypothesized as being responsible for improvement in adhesion properties between fiber and matrix [214,221]. A number of factors were thought to play a role in dictating the polymer crystallization across the substrate, which may provide a suitable condition for directional solidification to form a transcrystalline zone.

However, Keller [222] and Gray [205] have shown that the mechanical stress in a polypropylene melt near glass air surfaces gives rise to surface nucleation and subsequent transcrystalline morphology. In practice, the pressure applied during the compression molding of composites could provide the needed mechanical stress to create this peculiar morphology [221]. Sengupta et al. [221] observed crystallization behavior of polypropylene in the vicinity of carbon fibers using a polarizing light

microscope with a hot stage. They concluded that some interfacial stress is essential for developing surface nucleated transcrystalline morphology. Although fracture surface study indicated adherence of the transcrystalline polymer to the fiber surface, exactly how this morphology acts to improve the interfacial bond strength has yet to be explained.

Selective adsorption could play a significant role in the solution prepregging process of thermoplastic matrix composites. In the solution state where molecular mobility is high, larger polymer chains could segregate from the shorter chains and diffuse to the interface area.

The Effect of the Interface on Composite Properties

Longitudinal Strength and Stiffness

The longitudinal stiffness of a two-component composite (fiber plus matrix) is usually given by a rule-of-mixtures equation (Eq 1), assuming the absence of an interphase. Theocaris [223] modified this equation by introducing a third term to account for the modulus (E_i) of a third phase, the interphase:

$$E_c = E_f V_f + E_m V_m + E_i V_i \qquad \text{(Eq 18)}$$

In a typical composite, both V_i and E_i are expected to be rather small and therefore the longitudinal stiffness of the composite should be rather insensitive to the interface properties.

As pointed out by Swain et al. [1], most of the models developed for describing the micromechanics of strength normally ignored the effect of interphase properties. Perfect bonding between the fiber and the matrix is assumed in these models, which include Rosen's chain-of-bundles [224], Hedgepth and Van Dyke's shear lag model [225], and Linear Elastic Fracture Mechanics [226,227].

According to the chain-of-bundles model, if a fiber fractures, the matrix translates the load to the neighboring fibers in the composite. The stress concentrations at the broken fiber ends, unless dissipated properly, may induce failure in adjacent fibers and precipitate catastrophic failure of the composite. Interfacial debonding, caused by the large shear stress at the tip of the fiber end, can act to relieve the local stress. Therefore, the bond between fiber and resin must be sufficiently weak to fail in shear to relieve local stress concentrations near a fiber break to avoid brittle failure. On the other hand, the interfacial bond must be sufficiently high to allow effective stress transfer from the resin to the fiber in order to achieve a maximum tensile strength in composites.

Nairn [227] proposed that, in order to induce longitudinal splitting, the ratio of the critical strain energy release rates for splitting parallel to the fiber (G_{Lc}) versus cracking perpendicular to the fibers (G_{fc}) must be less than half of the square root of the ratio of composite shear modulus (G_c) to composite axial modulus (E_c):

$$\frac{G_{Lc}}{G_{fc}} < \frac{1}{2}\left(\frac{G_c}{E_c}\right)^{1/2} \qquad \text{(Eq 19)}$$

A low value of G_{Lc} can be achieved through low matrix shear modulus or low interphase modulus in a composite where $E_m < E_f$. This would encourage longitudinal splitting to avoid crack growth perpendicular to the fiber direction.

Transverse Stiffness and Strength

Tandon and Pagano [228] have obtained both the upper bound U and the lower bound L of the effective transverse elastic stiffness. In this approach, U is obtained by assuming displacement boundary conditions in an elasticity analysis, while L is calculated by assuming a traction boundary condition. For a Nicalon-based system, the values of E_{22} are 152.8 GPa (2.22 Msi) (U) and 150.6 GPa (2.18 Msi) (L) when perfect bonding is assumed, whereas for the perfect debonded case the E_{22} values are 23.1 GPa (335 ksi) (U) and 2.8 GPa (40.6 ksi) (L). An order of magnitude difference in transverse stiffness between perfect bonding and "pure sliding" was also shown in a recent work by Tong and Jasink [229]. These studies demonstrated the importance of interfacial bond in dictating the transverse stiffness of a composite.

An early work by Adams and Doner [230] indicated that the application of tension transverse to the fibers induces a maximum tensile stress at the poles, a compressive stress at the equator of a cylindrical fiber and shear along the fiber axis. Chamis [231] used a finite element method (FEM) to indicate stress magnification at the interface and suggested fiber debonding as a probable cause of transverse crack initiation. Evidence was given to show that crack propagation was close to the fiber-resin boundary in a [0°/90°/0°] laminate under transverse loading [232]. Zimmerman and Adams [233] used FEM to study the transverse failure behavior of composites. They predicted matrix yielding in the region along the 0° direction, which eventually led to matrix cracking and apparent interfacial failure.

Janssens et al. [234] investigated the effect of novel fiber surface treatment on the transverse bending strength of aramid-epoxy composites. The composites with treated fibers showed a 50% increase in transverse bending strength over the untreated ones. The untreated fibers easily debonded from the matrix while the treated fibers tend to promote fiber defibrillation. Metallization of glass fibers with aluminum to weaken the interfacial bond with a polyester resin was found to cause premature transverse cracking, leading to a reduced transverse strength [235]. Tensile transverse strength of BMI composites containing plasma-treated carbon fibers increased when the treatment time increased [89-91]. Untreated fibers debonded and pulled away from the matrix resin when the composite was subjected to transverse tensile loading [89].

In-Plane and Interlaminar Shear Strength

The four-point bending strength of epoxy laminates was studied as a function of carbon fiber surface treatments by electrolysis [57,181]. Interfacial bond strength was measured by single fiber tests and was correlated with the four-point shear strength. The shear strength of composites containing untreated fibers was relatively low when the interfacial bond, as determined by single fiber tests, was low [181]. In this type of test, it is generally assumed that a condition of in-plane shear exists and that failure occurs by matrix shear yielding, interfacial debonding or some combination of both [57]. A low 4-point-bending shear strength of composite is not necessarily caused by a weak interfacial bond; the matrix properties could play a critical role. Other factors being constant, a weaker interfacial bond should result in a lower shear strength.

In the short beam shear test (ASTM D2344), the specimen could fail by fiber rupture, microbuckling or interlaminar shear cracking, or a combination of fracture modes. An apparent low strength value as measured by this test does not necessarily mean low interfacial strength. Again, the matrix properties could be important. The interlaminar shear strength of composites in many cases can be improved by increasing the matrix tensile strength as well as the matrix volume fraction. The importance of interfacial bonding is evidenced by the well-known fact that glass-epoxy composites have higher interlaminar shear strength (ILSS) than those of vinyl ester and polyester composites. A better bond between epoxy and glass fiber is the overriding factor for this difference.

Finally, it should be noted that the ILSS (τ_{yz}^U) is not the same as the in-plane shear strength D (τ_{xy}^U); the latter is measured by the $[\pm45°]_s$ shear test (ASTM D3518), 10° off-axis test, or rail-shear test. A systematic study of the in-plane shear strength in relation to the fiber-matrix interfacial bonding has yet to be reported.

Compressive Strength

The longitudinal compressive strength of a 0° laminate depends on the fiber type, fiber volume fraction, matrix modulus, matrix yield strength, fiber straightness, as well as fiber-matrix bond strength [236,237]. A widely accepted concept is that compressive strength is governed by laminate buckling stability. For carbon fiber-epoxy composites, correlations were found to exist between composite compressive strength and matrix modulus [236]: the stiffer the matrix the better the fiber support to prevent buckling. For polymeric fiber reinforced composites, evidence indicated the micro-buckling is controlled by the microstructure of the fiber [238]. The effect of matrix-fiber interfacial bond strength is difficult to isolate, although one may expect that a weak or debonded interface would permit easier fiber microbuckling.

References

1. R.E. Swain, K.L. Reifsnider, K. Jayaraman, and M. El-Zein, Interface/Interphase Concepts in Composite Material Systems, *J. Thermoplastic Compos. Mater.*, Vol 3, (No. 1), 1990, p 13-23
2. W. Bolman, *Crystal Lattice, Interfaces, Matrices; An Extension of Crystallography,* published by the author,

22, Chemin Vert, CH - 1234 Pinchot, Geneva, Switzerland, 1982

3. R.W. Ballufi, CSL/DSC Lattice Model for General Crystal—Crystal Boundaries and their Line Defects, *Acta Metall.*, Vol 30, 1982, p 1453

4. R. Bonnet, Computation of Coincident and Near-Coincident Cells for any Two Lattices—Related DSC-1 and DSC-2 Lattices, *Acta Crystallog. A*, Vol 33, 1977, p 850

5. J. Aveston, G.A. Cooper, and A. Kelly, Single and Multiple Fracture, *The Properties of Fiber Composites: Conf. Proc.*, National Physical Laboratory, Surrey, England, 1971, p 15

6. B. Budiansky, J.W. Hutchinson, and A.G. Evans, Matrix Fracture in Fiber-Reinforced Ceramics, *J. Mech. Phys. Solids*, Vol 34, 1986, p 167

7. Y. Korchynski, *J. Mater. Sci.*, Vol 16, 1981, p 1533

8. T. Faber, *Ceram. Eng. Sci. Proc.*, Vol 5, 1984, p 408

9. K.T. Faber and A.G. Evans, *Acta Metall.*, Vol 31, 1983, p 365

10. J.R. Porter, *MRS Symposia Proc.*, Vol 78, 1987, p 289

11. E. Fredrick, *Fiz. Chim. Obrab. Mater.*, Vol 4, 1981, p 615

12. V.M. Kiiko, *Mech. Compos. Mater.*, Vol 4, 1981, p 615

13. G.Z. Cooper, *J. Mech. Phys. Solids*, Vol 15, 1967, p 279

14. K.M. Prewo, Advanced Characterization of SiC Fiber-Reinforced Glass-Ceramic Matrix Composites, *ONR Report*, R85-Y16629-1, 1985

15. R.A. Lowden, *Ceram. Eng. Sci. Proc.*, 1988, p 705

16. R.G. Hill, *Proc. 16th Refractory Working Group Meeting*, Seattle, WA, 1978

17. T. Mah, *J. Mater. Sci.*, Vol 19, 1984, p 119

18. W. Kowbel, H.L. Liu, and H.T. Tsou, Fiber-Matrix Interactions in Brittle Matrix Composites, *Metall. Trans. A*, Vol 23, 1992, p 1051

19. R.A. Lowden, *ORNL/TM-10403*, 1987

20. N. Fretty, *Compos. Sci. Tech.*, Vol 37, 1990, p 177

21. K.M. Prewo, *J. Mater. Sci.*, Vol 15, 1980, p 463

22. B.A. Bender, *Ceram. Eng. Sci. Proc.*, Vol 5, 1984, p 513

23. R. Naslain, *J. Mater. Sci.*, Vol 19, 1984, p 2749

24. P.R. Smith and F.H. Froes, Developments in Titanium Metal Matrix Composites, *J. Met.*, March 1984, p 19-26

25. C. Jones, *J. Mater. Res.*, Vol 4, 1989, p 327

26. S. Ochiai, K. Osamure, and Y. Murakami, *Proc. ICCM-IV*, Tokyo, Japan, 1982, p 1331-1338

27. S. Fisherman and A. Dhingra, Ed., *Interfaces in Metal Matrix Composites*, The Metallurgical Society, Warrendale, PA, 1986

28. H. Ishida, Interface in Polymer, Metal, and Ceramic Matrix Composites, I, II, and III, *Proc. Intl. Conf. Composite Interface* (ICCI-1,2,3), Cleveland, OH, 1986, 1988, and 1990

29. E.J.H. Chen *et al.*, Tailored Interface in Composites, *Proc. MRS Symp.*, Boston, MA, Nov 1989

30. Intl. Conf. Interfacial Phenomena, Symp. sponsored by *Composites* and European Composites Association, Birmingham, England, July 1989

31. N.H. Sung, G. Dagli, and L. Ying, *37th Conf. Reinforced Plastics/Composite Inst. SPI*, Jan. 1982, Sec. 23-B, p 1

32. J. Delmonte, *Technology of Carbon and Graphite Composites*, Van Nostrand Reinholdt Co., 1980

33. R.V. Subramanian and J.J. Jukubowski, *Polym. Eng. Sci.*, Vol 18, 1978, p 590

34. L.J. Broutman and B.D. Agarwal, *Polym. Eng. Sci.*, Vol 14, 1974, p 581

35. J.H. William, Jr. and P.N. Kousiounelos, *Fibre Sci. Technol.*, Vol 11, 1978, p 83

36. D.G. Peiffer, *J. Appl. Polym. Sci.*, Vol 24, 1979, p 1451

37. R.G.C. Arridge, *Polym. Eng. Sci.*, Vol 15, 1975, p 757

38. J.L. Kardos, The Role of Interface in Polymer Composites—Some Myths, Mechanisms and Modifications, in *Molecular Characterization of Composite Interfaces*, H. Ishida and G. Kumar, Ed., Plenum Press, New York, 1985, p 1

39. P. Ehrburger and J.B. Donnet, *Philos. Trans. Roy. Soc. London A*, Vol 294, 1980, p 495

40. P.S. Theocaris, Definition of Interphase in Composites, in *The Role of the Polymer Matrix in the Processing and Structural Properties of Composite Materials*, J.C. Seferis and L. Nicolois, Ed., Plenum Press, New York, 1983

41. J.D. Muzzy and A.O. Kays, *Polym. Compos.*, Vol 5, 1984, p 169

42. J.T. Hartness, *SPE NATEC Proc.*, Bal Harbour, FL, Oct 1982

43. E.P. Plueddemann, *Silane Coupling Agents*, Plenum Press, New York, 1982

44. M.R. Wertheimer and H.P. Schreiber, *J. Appl. Polym. Sci.*, Vol 26, 1982

45. L.S. Penn and T.K. Liao, *Compos. Technol. Rev.*, Vol 3, 1984, p 133

46. E.M. Petrie and J.C. Chottiner, Plasma Treatment of Aramid Fibers for Improved Composite Properties, *SPE ANTEC*, 1982, p 777

47. R.E. Allred, E.W. Merrill, and D.K. Roylance, in *Molecular Characterization of Composite Interfaces*, H. Ishida and G. Kumar, Ed., Plenum Press, New York, 1985, p 333

48. T.S. Keller, A.S. Hoffman, B.D. Ratner, and B.J. McElroy, in *Physical Chemistry Aspects of Polymer Surfaces*, K.L. Mittal, Ed., Vol 2, Plenum Press, 1983, p 861

49. M. Takayanali, T. Kajiyana, and T. Katayose, *J. Appl. Polym. Sci.*, Vol 27, 1982, p 3903

50. L.S. Penn, F.A. Bystry, and H.J. Marchionni, *Polym. Compos.*, Vol 4, 1983, p 26

51. M. Thomson, M.D. Baker, A. Christie, and J.F. Tyson, *Auger Electron Spectroscopy*, Wiley Interscience, New York, 1985, p 217-244

52. F. Hopfgarten, *Biennial Conf. Carbon*, Vol 13, 1977, p 288

53. J.E. Castle and J.F. Watts, in *Interfaces in Polymeric, Ceramic, and Metal Matrix Composites*, H. Ishida, Ed., Elsevier Publishing, New York, 1988, p 57-71

54. K. Siegahn and G. Nordberg, *ESCA Applied to Free Molecules*, North Holland Press, Amsterdam, 1969

55. W.D. Bascom and L.T. Drzal, NASA Contractor Report 4084, N87-25434, 1987

56. M. Narkis, J.H. Chen, and R.B. Pipes, Review of Methods for Characterization of Interfacial Fiber-Matrix Interactions, *Polym. Compos.* Vol 9, (No. 4), 1988, p 245-251

57. B.Z. Jang, L.R. Hwang, and Y.K. Lieu, The Assessment of Interfacial Adhesion in Fibrous Composites, in *Interfaces in Metal-Matrix Composites*, A.K. Dhingra and S.G. Fishman, Ed., Metallurgical Soc. of the AIME, 1986, p 95-109

58. R.L. Lehman, Ceramic Matrix Fiber Composites, *Treatise Mater. Sci. Technol.*, Vol 29, 1989, p 229-291

59. A.W. Adamson, *Physical Chemistry of Solids*, John Wiley and Sons, New York, 1976, p 12 and 15

60. B.J. Carroll, *J. Colloid Interface Sci.*, Vol 57, 1976, p 488-495

61. J.B. Donnet and R.C. Bansal, *Carbon Fibers*, Marcel Dekker, New York, 1984

62. L.T. Drzal, *Carbon*, Vol 15, 1977, p 129

63. L.T. Drzal, J.A. Meschner, and D.L. Hull, *Carbon*, Vol 17, 1979, p 375

64. J.M. Dale, L.D. Hulett, and T.M. Rosseel, *J. Appl. Polym. Sci.*, Vol 33, 1987, p 3055-3067

65. Also, see review in Ref 55
66. F.T. Tunistra and J.L. Koening, *J. Compos. Mater.*, Vol 4, 1970, p 402-409
67. E. Fitzer, E. Gantner, F. Rozploch, and D. Steinert, *High Temp. High Press.*, Vol 19, (No. 5), 1987, p 537-544
68. A.M. Zapper, A. Cholli, and J.L. Koenig, in *Molecular Characterization of Composite Interfaces*, H. Ishida and G. Kumar, Ed., Plenum Press, New York, 1985, p 299-312
69. D.E. Leyden and D.E. Williams, in *Molecular Characterization of Composite Interfaces*, H. Ishida and G. Kumar, Ed., Plenum Press, New York, 1985, p 377-386
70. L.W. Burggraf and D.E. Leyden, *Anal. Chem.*, Vol 53, 1981, p 759
71. D. Briggs, *Practical Surface Analysis by Auger and X-ray Photoelectron Spectroscopy*, John Wiley and Sons, Chichester, England, 1983
72. W. Kowbel, *J. Phys. Chem. Solids*, Vol 49, 1988, p 1279
73. R. Chaim and A.H. Heuer, *J. Am. Ceram. Soc.*, Vol 71, 1988, p 960
74. W.E. Brower, P. Tlomak, and W. Kowbel, *MRS Proc.*, Vol 80, 1988, p 246
75. W. Kowbel, *Ultramicroscopy*, Vol 29, 1989, p 71
76. F.S. Honecy, *NASA TM-100892*, 1988
77. D.V. Badami, J.C. Joiner, and G.A. Jones, *Nature*, Vol 215, 1967, p 386
78. B. Rand and R. Robinson, *Carbon*, Vol 15, 1977, p 257-263
79. H.P. Boehn, E. Diehl, W. Heck, and R. Sappock, *Agnew. Chem.*, Vol 3, 1964, p 669
80. F. Hopfgarten, *Fiber Sci. Technol.*, Vol 11, 1978, p 67
81. G.E. Hammer and L.T. Drzal, *Appl. Surf. Sci.*, Vol 4, 1980, p 350
82. A. Ishitani, *Carbon*, Vol 19, 1981, p 269
83. J.E. Bailey, P.J. Curtis, and A. Poruizi, *Proc. Royal Soc. London A*, Vol 366, 1979, p 599
84. G.A. Beitel, *Vacuum. Sci. Technol.*, Vol 8, 1971, p 647
85. C. Jones, *Compos. Sci. Technol.*, Vol 42, 1991, p 275-298
86. J.B. Donnet, *Carbon*, Vol 20, 1982, p 267
87. L.T. Drzal, M.J. Rich, D.J. Camping, and W.J. Park, *Proc. 35th Reinf. Plast./Compos. Conf.*, SPI, 1980, p 20-26
88. M.J. Rich and L.T. Drzal, *41st Ann. Conf. Reinf. Plast./Compos. Inst.*, SPI, Jan 1986, p 2f
89. T.C. Chang and B.Z. Jang, The Effects of Fiber Surface Treatments by a Cold Plasma in Carbon/BMI Composites, *MRS Symp Proc.*, Vol 170, C.G. Pantano and E.J.H. Chen, Ed., MRS, Pittsburgh, PA, 1989, p 321-326
90. T.C. Chang, C.A. Humphries, and B.Z. Jang, The Role of Interface in Polymer Composites, in *Advancements in Materials For Polymer Composites and Special Topics*, SPE/APC RETEC, Oct 1990, Los Angeles, p 247-254
91. T.C. Chang and B.Z. Jang, Plasma Treatments of Carbon Fibers in Polymer Composites, *SPE ANTEC*, Vol 36, 1990, p 1257-1262
92. W. Kowbel and S.H. Shan, *Carbon*, Vol 28, 1990, p 287
93. W.D. Wilson, *Phys. Rev. B.*, Vol 15, 1977, p 2458
94. E. Fotzer amd R. Weiss, in *Processing and Uses of Carbon Fiber Reinforced Plastics*, VDI-Verlag, 1981, p 45-65
95. J.V. Milewski and J.V. Anderle, Interstitial Growth of Silicone Carbide Whiskers in Carbon and Graphite Fibers, *GTC Tech. Report*, p 127.7-1, as cited in J.C. Goan and S.P. Prosen, in *Interfaces in Composites*, STP 452, ASTM, 1969, p 3
96. J.C. Goan and S.P. Prosen, in *Interfaces in Composites*, STP 452, ASTM, Philadelphia, p 3
97. R.V. Crane and V.J. Krukonis, *Ceram. Bull.*, Vol 54 (No. 2), 1974, p 184
98. J.N. Rabptnpv, B.V. Perov, V.G. Lutsan, T.G. Ssorina, and E.I. Stepanitsov, *Carbon Fibers—Their Place in Modern Technology*, The Plastics Institute, London, 1974, p 65
99. R.G. Shaver, *AICRE Mater. Conf.*, Philadelphia, Apr 1968
100. R.A. Simon and S.P. Bosen, *23rd Ann. Tech. Conf., SPI*, Reinf. Plastics Div., Feb 1968, Sec. 16B
101. R.E. Allred, E.L. Merrill, and D.K. Roylance, Amine Plasma Modification of Polyaramid Filaments, *Polym. Prep.*, Vol 24 (No. 1) 1983, p 223-224
102. V. Kishnamurthy and I. Kamel, Plasma Grafting of Allyamine Onto Glass Fibers, *33rd Intl. SAMPE Symp.*, Vol 33, 1988, p 560-570
103. I.H. Loh, R.E. Cohen, and R.F. Baddour, Modification of Carbon Surfaces in Cold Plasma, *J. Mater. Sci.*, Vol 22, 1987, p 2937-2947
104. C.A. Alsup, "Interface Control in Fiber Reinforced High-Temperature Polymer Composites," M.S. Thesis, Auburn University, AL, Mar 1991
105. B.Z. Jang, H. Das, L.R. Hwang, and T.C. Chang, in *Interfaces in Polymer, Ceramic, and Metal Matrix Composites*, H. Ishida, Ed., Elsevier Publishing, 1988, p 319-333
106. R.E. Allred and L.A. Harrah, Effect of Plasma Treatment on Carbon/BMI Interfacial Adhesion, in *Proc. 34th Intl. SAMPE Symp.*, 1989, p 2559-2569
107. G. Dagli and N.H. Sung, Properties of Carbon/Graphite Fibers Modified by Plasma Polymerization, *Polym. Compos.*, Vol 10, 1989, p 109-116
108. G. Dagli and N.H. Sung, Surface Modification of Graphite Blocks and Fibers by Plasma Polymerization, *ACS PMSE*, Vol 56, 1987, p 410-415
109. P.W. Rose, O.S. Kolluri, and R.D. Cormia, Continuous Plasma Polymerization Onto Carbon Fiber, in *Proc. 34th SAMPE Symp.*, Reno, NV, 1989
110. J.W. Herrick, P.E. Griiber, and F. T. Mansur, "Surface Treatments for Fibrous Carbon Reinforcements," AFML-TR-66-178, Part I, Air Force Wright Materials Laboratory, July 1966
111. D.W. McKee and V.J. Mimeault, *Chemistry and Physicals of Carbon*, Vol 88, P.L. Walker and P. Thrower, Ed., Marcel Dekker, New York, 1973
112. D.L. Schmidt and H.T. Hawkins, "Filamentous Carbon and Graphite," AFML-TR-65-160, Air Force Materials Laboratory, Aug 1965
113. B.S. Marks, R.E. Mauri, and W.G. Bradshaw, *12th Bienn. Conf. Carbon*, Pittsburgh, PA, 1975, p 337
114. E. Fitzer, W. Hüttner, and D. Wolter, *13th Bienn. Conf. Carbon*, Irvine, CA, 1977, p 180
115. J.J. Gebhardt, *14th Bienn. Conf. Carbon*, Penn State University, State College, PA, 1979, p 232
116. C.S. Brooks, G.S. Golden, and D.A. Scola, *Carbon*, Vol 12, 1974, p 609
117. J. C. Goan, *Graphite Fiber Oxidation*, Naval Ordnance Laboratory, TR-153, Nov 1969
118. J.W. Herrick, 23rd Ann. Tech. Conf., Reinf. Plast./Compos. Div., SPI, Feb 1968, Sec. 16A
119. E. Fitser, K.H. Geigel, W. Hüttner, and R. Weiss, *14th Bienn. Conf. Carbon*, Penn State University, State College, PA, 1979, p 220 and 236
120. J. Cziollek, E. Fitzer, and R. Weiss, *Carbon '80: 3rd Intl. Carbon Conf.*, Baden-Baden, 1980, p 632
121. L.M. Manocha, *Carbon '80: 3rd Intl. Carbon Conf.*, Baden-Baden, 1980, p 633
122. O.P. Bahl, R.B. Mathur, R.L. Seth, and T.L. Dhami, *Carbon '80: 3rd Intl. Carbon Conf.*, Baden-Baden, 1980, p 602

123. J.B. Donnet, E. Papirer, and H. Danksch, in *Carbon Fibers—Their Place in Modern Technology*, The Plastics Institute, London, 1974, p 58

124. B. Rard and R. Robinson, *Carbon*, Vol 15, 1977, p 311

125. R.C. Bansal, P. Chhabra, and B.R. Puri, *Indian J. Chem.*, Vol 19A, 1980, p 1149

126. D.W. McKee, *Carbon*, Vol 8, 1970, p 131

127. D.W. McKee, *Carbon*, Vol 8, 1970, p 623

128. MAN AG, DBP 2163071, 1971

129. R.C. Bansal and P. Chhabra, *Indian J. Chem.*, Vol 20A, 1981, p 449

130. R.J. Babka and L.P. Lowell, "Integrated Research on Carbon Composite Materials," AFML-TR-66-310 Part I, Air Force Materials Laboratory, Oct 1960, p 145-152

131. H. Well and W.F. Colclough, U.S. Patent 3,657,082, Apr 1972

132. M.L. Druin and R. Dix, U.S. Patent 3,894,884, July 1975

133. L.A. Joo, R. Prescott, G.E. Sharper, and F.E. Snodgrass, U.S. Patent 3,989,802, Nov 1976

134. J.P. Rieux and J. Lehurreau, French Patent 72.14.95, 1972

135. Courtaulds, Ltd., British Patent 14.4341, 1964

136. P. Ehrburger, J.J. Hergue, and J.B. Donnet, *Proc. 4th London Intl. Carbon and Graphite Conf.*, Soc. Chem. Ind., London, 1976, p 201

137. H. Nyo, D.L. Heckler, and P.L. Hoernschemeyer, Characterizing the Structure of PAN-based Carbon Fibers, *SAMPE 24th Natl. Symp.*, Vol 24, 1979, p 51-62

138. H.P. Boehm, E. Diehl, and W. Heck, *Angew. Chem.*, Vol 76, 1964, p 742-751

139. R.V. Subramanian, V. Sundaram, and A.K. Patel, *33rd Ann. Tech. Conf.*, SPI, 1978, Sec. 20-F, p 1-8

140. R.V. Subramanian, *Adv. Polym. Sci.*, Vol 33, 1979, p 33-58

141. J.V. Larsen, "Surface Treatments and Their Effects on Carbon Fibers," Paper and Bibliography from Naval Ordnance Laboratory, Maryland, 1971

142. J. Delmonte, Technology of Carbon and Graphite Fiber Composites, Van Nostrand Reinholdt, New York, 1981, Chapter 7

143. J.P. Bell, J. Chang, H. Rhee, and R. Joseph, Application of Ductile Polymeric Coatings Onto Graphite Fibers, *Polym. Compos.*, Vol 8, 1987, p 41-49

144. N.H. Ladizesky, J.M. Ward, and L.N. Phillips, *Progr. Sci. Eng. Compos. Mater. 4th Conf.*, Vol 1, 1982, p 203-210

145. U. Gaur and T. Davidson, Interfacial Effects of Plasma Treatment on Fiber Pull-Out, in *Interfaces in Composites: MRS Symp. Proc.*, Vol 170, C.G. Pantano and E.J.H. Chen, Ed., MRS, Pittsburgh, PA, 1989, p 309-314

146. S.L. Kaplan, P.W. Rose, H.X. Nguyen, and H.W. Chang, Gas Plasma Treatment of Spectra Fiber, *SAMPE Q.*, Vol 19 (No. 4), July 1988, p 55-59

147. H.X. Nguyen, G. Riahi, G. Wood, and A. Poursartip, Optimization of Polyethylene Fiber Reinforced Composites Using a Plasma Surface Treatment, *Proc. 33rd Intl. SAMPE Symp.*, Anaheim, CA, Vol 33, 1988, p 1721-1729

148. O.S. Kolluri, S.L. Kaplan, and P.W. Rose, "Gas Plasma and the Treatment of Advanced Fibers," SPE/APC Regional Technical Conf., Los Angeles, CA, Nov 1988

149. S.P. Wesson and R.E. Allred, Surface Energetics of Plasma Treated Carbon Fiber Reinforcements, *Proc. ACS Div. Polym. Mater. Sci. Eng.*, Vol 58, 1988, p 650-654

150. M. Sun, B. Hu, Y. Wu, Y. Tang, W. Huang, and Y. Da, The Surface of Carbon Fibers Continuously Treated by Cold Plasma, *Compos. Sci. Technol.*, Vol 34, 1984, p 353-364

151. J. Su, X. Tao, Y. Wei, Z. Zhang, and L. Liu, The Continuous Cold-Plasma Treatment of the Graphite Fiber Surface and the Mechanism of the Modification of the Interfacial Adhesion, in *Interfaces in Polymer, Ceramic and Metal Matrix Composites*, H. Ishida, Ed., Elsevier Publishing, New York, 1988, p 269-277

152. J.B. Donnet, Carbon Fibers Electrochemical and Plasma Surface Treatment and Its Assessment, in *Interfaces in Polymer, Ceramic, and Metal Matrix Composites*, H. Ishida, Ed., Elsevier Publishing, New York, 1988, p 35-42

153. J.B. Donnet, M. Brendle, T.L. Dhami, and O.P. Bahl, Plasma Treatment Effect on the Surface Energy of Carbon and Carbon Fibers, *Carbon*, Vol 24, 1986, p 757-770

154. J.B. Donnet, T.L. Dhami, S. Dong, and M. Brendle, Microwave Plasma Treatment Effect on the Surface Energy of Carbon Fibers, *J. Appl. Phys.*, Vol 20, 1987, p 269-275

155. I.K. Ismail and M.D. Vangsness, On the Improvement of Carbon Fiber/Matrix Adhesion, *Carbon*, Vol 26, 1988, p 749-751

156. C. Jones and E. Sammann, The Effect of Low Power Plasmas on Carbon Fiber Surfaces, *Carbon*, Vol 28, 1990, p 509-519

157. Y.Z. Wei, Z.Q. Zhang, Y. Li, Z.H. Guo, and B.L. Zheng, A Study of the Interfacial Bonding Between Carbon Fiber and PMR-15 Resin, in *Controlled Interphases in Composite Materials: Proc. ICCI-3*, H. Ishida, Ed., Elsevier Publishing, New York, 1990, p 167-174

158. R.J. Kerans, R.S. Hay, N.J. Pagano, and T.A. Parthasarathy, *Ceram. Bull.*, Vol 68, 1989, p 429-442

159. F.J. McGarry and D.W. Marshall, STP 327, ASTM, Philadelphia, PA, 1963, p 133-145

160. L. J. Broutman, in *Interfaces in Composites*, STP 452, ASTM, Philadelphia, PA, 1969, p 27-41

161. D.W. Richardson, *Ceram. Eng. Sci. Proc.*, Vol 12, 1988, p 671-678

162. D.C. Cranmer, *MRS Symp Proc.*, Vol 120, 1988

163. L.B. Gresczuk, in *Interfaces in Composites*, STP 452, ASTM, Philadelphia, PA, 1969, p 42-58

164. A. Takaku and R.G.C. Arridge, *J. Phys. D*, Vol 6, 1973, p 2038-2047

165. P. Bartos, *J. Mater. Sci.*, Vol 6, 1980, p 3122-3128

166. R.J. Gray, *J. Mater. Sci.*, Vol 19, 1984, p 861-870

167. B. Miller and P. Muri, *Compos. Sci. Technol.*, Vol 28, 1986, p 17

168. B. Miller, P. Muri, and L. Rebenfeld, *Compos. Sci. Technol.*, Vol 28, 1987, p 17-32

169. J.W. Laughner, N.J. Shaw, R.T. Bhatt, and J.A. Dicarlo, *Ceram. Eng. Sci. Proc.*, Vol 7, 1986, p 932

170. J.F. Mandell, J.H. Chen, and F.J. McGarry, *J. Adhesion and Adhesives*, Vol 1, 1988, p 40

171. J.F. Mandell, D.H. Grande, T.H. Tsiang, and F.J. MeGarry, in *Composite Materials: Testing and Design*, STP 893, ASTM, Philadelphia, PA, 1986, p 87

172. J.H. Chen and J. Young, *Compos. Sci. Technol.*, Vol 42, 1991

173. D.B. Marshall, An Indentation Method for Measuring Matrix-Fiber Frictional Stresses in Ceramic Components, *Comm. Am. Ceram. Soc.*, Dec 1984, p C259-60

174. D.B. Marshall and W.C. Oliver, Measurement of Interfacial Mechanical Properties in Fiber-Reinforced Ceramic Composites, *J. Am. Ceram. Soc.*, Vol 70 (No. 8) 1987, p 542-548

175. L.R. Hwang, "SiC-Reinforced Ceramic Matrix Composites," Ph.D. Dissertation, Auburn University, Sept 1990

176. R.A. Lowden and D.P. Stinton, The Influence of the Fiber-Matrix Bond on the Mechanical Behavior of Nicalon/SiC Composites, *ORNL/TM—10667*, DE88 006155, Dec 1987

177. F.J. McGraay and M. Fujiwara, *Modern Plast.*, July 1968, p 143

178. N.J. Wadsorth and I. Spilling, *British J. Appl. Sci.*, Vol 2, 1968, p 1049

179. P.E. McMahon, *SAMPE Q.*, Vol 6, Oct 1974

180. W.A. Fraser, F.H. Ancker, A.T. Dibenedetto, and B. Elbrirli, *Polym. Compos.*, Vol 4, 1983, p 238

181. L.T. Drzal, M.T. Rich, and P.F. Lloyd, *J. Adhesion*, Vol 16, 1982, p 1

182. A. Kelly and W.R. Tyson, *J. Mech. Phys. Solids*, Vol 13, 1965, p 329

183. P.S. Chua, *SAMPE Q.*, Vol 18, Apr 1987, p 10-15

184. C. Galiotis, R.J. Young, and D.N. Batchelder, *J. Mater. Sci.*, Vol 19, 1984, p 3640

185. I.M. Robinson, R.J. Young, C. Galiotis, and D.N. Batchelder, *J. Mater. Sci.*, Vol 22, 1987, p 3642

186. C. Galiotis, I.M. Robinson, R.J. Young, and D.N. Batchelder, in *Engineering Applications of New Composites*, S. Paipetis, Ed., Omega Scientific, Wallingford, 1988, p 409

187. H. Jahankhani, C. Vlattas, and C. Galiotis, in *Interfacial Phenomena in Composite Materials '89*, F.R. Jones, Ed., Butterworths, London, 1989, p 125-131

188. R.J. Day, M. Zakikhani, P.P. Ang, and R.J. Young, in *Interfacial Phenomena in Composite Materials '89*, F.R. Jones, Ed., Butterworths, London, 1989, p 121-124

189. D.N. Batchelder and D. Bloor, *J. Polym. Sci., Polym. Phys. Ed.*, Vol 17, 1979, p 569

190. C. Galiotis, R.J. Young, and D.N. Batchelder, *J. Polym. Sci., Polym. Phys. Ed.*, Vol 21, 1983, p 2483

191. I.M. Robinson, M. Zakikhani, R J. Day, R.J. Young, and C. Galiotis, *J. Mater. Sci. Letters*, Vol 6, 1987, p 1212

192. R.J. Young, R.J. Day, M. Zakikhani, and I.M. Robinson, *Compos. Sci. Technol.*, Vol 34, 1989, p 243

193. C. Galiotis, J. A. Peacock, D.N. Batchelder, and I. M. Robinson, *Composites*, Vol 19, 1988, p 321

194. C. Galiotis, *Compos. Sci. Technol.*, Vol 42, 1991, p 125-150

195. F.M. Underwood, D.N. Batchelder, and D.J. Sharpe, in *Interfacial Phenomena in Composite Materials '89*, F.R. Jones, Ed., Butterworths, London, 1989, p 69-73

196. R.L. Brady, R.S. Porter, and J.A. Donovan, in *Interfaces in Polymer, Ceramic and Metal Matrix Composites*, H. Ishida, Ed., Elsevier Publishing, New York, 1988, p 463-466

197. A.S. Argon, V. Gupta, H.S. Landis, and J.A. Cornie, Intrinsic Toughness of Interfaces between SiC Coatings and Substrates of Si or C Fiber, *J. Mater. Sci.*, Vol 24, 1989, p 1207-1218

198. A.S. Argon, V. Gupta, H.S. Landis, and J.A. Cornie, Intrinsic Toughness of Interfaces, *Mater. Sci. Eng.*, Vol A107, 1989, p 41-47

199. J.L. Kardos, *Trans. NY Acad. Sci. II*, Vol 35 (No. 2), 1973, p 136

200. Yu S. Lipatov, V.F. Babich, and V.F. Rosovizky, *J. Appl. Polym. Sci.*, Vol 20, 1976, p 1787

201. C.A. Kumins and J. Roteman, *J. Polym. Sci.*, Vol A1, 1963, p 527

202. T.K. Kwei, *J. Polymer Sci.*, Vol A3, 1965, p 3229

203. J.L. Racich and J.A. Koutsky, *J. Appl. Polym. Sci.*, Vol 20, (No. 8), 1976

204. L.A. Sukkareva, V.A. Voronkov, and PI. Zubov, *Vysokomol. Soedin*, Vol A11, 1969, p 407

205. D.G. Gray, *J. Polym. Sci., Polym. Letter Ed.*, Vol 12, 1974, p 645

206. J.J. Bickerman, *J. Adhesion*, Vol 3, 1972, p 333

207. L.H. Sharpe, *J. Adhesion*, Vol 4, 1972, p 51

208. R.J. Good, *J. Adhesion*, Vol 4, 1972, p 133

209. W.D. Bascom, C.O. Timmons, and R.L. Jones, *J. Mater. Sci.*, Vol 10, 1975, p 1037

210. L.T. Drzal, M.J. Rich, P.F. Lloyd, and M.F. Koenig, *J. Adhesion*, Vol 16, 1983, p 133

211. Yu.S. Lipatov, in *Interfaces in Polymer, Ceramic, and Metal Matrix Composites*, H. Ishid, Ed., Elsevier Publishing, New York, 1988, p 227-237

212. D.G. Legrand and G.L. Gaines, *J. Colloid Interface Sci.*, Vol 31, 1969, p 162

213. H.C. Tsai, A.M. Arocho, and L.W. Gause, "Prediction of Fiber/Matrix Interphase Prop. and Their Influence on Interface Stress, Displacement and Fracture Toughness of Comp. Materials," NADC Tech. Report, Naval Air Development Center, PA, 1990

214. S. Y. Hobbs, *Nature*, Vol 234, 1971, p 12

215. D.G. Gray, *J. Polym. Sci., Polym. Letter Ed.*, Vol 12, 1974, p 509

216. K. Thomas and D.E. Meyer, *Plast. Rubber Mater. Appl.*, Vol 1, 1976, p 35

217. T. Basell and J.B. Shortall, *J. Mater. Sci.*, Vol 10, 1975, p 2035

218. P.O. Frayer and J.B. Lando, *J. Polym. Sci., Polym. Letter Ed.*, Vol 10, 1972, p 29

219. F. Tunistra and D. Baer, *J. Polym. Sci., Polym. Letter Ed.*, Vol 8, 1970, p 861

220. F.N. Cogswell, *28th Natl. SAMPE Symp.*, 1983, p 528

221. P. K. Sengupta, D. Mukhopadhyay, and S.F. Xavier, in *Interfacial Phenomena in Composite Materials '89*, F.R. Jones, Ed., Butterworths, London, 1989, p 111-117

222. A. Keller, *Rep. Prog. Phys.*, Vol 31 (No. 2), 1968, p 623

223. P.S. Theocaris, *The Mesophase Concept in Composites*, Springer-Verlag, 1987

224. B.W. Rosen, *Fiber Composite Materials*, ASM, Metals Park, OH, 1965, p 37-75

225. J.M. Hedgepth and P. Van Dyke, *J. Compos. Mater.*, Vol 1, 1967, p 294-309

226. C.W. Smith, in *Inelastic Behavior of Composite Materials*, C.T. Herakovich, Ed., Appl. Mech. Div., ASME, 1975, p 157-175

227. J.A. Nairn, *J. Compos. Mater.*, Vol 22, 1988, p 561-588

228. G.P. Tandon and N.J. Pagano, *Proc. Fourth Japan-U.S. Conf. Comp. Mater.*, Technomic, 1989, p 191-200

229. Y. Tong and I. Jasink, in *Interfaces in Polymer, Ceramic, and Metal Matrix Composites*, H. Ishida, Ed., Elsevier Publishing, New York, 1988, p 757-764

230. D.F. Adams and D.R. Doner, *J. Compos. Mater.*, Vol 1, 1967, p 4-17

231. C.C. Chamis, in *Composite Materials*, Vol 5, L.J. Broutman and R.H. Krock, Ed., 1974

232. J.E. Bailey, P.J. Curtis, and A. Poruizi, *Proc. Royal Soc., London A*, Vol 366, 1979, p 599

233. R.S. Zimmerman and D.F. Adams, NASA CR-177970, Dec 1985

234. W. Janssens, L. Doxee, Jr., I. Verpoest, and P. De Meester, in *Interfacial Phenomena in Composite Materials '89*, F.R. Jones, Ed., Butterworths, London, 1989, p 147-154

235. J. Izbicka, in *Interfacial Phenomena in Composite Materials '89*, F.R. Jones, Ed., Butterworths, London, 1989, p 228-232

236. H.T. Hahn and J.G. Williams, NASA TM-85634, August 1984

237. P.T. Curtis and J. Marton, *Prog. Sci. Eng. Compos.: ICCM-IV*, Tokyo, 1982

238. S.J. Deteresa, R.S. Porter, and R.J. Farris, *J. Mater. Sci.*, Vol 20, 1985, p 1645

Chapter 4

Processing and Fabrication
of Polymer Composites

Efforts to ensure the manufacturing reliability and reproducibility of composite parts must start with the implementation of proper quality assurance and processability testing procedures. These procedures determine the curing kinetics, thermal stability, and optimum cure cycles of a matrix resin. Various methods feasible for in-process cure monitoring and management will be discussed in this chapter. A sensor to monitor the cure state in real-time during composite curing is important to the future automation of composite manufacturing. Nondestructive evaluation (NDE), performance and proof testing of the fabricated composite parts for quality control are also covered. The issues of automation of composite manufacturing are presented with an example of the cure modelling environment approach. Concepts of knowledge-based expert systems and potential process control methodologies for autoclave curing are also briefly covered.

Composite Manufacturing Technology

Thermoset Resin Composites

Composite material markets have been dominated by thermoset matrix composites over the past three decades, and this dominance will likely continue for at least another decade. Consequently, the manufacturing technology for thermoset composites is much more mature than that for thermoplastic, metal, and ceramic matrix composites. Thermoset composite manufacturing methods principally involve placing the uncured material into or onto a mold so that the material can be shaped into the final part. Today, after more than three decades of intensive thermoset development efforts, manual lay-up operations are still commonly used in many composite manufacturing facilities.

Wet Lay-up Method. In wet lay-up, fabric or mat is laid into or onto a mold, and then catalyzed resin is poured, brushed, or sprayed over the reinforcement. The wet composite is rolled by hand or with a proper tool to evenly distribute the resin and to remove air pockets. Another layer of reinforcement is placed on top, after which more resin is applied. This sequence is repeated until the desired thickness is obtained. The reinforcement is saturated or impregnated with liquid resin during the lay-up operation. The layered structure is then allowed to harden or cure, which at the molecular level involves chain extension, branching, and cross-linking. This method is used more extensively with glass fiber-reinforced polyester.

There are many variations or modifications to the above sequence. For instance, better resin uniformity can be achieved if the reinforcement is pre-wetted with the resin before being laid into the mold. The dry fabric and resin can be weighed to ensure desirable fiber/resin ratios. On a flat surface, the resin can then be rolled or squeezed into the fabric more uniformly than on a tool because resin-rich or resin-deficient zones can be more clearly spotted. The wet lay-up method is conceptually simple and does not require special handling of wet fabrics. Some pressure is normally applied by hand rolling, wiping with a squeegee or using vacuum bagging to remove the entrapped air and give better uniformity. However, the gradual or rapid curing of pre-catalyzed resins prior to completion of the lay-up procedure can cause poor wet-out or resin runoff problems.

Prepreg Lay-up Method. Prepreg is a reinforcement that has been preimpregnated with resin, which is cured slightly to increase viscosity to facilitate easy handling and lay-up. A somewhat superior composite part can be produced, with less resin and fiber handling difficulty, if a prepreg lay-up method is used. The prepreg is normally produced at a facility dedicated to the manufacturing of prepreg by using a method that permits better control of the resin/fiber ratio. The fibers are usually arranged in a unidirectional tape or a woven fabric, impregnated with catalyzed resin, partially cured and then rolled up for shipment to the composite manufacturing sites. Prepregs are used in applications where part performance is critical, since the prepreg lay-up method is more precise than the wet lay-up method.

At a composite manufacturing site, prepreg tapes are profiled and cut to specific dimensions and shapes, laid up ply-by-ply into a mold and then cured. This method usually requires vacuum bagging and often autoclaving. The prepregs typically are leathery and have a slight tackiness so that the layers will not slide over one another during lay-up. They should also be conformable to the mold so that parts with complex shapes may be produced. This ability to conform is commonly referred to as the material's drape. Drape and tack are closely related; both depend on resin type and degree of resin curing and handling temperature.

Fabric prepregs, as opposed to unidirectional tape, are generally preferred on complex contours because of their better ability to drape and conform to the mold. Conventional thermoplastic tapes produced by solution or melt prepregging methods are often stiff at room temperature and have almost no tack or drape. However, some tack and drape are obtained with the so-called "tow-preg" prepared by proprietary powder or slurry techniques. As a result of crimping and mechanical damage of the weaving process, composites made of woven reinforcements are not as strong as the corresponding cross-plied laminates made of unidirectional tape.

Prepreg Manufacture. Solvent impregnation and direct or hot melt resin impregnation represent the two most common prepreg tape manufacturing methods. In either process, the individual fiber bobbins are placed on a pay-off device such as a creel and then collimated into a ribbon prior to contact with the resin. Alternatively, a reinforcement fabric can be used for impregnation. In the solvent process, the dry fibers are passed through a resin/solvent bath where the fibers pick up some resin/solvent mixture which is squeezed to the correct fiber/resin ratio by a series of nip rollers. These rollers serve to work the resin into the fiber bundles for better wet-out. Solvent is used to reduce the resin viscosity for easier fiber wet-out. The impregnated ribbons or fabrics are then passed through a heated drying chamber where much of the solvent is removed. The prepreg is then cured slightly, in some cases to the B stage, sandwiched between a plastic film and a release paper, and wound on a drum. The prepreg rolls are then stored at low temperatures (preferably <-10 °C or 14 °F) to prevent further cure reactions.

In the hot-melt process, the resin is normally calendered or cast into a film prior to impregnation. The reinforcement ribbons or fabrics pass through a hot roller mechanism which applies the resin in the form of a dry resin film. Heat and pressure are then applied to melt the resin and lower its viscosity to wet the fibers. The material is cooled back to room temperature and the tape is wound up, packaged, and stored [1].

Automated Ply Profiling and Tape Lamination. A major portion of the time involved in composite part

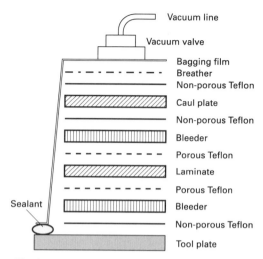

Fig. 1 A possible vacuum bag assembly

fabrication by a manual lay-up technique is taken up by the ply profiling operation. Software can be used to nest the ply requirements together to ensure a minimum of material waste. A device can then be used to profile plies at high speed from a length of material to the desired configuration. The devices that have been proposed to accomplish this task include laser cutting, water jet cutting, reciprocating knife, ultrasonic knife and preset steel rule die [2].

A tape laying machine is designed both to profile the plies and to lay them into a laminate simultaneously. Much development effort has been expended on the design of this type of tape laying machine. At the present time, the machines are available that can lay tape directly onto curved tools with up to 30° of curvature with reasonable deposition rates for the prepregs. However, there is room for improvement in the level of curvature which a laying machine can handle.

Vacuum Bagging and Autoclave Curing. The vacuum bagging process is used predominantly in the aerospace industry where composite part performance rather than high production rate is a major consideration. The starting material is normally a prepreg although the wet lay-up laminates are also vacuum-bag molded. Figure 1 shows a typical vacuum bag configuration. The mold surface is covered with a non-sticking glass fabric separator (usually Teflon-coated) on which the prepreg plies are laid up. The lay-up procedure is followed by the placement of a porous release cloth and a few layers of bleeder paper on top of the prepreg stack. The bleeder paper absorbs the excess resin flowing out during the molding procedure. The whole stack is then covered with another sheet of fabric separator, a caul plate, a plastic breather and a thin heat-resistant vacuum bag (a nylon film for epoxy-based laminates). In a common aerospace process, the entire assembly is placed inside an autoclave where a combination of heat, pressure, and vacuum consolidates and cures the laminate.

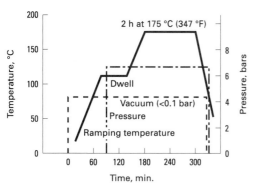

Fig. 2 A traditional two-stage cure cycle for a carbon fiber-epoxy prepreg

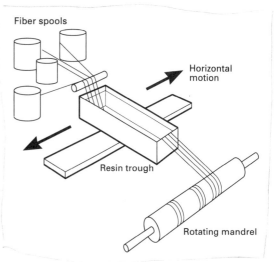

Fig. 3 A filament winding machine

The autoclave provides pressure beyond what can be achieved with vacuum only, and therefore imparts greater compression and void elimination. Curing in an autoclave produces a part that is superior to one produced through non-pressure curing.

Figure 2 shows a typical two-stage cure cycle for a B-staged carbon fiber-epoxy prepreg. The first stage involves increasing the temperature up to 130 °C (266 °F) and dwelling at this temperature for 60 min. During the heat-up process the resin viscosity in the prepreg first decreases, attains a minimum (during the period of temperature dwell), and then increases rapidly as gelation takes place and cross-linking proceeds toward completion. At some moment during temperature dwell, an external pressure is applied on the prepreg stack to squeeze out the excess resin. This resin flow permits the removal of entrapped air and volatiles from the prepreg and thus reduces the void content in the cured part. At the second stage, the autoclave temperature is increased to the actual temperature for the resin. This two-step heating process reduces the risk of locally overheating the prepreg because of heat liberated by the cure reactions. The cure temperature and pressure are maintained for 2 h or more until a predetermined degree of cure has been achieved. The temperature is then gradually decreased while the part is still under pressure. At the end of the cure cycle, the part is removed from the vacuum bag and, if necessary, postcured at an elevated temperature in an oven.

The processing temperature and pressure are chosen to ensure that the resin is cured uniformly in the shortest possible time. The temperature at any portion of the prepreg must not exceed a safe limit during the cure. The pressure must be sufficiently high to remove the excess resin from every ply prior to the gelation of resin at any location inside the prepreg [3]. The thermomechanical processes involved in a vacuum bag have been studied by many researchers [4,5].

Loos and Springer [5] have concluded that the maximum temperature inside the lay-up depends on the maximum cure temperature, the heating rate and the lay-up thickness. At high heating rates with thick laminates, the cure-generated heat can cause local "overshoot." Resin flow in the lay-up depends on the maximum pressure,

lay-up thickness, heating and pressure application rates. Thicker laminates generally require greater pressure to remove the excess resin thoroughly. If the heating rate is too high, the resin may begin to gel before the excess resin is adequately removed. Therefore, the maximum cure pressure should be applied just before the resin viscosity in the top ply becomes sufficiently low for the resin flow to occur. If pressure is exerted too early, excess resin loss would occur since the resin viscosity in the pre-gelation period is very low. Conversely, if the pressure is applied after gelation, the high resin viscosity may prevent the resin from flowing into the bleeder cloth. Thus, the pressurization timing is an important processing parameter in a vacuum bag/autoclave process.

In filament winding, a continuous band or tape of resin-impregnated fibers is wrapped over a mandrel to produce a hollow part. As shown in Fig. 3, a large number of fiber rovings are pulled from a series of payoff devices, collimated into a band through the use of a textile thread board or a stainless steel comb, passed into a liquid resin bath and a wiping device, and then wrapped over a mandrel. Either the mandrel or the application head (carriage) can rotate to give the fiber coverage over the mandrel. Although the rotating mandrel is far more common, simultaneous rotation/translation of both mandrel and application head permits filament winding of complex non-uniform shapes. The process which incorporates the resin impregnation as part of the winding process is called wet filament winding or wet winding. The process which uses prepreg tow as the winding medium is called dry filament winding.

In wet winding, fiber tension is controlled by the fiber guides or scissor bars located between each payoff device (creel) and the resin bath. A wiping device is normally a set of squeeze rollers where the clearance gap between rollers can be adjusted to control the resin content. The traversing speed of the carriage and the winding speed of the mandrel are controlled to provide the desired winding angle patterns. A helical winding pattern is created with a

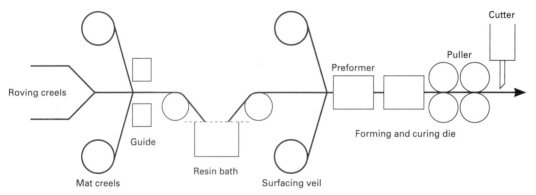

Fig. 4   A pultrusion process

rotating mandrel and a translating carriage. By adjusting the carriage speed and the mandrel rotation rate, any wind angle between near 0° and near 90° can be achieved. For a circular mandrel rotating with a constant rotational speed (N revolutions per minute) and a constant carriage feed of V, the wind angle is given by $\theta = 2\pi Nr/V$, where r is the radius of the mandrel. A constant θ can be maintained in a thick part only if the ratio N/V is adjusted from layer to layer.

Most standard matrix resins can be used for filament winding, provided that they are low in volatile content and have proper viscosity characteristics. Viscosity of the resin should be high enough so that resin dripping in the mandrel can be avoided and yet low enough so that good fiber wet-out can be achieved. Resin viscosities in the range of 350 to 1500 Pa · s have been found to satisfy these requirements [1]. Epoxy, polyester, phenolic, some imides, silicone and thermoplastic can be used for filament winding. Most common continuous reinforcements for composites have been proven suited to filament winding.

The parts can be left on the mandrel and a mold is placed over the part for autoclave curing. In some cases, parts can be simply oven-cured. The effects of gravity may be minimized if the part is rotated while being cured. Microwave or other dielectric heating methods can be used to cure the filament-wound parts with shorter cure cycles.

Some applications of filament winding include pressure vessels, fuel and water tanks, rocket motor cases, pipelines, automotive drive shafts, and helicopter blades. In sporting goods, the filament winding process is used to manufacture tennis rackets from prepreg sheets. The sheets are formed by slitting the wound shape parallel to the mandrel axis or at an inclined angle to provide adequate normal and shear strengths to the tennis rackets.

The advantages of filament winding include [1]:

- Parts of widely varying size may be produced
- Non-cylindrical shapes can be made
- Panels and fittings for reinforcement or attachment can be easily included in the winding process

- Low void content and good fiber/resin ratio can be achieved
- Parts with high pressure ratings can be fabricated.

But resin viscosity and pot life must be carefully chosen and monitored. Control of key parameters such as fiber tension, fiber wet-out, and resin content is important. Not all shapes can be reasonably made by filament winding.

Pultrusion is a continuous process used to produce fiber-reinforced plastic structural shapes. The process involves pulling resin-impregnated fiber reinforcements through a preformer and a heated die to cure the resin (Fig. 4). Pultrusion differs from filament winding in that filament winding places the primary reinforcement in the hoop direction, while pultrusion has the primary reinforcement in the longitudinal direction. Pultrusion can produce a variety of reinforced solid, tubular, or structural profiles. This process normally manufactures 2 to 200 m/h (2.2 to 220 yd/h) depending on the shape and the resin used.

Guide plates are used to position the longitudinal and transverse reinforcements in their designed locations in the shape being pultruded. Preformers are utilized to gently and gradually bend the impregnated reinforcements to form the shape being pultruded and to minimize the amount of excess resin present during the cure cycle. The shape is then cured within a heated die which can range in length from 30 to 155 cm (12 to 60 in.), depending on the resin used and the desired line speed.

Most of the commonly used continuous reinforcements such as glass fiber, carbon and aramid work with pultrusion. A variety of reinforcement forms can be used, including roving, mat, and fabric. All resins can be chosen to work with pultrusion provided they meet certain requirements. The resin needs to cure quickly because of the high-speed continuous nature of the process. Viscosity values around 500 mPa · s are typical for protruded resins. Polyester is a popular resin for pultrusion because it shrinks slightly on curing and thus easily releases from a die. In contrast, epoxy resins tend to stick to the die and other parts of the pultrusion equipment. Release agents, viscosity modifiers and non-sticking cloth represent possible means to prevent sticking.

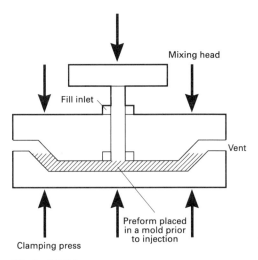

Mixing head

Fill inlet

Vent

Preform placed
in a mold prior
to injection

Clamping press

Fig. 5 A RTM process

The knowledge and technology about the operations of pultrusion equipment were gained mainly through trial-and-error experimentation during the period of 1951 to 1980 [6]. Price [7] made the first attempt to study the fundamentals of pultrusion. Price treated the die as a combination of a reactor and a rheometer. The "reactor" contains a low-molecular-weight resin which undergoes a chemical reaction to become a higher molecular-weight resin which eventually forms the cross-linked matrix. The "slit rheometer" involves a volume of material entering the die and undergoing changes which produce pressures and resisting forces. The parameters that influence the pultrusion process were identified by Sumerak [8]. These include internal pressure, pull force, product and surface temperature, mold temperature profile, internal product exotherm, viscosity and pull speed.

By using a numerical simulation approach to solve a heat transfer equation, Tulig [9] included the effect of heat input from heaters and convection heat input from heaters and convection heat loss to surroundings as a part of the boundary condition. Han *et al.* [10] selected an empirical kinetic model and a prescribed wall temperature profile in their pultrusion model. The model considered the effect of initiator type, fiber reinforcement, resin type, and the pulling speed. Batch and Macosko [11] incorporated a heat transfer model, a resin pressure model and a pulling force model to formulate a pultrusion model based on a mechanistic kinetic approach. The model is capable of predicting the effect of changing initiator, monomer or inhibitor concentration on reactivity. By assuming the reinforcement as an anisotropic porous medium, Batch and Macosko used Darcy's law as a basis to express the dependence of fluid pressure on the changes of resin density, viscosity and the volume fraction of fibers.

Wu and Joseph [12] studied the unsteady state modeling of the pultrusion process and the use of a mathematical model in a knowledge-based decision support system. The unsteady state simulation of pultrusion can be used to study the transient conditions within the die

and the composite material during the cold die start-up or during a change of operation conditions. This information is valuable in the design stage when the power requirement of the heater is specified. The work [12] emphasized the integration of the operator's experience and a steady-state simulation module into a knowledge-based system for the operation of pultrusion processes. Given a set of raw material specifications, this system can be used off-line to generate a suitable set of operating conditions. The system can also be included in the control loop to provide on-line advice to an operator regarding minor changes to the operating parameters [12]. Additional studies on the process parameter definition and potential closed-loop control of pultrusion processes would aid in further improving process reliability.

Liquid composite molding (LCM) embraces both resin transfer molding (RTM) and its faster counterpart, structural reaction injection molding (SRIM). LCM is generally considered to be the future process of choice for structural components in a variety of automotive, marine, and aerospace applications [13]. The recent development of snap-cure SRIM resin systems, low-viscosity RTM systems and the associated preform technology has paved the way for LCM to possibly enter the realm of large-quantity structural parts.

In RTM, a reinforcement preform is placed in a mold, the mold is closed and resin is then injected into it (Fig. 5). A dry reinforcement material, originally in the forms of roving, mat, fabric, or a combination, is cut and shaped into a preform. The preform is normally pre-rigidized by using a small amount of fast-curing resin before being placed in a RTM mold. The preform must not extend beyond the desired seal or pinch-off area in the mold to permit the mold to close and seal properly. Resin is injected into the mold cavity where it flows through the reinforcement preform, expelling the air in the cavity and impregnating the reinforcement. When excess resin starts to flow from the vent areas of the mold, the resin flow is stopped and curing begins. Completion of curing can take from several minutes to several hours. When the cure process ends, the part is removed from the mold, which is prepared to accept another preform. The cured part may require a postcure to complete the resin reactions.

As shown in Fig. 6, SRIM is a very similar process. However, mold release and reinforcement sizings are varied to optimize their chemical characteristics for the SRIM chemistry. Once the mold has been closed, the resin is rapidly injected into the mold and reacts quickly to cure fully within a few seconds. This chemical reaction proceeds as the resin penetrates through the preform and, therefore, SRIM requires fast fiber wet-out and air displacement. Thorough impregnation of reinforcement is quickly followed by complete cure of resin. The resin viscosity rapidly becomes too thick to permit resin flash through vents. SRIM parts normally do not require postcuring [14].

RTM resins usually have two components mixed at a ratio of approximately 100:1 at low pressure. The reaction rates are much slower to permit resins with viscosities of 100 to 1000 cp to fully impregnate a large preform.

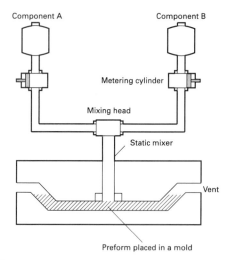

Component A   Component B

Metering cylinder

Mixing head

Static mixer

Vent

Preform placed in a mold

Fig. 6 A SRIM process

SRIM resins are highly reactive and require very fast, high-pressure impingement mixing to achieve thorough mixing before injection. The two components are mixed at a nearly 1:1 ratio to facilitate rapid impingement mixing and fast reaction. While RTM resins are being pushed to achieve cure as fast as 20 s, some SRIM resins are being slowed to allow filling of large molds and impregnation of higher content reinforcements. This trend is also leading to a convergence in resin dispensing technology; some SRIM metering systems are being adopted for fast RTM dispensing [14].

A variety of major RTM and SRIM prototype parts are being developed and will most likely become commercialized soon. These include pick-up truck boxes, bumper beams, auto instrument-panel retainers and floor pans, horizontal auto panel cross members, missile and helicopter components, satellite dishes, boat hulls and electrical transformer cases. The LCM process is characterized by low temperatures, low pressures and therefore, lower equipment costs. The reinforcements can be placed exactly where they are needed and complex sub-assemblies can be consolidated into one-shot molding.

The use of LCM to produce structural composites has gained considerable attention over the last decade. One barrier to further acceptance has been the lack of adequate knowledge and expertise in the cost-effective production of reinforcement preforms. If the LCM process is to remain economically viable, low-cost methods of preform production must be further advanced. At present, two basic input forms of fiberglass are available to the LCM molder for producing a stiff 3-D preform. These are a thermoformable continuous strand mat and a multi-end roving. Furthermore, either form of reinforcement can be combined with unidirectional or bi-directional forms of reinforcement in selected areas during preform fabrication [7]. The specialty reinforcements are added to impart additional strength to areas where the composite may experience higher stresses.

There are three basic routes of fabricating the LCM preforms from the two basic input forms of reinforcement (mats and rovings). These are cut-and-sew preforming, directed fiber spray-up, and stamping of thermoformed mats. Cut-and-sew preforming is utilized in aerospace and low-volume applications. In a cut-and-sew preform, areas of material are defined based on the requirements determined in a finite element analysis. In this process, the general size and shape of each area is cut from a conformable material and fit to the part mold or a part model; this is then cut, trimmed and sewn to fit the desired dimensions. A final template is built and the actual reinforcement is cut and sewed on the preform. This process is slow and labor-intensive, but lends itself to easy translation from finite element modeling to componentry [15].

The directed fiber spray-up process utilizes an air-assisted chopper/binder gun which conveys glass and binder to a perforated metal screen shape identical to the part to be molded. The chopped fibers are held in place on the screen by a large blower drawing air through the screen (Fig. 7). Once the desired thickness of reinforcement has been achieved, the chopping system is turned off and the preform is formed by polymerizing or curing the binder. Once stabilized, the preform is cooled and removed from the screen.

The thermoformed mat process requires an oven to heat the mat, a frame to hold it while being stretched into shape, and a tool to form the mat into a preform [16]. In a typical process, several plies of mat would be cut to the approximate desired shape of the molded part, allowing extra material to be held in a frame. The frame containing the material is then placed in an oven to be heated (up to 170 °C, or 338 °F) and then quickly transferred to the forming tool. The tool is closed, forming and cooling the mat for a short period of time. After removing the frame and trimming the waste fibers clamped in the frame, the preform is ready for molding. Both thermoplastic and thermoset binder systems are available to retain the formed shape.

Many recent efforts have been directed to improve the processes involved in LCM [13,14,16-28]. Major advantages of LCM include [1]:

- Very large and complex shapes can be made efficiently and inexpensively
- Production times are much shorter than lay-up
- Inserts and special reinforcements can be added easily

However, at the present stage of development, LCM still suffers from the following major disadvantages:

- The mold design is critical and requires great skill
- Properties of LCM parts are not generally as good as those produced by vacuum bagging, filament winding or pultrusion
- Control of resin uniformity and elimination of reinforcement movement during resin injection are still quite difficult

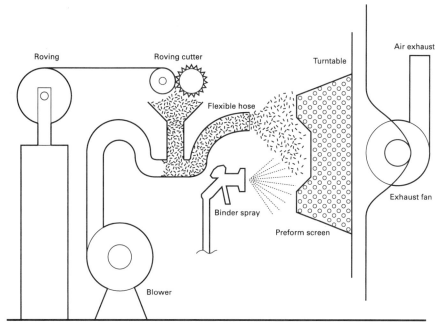

Fig. 7 A directed-fiber spray-up process for preform fabrication

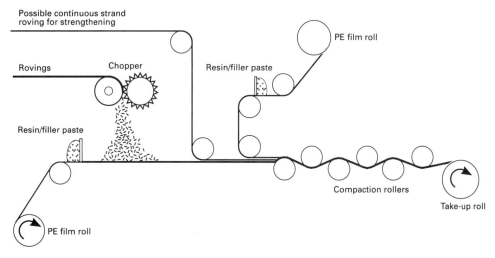

Fig. 8 A SMC process

Matched-Die Molding. Sheet molding compounds (SMCs) and bulk molding compounds (BMCs) utilize the same type of high-pressure molding equipment (compression molding press), but differ in the form of the material that is placed in the mold to form the part. Commonly used materials in these techniques are fiberglass-reinforced polyester and epoxy, plus a sizable amount of fillers such as clay, alumina, or calcium carbonate. A typical SMC composition would contain 30 to 50% fibers (generally shorter than 75 mm, or 3.0 in.),

25% resin, plus the balance amount of fillers, viscosity modifier, and low-profile agents.

As shown in Fig. 8, a SMC process consists of chopping glass fibers onto a sheet of plastic film (usually polyethylene) on which a mixture of resin-catalyst-fillers has been doctored. A predetermined amount of this paste mixture, placed on top of another film, is then conveyed forward to receive the dropping chopped fibers. The "sandwich" of resin mixture and chopped glass fibers is passed between compaction rolls to wet the fibers and thoroughly mix the ingredients. The mixture is then cured slightly (called aging, maturing, or B-staging) to produce a leather-like texture, and rolled up for shipment.

At the matched-die molding facilities, SMC is cut off the roll, the plastic film backing removed, and loaded into the molding presses. Molding temperatures and pressures can be as high as 160 °C (320 °F) and 30 MPa (4350 psi), respectively, depending on the resin chemistry and filler content.

BMC usually has a composition similar to SMC, but the fiberglass content is slightly lower and lengths are shorter. BMC is usually mixed in bulk rather than as a sheet and is sold in a log or rope form. Such a premix can also be conveniently prepared at the molding facilities. Molding procedures and conditions are similar to those for SMC. The processing technology, material compositions, and mechanical properties of SMC, BMC, and other types of molding compounds have been discussed in detail [29,30].

Thermoplastic Matrix Composites

Thermoplastics offer a wide range of advantages over thermosetting matrices, including unlimited shelf life, short process cycle times, the ability to be reformed, increased moisture resistance, increased toughness and impact resistance, and the potential for improved aircraft system supportability. However, there are several disadvantages associated with thermoplastic matrix composites. The processing temperatures are higher than those required to cure epoxy-based composites. The present cost of the thermoplastic materials and preforms for advanced composites is higher than that of the conventional thermosets. The manufacture of thermoplastic prepreg or other type of preform is much more difficult.

Manufacture of Thermoplastic Prepreg. Mixing high-performance thermoplastics with continuous fiber tows is a difficult task, mainly because of the high viscosity of polymer melts leading to high shear stresses and long residence times to permeate and wet out a fiber bundle. Greatly increasing the melt temperature to lower the resin viscosity is not always possible, because decomposition can begin to occur before viscosity reaches a reasonable value. These considerations have made it difficult to fabricate thermoplastic prepreg materials using a conventional hot-melt process. Combined temperature and shear thinning has proven to be effective in reducing the resin viscosity. Pseudo-thermoplastics may have lower viscosities as they are typically not fully polymerized at the time of impregnation. Examples include Torlon, Avimide K-III, and LaRC-TPI (in its precursor poly(amic acid) form). The solution prepregging methods overcome this viscosity problem, but introduce the complication of residual solvent which is hard to remove. In addition to the conventional solution and hot melt impregnation processes, more exotic processes such as emulsion/slurry, dry powder, and fiber commingling and co-weaving methods have recently been developed for producing prepreg, "towpreg", and other styles of preform. In laboratories, film stacking [31,32] and fiber

surface polymerization techniques are also used for fabrication of thermoplastic composite parts.

The "FIT" Process. A recent paper [33] discusses the processing parameters of PEEK/graphite composites prepared from the "FIT" process originally developed by ATOCHEM, France. In this process, fiber tow is spread, passes through a fluidized bed of powders less than 10 microns in size and then is extrusion coated to encapsulate the powder and the tow in a film sheath. The product can be readily woven and used in fabric form. However, the composite laminate produced from the encapsulated tows may contain excessive resin-rich areas due to the presence of the sheath layer. A good fiber-matrix distribution may be difficult to achieve. The FIT process requires that the polymer be suitable for conversion into a fine powder, which can be very expensive. The polymer must be suited for extrusion, which requires good thermal stability in the melt. Nonetheless, this process represents a new milestone in the developments of thermoplastic composite technology.

Wet Powder Technology. A thermoplastic powder technology to impregnate fiber reinforcements was discussed by Clemans and Hartness [34,35], who did not provide details on how impregnation was physically done. This process is believed to involve dispersing the powder in an aqueous suspension to form a slurry, which is then used to impregnate the fibers. Various thermoplastic powders can be combined with the fibers in a product that exhibits good tack and drape. Because of the low shear viscosity of the slurry, woven fabrics as well as separate fiber tows can be impregnated directly as needed. The water is then driven off during a separate heating step.

One important advantage of powder impregnation technology is its ability to process matrix systems with a very high melt viscosity and a high melt temperature. It was claimed [34,35] that even PMR-15 and PEEK of high viscosity grades can be processed using this approach. This process again requires a step to obtain the particle size necessary to match the fiber dimensions. The unidirectional carbon prepreg produced by this method lacks sufficient transverse strength. Therefore special care and effort are required in handling the prepreg during lay-up to form a complex-shaped part. It may also be necessary to remove the polymeric binding agent added in the slurry. How one can achieve adequate interfacial adhesion between fibers and matrix in this process remains questionable. Nevertheless, this process technology appears to hold a great potential.

Dry Powder Technology. Electrostatic fluidized bed coating has been applied to impregnating fiber tows with thermoplastic powder (Fig. 9) [36]. Both PEEK and LaRC-TPI have been successfully formed into "towpregs" for conversion into unidirectional fabric. The powder is electrostatically charged and the tow grounded to promote powder pickup. The powders are then partially melted to allow the polymer to "glue" to the fiber, yet maintain isolated particles. This process appears to be capable of producing towpreg rapidly without imposing severe stresses on the fibers or requiring excessively long

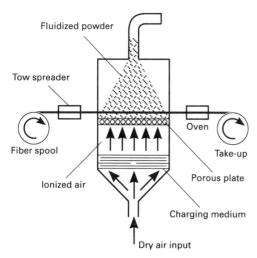

Fluidized powder

Tow spreader

Oven

Fiber spool

Take-up

Ionized air

Porous plate

Charging medium

Dry air input

Fig. 9 Electrostatic fluidized bed powder coating process for continuous fiber tow (the GaTech process)

high-temperature residence times for the polymer. This process looks promising but is still in the laboratory stage. Much developmental work is required before this process can be scaled up. A similar dry powder process without ion charging was developed at NASA LaRC [37].

Commingled Yarn Technology. The "hybrid yarn" concept represents a major step forward in the developments of thermoplastic composite processing technology. This technology originated as a NASA contract for commingling Celion carbon fibers with polymeric filaments made from PBT, PET and liquid crystal polymers. The U.S. Air Force then supported work on commingling of carbon fibers with PEEK filaments. The work has been extended in various research and development establishments around the world to commingle various types of reinforcement fibers with newly developed high-performance polymers in filament form [38-40].

The commingling process requires a texturing procedure to open the yarn bundle without damaging the brittle fibers. This may be difficult to accomplish for a high filament count carbon tow (such as 6K or 12K). There is certainly an added cost to produce filaments from the matrix polymers. Polymer fibers tend to be much larger in diameter than carbon fibers so that there are many resin-rich areas in the molded composites. Some difficulty exists for the commingled yarns to be utilized in high-speed process like filament winding possibly because of the lack of bundle integrity. Since no fusion of the matrix on the reinforcement fiber occurs, the commingled yarns are flexible enough for use in braiding and weaving, two important processes for composite preform fabrication.

Alternatively, tows of reinforcing fiber and polymer fiber can be co-woven together [39]. This is a simple textile operation, and the produced fabric has drapeability. Again, the drawbacks are the necessity to produce the polymer in fiber form and that impregnation takes place during the processing or fabrication stage. Full impregna-

tion is more difficult to achieve with co-woven fabrics because of a lesser degree of mixing as compared to the commingled materials.

The HELTRA Process. The commingled process, as described in [38], tends to suffer from poor dispersion of matrix fibers in reinforcing fibers leading to composites with resin-rich areas. As carbon fiber tows heavier than 3K are used, this process does not allow the production of very light-weight, thin prepreg fabrics. These concerns motivated the engineers in Courtaulds to develop the HELTRA process [31], which produces fine staple yarns from continuous filament tows. These yarns consist of discontinuous fibers bound together into a well oriented coherent bundle by the insertion of a degree of twist. Co-spun intimate blends of reinforcing and thermoplastic matrix filaments can be fabricated, which are well suited to textile processing such as weaving, braiding, and knitting. In principle, hybrid yarns with more than one type of matrix or reinforcing fibers can be blended to produce hybrid composites.

The "Cyclics" Technology. GE Plastics has recently developed a new route to synthesize/process thermoplastic resins and their fiber composites [41]. This new technology takes advantage of the ability of ring-opening polymerization reactions to convert several classes of low-molecular-weight species (intermediate cyclic oligomers) to high-molecular-weight polymers without the formation of by-products. By using an interfacial, amine-catalyzed hydrolysis/condensation of aromatic bischloroformates, GE scientists have obtained cyclic, oligomeric carbonates. These low molecular weight species, exhibiting low viscosity, can be fabricated into final geometry (dimensions and shape) by a variety of reactive processing techniques including pultrusion, RTM, and other in-mold cross-linking processes. By using a proper catalyst, and with the assistance of heat, the cyclic intermediates can be readily converted to very high-molecular-weight polycarbonates.

Processing of Thermoplastic Composites. Wet lay-up methods, in general, are not applicable for thermoplastic matrix composites. Thermoplastic prepregs prepared by hot melt or solution processes are stiff and boardy with no drape or tack. The tack and drape can be improved by slightly melting the resin during the hand lay-up procedure. Bagging materials must be able to withstand the higher temperatures and pressures than normally required for thermoset composites. Some machines, with a heated shoe attached to the applicator head, have been developed for the automatic lay-down of thermoplastic prepreg. Thermoplastic prepreg tow or "towpreg" can be filament wound. Some preheating is usually necessary to obtain good lay-down on the mandrel. Consolidation is a difficult step to accomplish with thermoplastic composites.

Pultrusion can be used to manufacture thermoplastic composites into standard shapes such as rods, channels, I-beams, hat structures, and plates. These shapes are then postformed to create various types of structural members. Because of the very high melting points and the high viscosities, high-performance thermoplastics such as PEEK, PES, TPI, etc. have not been used in LCM proc-

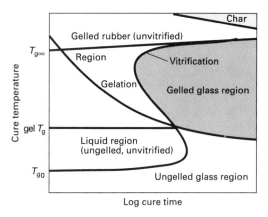

Fig. 10 Time-temperature-transformation (TTT) diagram for isothermal cure of a thermosetting resin (after [44])

esses and the like. Traditional matched-die molding, when combined with a preheating step, can be used for "thermoforming" thermoplastic composites. When one of the dies is flexible, the process is called hydroforming. Fabrication technology of thermoplastic composites have recently been reviewed by Leach [42].

Processing Science of Composites

Manufacturing reliability, reproducibility, and cost effectiveness are the three major ingredients in the future wide-spread use of composites in primary structures. Not only is a high production rate required, but also a uniform part-to-part quality is essential. With a better understanding of the cure chemistry as well as the introduction of automation, fast-curing resins, innovative processing techniques, and adequate quality assurance and control tools, the manufacturing technology for fiber composites is advancing rapidly.

Kaelble [43] suggested that a comprehensive composites manufacturing facility should possess the following capabilities:

- Chemical quality assurance testing of the starting materials such as resins, fibers, and prepreg
- Processability testing of select batches of these materials
- Monitoring and management, preferably automatic control, of the cure cycles
- Nondestructive evaluation (NDE) of the produced parts
- Performance and proof testing of select parts
- Durability analysis and service life prediction

Major characterization methodologies required for ensuring the manufacturing reliability of composites will be discussed in the subsequent sections.

Phase Transitions During Curing

During processing of composites, epoxy resins are exposed to various levels of heat treatment. Two major physical transitions are usually observed as cure proceeds. At the molecular level, gelation corresponds to a transition from linear or branched molecules to an infinite network of chains. Vitrification represents a transition from a rubbery to a glassy state. At the gel point, the weight-average molecular weight of a resin approaches infinity. Further increase of the degree of cure beyond this gel point causes a rapid rise in the insoluble fraction (termed gel) formed by the crosslinking of these large molecules. More and more chains will be incorporated in the cross-linked network, and the proportion of the soluble molecules (termed sol) of the resin will diminish. To describe the phase transition behavior of a curing resin, the definition of the following terms is required (Fig. 10):

- T_{g0} = the glass transition temperature of the monomers
- $T_{g\infty}$ = the glass transition temperature of the fully cured resin
- $G^* = G' + iG''$ = complex shear modulus,
- G' = storage shear modulus
- G'' = loss shear modulus
- $\tan \delta = G''/G'$ = loss tangent

As shown in the time-temperature-transformation (TTT) diagram (Fig. 10) [44], the resin will remain as an ungelled glass if stored at a temperature lower than T_{g0}. At a cure temperature $T_{cure} = T_A$ slightly higher than T_{g0}, the resin reactions (mainly chain extension and possibly some branching) will proceed at a low rate to gradually increase the resin T_g until it reaches and exceeds the T_{cure}. At this point, the resin is said to vitrify (frozen to become a glass). The resin may remain glassy and may never gel. At a higher cure temperature, $T_{cure} = T_B$, the resin reactions can proceed rapidly to reach a gel point, where the resin changes from a liquid (ungelled and unvitrified) to a gelled rubber (gelled but not vitrified). At this point and slightly beyond, the T_g of the resin is still lower than T_{cure}, but is rapidly catching up as the fraction of gels is increasing quickly at the expense of sol. The resin will eventually reach the vitrification point where $T_g = T_{cure}$, beyond which T_{cure} becomes lower than the actual T_g of the resin and the resin is vitrified. After vitrification, it is difficult to proceed further with chemical reactions; a higher temperature would be required. This is the main reason why a post-cure is normally needed with most composite matrix resins. Bubbles, entrapped air, and unwetted interface all must be eliminated in the flow state and prior to gelation. This is essential because a gelled polymer, with molecular weight growing closer to infinity will have a viscosity approaching infinity, and will not flow. At the gel point or beyond, only the gelled rubber and gelled glass states exist.

The rate of viscosity (and also G') increase is low at the early stages of curing. After a threshold degree of cure is achieved, the resin viscosity increases rapidly. The time at which this occurs corresponds to the gel point, where the tan δ versus time curve would normally exhibit a peak. The tan δ versus time curve will also peak at the point of gelation. Therefore, dynamic mechanical techniques which are capable of measuring G', G'', and tan δ will prove useful for cure monitoring. A B-staged or a thickened resin has a much higher viscosity than the neat resin at all stages of curing. The addition of fillers such as $CaCO_3$ will raise the resin viscosity and the rate of viscosity increase during curing. The addition of thermoplastic additives decreases the rate of viscosity increase during curing, however. The increase in viscosity with cure time is less if the shear rate is increased.

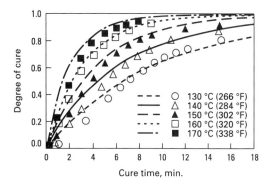

Fig. 11 Degree of cure of an epoxy resin as a function of cure time at various cure temperatures

Degree of Cure

The heat liberated in a curing reaction is related [4,45-47] to the degree of cure achieved at any time during the curing process. A differential scanning calorimeter (DSC) is commonly used to measure the heat released. The instrumentation in DSC monitors the rate of heat generation as a function of time when a small sample is heated either isothermally or dynamically. The total heat liberated to complete a cure reaction is equal to the area under the curve of heat generation rate versus time obtained in a dynamic heating experiment (uniform temperature ramping). Therefore, in dynamic heating:

$$H_R = \int_0^{t_f} \frac{(dQ)}{(dt)_d}\, dt \qquad \text{(Eq 1)}$$

where H_R = the heat of reaction, $(dQ/dt)_d$ = the rate of heat generation in a dynamic test, and t_f = the time required to complete the reaction.

Under isothermal conditions, the amount of heat released in time t at a given temperature T is expressed as:

$$H = \int_0^t \frac{(dQ)}{(dt)_i}\, dt \qquad \text{(Eq 2)}$$

where H = the amount of heat released in time t and $(dQ/dt)_i$ = the rate of heat generation in an isothermal test. The degree of cure at any time t is then defined as

$$\alpha_c = H(t)/H_R \qquad \text{(Eq 3)}$$

Figure 11 shows a number of curves relating the degree of cure to the cure time for an epoxy resin cured at different temperatures [48]. It is clear that α_c increases with both cure time and temperature. If the cure temperature is too low, α_c may not reach a 100% level within a reasonable length of time.

Cure modeling involves the study of cure reaction kinetics, physical property correlation, and cure variables analysis (resin transport, temperature distribution, and voids control, and so on). Several rate equations have been derived to describe the extent of reaction, reaction rate and reaction order for various thermoset systems [49-56]. The experiments were conducted generally by using a combination of DSC, FTIR, and other thermomechanical techniques. Theoretical expressions to predict the physical properties of the polymeric matrix (such as dynamic moduli, dielectric permittivity or viscosity) as a function of cure temperature, time or degree of cure have been discussed by several authors [57-62].

The studies on reaction kinetics of commonly used matrix resins have led Kamal and Sourour [63] to propose that the reaction rates of thermosetting resins can be expressed as

$$d\alpha_c / dt = (K_1 + K_2\alpha_c)^m (1 - \alpha_c)^n \qquad \text{(Eq 4)}$$

where K_1 and K_2 are reaction rate constants, and m and n are constants describing the order of reaction. The constants m and n do not vary much with T, for a second order reaction, $m + n = 2$.

Standard Test Method for Gel Time

ASTM Standard ASTM D3532-76 has been developed for the determination of gel time of carbon fiber-epoxy tape and sheet. This method is suitable for the measurement of gel time of resin systems having either high or low viscosity. In this method, a specimen of prepreg material approximately 6.3 mm (0.25 in.) square is placed between microscope cover glasses on a hot plate preheated to either 121 or 177 °C (250 or 350 °F). Pressure is applied to the specimen through the cover glass with a wood pick to form a bead of resin at the edge of the cover glass. The time from the application of heat until the resin ceases to form strings by contact with the pick is noted and reported as the gel time.

This method can be used to obtain the gel time of resin squeezed from prepreg tape or sheet material. The gel time will vary with the test temperature. The temperatures specified in this standard are two of many temperatures often used in processing epoxy prepreg material. If other test temperatures are employed, this is to be clearly noted in the report. Carbon fiber-epoxy prepreg tape should be stored at low temperatures, approximately –18 °C (0 °F), to prolong the usefulness of the material. Allow the sealed packages of material to warm to ambient

Table 1 Techniques for quality assurance analysis of starting materials [43]

Technique	Advantages	Limitations
HPLC (High-performance liquid chromatography)	Quick separations; very high resolution; simple procedures; very small sample size; largely automated	Requires high-resolution column and high-pressure pump; instruments are expensive; extensive experience required; limited to soluble materials
GC/MS (Gas chromatography/mass spectroscopy)	Direct chemical separation and analysis of insoluble and cross-linked polymers; profile of gas chromatographic peaks for mass spectrum analysis	Expensive
FTIR (Fourier transform infrared spectroscopy)	High resolution, rapid scanning of the IR spectra; computer assisted; versatility	Few limitations
NMR (Nuclear magnetic resonance)	Defining short range stereochemical structure; tacticity, comonomer sequence, cis-trans isomerism, branching and cross-linking; high resolution	Not amenable to routine use; expensive
Elemental analysis	Identifying the organic chemistry of composite materials; automated; highly sensitive; useful in detecting organometallic catalysts which modify the processability of thermosetting resins	No direct information on molecular structure; must be supplemented by other analysis
Wettability, SEA via scanning	Measures fiber circumference; estimates surface roughness; indirectly measures surface energies and environmental durability	Sensitive to swelling; liquids analysis complicated; gives no direct chemical information
SEM plus EDAX	High resolution; great depth of focus; low magnification; can bypass metal coatings	Requires high vacuum; may cause charging; signal from thin coatings too weak for analysis
ESCA	Small sampling depth; X-ray does not damage the surface; coating thickness can be evaluated	Requires high vacuum; depth profiling restricted
ASTM adhesion tests	Direct measure of apparent bond strength; durability	Measures a system response; no direct chemical information
FTIR, internal reflection and ATR	Direct chemical identification; extended data processing; rapid multiscans; isolates chemistry of thin coatings	Limited penetration depth
Optical microscopy	Straightforward sample preparation; records color and birefringence; can scan large areas	Low resolution, limited depth of focus; does not resolve curved surfaces nor show evenly distributed coatings; no chemical information
SIMS (Secondary ion mass spectroscopy)	Small sampling depth; potentially broad range of secondary ions	High vacuum required; degradation and pyrolysis of samples; nonoperative on polymeric fibers
Laser microprobe mass analysis	Analysis of a small fiber area, possible to map surface; small sample volume	High vacuum; disintegration of the fiber
Raman microscopy	Small area; does not require high vacuum	Requires strong Raman scatterers in coating and weak scatterers in fiber

temperature before the seal is opened to ensure that the material does not absorb moisture from the atmosphere.

Chemical Quality Assurance of Prepregs and Resins

The first important step in ensuring the composite part reliability is to verify the lot-to-lot reproducibility of material constituents in prepregs before massive composite lay-up operations are attempted. The primary purpose is to detect the possible variations in the degree of curing among batches of prepreg so that the subsequent processability testing and curing schedule can be adjusted accordingly. Each prepreg chemistry may require a complete development of a chemical quality assurance pro-

gram. Many techniques have been proposed for use to analyze the chemical compositions and the degree of reaction advancement in a resin or prepreg. The advantages and limitations of these techniques have been summarized by Kaelble [43] and are presented in Table 1.

Quality assurance methods such as HPLC, FTIR, and DSC are being implemented by industry to supplement the traditional mechanical tests and gel time or resin flow tests. Many other techniques need improvement or development. It is necessary to develop, optimize, and implement standard test methods for materials characterization for current and future prepregs, fibers, and resins to ensure uniformity of data and unification of materials specifications. Development of techniques for 100% inspection of prepregs is required for monitoring the condition of materials prior to manufacture of composites. To

Table 2   Techniques for composite processability testing [43,64]

Technique	Advantages	Limitations
DTA (Differential thermal analysis)	Measures the temperature difference between sample and reference under programmed thermal scan; high sensitivity, wide range of high temperature and pressure, small sample size, and measurement simplicity	Calibrations and data processing required
DSC (Differential scanning calorimetry)	Directly measures the rate of heat release; quantitative measure of heat of reaction and heat capacity; directly measures thermal state of cure, cure kinetics and cure effects on glass transition	Less sensitive and more limited temperature and pressure ranges than DTA
TMA (Thermal mechanical analysis)	Measures thermal expansion; high displacement sensitivity	Physical limitations of precision measurement
TGA (Thermal gravimetric analysis)	Measures weight changes and the underlying degradation reactions; high precision and sensitivity	Weight change measurements need additional modes of characterization
SEA (Surface energy analysis)	Tests for bonding ability; automatic measure of advancing and receding contact angle on fibers; adhesive bonding and interface durability parameters can be calculated	Limited to continuous surfaces; high microroughness introduces contact angle hysteresis
DMA (Dynamic mechanical analysis)	Both the storage and loss components of rheological properties; applicable to unsupported or supported polymer; highly sensitive to both flow and glass transitions at all states of cure	Multiple DMA required; requires concurrent chemical analysis
MIA (Mechanical impedance analysis)	A more versatile version of DMA; also useful for detecting defects, condition monitoring and cure monitoring in real time	Developmental technique

improve the reliability of data, automated methods and statistical process control techniques must be used. New NDE techniques are needed for assessing the moisture content in fibers and prepregs to improve reliability and durability of composites. A need also exists for developing shelf-life indicators for prepregs and materials in repair kits to ensure processability and structural integrity.

Processability Testing

Curing of the matrix resin is a key step in the fabrication of fiber-reinforced thermoset composites. The product quality of composites is controlled to a great extent by the cure cycle parameters such as time, temperature, pressure and their combination sequence. Traditionally a downstream composite manufacturer just follows the cure schedule suggested by the prepregger or the resin supplier. This schedule was usually determined in an ideal situation where the resin was still "fresh" and may not represent the actual cure state of the resin in a composite manufacturing facility. The composite manufacturer must study the cure behavior of the matrix resin just prior to autoclave or press curing. This is necessary for processability verification and cure cycles optimization especially if the resin has been shelved for some time.

Analytical instruments such as dynamic mechanical analyzer (DMA), dielectric thermal analyzer (DETA), differential scanning calorimeter (DSC), Fourier transform infrared spectrometer (FTIR), and high-performance liquid chromatography (HPLC) can be employed to analyze the cure reaction chemistry. These instruments can be used in a laboratory environment to study the reaction kinetics and the phase transitions in a thermoset system and therefore are useful for not only quality assurance but also processability verification.

Laboratory tests for processability are aimed at determining how a sample performs during a simulated manufacturing cure cycle. These tests use small quantities of material, are fully instrumented, and are operated by programmed scanning of temperature at a constant scan rate. The function of processability testing is to define the kinetics of curing, the limits of thermal stability, and the optimum cure cycle which leads to high performance and durability. Very often, procedures of chemical analysis are implemented to verify thermal analysis data and to define the chemical mechanisms of curing. Matrix resins are commonly classified in terms of their processing temperature range or the service ceiling temperature for environmental stability. DSC and thermal mechanical analysis (TMA) are central components of processability testing. These combined tests characterize the degree of cure and the effect of cure on the melt temperature and glass temperature. The function of processability testing is to find the optimum processing "window", which is a combination of processing times, temperatures, and pressures that consolidate, form by flow, and chemically cure the composite laminate. These processability studies should always be accompanied by relevant chemical analysis [43].

The processing of laminate composites inevitably involves interface bonding during the process cycle. Wet-

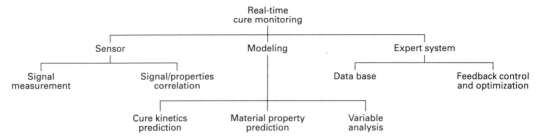

Fig. 12 Science-based real-time cure control for composite fabrication

tability tests and surface energy analysis (SEA) can be quite helpful in this regard. SEA involves the testing of solid surface wettability by contact angle measurements with a selected set of test liquids. The analysis which accompanies SEA testing gives predictions of bonding ability and bond durability of the composite interface. Brief descriptions of the commonly used processability test methods are presented in Table 2 [43].

Cure Monitoring and Management

Future growth of advanced polymer composite applications to large primary structures will require development of a scientific foundation for process mechanics and control. Affordable manufacturing procedures will determine the next stage of growth for the industry. The ability to build in quality with sensors that monitor the cure chemistry in real time is essential to the success of composite manufacturing.

One way to maintain the product quality is to monitor the manufacturing process variables by curing test coupons simultaneously with the corresponding parts. In principle, the cure states of these coupons can be characterized possibly in real time by using the above-mentioned analytical tools. The part quality is usually measured by a number of destructive testing techniques. Such methods have been shown to provide only limited assurance on the quality of the fabricated composite components [65]. Therefore, the utilization of a non-destructive evaluation technique for real-time monitoring of the cure state and for feed-back control of the composite fabrication process is highly desirable.

An ultimate real-time feed-back control incorporating an NDE monitoring system for high-performance composite manufacture involves the following three main stages (Fig. 12) [66]:

- Selection of a cure monitoring sensor
- Cure modeling
- Development of an expert system

A direct measurement sensor for the engineering parameters (dielectric or mechanical properties) or, if possible, for the cross-linking reaction (to assess the degree of cure) is required for an NDE technique to become an integral part of a manufacturing process. Normally the sensor will monitor the global property changes due to the effect of cure temperature and pressure. Theoretical cure modeling is essential to the prediction of the cure reaction. The complex cure variables require the development and application of an expert control system to manipulate the interrelated parameters in curing.

Cure Monitoring Sensor Development

An *in situ* measurement sensor is important for an automated control of composite fabrication. Chemical analysis devices such as DSC, FTIR, HPLC and GPC (gel permeation chromatography), which are generally used for studying cure kinetics and monitoring the onset of gelation and vitrification in composite resin, are not practical for direct usage as a real-time sensor in an autoclave environment during the composite fabrication process. Other commercial instruments including rheometers, dynamic mechanical analyzers (DMA), torsional braid analyzers (TBA) and dynamic dielectric thermal analyzers (DETA), which can be used for curing reaction studies, also cannot be used for *in situ* cure monitoring because they are destructive. They also impose constraints in sample geometry and dimensions which prevent them from being effective when applied to composite manufacturing [43,67].

Currently, very few direct measurement sensors can successfully monitor the complete curing cycle of composite components in the autoclave and press curing situations. Most techniques are still under development. These techniques include microdielectrometry, acoustic emission/ultrasonics, fluorescence, fiber optics, and dynamic or vibration-based methods.

Microdielectrometry. Dielectric analysis has been in use for years [68-84]. Dielectric measurements for cure monitoring are normally done with sinusoidal excitations at specific frequencies. By applying a step-change voltage to the material, the resulting current waveform can be obtained. The frequency-dependent dielectric properties therefore can be calculated from the Fourier transform of the current waveform. The weak points of the dielectric analysis include the empirical nature of the research, inadequate models available to interpret data, and inadequate correlation between dielectric properties and other materials characteristics such as mechanical properties.

The permittivity and the loss factor are the two dielectric properties that originate from dipole motion and ionic conduction. The measurements of these dielectric prop-

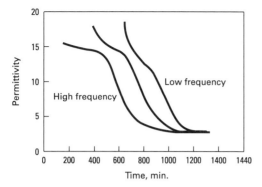

Fig. 13 A typical change of permittivity during the cure of epoxy as a function of time and frequency

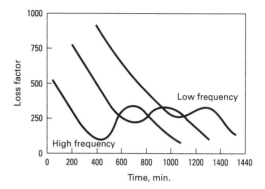

Fig. 14 A typical change of loss factor during the cure of epoxy as a function of time and frequency

erties can be affected by electrode polarization. The changes in permittivity during cure reaction are a function of dipole motion. The motion of dipoles will be hindered and therefore the permittivity will decrease as the reaction proceeds. Typical permittivity versus time curves from the epoxy curing process are shown in Fig. 13. However, the changes of dielectric loss factor are a function of both dipole motion and ionic conduction. The contribution from dipole motion is relatively small compared to ionic conduction. Since ionic conduction is an indication of the mobility of ions in the resin, it can be correlated to the viscosity and the degree of cure in the resin. The ionic conduction is inversely proportional to frequency and usually changes by several orders of magnitude during the cure reaction. Therefore, ionic conduction always dominates the loss factor, especially at low frequencies. The dielectric loss factor is thus very sensitive to the degree of cure throughout the cross-linking reaction. Fig. 14 shows the loss factor as a function of time obtained from the cure of epoxy resin at different frequencies.

At the present time, two major types of dielectric sensors are available. They are parallel plate and comb electrodes. The Hewlett Packard HP4192A Low Frequency Impedance Analyzer and Polymer Laboratories Dielectric Thermal Analyzer (DETA) are two of the commercial plate-type sensors that can be applied in cure reaction studies by measuring the admittance. The effectiveness of these instruments for dielectric cure monitoring depends upon sensitivity and accuracy. The sensitivity depends on the frequency, admittance, cabling and shielding, and electrical noise. The dynamic dielectric analysis (DDA) using a HP4192A can be employed as an *in situ* nondestructive method for cure monitoring in thermosets and phase changes in thermoplastics [74]. The cure rate variations within a thick part in an autoclave can also be monitored by DDA.

The dielectric response from Polymer Laboratories DETA of a real-time cure process can be correlated to the mechanical property changes of a resin during cure [75]. This dielectric sensor can be used to optimize the cure cycles of a given resin. Dielectric thermal analysis, in principle, can be utilized to assess the viscosity and the cure state during the cross-linking reaction of epoxy [76]

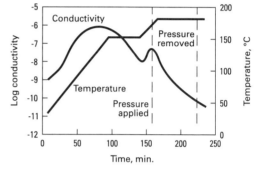

Fig. 15 Ion conductivity variations during the cure of a graphite/epoxy laminate using the parameter control method (after [81])

and other thermosetting resins. Nevertheless, the way the instrument is configured presently precludes the Polymer Laboratories DETA from being used as an in-process cure monitor of composites.

The microdielectrometer sensor combines a comb electrode with a pair of field-effect transistors in a silicon integrated microcircuit to achieve sensitivities comparable with parallel plate electrodes, but retaining the reproducible calibration features of comb [77]. The corresponding permittivity and the loss factor can be obtained from a contour plot [78] if the magnitude and phase of a transfer function are given.

The Eumetric System II Microdielectrometer by Micromet Instruments became commercially available in 1983 [79] . The use of charge measurement rather than the admittance measurement allows the dielectric cure monitoring to be performed at lower frequencies. The dielectrometer generates an excitation signal at a given frequency. The change in gain and phase of the response signal is then recorded and converted into permittivity and loss factor. Temperature is measured by a thermal diode.

One of the drawbacks of the conventional dielectrometry is that changes in spacing between electrodes may modify the dielectric output. The microdielec-

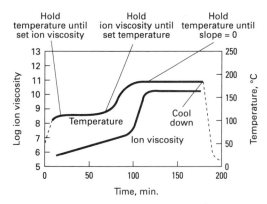

Fig. 16 Closed-loop cure control of graphite/epoxy composites (after [81])

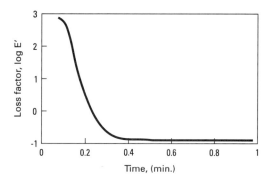

Fig. 18 The change of loss factor with respect to time in a fast RIM system (after [83])

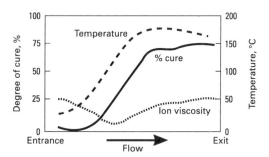

Fig. 17 Cure monitoring in pultrusion process (after [82])

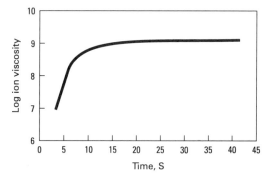

Fig. 19 Ion viscosity versus time of PUR/urea cured in RIM process (after [83])

trometry technique eliminates this deficiency. This microsensor shows good reproducibility. A very detailed review on the dielectric analysis was made by Senturia and Sheppard [80].

The utilization of the dielectric response for closed loop composite process control has been described [81]. When the chemical reaction starts, the loss factor increases gradually until it reaches a maximum (a viscosity minimum). The onset of cross-linking reaction is indicated by the rapid decrease in the loss factor resulting from the decrease in ionic mobility. When the first derivative or slope of the loss factor is zero, this would indicate that the cure is close to completeness. The loss factor measured from the microdielectrometer can be converted to the ionic conductivity. The process under control may be triggered by matching the values of the first and second derivative of the ionic conductivity with the preset setting which, with option, is confined in time and temperature windows (Fig. 15). Since the fluid viscosity is inversely proportional to the ionic conductivity, an ionic conductivity or fluidity maximum is also the viscosity minimum. Often pressure is applied at the viscosity minimum to ensure good flow and fewer voids. This action can be triggered by detecting the zero slope (first derivative) and negative trend (second derivative) of the conductivity in preset time and temperature windows. The

applied pressure may be removed when the end point is detected.

As shown in Fig. 16, an ideal closed loop process control includes the measurements of dielectric properties and temperature, the automatic application and removal of the pressure, and the automatic temperature control and shut off when the change of ion viscosity with respect to time reaches zero. This may be achieved by a well automated microdielectrometer and intelligent control system. Dielectric data can be correlated with the viscosity and with the degree of cure. This microdielectric method can be applied to monitoring and control of both conventional batch thermoset cures as well as continuous pultrusion curing as shown in Fig. 17 [82].

In reaction injection molding, the gelation time is as short as seconds. Traditional cure monitoring methods are not fast enough in this high speed system. Recent developments in high speed microdielectrometry as shown in Fig. 18 and 19 [83], which have resulted in taking dielectric data every 8.3×10^{-3} s, will alleviate this problem.

Variations in composition and aging from batch to batch in prepreg materials may affect material properties after cure [84]. The viscosity of the prepreg materials can change several times with changes in aging or solvent content. The viscosity variations can cause significant

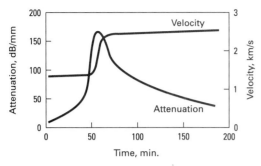

Fig. 20 Typical attenuation-time and velocity-time spectra of ultrasonic measurement for the cure of thermoset materials (after [93,96])

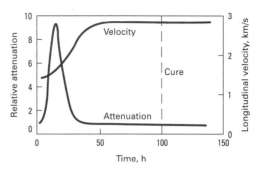

Fig. 21 Signal attenuation and velocity data as a function of cure time for longitudinal wave (after [91])

fluctuations in flow and consolidation properties, leading to inconsistency in the part quality and reproducibility. The increase in viscosity caused by aging can be measured by dielectric analysis at room temperature. This ensures the quality of prepreg materials before and after the prepreg lay-up.

Acoustic emission (AE) is a pseudo-nondestructive testing technique which shows great promise in the area of composite materials. When a material is loaded, it emits mechanical stress waves. These stress wave emissions are commonly called acoustic emissions. In a composite material, several mechanisms which can produce acoustic emission include fiber breakage, fiber-matrix debonding, delamination, matrix plastic deformation and cracking, and possibly plastic deformation of the fiber. These acoustic emission signals are transformed from mechanical energy to electrical energy by means of a piezoelectric transducer. The signal from the transducer is filtered and then amplified. AE characteristics are number of events, ringdown count (RDC), amplitude (the peak height of the wave form), rise time (the time period for the signal to arrive at its peak amplitude), event duration (the time from first to last threshold crossing), slope, frequency, and energy (velocity and amplitude) emitted by the signal. Each AE-generating mechanism will send out signals with different characteristics.

Acoustic-based techniques (including AE) have been used as tools to study materials. A description of the background and history of acoustic technique application was compiled by Harrold and Sanjana [85]. Acoustic monitoring of the cure of resin was first investigated in 1955. Feasibility of monitoring the acoustic wave propagation (including AE waves) through materials during curing has been demonstrated [82-90]. Since the transducers are acoustically coupled to the exterior surface of a material, dispersion and scattering of acoustic waves cause difficulty in interpreting the results. A common problem is the low signal-to-noise ratio. Correlations were found between the acoustic emission activity, the DC resistance and the resultant residual stresses [87]. However, further studies are still being undertaken in correlating AE data with material properties. The combination of acoustic emission technique and ultrasonic

measurement will be a better tool for an *in situ* cure monitoring sensor.

Ultrasonics. Investigations of the application of ultrasonic techniques to monitor the curing of epoxy resin were reported in several papers [91-98]. Both the phase velocity and the attenuation of ultrasonic waves were measured as a function of time during the reaction of epoxy resin [91,95]. Physical transitions in resin during curing can be identified. Halmshaw [97] gave a detailed description of the ultrasonic testing of materials.

Sofer and Hauser [93] first applied the bulk longitudinal ultrasonic waves to study the curing process of thermosetting polymers by the measurement of attenuation and velocity. Papadakis [92] used ultrasonic methods to measure the elastic moduli of polymers during the solidification process. Lindrose [91] found that the shear and bulk relaxation moduli changed during cure. Rokhlin *et al.* [96] studied the real-time frequency dependence of attenuation and velocity of ultrasonic waves during the curing of epoxy resin. He also investigated the temperature effect on the velocity and attenuation of longitudinal waves at different stages of the cure reaction of thermosetting resins. A model consisting of springs and dashpots was proposed by Hahn [98] to explain the experimental data on the wave speed and attenuation of ultrasonic waves.

The ultrasonic waves passing through a composite specimen interacts with the various defects. These defects in turn affect the wave velocity and attenuation. The wave velocity is directly related to the stiffness, and attenuation is a measure of the damping characteristics of the material. Significant changes in attenuation and velocity occur beyond the gel point.

Since the bulk modulus of the material is proportional to the density of the medium and the ultrasonic wave velocity squared, the modulus will increase dramatically during gelation and then finally level off at vitrification. The attenuation, which is indicative of the damping factor, will increase first then decrease rapidly at gelation and die out or minimize at vitrification. Typical attenuation-time and velocity-time spectra are shown in Fig. 20 [93,96].

Lindrose [91] measured the ultrasonic attenuation and velocity as a function of cure time for both longitudi-

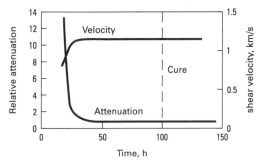

Fig. 22 Signal attenuation and velocity data as a function of cure time for shear wave (after [91])

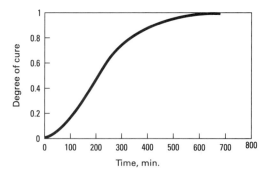

Fig. 24 The relationship between the degree of cure and cure time for epoxy (after [95])

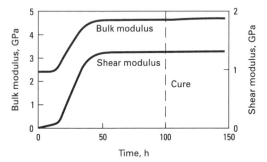

Fig. 23 The changes of apparent elastic moduli as a function of cure time (after [91])

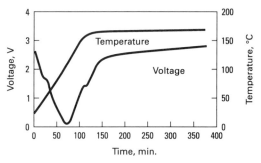

Fig. 25 A typical acoustic response for the cure of graphite/epoxy composites (after [103])

nal and shear waves (Fig. 21 and 22). The velocity-time data were then correlated to the apparent elastic moduli versus time data (Fig. 23). The degree of cure as a function of cure time for epoxy from the ultrasonic longitudinal velocity was illustrated by Winfree and Parker [95] (Fig. 24). An *in situ* cure monitor is essential to the implementation of an automatic control system. This intelligent system will trigger the application of pressure, and will adjust temperature by comparing the measured data with the preset or calculated values. A new development in the acoustic/ultrasonic technique for cure monitoring applications was the integration of acoustic waveguides into the composites [85,99,100]. These acoustic waveguide systems are extremely sensitive to the pressure on the waveguide surface and the change in the acoustic impedance of the surrounding media.

Acousto-ultrasonics is a contraction of acoustic emission simulation with ultrasonic sources. The acoustic emission method depends on externally applied loading to excite spontaneous stress waves. Acousto-ultrasonics differs from acoustic emission mainly in that the ultrasonic waves are generated externally by a pulser or piezotransducer.

Vary [101] used a piezoelectric transducer to generate stress waves in a graphite fiber/epoxy laminate with a sensitive piezoelectric transducer that monitored the waves on the same surface of the laminate. Acousto-ul-

trasonic signals are characterized by a stress wave factor (SWF). The SWF is based directly on ringdown count, peak voltage, and energy. Lower values of SWF usually indicate regions of higher attenuation. The magnitude of attenuation in turn depends on material factors such as morphology, bond quality, microcracks, microstructure, porosity, cure state, and so on [102]. The changes of viscosity with time and temperature are closely related to the resin flow, devolatilization, and fiber compaction during autoclave fabrication of composites.

An acousto-ultrasonic sensor was used to monitor the cure of graphite/epoxy composites [103] and carbon/phenolic composites [90]. A broadband ultrasonic transducer sends ultrasonic pulses into the composite material. Each pulse produces stimulated stress waves that resemble acoustic emission events. The receiving acoustic emission transducer receives acoustic signals which will be processed thereafter. A general correlation between acoustic signals and viscosity or degree of cure was developed. A typical acoustic response curve for graphite/epoxy is shown in Fig. 25. The acoustic attenuation decreases first to a minimum (viscosity minimum), then increases dramatically and finally levels off upon completion of cure. The acoustic response can be further correlated with the viscosity of the resin. A representative viscosity-time curve obtained from the acoustic emission sensor during the autoclave process is shown in Fig. 26

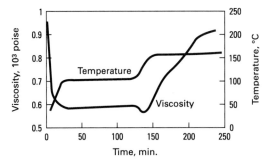

Fig. 26 A representative viscosity-time curve obtained from the acoustic emission sensor during autoclave process (after [103])

[103]. The *in situ* acoustic cure monitoring is feasible based on the viscosity-time-temperature profile. This online control system will trigger the application of pressure as the viscosity reaches the preset value and will automatically adjust the temperature in order to achieve optimal cure.

Fluorescence Techniques. The fluorescence quantum yields of certain compounds are known to exhibit a strong dependence on the viscosity of the medium [104]. Loutfy [105,106] was among the first to exploit this effect for monitoring polymerization reactions. Fluorescence spectroscopy has also been applied to cure monitoring of epoxy resin [104,107-112]. Two such techniques are currently available to monitor the variations in local mobility for the added organic dye in an epoxy resin system [113].

One method is based on the premise that the fluorescence intensity of certain dye molecules is sensitive to their local mobility. A reduction in local mobility will hinder the local rotational motions, which act as alternative energy losses. The energy that cannot be released through rotation appears as fluorescence, leading to an increase in fluorescence intensity. To avoid the need for absolute measurements and to minimize other factors that affect the fluorescence intensity, Wang *et al.* [108] used a viscosity-insensitive internal standard. The ratio of the fluorescence intensities for the two dyes is treated as a measure of the local microviscosity. To facilitate remote sensing of composite cure several different types of optical fiber sensors may be used.

The second method of cure monitoring using fluorescence spectroscopy relies on the recovery of fluorescence after photobleaching. The probe molecules in a well-defined small region of the sample are first destroyed by a high intensity pulse of light. The rate of fluorescence return due to the diffusion of unbleached probe molecules from outside the region is measured and correlated with the translational diffusion coefficient of the dye in the resin [111,113]. Since the above two techniques are insensitive to the sample geometry, they can be adopted for *in situ* monitoring of the cure of composite structures in a factory environment.

Fiber Optics Techniques. One practical way for *in situ* monitoring of polymers and composites processing

is through the use of fiber optics because the light path can be extended outside of the instrument. Light is carried from the spectrophotometer to the point of measurements through optical fibers/cables. The probes at the end of the cable experience the process conditions while the spectrophotometer is located some distance away from a more protected area. This arrangement allows direct measurements even under the extreme processing conditions (e.g., high temperatures and pressures) [114]. The fiber-optic sensors could be fluorescence-based [113,115] or infrared-based [116] with their signals processed by a photodetector or Fourier transform infrared spectrometer.

Mechanical Impedance Analysis. Vibration techniques have been used in nondestructive testing for years [117-119]. Recently, Jang [67] proposed the idea that the mechanical impedance analysis (MIA) can be applied to monitor the cure reaction in thermoset resins. This MIA technique is expected to work as a smart sensor for *in situ* cure monitoring in composite fabrication.

The mechanical impedance analysis technique basically involves quantitative measurement of the effect of a vibratory force on a structure. The dynamic behavior of materials can be characterized by measuring the ratio of motion to force as a function of frequency. The point impedance, Z, is defined as F/V, where F is the harmonic force input to the structure and V is the resultant velocity of the structure at the same point.

The impedance spectrum, when plotted over a frequency range, contains all the information about the dynamic characteristics of the materials. The mechanical impedance analysis for other boundary conditions such as clamped-free, and clamped-clamped modes can be achieved by the similar approach in free-free mode.

For a liquid-like matrix material, wave measurements will fail due to extreme damping of the solution. In this case, mechanical impedance analysis of a moving element immersed in the resin solution can be used to obtain the complex shear modulus of the resin. The principle for the mechanical impedance measurement of a moving mass was proposed by Ferry *et al.* [120-124]. The force and motion are both measured on the moving element [125,126].

Figure 27 shows the apparatus setup for the cure monitoring by the MIA technique [66]. The frequency response function (FRF) is found from the ratio of the Fourier transforms of excitation and response signals (velocity/force or acceleration/force). These FRF spectra are obtained from the Fast-Fourier-Transform (FFT) analyzer while the specimen is continuously excited by an electromagnetic shaker with a random input. The random signal was generated by the noise source of the FFT analyzer. During the phase transformation of the material, all resonant frequencies can be excited simultaneously by the random excitation. Excitation and response signals are fed into the FFT analyzer and the FRF spectrum of the desired frequency span is displayed by a computer in a real-time manner. A data acquisition system stores all the spectrum outputs from the analyzer for further calculations.

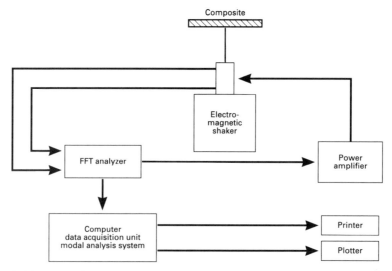

Fig. 27   Setup for MIA cure monitoring

A typical FRF spectrum, which is expressed in terms of inertance (acceleration/force) versus frequency, shows several resonant peaks for elastic materials. For viscoelastic materials, the resonant peaks are broadened due to the damping of the vibration wave. From the FRF spectrum, resonant frequency and damping ratio calculations can be carried out by modal analysis.

Random excitation is preferred to sinusoidal signals in the application for cure monitoring. It is difficult to tune in a suitable frequency range, if sinusoidal signals are used, to include the resonant response continuously of a composite structure whose natural frequency varies as curing proceeds. In contrast, a broad excitation band from random signals provides a feasible measurement, with all the resonant responses being excited simultaneously. This band would cover a wide range of damping values from very high damping of an uncured resin to relatively low damping of a cured one.

The FRF spectra for the Kevlar/epoxy laminates isothermally cured at 120 °C (248 °F) show two resonant peaks in the frequency range which correspond to the first and second mode of vibration respectively at the early curing stage where the internal portion of the sample has not yet reached thermal equilibrium (Fig. 28). Then the sample viscosity continues to decrease because of the increasing segmental mobility at elevated temperature prior to gelation. Since the storage Young's modulus is proportional to the square of resonant frequency and the loss tangent is proportional to the half-power band width of the resonant peak, the first peak shifts to lower frequency indicating a lower modulus while the second peak damped out indicating a higher damping factor. The molecular weight of the resin will build up quickly after cross-linking reaction starts and so will the material modulus. The first peak will shift to higher frequency because of higher storage modulus, and become sharper due to lower damping. The second peak, which is not

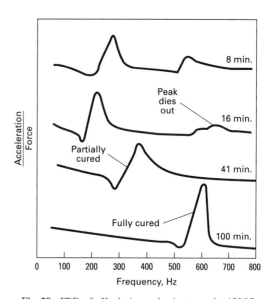

Fig. 28   FRFs of a Kevlar/epoxy laminate cured at 120 °C (248 °F)

seen, now moves beyond 800 Hz. Similar results from other materials such as graphite/epoxy and glass/epoxy laminates were obtained [66].

Rapid data acquisition and reduction in this MIA system provide practically real-time monitoring of the continuous changes in the FRF spectra. The dynamic modulus E' as well as the loss tangent (tan δ) can be calculated. These direct material properties are used to characterize the cure states of a thermosetting resin during composite fabrication. These properties can be obtained at any stage of curing and, therefore, can be expressed as a function of cure time at any cure temperature.

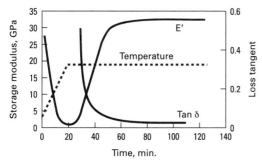

Fig. 29 Dynamic mechanical properties, obtained from MIA, of a graphite/epoxy laminate cured at 140°C (284 °F)

In situ cure monitoring of a free-free vibrating beam of the graphite/epoxy by MIA is shown in Fig. 29. The cure reaction is monitored by the variations in dynamic characteristics. The storage modulus reaches a minimum (viscosity minimum) and then increases dramatically due to gelation. It finally levels off after vitrification. These variations are attributed to the phase transformation of the thermosetting resin from a liquid phase through a rubbery and eventually a glassy phase. Onset of the cross-linking reaction can be observed by the dramatic increase in modulus and a significant decrease in loss tangent. The MIA technique has been successfully applied to a real compression molding environment. Its application toward the autoclave cure situation is under investigation.

This MIA technique offers the advantage that no geometry constraint exists, and it can be applied for cure state characterization and condition monitoring of composite structures [125-128]. This technique may serve as an effective, real-time sensor that can be integrated in a closed-loop feedback control system for the fabrication of large-scale composite structures.

A parameter-controlled process depends upon the measurement of control parameters such as temperature, pressure and vacuum. In a conventional closed-loop control of cure process the adjustment of heating/cooling rates, pressure and vacuum relies upon the continual comparison between the actual and specified parameters. Matchup of these parameters in a manual controlled system does not necessarily assure that the resin has properly reacted. The cure monitoring and control of composite fabrication using a MIA sensor is a property-controlled technique. MIA differs from microdielectrometry in the use of control parameters. The former uses dynamic material properties; the latter uses dielectric properties.

A feasible software flow chart for real-time cure monitoring in conjunction with the MIA setup is shown in Fig. 30. Three major steps are involved. First, the impedance spectrum from a wide band resonant excitation is recorded by the data acquisition software package. Second, *in situ* dynamic responses based on the modal fit data are estimated according to the different types of test coupon setup. Finally, control commands as well as the cure control model are extended to trigger the tempera-

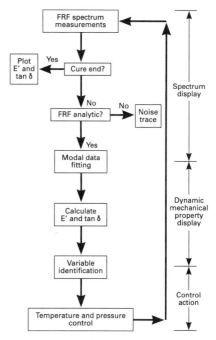

Fig. 30 Methodology for real-time cure monitoring and control of the MIA technique

ture and pressure control devices. Since the vibrational test is extremely sensitive in detecting all the signals in conjunction with the test structure, tracing the noise source is needed. The operation of this MIA technique in composite structure curing will require a skillful operator in vibrational response analysis.

In summary, the MIA technique can be employed to directly measure the storage and loss components of the dynamic response of a viscous liquid or a viscoelastic solid. It is highly sensitive for characterizing both the molecular flow behavior and the cure states of a thermosetting resin and composite. This technique can be used for cure cycle design and optimization for any developmental or new thermosetting resin and can also be used for quality assurance of incoming materials. When applied to composite fabrication, MIA can serve as a real-time *in situ* dynamic mechanical sensor for the cure state characterization and is not subject to any constraints on the test specimen geometry and dimensions.

Comparison of Techniques

Three major candidates for *in situ* cure monitoring techniques (microdielectrometry, acousto-ultrasonics, and mechanical impedance analysis) have been reviewed. The measured signals are permittivity and loss factor for microdielectrometry, attenuation and velocity for acousto-ultrasonics, and mechanical impedance for MIA, respectively. These measured data are usually correlated to other physical properties of the composites during the cure process for the assessment of cure state. Fluorescence and fiber optics techniques are also being

Table 3 Useful NDE techniques for quality control of composites [43,64]

Technique	Advantages	Limitations
Ultrasonic NDE	Uses pulsed ultrasound; both contact and immersion techniques employed	Suffers from destructive wave interference
Fokker Bondtester	Based resonance impedance; can detect porosity; measures both amplitude and frequency shift	Low spatial resolution; must be manually scanned
NDT-210 Bondtester	Operation similar to Fokker bondtester	Interpretation difficulty
Neutron radiography	Particularly useful when bonding components are not X-ray opaque; can be used to detect voids and porosity	Requires radiographic facility
X-ray radiography	Maximum contrast; effective for complex geometries difficult to inspect ultrasonically; can be used to detect water intrusion	Requires X-ray opaque materials as adhesive and matrix components
Coin tap test	Useful in locating large voids and disbonds; applicable for metal-metal or thin skin-honeycomb assemblies	Limited to the outer ply disbonds; method is subjective and may yield variation in test results
MIA (Mechanical impedance analysis)	Global technique for detecting defects; both resonant frequency and damping can be obtained; useful for condition monitoring; may also be used for processability study and cure monitoring in real time	Developmental technique
Acoustic emission	Detects wet interface corrosion delamination	Detector transducer must be placed over the corroded area
Thermography	Locates stress concentrations or structural defects	Requires physical interpretation; not applicable to metal skin laminates

developed and they also appear to have good potential for use as *in situ* cure sensors. Comparison among different techniques is difficult at this stage of development. Although the microdielectic instruments specifically designed for composite cure monitoring have been commercially available for some time, there remains a great deal of room for improvement in this technique. This technique has not yet been incorporated in a closed loop control system in a real composite manufacturing facility. All the other techniques are still under active development.

The main difficulties in all these techniques are to relate accurately the measured signals or data to mechanical, dielectric, and rheological properties and cure state (degree of cure) based on the cure process variables; and to control the process variables precisely so that an ultimate cure state will be reached efficiently. Therefore, two key future research directions in this field of real-time cure monitoring and control are to establish well-defined cure models to interpret the measured data and to develop a knowledge base or expert system to manipulate the large numbers of process variables.

Post-Fabrication Quality Control

Nondestructive evaluation (NDE) techniques for the assessment of damage in composites subjected to various loading conditions include ultrasonics, radiography, IR thermography, vibration-based techniques, etc. These methods can also be applied for the detection of manufacture-induced flaws (Table 3). By adopting new modes of computer-controlled 3-D scanning, the ultra-sonic response of parts with complex surface curvatures can be automatically mapped for flaws. Computed tomography (CT or CAT-SCAN) has also proven to be a powerful but expensive technique for flaw detection in composites. At the present time, no nondestructive testing method can reliably detect poor interfacial adhesion, which could result in low strength and durability in composites.

Some new methods provide a viable reliability and durability test methodology for adhesive bonded structures. Surface NDE methods (Table 4) in general are modifications to the tools of surface characterization to permit automation, rapid surface property mapping, and computerized data acquisition and processing. Surface NDE is designed to perform a final inspection of surfaces to be bonded and to make accept-reject decisions on whether the surface will form a reliable, durable bonded joint or whether rejection and recycling through surface treatment is required [43]. The combination of surface NDE and surface chemical analysis provides a valid approach to reliability and durability analysis of structural adhesive bonding. Surface NDE needs further development and integration of measurement and analysis methodologies to provide quantitative predictions for reliability and durability of composites.

Performance and proof testing of composite reliability can be conducted by various standard ASTM methods, which fall into six categories of response: (1) Processing, (2) Mechanical Properties, (3) Thermal Properties, (4) Electrical Properties, (5) Optical Properties, and (6) Environmental Properties. These ASTM tests are commonly accepted and used, providing a common fund of characterization data. The utilization of standard

Table 4 Surface NDE techniques for bonding applications [43,64]

Technique	Advantages	Limitations
Ellipsometry	Noncontacting and nondestructive; automated and developed for rapid computer controlled surface mapping	Sensitivity is limited by difference in refractive index
SPD (Surface potential difference)	Noncontacting and nondestructive; commercial instruments are available and computerized surface mapping has been developed	Requires other measurements to make a physical interpretation of data
PEE (Photoelectron emission)	Sensitive to both substrate and surface properties; extremely sensitive to thickness effects	Requires intense ultraviolet light source
SRP (Surface remission photometry)	Permits surface spectral analysis; influence of surface roughness is small	Requires use of a light integrating sphere and twin beam optics

ASTM tests to establish the isothermal stress-strain-time response is essential to fully define the mechanical responses of composites. However, ASTM tests used alone could be expensive and may not be adequate for testing composite reliability and durability. Contrarily, extensive quantitative characterization without ASTM testing provides a data base without a general technology reference. The ability to conduct a minimal but adequate amount of performance and proof testing requires a full understanding of the stress- and environment-dependent glass temperature and flow temperature of composites.

Life Monitoring Measurement Science. Reliability of a composite structure can be further enhanced if life monitoring measurement science is fully developed. Future research and development efforts in this aspect should include evaluation of "sensible" composite with an internal "nervous" system, demonstration of life-time monitoring potential for stress, strain, impact, and fatigue real-time measurement, development of science-based quantitative material characterization, and demonstration of sensors in prototype full-scale applications. Hopefully these efforts will result in the selection of feedback sensors, smart materials with life-remaining gages and instantaneous damage warning and cockpit readout of quantifiable structural condition and performance.

Cure Modeling and Expert Systems

Cure modeling is the study of cure reaction kinetics, physical property correlation, and cure variables analysis (resin transport, temperature distribution, void control, etc). Several rate equations have been derived to describe the extent of reaction, reaction rate and reaction order for various thermoset systems [49-56]. The experiments were conducted generally by using a combination of DSC, FTIR, and other thermomechanical techniques. Theoretical expressions to predict the physical properties of the polymeric matrix (such as dynamic moduli, dielectric permittivity or viscosity) as a function of cure temperature, time or degree of cure have been discussed by several authors [57-62].

Analysis of a standard commercial cure cycle in an autoclave/vacuum bagging method has been conducted by Champbell *et al*. [129,130]. The cure processing pa-

rameters and cure cycles currently used in the fabrication of composite structures are usually derived from experience, primarily as a result of interactive experimental testing. Two major drawbacks of the current process development approaches are the need for extensive and expensive experimental testing to determine the proper cure cycles, and the fact that a cure cycle found to be suitable for a given material for one application may not be applicable for a different material, or for the same material in a different part geometry [131].

Cure models for calculating the cure process variables (such as void growth, resin transportation, cure pressure effect, fiber deformation and local temperature distribution) in conjunction with the extent of reaction and viscosity have been discussed in several publications [4,45,131-133]. Recent advancements in cure simulation and instrumentation technology offer the opportunity to develop a system potentially capable of automating the cure process control. Mallow [134] proposed a so-called "Science-Based Cure Model", in which the composite curing process is simulated by a computer program which predicts the final resin content, degree of cure and part quality based upon autoclave operating parameters and material physical properties (resin viscosity for any time-temperature cycle). This cure process model would allow the automated control of autoclaves to be integrated into a computer control manufacturing process. Thus far, all the available models have only had limited success. Most models were built on the basis of very simple laminate geometry and autoclave configuration.

Process variables as shown in Fig. 31 led to the development of the expert system concept for the autoclave cure process [135]. The knowledge base for the expert system in cure process applications contains rules concerning interpretation of sensor inputs and control decision making [43,135-137]. The principles of the autoclave cure process and the constraints of its operation, which were derived from the cure model prediction, are the underlying criteria for decision making. Major refinements of input interpretation and decision making to the existing expert systems are needed before they will become more adaptable to real curing environments.

Wu and Joseph [138] demonstrated that the recent advancements in processing knowledge of composites, computer hardware, artificial intelligence, and control

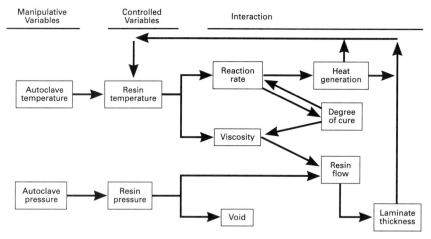

Fig. 31   Process variables for the cure of thermoset-based composite materials (after [135])

theory can be combined to generate knowledge-based control algorithms for automation of autoclave composite manufacturing. The same authors [12] have also investigated model-based and knowledge-based control of pultrusion processes.

A Cure Modeling Approach to Automation (An Example) [139]. To control the overall composite process, industry has established a set of processing windows to keep the materials in an "acceptable" processing state. These windows (out-time, expiration dates, re-test, shelf life, etc.) are very restrictive and often do not lead to the desired results. A batch of resin can be purchased that marginally passes the test criteria of incoming inspection, and is then given the normal shelf life. This resin may not be processable using the standard set of cure conditions established for this resin after any significant shelf life. An alternative to this parametric or processing window environment is the cure modeling environment. Cure modeling will account for the actual state of the resin and adjust the cure cycles accordingly.

The majority of the final properties of a composite are obtained during the cure of the matrix material of the composite. During the cure cycle, chemical reactions occur, which cause the matrix material (resin) to polymerize and obtain the desired properties of the material. The cure of the resin can be considered the most important phase of the manufacturing process. Before a resin cure phase, a considerable investment has been made into a composite component. If the part is incorrectly cured, then the outcome of the cure is irreversible and the part is usually scrapped. No amount of work before cure can ensure that an incorrectly cured part will be usable. It is apparent that the most critical step in obtaining good composites parts is the cure cycle.

Some of the cure monitoring systems have been coupled to automated computer control systems and used to trigger different steps within the cure cycle. The steps are generally pressure application, beginning of heat-up sequences and thermal soaks. The sensor data is required to correctly and reliably cure the resin system. The cure

of the system relies on many conditions including time at temperature, heat rate, and humidity exposure. One way to determine how the cure cycle is affecting the resin is to use sensors during the cure cycle. With this type of data, the cure cycle can be adaptively adjusted to conform to the resin requirements instead of parametrically cured as it is today. The success rate of the cure cycle process will be increased because the sensor approach will eliminate "marginal" cures, where the parametric cure cycle does not adequately account for the condition of the resin system (e.g., excessive out-time, high humidity exposure, marginal resin batch chemistry, etc.). It also has the potential of increasing autoclave throughput by shortening and lengthening cure cycles as appropriate, depending on the resin requirements.

The current in-bag monitoring devices are somewhat cumbersome in the production environment. The sensors could be sensitive to application method and are still relatively expensive. If not used properly, the current cure monitors are prone to damage and can fail to trigger the desired action. Sensors require additional penetration in the vacuum bag and could generate additional potential leak paths. Even when they are used, fail-safe thermal triggers are placed within the cure profile in case the sensor does not signal before a certain time-temperature relationship exists. The sensors at the present stage of development are not yet optimal for the production environment.

Similar data can be obtained without the use of sensors in production by the use of the cure modeling technique [139]. To model a cure, mathematical models are used to predict the current state of the resin system. The resin state prediction is related to the resin degree of cure α_c and can be modeled before, during, and after the cure process. The resin degree of cure indicates how far along the reaction path the resin system is at a given set of temperature and time conditions. The implementation of any model-based automated composites cure control system depends on the ability to accurately predict the degree of cure at any point of time.

The Automated Composites Cure Control (ACCC) system introduced by Roberts [139] involved the following elements:

- Theoretical model—resin degree of cure (kinetics and visco-elastic flow state)
- Meaningful battery of incoming material tests
- Basic cure model for each resin system
- Material database (material properties, material history, and in-process material tracking)
- Model validation requirements
- Software (model, expert shell, autoclave control unit, periodic revalidation of model, and data storage methods for historical purposes)

The physical characteristics that an acceptable composite laminate possesses are:

- Low to no void/porosity content
- Correct degree of cure
- Correct fiber volume/resin content
- Correct fiber orientation

Each element of the Automated Composite Cure Control System relates the processing to one or all of these goals to produce a good laminate.

The cure modeling environment appears to be capable of fully utilizing the power of the computer to accurately trace and monitor raw materials, from entry into the processing facility through cure, and to use this resin condition data to predict the cure requirements interactively during the cure cycle. This approach provides an alternative route to sensor monitoring for ensuring that the resin of the composite is fully cured.

References

1. A.B. Strong, *Fundamentals of Composites Manufacturing*, SME, Dearborn, MI, 1989
2. D. Woolstencroft, Advanced Composite Fabrication, in *Advanced Composites*, I.K. Partridge, Ed., Elsevier Publishing, London, 1989, p 251-268
3. P.K. Mallick, *Fiber Reinforced Composites*, Marcel Dekker, Inc., New York, 1988
4. A.C. Loos and G.S. Springer, Calculation of Cure Process Variables During Cure of Graphite/Epoxy Composite, *Composite Materials: Quality Assurance and Process*, STP 797, C.E. Browning, Ed., ASTM, 1983, p 110-118
5. A.C. Loos and G.S. Spring, *J. Compos. Mater.*, Vol 17, 1983, p 135
6. J.E. Sumerak and J.D. Martin, *Proc. 41st Annual Conf., Reinforced Plastics/Composites Institute*, SPI, 1986
7. H.L. Price, "Curing and Flow of Thermosetting Resins for Composite Material Pultrusion," Ph.D. thesis, Old Dominion University, 1979
8. J.R. Sumerak, *Modern Plast.*, March 1985, p 58
9. T.J. Tulig, *Proc. AIChE*, 1985
10. C.D. Han, D.S. Lee, and H.B. Chin, *Polym. Eng. Sci.*, Vol 26, 1986, p 393
11. G.L. Batch and C.W. Macosko, *Proc. 42nd Annual Conf. Reinforced Plastics/Composite Institute*, SPI, 1987
12. H.T. Wu and B. Joseph, *SAMPE J.*, Vol 26, No. 6, 1990, p 59-70
13. J.K. Rogers, *Plast. Technol.*, March 1989, p 50-58
14. H.S. Kliger and B.A. Wilson, Resin Transfer Molding, in *Composite Materials Technology*, P.K. Mallick and S. Newman, Ed., Hanser Publishing, Munich, 1990, p 149-178
15. P.K. Mallick and S. Newman, Ed., *Composite Materials Technology*, Hanser Publishing, Munich, 1990
16. E.P. Carley, J.F. Dockum, and P.L. Schell, in *Proc. 5th Annual Conf. Advanced Composites*, ASM International, Sept 1989, p 259-273
17. C.F. Johnson and N.G. Chavka, in *Proc. 4th Annual Conf. Advanced Composites*, ASM International, Sept. 1988
18. B.C. Mazzoni, in *Proc. 43rd Annual Conf., Reinforced Plastics/Composites Institute*, SPI, Feb 1988, Sec.11-F
19. S.G. Dunbar, in *Proc. 43rd Annual Conf., Reinforced Plastics/Composites Institute*, SPI, Feb 1988, Sec.4-D
20. M.J. Cloud, in *Proc. 4th Annual Conf. Advanced Composites*, ASM International, Sept 1988
21. E.P. Carley, J.F. Duckum, and P.L. Schell, in *Proc. 44th Annual Conf. Reinforced Plastics/Composites Institute*, SPI, Feb 1989, Sec.10-B
22. B. Miller, *Plast. World*, Jan 1988, p 23
23. M. Kallaur, in *Proc. 43rd Annual Conf. Reinforced Plastics/Composites Institute*, SPI, Feb 1988, Sec.22-A
24. J. Scirvo, in *Proc. 43rd Annual Conf., Reinforced Plastics/Composites Institute*, SPI, Feb 1988, Sec.22-B
25. S.L. Voeks et al., in *Proc. 43rd Annual Conf., Reinforced Plastics/Composites Institute*, SPI, Feb 1988, Sec.4-E
26. C.D. Shirrel, *AutoCom '88 Conf.*, SME, May 1988, Paper No. EM88-228
27. P. Emrich, in *Proc. 3rd Annual Conf. Advanced Composites*, ASM International, Sept 1987
28. D.A. Kleymeer, in *Proc. 4th Annual Conf. Advanced Composites*, ASM International, Sept 1988, p 289-295
29. R.W. Meyer, *Handbook of Polyester Molding Compounds and Molding Technology*, Chapman and Hall, New York, 1987
30. P.K. Mallick, Sheet Molding Compounds, in *Composite Materials Technology*, P.K. Mallick and S. Newman, Ed., Hanser Publishing, Munich, 1990, p 25-65
31. R.J. Coldicott, T. Longdon, S. Green, P.J. Ives, HELTRA—A New System for Blending Fibers and Matrix for Thermoplastic-Based High Performance Composites, *34th Intl. SAMPE Symp.*, May 1989, p 2206-2216
32. P.A. Hogan, *Proc. 14th SAMPE Tech. Conf.*, New York, 1980
33. M. Thiede-Smet, M. Liu, and V. Ho, Study of Processing Parameters of PEEK/Graphite Composites Fabricated with "FIT" Prepreg, *34th Intl. SAMPE Symp.*, May 1989, p 1223-1235. See also US Patent 4,614,678, 1986
34. T. Hartness, Thermoplastic Powder Technology for Advanced Composites Systems, *J. Thermoplast. Compos. Mater.*, Vol 1, 1988, p 210-220
35. S. Clemans and T. Hartness, Thermoplastic Prepreg Technology for High Performance Composites, *SAMPE Q.*, July 1989, p 38-42. Also see International Patent Application PCT/US87/02867, 1987
36. J. Muzzy, B. Varughese, V. Thammongkol, and W. Tincher, Electrostatic Prepregging of Thermoplastic Matrices, *34th Intl. SAMPE Symp.*, May 1989, p 1940-1950
37. R.M. Baucom and J.M. Marchello, *SAMPE Q.*, July 1990, p 14-19

38. A.C. Handermann, Advances in Commingled Yarn Technology, *20th Intl. SAMPE Conf.*, Sept 1988, p 681-688

39. S.R. Clemans, E.D. Western, and A.C. Handerman, *Mater. Eng.*, March 1988, p 27-30

40. E.M. Silverman and R.J. Jones, *SAMPE J.*, July/August 1988, p 33-40

41. D.J. Brunelle, T.L. Evans, T.G. Shannon, E.P. Boden, K.R. Steward, L.P. Fontana, and D.K. Bonauto, Preparation and Polymerization of Cyclic Oligomeric Carbonates, *Polymer Preprints*, ACS, Vol 30, No. 2, 1989, p 569-570

42. D.C. Leach, in *Advanced Composites*, I.K. Partridge, Ed., Elsevier Publishing, London, 1989, p 43-110

43. D. H. Kaelble, *Computer-Aided Design of Polymers and Composites*, 1985, Chapter 2, Marcel Dekker, 1985

44. J.K. Gillham, in *Developments in Polymer Characterization—3*, J.V. Dawkins, Ed., Applied Science Publishing, London, 1982

45. G.S. Springer, Modeling the Cure Process of Composites, *SAMPE J.*, Sept/Oct 1986, p 22-27

46. W.I. Lee, A.C. Loos, and G.S. Springer, *J. Compos. Mater.*, Vol 16 , 1982, p 510

47. K.W. Lem and C.D. Han, *Polym. Eng. Sci.*, Vol 24, 1984, p 175

48. C.P. Chung, "Cure Modeling and Monitoring of Thermosetting Resins," MS Thesis, Auburn University, Aug 1989

49. M.T. Aronhime and J.K. Gillham, TTT Cure Diagram of Thermosetting Polymeric Systems, *Advances in Polym. Sc.*, K. Dusek, Ed., Springer-Verlag, 1986, p 84-112

50. R.E. Wetton, M.R. Stone, and J.W.E. Gearing, Absolute Properties of Thermoset during Curing, *Progress in Advanced Materials and Process*, G. Bartelds, Ed., Air Force Wright Materials Laboratory, 1985, p 293-296

51. R.E. Wetton, Dynamic Mechanical Thermal Analysis of Polymers and Related Systems, *Development of Polymer Technology*, Vol 5, 1986, p 179

52. V.B. Gupta, L.T. Drzal, C.Y.-C. Lee, and M.J. Rich, The Temperature Dependance of Some Mechanical Properties of a Cured Epoxy Resin System, *Polym. Eng. Sci.*, Vol 25, No. 13, 1985, p 812

53. L.T. Pappalardo, DSC Evaluation of Epoxy and Polyamide-Impregnated Laminates, *J. Appl. Polym. Sci.*, Vol 21, 1977, p 809-820

54. R.B. Prime, Differential Scanning Calorimetry of the Epoxy Cure Reaction, *Polym. Eng. Sci.*, Vol 13 (No. 5), Sept 1973

55. C.R. Carmen and R.J.J. Williams, A Kinetic Scheme for an Amine-Epoxy Reaction with Simultaneous Etherification, *J. Appl. Polym. Sci.*, Vol 32, 1986, p 3445-3456

56. A. Moroni, J. Mijovic, E. Pearce, and C.C. Foun, Cure Kinetics of Epoxy Resins and Aromatic Diamines, *J. Appl. Polym. Sci.*, Vol 32, 1986, p 3761-3773

57. J.B. Enns and J.K. Gillham, TTT Diagram: Modeling the Cure Behavior of a Thermoset, *J. Appl. Polym. Sci.*, Vol 28, 1983, p 2567-2591

58. S. Sourour and M.R. Kamal, DSC of Epoxy Cure: Isothermal Kinetics, *Thermochemica Acta*, Vol 14, 1976, p 41-59

59. J. W. Lane, M. A. Bachmann, and J. C. Seferis, Monitoring of Matrix Property Changes during Composite Processing, *SPE Technical Papers*, Vol 31, 1985, p 318-320

60. S.D. Senturia and N.F. Sheppard, Dielectric Analysis of Thermoset Cure, *Advances in Polymer Science*, K. Dusek, Ed., Vol 80, Air Force Wright Materials Laboratory, 1986, p 1-48

61. M.R. Dusi, C.A. May, and J.C. Seferis, Predictive Models as Aids to Thermoset Resin Processing, *Chemorheology of Thermosetting Polymers, ACS Symposium Series 227*, 1983, p 301-318

62. W.M. Sanford and R.L. McCullough, Modeling the Viscosity and Dielectric Behavior During the Cure of Epoxy Matrix Composites, Proc. Am. Soc. Compos. Conf., 1987, p 21-30

63. M.R. Kamal and S. Sourour, *Polym. Eng. Sci.*, Vol 13, 1973, p 59

64. D.H. Kaelble and R.J. Shuford, *US Army ManTech J.*, Vol 10, No. 2, 1985, p 17-26

65. A. Mahoon, Non-Destructive Testing of Aircraft Composite Structure, Proc. Int. Conf. Testing, Evaluation and Quality Control of Composites, ASTM, 1983, p 1-29

66. H.B. Hsieh, "Real-Time Cure Monitoring of Composite Using Mechanical Impedance Analysis," M.S. thesis, Auburn University, 1989

67. B.Z. Jang and G.H. Zhu, Monitoring the Dynamic Mechanical Behavior of Polymers and Composites Using MIA, *J. Appl. Polym. Sci.*, Vol 31, 1986, p 2627-2646

68. R.E. Wetton, M.R. Morton, A.M. Rowe, and T.N. Tweeten, Comparison of Information Obtained During Epoxy Cure by DMTA and DETA Techniques, *Proc. Ann. Tech. Conf. ANTEC SPE*, 1986, p 439-440

69. R.D. Hoffman, J.J. Dodfrey, D.E. Kranbuehl, L. Weller, and M. Hoff, Dynamic Dielectric Analysis: A Nondestructive Quality-Assurance Monitor of Resin Processing Properties, *41st Ann. Conf., Reinforced Plastics/Composites Institute*, The Society of the Plastics Industry, Inc., Jan 27-31, 1981

70. D.E. Kranbuehl, S.E. Delos, and P.K. Jue, Dynamic Dielectric Characterization of the Cure Process: LARC-160, *28th Natl. SAMPE Symp.*, April 12-14, 1983, p 608

71. S.D. Senturia and N.F. Sheppard, Cure Monitoring and Control With Combined Dielectric/Temperature Probes, *28th Natl. SAMPE Symp.*, April 12-14, 1983, p 851

72. Z.N. Sanjana and R.L. Selby, Monitoring Cure of Epoxy Resins Using a Microdielectrometer, *Proc. SAMPE Symp.*, Vol 29, 1984, p 1233-1242

73. R. Kienle and H. Race, The Electric, Chemical and Physical Properties of Alkyd Resins, *Trans. Electrochem. Soc.*, Vol 65, 1934, p 87

74. D.E. Kranbuehl, S.E. Delos, M.S. Hoff, M.E. Whitham, L.W. Weller, and P.D. Harverty, Dynamic Dielectric Analysis for Nondestructive Cure Monitoring and Process Control, *Advanced Composites: The Latest Developments*, ASM International, 1986, p 61

75. R.E. Wetton, G.M. Foster, J.W.E. Gearing, and J.S. Fisher, Thermal Scanning Dielectric Instrumentation—The PL-Dielectric Thermal Analyzer, in *Dielectric Materials: Measurements and Applications*, The Institution of Electrical Engineers, Chameleon Press, London, 1988

76. R.E. Wetton, G.M. Foster, V.R. Smith, J.C. Richmond, and J.T. Neill, Dielectric Monitoring of Epoxy Cure—Detailed Analysis, *33rd Intl. SAMPE Symp.*, March 7-10, 1988, p 1285

77. N.F. Sheppard, Jr., S.L. Garverick, D.R. Day, and S.D. Senturia, Microdielectrometry: a New Method for *in-situ* Cure Monitoring, *26th SAMPE Symp.*, 1981, p 65

78. M.C.W. Coln and S.D. Senturia, The Application of Linear System Theory to Parametric Microsensors, *Proc. Transducers '85*, 1985, p 118

79. S.D. Senturia, N.F. Sheppard, Jr., H.L. Lee, and S.B. Marshall, Cure Monitoring and Control with Combined Dielectric/Temperature Probes, *SAMPE J.*, Vol 19, 1983, p 22

80. S.D. Senturia and N.F. Sheppard, Dielectric Analysis of Thermoset Cure, in *Epoxy Resins and Composites IV*, K. Dusek, Ed., Springer-Verlag, Berlin 1986, p 1

81. D. R. Day, Thermoset Cure Process Control for Utilizing Microdielectric Feedback, *33rd Intl. SAMPE Symp.*, March 7-10, 1988, p 594

82. M.L. Bromberg, D.R. Day, H.L. Lee, and K.A. Russell, "New Applications for Dielectric Monitoring and Control," Sec. 22-E, 42nd Annual Conf. Proc., Composites Inst. Soc. of Plastics Industry, Feb 1987, p 1-5

83. D.D. Sheppard, H.L. Lee, and D.R. Day, In-Process RIM Analysis with Microdielectric Sensors, *34th Intl. SAMPE Symp.*, May 8-11, 1989, p 407

84. D.R. Day and D.D. Shepard, Effect of Advancement on Epoxy Pre-Preg Processing—A Dielectric Analysis, *ANTEC SPE*, 1989, p 1534

85. R.T. Harrold and Z.N. Sanjana, Acoustic Waveguide Monitoring of the Cure and Structural Integrity of Composite Materials, *Polym. Eng. Sci.*, Vol 26, No. 5, 1986, p 367-372

86. E.M. Woo and J.C. Seferis, Acoustic Cure Monitoring of Epoxy Matrixes and Composites, *ANTEC SPE*, 1986, p 375-379

87. Y.L. Hinton, R.J. Shuford, and W.W. Houghton, Acoustic Emission During Curing of Fiber Reinforced Composites, *Proc. Critical Review: Techniques for the Characterization of Composite Materials*, AMMRC MS 82-3, Army Materials and Mech. Res. Center, May 1982, p 25-36

88. W.W. Houghton, R.J. Shuford, and J.F. Sprouse, Acoustic Emission as an Aid for Investigating Composite Manufacture Processes, *New Horizons—Materials and Processes for the Eighties*, Vol 11, National SAMPE Technical Conference Series, 1979, p 131-150

89. J.R. Mitchell, "Cure Monitoring of Composite Materials Using Acoustic Transmission and Acoustic Emission," Physical Acoustic Corp., 1987, TR-103-60D-7/87

90. S.C. Brown, P.K. Jarrett, and T. Rose, *"In Situ* Cure Monitoring of 2D Carbon Fiber Reinforced Phenolic Composite," Physical Acoustic Corp., 1986, TR-103-60A-9/86

91. A.M. Lindrose, Ultrasonic Wave and Moduli Change in a Curing Epoxy Resin, *Experimental Mechanics*, July 1978, p 227

92. E.P. Papadakis, Monitoring the Moduli of Polymers with Ultrasound, *J. Appl. Phys.*, Vol 45, No. 3, March 1974, p 1218-1222

93. G.A. Sofer and E.A. Hauser, A New Tool for Determination of the Stage of Polymerization of Thermosetting Polymers, *J. Polym. Sci.*, Vol 8, No. 6, p 611-620

94. E.J. Tuegel and H.T. Hahn, Ultrasonic Cure Characterization of Epoxy Resins: Constitutive Modeling, in *Advances in Modeling of Composites Processes*, G. Bartelds, Ed., Air Force Wright Materials Laboratory, 1985, p 129-136

95. W.P. Winfree and F.R. Parker, F.R., Measurement of the Order of Cure in Epoxy with Ultrasonic Velocity, *Review of Progress in Quantitative NDE*, 1985, p 1-8

96. S.I. Rokhlin, D.K. Lewis, K.F. Graff, and L. Adler, Real-Time Study of Frequency Dependence of Attenuation and Velocity of Ultrasonic Waves during the Curing Reaction of Epoxy Resin, *J. Acoust. Soc. Am.*, Vol 76, No. 9, June 1986, p 1786-1793

97. R. Halmshaw, Chapter 4, *Nondestructive Testing*, Edward Arnold Publishers, London, 1987

98. H.T. Hahn, Application of Ultrasonic Technique to Cure Characterization of Epoxies, in *Nondestructive Methods for Material Property Determination*, C.O. Ruud and R.E. Green, Ed., Plenum Press, New York, 1983

99. R.T. Harrold and Z.N. Sanjana, Z.N., Theoretical and Practical Aspects of Acoustic Waveguide Cure Monitoring of Composites and Materials, *31st Intl. SAMPE Symp.*, April 1986, p 1713-1721

100. R.T. Harrold and Z.N. Sanjana, Material Cure and Internal Stresses Monitored via Embedded Acoustic Waveguides, *Intl. Congr. Technology and Technology Exchange*, Oct. 1986, Paper AM-2.3, p 30-33

101. A. Vary and K.J. Boeles, An Ultrasonic-Acoustic Technique for Nondestructive Evaluation of Fiber Composite Quality, *Polym. Eng. Sci.*, Vol 19, 1979, p 373

102. J.C. Duke, Jr., *Acousto-Ultrasonics—Theory and Application*, Plenum Press, New York, 1988

103. S.S. Saliba, T.E. Saliba, and J.F. Lanzafame, Acoustic Monitoring of Composite Materials During the Cure Cycle, *34th Intl. SAMPE Symp.*, 1989, p 397

104. R.L. Levy and D.P. Ames, Monitoring Epoxy Cure Kinetics with a Viscosity-Dependent Fluorescence Probe, *Organic Coatings* and *Appl. Polymer Sci.* Preprints, ACS, Vol 48, 1983, p 116-120

105. R.O. Loutfy, High-Conversion Polymerization Fluorescence Probes. 1. Polymerization of Methyl Methacrylate, *Macromolecules*, Vol 14, 1981, p 270-275

106. R.O. Loutfy, Fluorescence Probes for Polymerization Reactions: Bulk Polymerization of Styrene, N-Butyl Methacrylate, Ethyl Methacrylate, and Ethyl Acrylate, *J. Polym. Sci., Polym. Phys. Ed.*, Vol 20, 1982, p 825-835

107. F.W. Wang, R.E. Lowry, and W.H. Grant, Novel Excimer Fluorescence Method for Monitoring Polymerization: 1. Polymerization of Methyl Methacrylate, *Polymer*, Vol 25, 1984, p 690-692

108. F.W. Wang, R.E. Lowry, and B.M. Franconi, Cure Monitoring of Epoxy Resins by Fluorescence Spectroscopy, *ACS Polym. Mater. Sci. Eng.*, Vol 53, 1985, p 180-184

109. A. Bur, F.W. Wang, and R. Lowry, Fluorescence Monitoring of Polymer Processing: Mixing and Zero-Shear Viscosity, *ANTEC SPE*, 1988, p 1107-1110

110. A. Stroeks, M. Shmorhun, A.M. Jamieson, and R. Simha, Cure Monitoring of Epoxy Resins by Excimer Fluorescence, *Polymer*, Vol 29, 1988, p 467-470

111. F.W. Wang and E.S. Wu, Cure Monitoring of Epoxy Resins by Fluorescence Recovery After Photobleaching, *Polymer*, Vol 28, May 1987, p 73-75

112. F.W. Wang, R.E. Lowry, and B.M. Faconi, Novel Fluorescence Method for Cure Monitoring of Epoxy Resins, *Polymer*, Vol 27, 1986, p 1529-1532

113. S.S. Chang, F.I. Mopsik, and D.L. Hunston, Correlation of Cure Monitoring Techniques, *19th Intl. SAMPE Tech. Conf.*, Oct 1987, p 253-264

114. R.E. Shirmer and A.G. Gargus, Monitoring Polymer Processing Through Fiber Optics, *Am. Lab.*, Nov 1988, p 37-43

115. R.L. Levy and S.D. Schwab, Monitoring the Composite Curing Process with a Fluorescence-Based Fiber-Optic Sensor, *ANTEC SPE*, 1989, p 1530-1533

116. P.R. Young, M.A. Druy, W.A., Stevenson, and D.A. Compton, *In situ* Composite Monitoring using IR Transmitting Optical Fibers, *20th Intl. SAMPE Tech. Conf.*, Sept 1988, p 336-347

117. M.S. Vratsanos and R.J. Farris, A New Method for Determining Shrinkage Stresses and Properties of Curing Thermoset, in *Composite Interfaces*, H. Ishida and J. L. Koenig, Ed., North-Holland, 1986, p 71-80

118. B.Z. Jang and M. Rao, Cure Monitoring of Composite Structure Using Mechanical Impedance Analysis, *Nondestructive Characterization of Material II*, J.F. Bussiere et al., Ed., Plenum Publishing, 1987

119. P. Cawley and R.D. Adams, Vibration Technique, *Nondestructive Testing of Fiber-Reinforced Plastic Composites*, J. Summerscales, Ed., Elsevier Appl. Sci., 1987

120. J.D. Ferry, *Viscoelastic Properties of Polymers*, John Wiley and Sons, Inc., New York, 1970, p 125-157

121. T.L. Smith, J.D. Ferry, and F.W. Schremp, Measurements of the Mechanical Properties of Polymer Solution by Electromagnetic Transducers, *J. Appl. Phys.*, Vol 20, Feb 1949, p 144-153

122. J.H. Dillon, I.B. Prettyman, and G.L. Hall, Hysteretic and Elastic Properties of Rubberlike Materials under Dynamic Shear Stresses, *J. Appl. Phys.*, Vol 15, April 1944, p 309-323

123. M.H. Birnboim and J.D. Ferry, Method for Measuring Dynamic Mechanical Properties of Viscoelastic Liquids and Gels, *J. Appl. Phys.*, Vol 32, No. 11, Nov 1961, p 2305-2313

124. W. Philippoff, Mechanical Investigation of Elastomers in a Wide Range of Frequency, *J. Appl. Phys.*, Vol 24, No. 6, June 1953, p 685-689

125. B.Z. Jang, H.B. Hsieh, and M.D. Shelby, in *ANTEC SPE*, 1988, p 1189-1191

126. B.Z. Jang, H.B. Hsieh, and M.D. Shelby, Vibration-Based Cure Monitoring for Composites, *ANTEC SPE*, 1989, p 1527-1529

127. B.Z. Jang, M.D. Shelby, H.B. Hsieh, and T.L. Lin, *19th Intl. SAMPE Tech. Conf.*, Crystal City, VA, Oct 13-15, 1987, p 265-276

128. B.Z. Jang, H.B. Hsieh, M.D. Shelby, and Y.J. Paig, *14th Conf. Production and Research Technology*, SME-NSF, Ann Arbor, MI, 1987, p 427-432

129. F.C. Champbell, A.R. Mallow, and K.C. Amuedo, "Computer Aided Curing of Composites," McDonnell Aircraft Company, 5th Interim Report, 4/1/85/-6/31/85, Contract No. F33615-83-C-5088, AFWAL/MLBC, Wright-Patterson AFB, OH 45433

130. F.C. Champbell, A.R. Mallow, and K.C. Amuedo, "Computer Aided Curing of Composites," McDonnell Aircraft Company, 6th Interim Report, 7/1/85-9/1/85, Contract No. F33615-83-C-5088, AFWAL/MLBC, Wright-Patterson AFB, OH 45433

131. R. Dave, J.L. Kardos, and M.P. Dudukovic, Process Modeling of Thermosetting Matrix Composites: A Guide for Autoclave Cure Cycle Selection, *Advances in Modeling of Composite Processes*, G. Bartelds, Ed., Air Force Wright Materials Laboratory, 1985, p 137-153

132. J.M. Tang, W.T. Lee, and G.S. Springer, Effect of Cure Pressure on Resin Flow, Voids and Mechanical Properties, *J. Compos. Mater.*, Vol 21, May 1987, p 421-440

133. T.G. Gutowski, A Resin Flow/Fiber Deformation Model for Composite, *SAMPE Q.*, July 1985, p 58-64

134. A.R. Mallow, F.R. Muncaster, and F.C. Champbell, Science Based Cure Model for Composites, *Advances in Modeling of Composite Processes*, G. Bartelds, Ed., Air Force Wright Materials Laboratory, 1985, p 171-185

135. C.W. Lee, Composite Cure Control by Expert System, *Advances in Modeling of Composite Process*, G. Bartelds, Ed., Air Force Wright Materials Laboratory, 1985, p 187-194

136. S.R. Le Clair, Sensor Fusion: The Application of Artificial Intelligence Technology to Process Control, *1986 Rochester FORTH Conf.*, Rochester, NY, June 11-14, 1986

137. R.A. Servais, C.W. Lee, and C.E. Browning, Intelligent Processing of Composite Materials, *31st Intl. SAMPE Symp.*, Las Vegas, NV, April 7-10, 1986

138. H.T. Wu and B. Joseph, *SAMPE J.*, Vol 26, No. 6, 1990, p 39-54

139. R.W. Roberts, Automated Composites Cure Control Implementation: A Cure Model Approach to Automation, *SAMPE J.*, Sept/Oct 1987, p 29-33

Part II

Mechanics,
Mechanical Properties,
and Design Considerations

Chapter 5

Design and Prediction for Elastic Constants of Composites

In this chapter, we will learn how to predict the elastic constants of a composite, given the constituent properties and proportions. We will come to realize that, in general, the longitudinal elastic constants (e.g., modulus along the fiber axial direction) of a composite are dictated by the fiber properties. However, under transverse and shear loading conditions, the matrix also plays a key role in determining the properties of the composite. To be considered in this chapter are different types of composites, schematically shown in Fig. 1, that are loaded along various directions (longitudinal, transverse, or at any arbitrary angle) under normal or shear stress conditions.

Unidirectional Continuous Fiber Laminas or Laminates

Prediction of Elastic Constants from Constituent Properties

The mechanical behavior of fiber composites can be studied from two perspectives: micromechanics and macromechanics. Micromechanical approaches are used to predict the properties of composites in terms of the properties and interactions of the fiber and matrix. The microstructure of a composite is simulated by an approximate model that predicts the "average" properties of the composite. The composite is generally in the form of unidirectional plies laminated together at various orientations. Macromechanics is aimed at designing or predicting the behavior of a composite structure on the basis of the "average" properties of the unidirectional plies. The properties of interest include the longitudinal modulus, E_1; the transverse modulus, E_2; the major Poisson's ratio, ν_{12}; the in-plane shear modulus, G_{12}; and various strength values.

To illustrate the roles of fibers in determining the elastic constant of a unidirectional ply or lamina, let us take the mechanics-of-materials approach, one of the

micromechanics analyses, as an example. Assume that a lamina (Fig. 2a) can be modeled as a simplified two-dimensional element (Fig. 2b) that is subjected to a load P_1. Assuming an iso-strain condition, one can easily show that

$$E_1 = E_f V_f + E_m V_m \qquad \text{(Eq 1)}$$

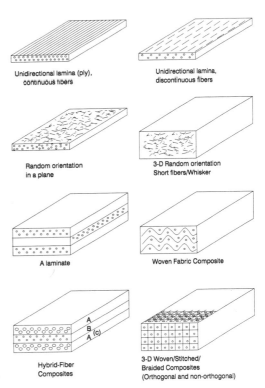

Unidirectional lamina (ply), continuous fibers

Unidirectional lamina, discontinuous fibers

Random orientation in a plane

3-D Random orientation Short fibers/Whisker

A laminate

Woven Fabric Composite

Hybrid-Fiber Composites

3-D Woven/Stitched/ Braided Composites (Orthogonal and non-orthogonal)

Fig. 1 Schematic of several different types of composites considered in this chapter

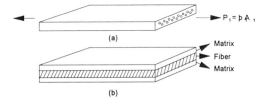

Fig. 2 A unidirectional ply or lamina modeled as a simplified two-dimensional element using the mechanics of materials approach

This is the well-known "rule of mixtures" equation. In an advanced composite, typically $E_m << E_f$ and $V_f \geq V_m$, so we have

$$E_1 \approx E_f V_f \qquad \text{(Eq 2)}$$

This suggests that the longitudinal modulus of a unidirectional composite is practically dictated by the axial modulus of the fibers. The significance of this observation can be enlarged by considering the load shared by the fibers (P_f) with respect to the total load (P_1):

$$\frac{P_f}{P_1} = \frac{\sigma_f A_f}{\sigma_f A_f + \sigma_m A_m} = \frac{E_f \varepsilon A_f}{E_f \varepsilon A_f + E_m \varepsilon A_m}$$
$$= \frac{1}{1 + \dfrac{E_m A_m}{E_f A_f}} = \frac{1}{1 + (E_m / E_f)(1 - V_f) / V_f} \qquad \text{(Eq 3)}$$

For a graphite-fiber-reinforced glass with $V_f = 0.6$, $E_f = 300$ GPa (44×10^6 psi), and $E_m = 70$ GPa (10×10^6 psi), Eq 3 gives (P_f / P_1) = 0.87. This implies that the fibers carry 87% of the total load. For a graphite-fiber-reinforced epoxy where $V_f = 0.6$, $E_f = 390$ GPa (57×10^6 psi), and $E_m = 3$ GPa (435 ksi), we obtain (P_f / P_1) = 0.995, suggesting an even greater role played by the fibers in carrying the load.

When loaded transversely, the fiber and matrix can be assumed to act in series, both carrying the same applied stress (iso-stress condition). It can be readily shown that the transverse modulus of the composite, E_2, is given by

$$\frac{1}{E_2} = \frac{V_f}{E_{ft}} + \frac{V_m}{E_m} \qquad \text{(Eq 4)}$$

where E_{ft} is the transverse modulus of the fiber. Although usually not as high as the longitudinal modulus E_f, the magnitude of E_{ft} generally is still much higher than that of E_m. Equation 4 is therefore reduced to $E_2 \approx E_m/V_m$, implying that the fibers have only a small effect on E_2. It is of interest to note that $E_f \approx E_{ft}$ for glass fibers.

Following similar considerations, one can prove that

$$\nu_{12} = \nu_f V_f + \nu_m V_m \qquad \text{(Eq 5a)}$$

$$\frac{1}{G_{12}} = \frac{V_f}{G_f} + \frac{V_m}{G_m} \qquad \text{(Eq 5b)}$$

The major Poisson's ratio (ν_{12}) is defined by $\nu_{12} = -E_2/E_1$, where the only applied stress is σ_1. Again, the value of G_m has the major effect on G_{12}, the in-plane shear modulus, for a typical advanced composite, because G_f is much greater than G_m. Although Eq 1 to 5 provide a good estimate of the various elastic constants, more accurate predictions require refinements to the mechanics-of-materials approach, which can be found in several textbooks on composites mechanics [1-3].

The linear coefficients of thermal expansion (CTEs) for unidirectional continuous fiber laminas have been developed by Schapery [4] using the energy principles. The CTEs in the 0° and 90° directions can be calculated from the following equations:

$$\alpha_{11} = \frac{\alpha_{fl} E_f V_f + \alpha_m E_m V_m}{E_f V_f + E_m V_m}$$

$$\alpha_{22} = (1 + \nu_f) \frac{(\alpha_{fl} + \alpha_{fr})}{2} V_f$$
$$+ (1 + \nu_m) \alpha_m V_m - \alpha_{11} \nu_{12} \qquad \text{(Eq 6a)}$$

where $\nu_{12} = \nu_f V_f + \nu_m V_m$, α_{fl} = the CTE for the fiber in the longitudinal direction, α_{fr} = the CTE for the fiber in the radial direction, and α_m = the CTE for the matrix.

If the fibers of the lamina are at an angle with the x-direction, the CTEs in the x- and y-directions can be obtained using α_{11} and α_{22}:

$$\alpha_{xx} = \alpha_{11} \cos^2\theta + \alpha_{22} \sin^2\theta$$
$$\alpha_{yy} = \alpha_{11} \sin^2\theta + \alpha_{22} \cos^2\theta$$
$$\alpha_{xy} = 2 \sin\theta \cos\theta \cdot (\alpha_{11} - \alpha_{22}) \qquad \text{(Eq 6b)}$$

where α_{xy} is the coefficient of shear expansion. It may be noted that the CTE of a unidirectional lamina (or laminate) is greater in the transverse direction than in the longitudinal direction ($\alpha_{22} > \alpha_{11}$). This is because the fibers, which usually have a smaller coefficient than that for the polymer matrices, impose a mechanical restraint on the matrix material. Unless $\theta = 0°$ or $90°$, a change in temperature generates a shear strain because of the existence of α_{xy}. The other two "off-axis" CTEs, α_{xx} and α_{yy}, produce extensional strains in the x- and y-directions, respectively.

A closer examination of the above equations leads to the following conclusions:

- The longitudinal modulus (E_1) is always greater than the transverse modulus (E_2). The major Poisson's ratio (ν_{12}) is always greater than the minor Poisson's ratio (ν_{21}); only one of them can be considered independent.

- The fibers contribute more to the longitudinal modulus of a composite, whereas the matrix contributes more to the transverse modulus and the shear modulus.

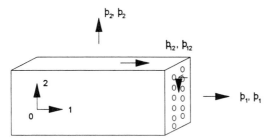

Fig. 3   Material axes for a single lamina or a unidirectional laminate

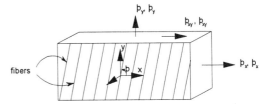

Fig. 4   Structural axes for a single lamina or a unidirectional laminate loaded at an angle θ with respect to the fiber axial direction

- Four independent elastic constants, E_1, E_2, ν_{12}, and G_{12}, are required to describe the in-plane elastic behavior of a lamina.

Stress-Strain Laws and Elastic Constants in the Material Axes

In the above paragraphs we have learned how to predict the elastic constants of a single lamina (ply) or a unidirectional laminate from the constituent material constants. In the structural mechanics of composites we have found it convenient to express the elastic constants in terms of the reduced stiffness coefficients and to discuss the stress-strain relationships in matrix forms. Consider a rectangular element of a single lamina with the sides of the element parallel and perpendicular to the fiber direction (Fig. 3). Let us choose a cartesian set of "materials axes" 1-2-3 with the 1-axis in the fiber direction, the 2-axis perpendicular to the fibers and in the lamina plane, and the 3-axis perpendicular to the lamina plane. We are particularly interested in the behavior of the lamina when it is subjected to stresses acting in its plane (plane stress conditions).

Under plane stress conditions, the stresses in the material axes will be denoted by σ_1, σ_2, and τ_{12} and the associated strains by ε_1, ε_2, and γ_{12}. A scrutiny of the rectangle in Fig. 3 indicates the existence of three planes of material symmetry: O12, O23, and O31. A material possessing three mutually orthogonal planes of symmetry is known as "orthotropic." The stress-strain relationship for an orthotropic material under plane stress conditions can be expressed as [1-3]:

$$\begin{bmatrix} \varepsilon_1 \\ \varepsilon_2 \\ \gamma_{12} \end{bmatrix} = \begin{bmatrix} 1/E_1 & -\nu_{21}/E_2 & 0 \\ -\nu_{12}/E_1 & 1/E_2 & 0 \\ 0 & 0 & 1/G_{12} \end{bmatrix} \begin{bmatrix} \sigma_1 \\ \sigma_2 \\ \tau_{12} \end{bmatrix} \quad \text{(Eq 7)}$$

where E_1 and E_2 = Young's moduli in the 1- and 2-directions, respectively; ν_{12} = the major Poisson's ratio specifying the contraction in the 2-direction for a tension in the 1-direction; ν_{21} = the minor Poisson's ratio governing the contraction in the 1-direction for a tension in the 2-direction; and G_{12} = the in-plane shear modulus.

For an orthotropic lamina under plane stress conditions, there are four independent material constants, even though five constants appear in Eq 7. This is because of the following symmetry relation:

$$\nu_{12}/E_1 = \nu_{21}/E_2 \quad \text{(Eq 8)}$$

In many cases, it is more convenient to use the inverse form of Eq 7:

$$\begin{bmatrix} \sigma_1 \\ \sigma_2 \\ \tau_{12} \end{bmatrix} = \begin{bmatrix} Q_{11} & Q_{12} & 0 \\ Q_{12} & Q_{22} & 0 \\ 0 & 0 & Q_{66} \end{bmatrix} \begin{bmatrix} \varepsilon_1 \\ \varepsilon_2 \\ \gamma_{12} \end{bmatrix} \quad \text{(Eq 9)}$$

where the reduced stiffness coefficients, Q_{ij}'s, are given by

$$\begin{aligned}
Q_{11} &= E_1/(1 - \nu_{12}\nu_{21}) \\
Q_{22} &= E_2/(1 - \nu_{12}\nu_{21}) \\
Q_{12} &= \nu_{21}E_1/(1 - \nu_{12}\nu_{21}) \\
Q_{66} &= G_{12} \quad \text{(Eq 10)}
\end{aligned}$$

It may be noted that Eq 7 through 10 apply not only to a single ply, but also to a unidirectional laminate in which the fiber direction is the same in all of the plies. This set of four independent stiffness coefficients represents an alternative to the characterization of the elastic response of a lamina.

Stress-Strain Laws for Single Lamina in the Structural Axes; "Off-Axis" Laminates

In many laminates, individual plies may be composed of fibers that make some prescribed angle θ with respect to a stressing direction (e.g., x-direction in Fig. 4). It is of interest to understand the stress-strain relationship and the elastic constants of such an "off-axis" lamina. Once the elastic constants of various plies (both on-axis and off-axis) are obtained, we may use them to predict the overall response of a laminate using the classical lamination theory.

Consider Fig. 4, where the angle θ is measured from the x-axis (one of the structural axes) to the 1-axis (in the fiber axial direction; one of the material axes). The angle θ is considered positive when measured counterclockwise from the positive x-axis if the z-axis is normal to the lamina plane in the outward direction (pointing to the reader). The y-axis is perpendicular to the x-axis and in the plane of the lamina. The calculations in the lamination theory are normally made using the structural axes (x-y axes), so it is necessary to transform the stress-strain law from the material axes to the structural axes. The stresses

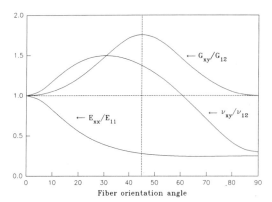

Fig. 5 Variation of calculated elastic constants of a continuous fiber lamina with the fiber orientation angle θ

in the structural axes (σ_x, σ_y, and τ_{xy}) can be related to those in the material axes (σ_1, σ_2, and τ_{12}) by the following transformation equations [1-3]:

$$
\begin{bmatrix} \sigma_x \\ \sigma_y \\ \tau_{xy} \end{bmatrix} = \begin{bmatrix} m^2 & n^2 & -2mn \\ n^2 & m^2 & 2mn \\ mn & -mn & m^2-n^2 \end{bmatrix} \begin{bmatrix} \sigma_1 \\ \sigma_2 \\ \sigma_3 \end{bmatrix} \quad \text{(Eq 11)}
$$

where $m = \cos\theta$ and $n = \sin\theta$. However, the strains in the material axes are related to those in the structural axes (ε_x, ε_y, and γ_{xy}) by

$$
\begin{bmatrix} \varepsilon_1 \\ \varepsilon_2 \\ \gamma_{12} \end{bmatrix} = \begin{bmatrix} m^2 & n^2 & mn \\ n^2 & m^2 & -mn \\ -2mn & 2mn & m^2-n^2 \end{bmatrix} \begin{bmatrix} \varepsilon_x \\ \varepsilon_y \\ \gamma_{xy} \end{bmatrix} \quad \text{(Eq 12)}
$$

By combining Eq 9, 11, and 12, we can obtain the stress-strain law in the structural axes:

$$
\begin{bmatrix} \sigma_x \\ \sigma_y \\ \tau_{xy} \end{bmatrix} = \begin{bmatrix} Q'_{11} & Q'_{12} & Q'_{16} \\ Q'_{12} & Q'_{22} & Q'_{26} \\ Q'_{16} & Q'_{26} & Q'_{66} \end{bmatrix} \begin{bmatrix} \varepsilon_x \\ \varepsilon_y \\ \gamma_{xy} \end{bmatrix} \quad \text{(Eq 13)}
$$

where the Q'_{xy}'s are related to the Q_{xy}'s by the following equations [2]:

$$
\begin{bmatrix} Q'_{11} \\ Q'_{12} \\ Q'_{22} \\ Q'_{16} \\ Q'_{26} \\ Q'_{66} \end{bmatrix} = \begin{bmatrix} m^4 & 2m^2n^2 & n^4 & 4m^2n^2 \\ m^2n^2 & m^4+n^4 & m^2n^2 & -4m^2n^2 \\ n^4 & 2m^2n^2 & m^4 & 4m^2n^2 \\ m^3n & -mn(m^2-n^2) & -mn^3 & -2mn(m^2-n^2) \\ mn^3 & mn(m^2-n^2) & -m^3n & 2mn(m^2-n^2) \\ m^2n^2 & -2m^2n^2 & m^2n^2 & (m^2-n^2)^2 \end{bmatrix}
$$

$$
\times \begin{bmatrix} Q_{11} \\ Q_{12} \\ Q_{22} \\ Q_{66} \end{bmatrix} \quad \text{(Eq 14)}
$$

From Eq 14, it appears that there are six elastic constants that govern the stress-strain behavior of an off-axis lamina (or "angle-ply"). However, a closer examination of this equation would indicate that Q'_{16} and Q'_{26} are linear combinations of the four basic reduced stiffness coefficients (or elastic constants), Q_{11}, Q_{12}, Q_{22}, and Q_{66}, and therefore are not independent. Further, elements in the $[Q'_{ij}]$ matrix are expressed in terms of the properties in the material axes (Q_{11}, Q_{12}, Q_{22}, and Q_{66}, or equivalently, E_1, E_2, G_{12}, and v_{12}), which can be either experimentally determined or predicted approximately from the constituent properties using Eq 1 to 6.

Although the above discussion has been related to a single lamina, it is equally valid for a laminate in which the fiber direction is the same in all laminas (plies). A unidirectional laminate in which the fiber direction makes a nonzero angle with the structural axis (x-axis) is known as an "off-axis" laminate. The elastic constants of an "angle-ply" lamina or an off-axis laminate can be calculated from the following equations:

$$
\frac{1}{E_{xx}} = \frac{m^4}{E_1} + \frac{n^4}{E_2} + \left(\frac{1}{G_{12}} - 2\frac{v_{12}}{E_1} \right) m^2 n^2 \quad \text{(Eq 15)}
$$

$$
\frac{1}{E_{yy}} = \frac{n^4}{E_1} + \frac{m^4}{E_2} + \left(\frac{1}{G_{12}} - \frac{2v_{12}}{E_1} \right) m^2 n^2 \quad \text{(Eq 16)}
$$

$$
\frac{1}{G_{xy}} = \frac{1}{E_1} + \frac{2v_{12}}{E_1} + \frac{1}{E_2}
$$
$$
- \left(\frac{1}{E_1} + \frac{2v_{12}}{E_1} + \frac{1}{E_2} - \frac{1}{G_{12}} \right)(m^2-n^2)^2 \quad \text{(Eq 17)}
$$

$$
v_{xy} = E_1 \left[\frac{v_{12}}{E_1} - \left(\frac{1}{E_1} + \frac{2v_{12}}{E_1} + \frac{1}{E_2} - \frac{1}{G_{12}} \right) m^2 n^2 \right] \quad \text{(Eq 18)}
$$

$$
v_{yx} = \frac{E_{yy}}{E_{xx}} v_{xy} \quad \text{(Eq 19)}
$$

Figure 5 shows the variation of E_{xx} as a function of fiber orientation angle θ for an angle-ply lamina. It is of interest to note that $E_{xx} = E_1$ at $\theta = 0°$, and that $E_{xx} = E_2$ at $\theta = 90°$. Depending on the values of the shear modulus G_{12}, E_{xx} can be either greater than E_1 or less than E_2 at some intermediate value of θ.

By combining Eq 14 and the inverse form of Eq 13, we obtain more explicit equations of the elastic stress-strain relations for a thin orthotropic lamina under plane stress conditions ($\sigma_z = \sigma_{xz} = \tau_{yz}$):

$$
\varepsilon_x = \frac{\sigma_x}{E_{xx}} - v_{yx}\frac{\sigma_y}{E_{yy}} - m_x \tau_{xy} \quad \text{(Eq 20)}
$$

$$
\varepsilon_y = -v_{xy}\frac{\sigma_x}{E_{xx}} + \frac{\sigma_y}{E_{yy}} - m_y \tau_{xy} \quad \text{(Eq 21)}
$$

$$\gamma_{xy} = -m_x \sigma_x - m_y \sigma_y + \frac{\tau_{xy}}{G_{xy}} \qquad \text{(Eq 22)}$$

where the coefficients of mutual influence, m_x and m_y are given by:

$$m_x = 2mn\left[\frac{v_{12}}{E_1} + \frac{1}{E_2} - \frac{1}{2G_{12}}\right.$$
$$\left. - m^2\left(\frac{1}{E_1} + \frac{2v_{12}}{E_1} + \frac{1}{E_2} - \frac{1}{G_{12}}\right)\right] \qquad \text{(Eq 23)}$$

$$m_y = 2mn\left[\frac{v_{12}}{E_1} + \frac{1}{E_2} - \frac{1}{2G_{12}}\right.$$
$$\left. - n^2\left(\frac{1}{E_1} + \frac{2v_{12}}{E_1} + \frac{1}{E_2} - \frac{1}{G_{12}}\right)\right] \qquad \text{(Eq 24)}$$

These new elastic constants represent the influence of shear stresses on extensional strains in Eq 20 and 21. Unlike isotropic materials, extensional and shear deformations are coupled in a general orthotropic lamina (i.e., shear stress causes both shear and normal strains, and normal stresses induce both normal and shear strains). The effects of such normal-shear coupling phenomena are demonstrated in Fig. 6. Note that no extension-shear coupling exists for $\theta = 0°$ and $90°$ where both m_x and m_y are zero. The laminas in which the principal material axes (1 and 2) coincide with the structural or loading axes (x and y) are called "specially orthotropic." It may be further noted that both m_x and m_y are functions of the fiber orientation angle θ and exhibit maximum values at an intermediate angle between $\theta = 0°$ and $90°$. This observation has an important implication, in that two neighboring plies having different values of m_x and m_y tend to result in high interlaminar stresses and hence great tendency to delaminate.

Elements in the $[Q']$ matrix can be expressed in terms of five invariant properties of the lamina [5]:

$$Q'_{11} = U_1 + U_2 \cos2\theta + U_3 \cos4\theta$$
$$Q'_{12} = Q'_{21} = U_4 - U_3 \cos4\theta$$
$$Q'_{22} = U_1 - U_2 \cos2\theta + U_3 \cos4\theta$$
$$Q'_{16} = \frac{1}{2} U_2 \sin2\theta + U_3 \sin4\theta$$
$$Q'_{26} = \frac{1}{2} U_2 \sin2\theta - U_3 \sin4\theta$$
$$Q'_{66} = U_5 - U_3 \cos4\theta \qquad \text{(Eq 25)}$$

where U_1 to U_5 represent angle-invariant stiffness properties of a lamina and are given as

$$U_1 = \frac{1}{8}(3Q_{11} + 3Q_{22} + 2Q_{12} + 4Q_{66})$$
$$U_2 = \frac{1}{2}(Q_{11} - Q_{22})$$
$$U_3 = \frac{1}{8}(Q_{11} + Q_{22} - 2Q_{12} - 4Q_{66})$$

$$U_4 = \frac{1}{8}(Q_{11} + Q_{22} + 6Q_{12} - 4Q_{66})$$
$$U_5 = \frac{1}{2}(U_1 - U_4) \qquad \text{(Eq 26)}$$

The equations are very useful in computing the reduced stiffness coefficients $[Q'_{ij}]$ for an angle-ply lamina.

Example 1. Determine the elements in the reduced stiffness coefficient matrix for an angle-ply lamina containing 60 vol% of AS-4 carbon fibers in a PEEK matrix. Consider fiber orientation angles of both +45 and −45°. Data: $E_f = 238$ GPa , $v_f = 0.2$, $E_m = 3.7$ GPa, and $v_m = 0.35$.

Solution: The first step is to calculate various elastic constants for a lamina of $\theta = 0°$ using Eq 1 to 7.

$$E_{11} = V_f E_f + V_m E_m$$
$$= 0.6 \times 238 + 0.4 \times 3.7 = 145.5 \text{ GPa}$$

$$E_{22} = \frac{E_f E_m}{E_f V_m + E_m V_f}$$
$$= \frac{238 \times 3.7}{238 \times 0.4 + 3.7 \times 0.6} = 9.04 \text{ GPa}$$

$$v_{12} = v_f V_f + v_m V_m = 0.2 \times 0.6 + 0.35 \times 0.4 = 0.26$$

$$v_{21} = \frac{E_{22}}{E_{11}} v_{12} = 0.016$$

Assuming isotropic relationships for both the fiber and the matrix, we estimate

$$G_m = \frac{E_m}{2(1 + v_m)} = \frac{3.7}{2(1 + 0.35)} = 1.37 \text{ GPa}$$

$$G_f = \frac{E_f}{2(1 + v_f)} = \frac{238}{2(1 + 0.2)} = 99.2 \text{ GPa}$$

(Shear modulus values of thin fibers are difficult to measure.) Therefore, we have

$$G_{12} = \frac{G_f G_m}{G_f V_m + G_m V_f}$$
$$= \frac{99.2 \times 1.37}{99.2 \times 0.4 + 1.37 \times 0.6} = 3.36 \text{ GPa}$$

The second step is to calculate Q_{ij}'s using Eq 10.

$$Q_{11} = \frac{E_{11}}{1 - v_{12}v_{21}} = \frac{145.5}{1 - 0.26 \times 0.016} = 146.11 \text{ GPa}$$

$$Q_{22} = \frac{E_{22}}{1 - v_{12}v_{21}} = \frac{9.04}{1 - 0.26 \times 0.016} = 9.08 \text{ GPa}$$

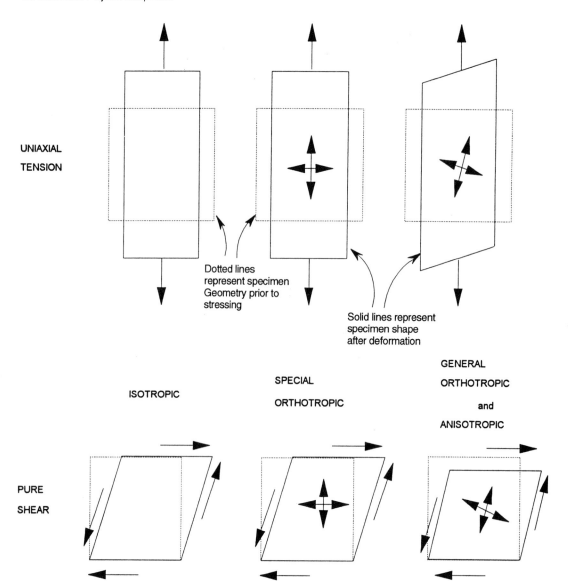

UNIAXIAL
TENSION

Dotted lines
represent specimen
Geometry prior to
stressing

Solid lines represent
specimen shape
after deformation

ISOTROPIC

SPECIAL
ORTHOTROPIC

GENERAL
ORTHOTROPIC
and
ANISOTROPIC

PURE
SHEAR

Fig. 6   Deformation of isotropic, special orthotropic, and anisotropic materials when subjected to either uniaxial tension or simple shear

$$Q_{12} = Q_{21} = \frac{v_{12}E_{22}}{1 - v_{12}v_{21}} = \frac{0.26 \times 9.04}{1 - 0.26 \times 0.016} = 2.36 \text{ GPa}$$

$$Q_{66} = G_{12} = 3.36 \text{ GPa}$$

The third step is to calculate the angle variants, U_i's, using Eq 26.

$$U_1 = \frac{1}{8}(3 \times 146.11 + 3 \times 9.08 + 2$$
$$\times 2.36 + 4 \times 3.36) = 60.47 \text{ GPa}$$

$$U_2 = \frac{1}{2}(146.11 - 9.08) = 68.52 \text{ GPa}$$

$$U_3 = \frac{1}{8}(146.11 + 9.08 - 2$$
$$\times 2.36 - 4 \times 3.36) = 17.13 \text{ GPa}$$

$$U_4 = \frac{1}{8}(146.11 + 9.08 + 6$$
$$\times 2.36 - 4 \times 3.36) = 19.49 \text{ GPa}$$

$$U_5 = \frac{1}{2}(60.47 - 19.49) = 20.49 \text{ GPa}$$

The fourth step is to calculate Q'_{ij} using Eq 25. For a lamina of $\theta = +45°$ we have

$$Q'_{11} = 60.47 + 68.52 \cos 90°$$
$$+ 17.13 \cos 180° = 43.34 \text{ GPa}$$

$$Q'_{12} = 19.49 - 17.13 \cos 180° = 36.62 \text{ GPa}$$

$$Q'_{22} = 60.47 - 68.52 \cos 90°$$
$$+ 17.13 \cos 180° = 43.34 \text{ GPa}$$

$$Q'_{16} = \frac{1}{2} \times 68.52 \cdot \sin 90°$$
$$+ 17.13 \cdot \sin 180° = 34.26 \text{ GPa}$$
$$Q'_{26} = \frac{1}{2} \times 68.52 \cdot \sin 90°$$
$$- 17.13 \cdot \sin 180° = 34.26 \text{ GPa}$$
$$Q'_{66} = 20.49 - 17.13 \cdot \cos 180° = 37.62 \text{ GPa}$$

Similarly, for a lamina of $\theta = -45°$, we have

$$Q'_{11} = 43.34 \text{ GPa}$$
$$Q'_{12} = 36.62 \text{ GPa}$$
$$Q'_{22} = 43.34 \text{ GPa}$$
$$Q'_{16} = -34.26 \text{ GPa}$$
$$Q'_{26} = -34.26 \text{ GPa}$$
$$Q'_{66} = 37.62 \text{ GPa}$$

Therefore, we obtain

$$[Q'_{ij}]_{0°} = [Q_{ij}] = \begin{bmatrix} 146.11 & 2.36 & 0 \\ 2.36 & 9.08 & 0 \\ 0 & 0 & 3.36 \end{bmatrix} \text{GPa}$$

$$[Q'_{ij}]_{45°} = \begin{bmatrix} 43.34 & 36.62 & 34.26 \\ 36.62 & 43.34 & 34.26 \\ 34.26 & 34.26 & 37.62 \end{bmatrix} \text{GPa}$$

$$[Q'_{ij}]_{-45°} = \begin{bmatrix} 43.34 & 36.62 & -34.26 \\ 36.62 & 43.34 & -34.26 \\ -34.26 & -34.26 & 37.62 \end{bmatrix} \text{GPa}$$

Laminates and the Lamination Theory

In the previous sections we learned several important concepts about the design of composite properties. We now know how to estimate the elastic constants of a unidirectional composite (a lamina, or a ply) from the properties of the constituents (fiber and matrix). We also understand how to convert the elastic constants into the reduced stiffness coefficients (Q_{ij}) and how to conveniently express the stress-strain relationships in a matrix form for a unidirectional lamina (or a laminate with all the fibers parallel to each other) in the natural material axes ($\theta = 0°$ or $90°$). For an angle-ply lamina or an off-axis laminate in which all the fibers make an angle θ ($0 < \theta < 90°$ or $-90° < \theta < 0$) with reference to one of the structural axes (loading directions), stress-strain laws and elastic constants (e.g., Q'_{ij}) have been developed and expressed in a proper form using matrix operations. The next logical step is to develop a methodology with which one can predict the properties of a general laminate containing laminas of various fiber orientations from the constituent lamina properties. The classical lamination theory provides such a methodology.

Examples of Commonly Used Laminates

Before embarking on an introduction to classical lamination theory, we should familiarize ourselves with a few special types of laminates and the standard lamination code.

- *Unidirectional laminates:* Fiber orientation angle is the same in all laminas (e.g., in a unidirectional $0°$ laminate, $\theta = 0°$ in all laminas).
- *Angle-ply laminates:* Fiber orientation angles in alternate layers follow the sequence [.../ θ/–θ/θ/–θ/...], where $\theta \neq 0$ or $90°$.
- *Cross-ply laminates:* Fiber orientation angles in alternate layers are [.../$0°$/$90°$/$0°$/$90°$/...].
- *Symmetric laminates:* The lamina fiber orientation is symmetrical about the center line of the laminate. For each lamina above the midplane, there is an identical lamina (identical in material, thickness, and fiber angle) at an equal distance below the midplane. One example is a stacking sequence of [0/+45/90/90/+45/0] = [0/+45/90]s, where the subscript indicates symmetry about the midplane.
- *Balanced laminates:* For every lamina of +θ orientation, there is an identical (equal in thickness and material) lamina of –θ orientation. The locations of these two laminas are arbitrary, e.g., [0/+60/–60/+60/–60/0]. It should be noted that a balanced laminate is not necessarily a symmetric laminate, or vice versa. However, one can have a laminate that is both symmetric and balanced, e.g., [0/+60/–60/–60/+60/0].
- *Quasi-isotropic laminates:* These laminates are made of three or more laminas of identical thickness and materials, with equal angles between each adjacent lamina. Thus, if the total number of laminas is n, the angles of the laminas are at increments of π/n. The resulting laminates typically exhibit an isotropic elastic behavior in the xy plane [6]. Examples include [+60/0/–60] and [90/+45/0/–45/90]. A widely used isotropic and symmetrical stacking sequence is [0/±45/90]s.
- *Hybrid laminates:* Laminas of different or identical orientation angles can be made up of different fiber-matrix combinations. For example, [0_K/0_K/+45_C/–45_C/90_G/90_G/–45_C/+45_C/0_K/0_K] = [0_{2K}/(±45)$_C$/90_G]s is a symmetrical, balanced hybrid laminate comprising Kevlar-, carbon-, and glass-fiber-based laminas.

The reader may also find the following examples of lamination code useful:

- [0/+45/90/+45/0] = [0/+45/90̄]s, where a bar on top of the 90 indicates that the plane of symmetry passes midway through this 90° lamina.

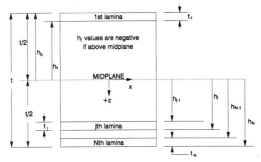

Fig. 7 Laminate configuration convention used in this book. Source: Ref 6

Fig. 8 In-plane bending and twisting loads applied on a laminate. Source: Ref 6

- [0/0/0/90/+60/–60/+60/90/0/0/0] = [0₃/90/±60]ₛ, where three adjacent +60° and –60° laminas are grouped as ±60.

  [0/0/0/90/+60/−60/+60/90/0/0/0] = $[0_3/90/\pm60]_s$, where three adjacent +60° and –60° laminas are grouped as ±60.

- $[0/45/-45/45/45/-45/-45/90/90/-45/+45/-45/+45/-45/+45/0] = [0/(\pm45)_3/90]_s$

- $[\theta/ -\theta/\theta/-\theta/ -\theta/\theta/-\theta/\theta] = [\theta/-\theta]_{2s} = [\pm\theta]_{2s}$ represents a symmetric angle-ply laminate where two adjacent ±θ plies on each side of the plane of symmetry are denoted by the subscript "2s."

The Lamination Theory [2,4,6,7]

The classical lamination theory is useful in calculating stresses and strains in each lamina of a thin laminate and in estimating the overall elastic constants of this laminate. This theory is based on the following assumptions:

- The laminate is thin and wide.
- A perfect interply bond exists between neighboring laminas.
- The strain distribution across the thickness (z-direction) is linear.
- All laminas are macroscopically homogeneous and behave in a linearly elastic manner.

In what follows, a cartesian coordinate system will be established in such a way that the geometric midplane of the laminate contains the x- and y- axes, and the z-axis is normal to this midplane. Assume that a laminate with a total thickness t is constructed from N laminas, of which the individual thicknesses are t_1, t_2, t_3, and so on (Fig. 7). With this coordinate system in place, we are ready to follow a typical sequence of steps involved in lamination theory:

1. Calculate the stiffness matrices for the laminate, given the stiffness matrices $[Q'_{ij}]$ and $[Q_{ij}]$.
2. Calculate the midplane strains and curvatures for the laminate caused by a given set of applied forces and moments.
3. Calculate the in-plane strains ε_{xx}, ε_{yy}, and γ_{xy} for each lamina.
4. Calculate the in-plane stresses σ_{xx}, σ_{yy}, and τ_{xy} in each lamina.
5. Calculate the more explicit elastic constants E_{xx}, E_{yy}, v_{xy}, v'_{yx}, and G_{xy} of the laminate.
6. Estimate the laminate strength by invoking a proper failure criterion (Chapter 6).

According to assumption 3, laminate strains are linearly related to the distance from the midplane as

$$\varepsilon_{xx} = \varepsilon_{xx}^0 + zK_{xx}$$
$$\varepsilon_{yy} = \varepsilon_{yy}^0 + zK_{yy}$$
$$\gamma_{xy} = \gamma_{xy}^0 + zK_{xy}$$

where ε_{xx}^0 and ε_{yy}^0 = midplane normal strains, γ_{xy}^0 = the midplane shear strain, K_{xx} and K_{yy} = bending curvatures, K_{xy} = the twisting curvature in the laminate, and z = the distance from the midplane in the thickness direction. Consider the in-plane, bending, and twisting loads exerted on the laminate, as shown in Fig. 8. One can easily show that the applied force and moment resultants on a laminate can be related to the midplane strains and curvatures by the following equations [2,4,6,7]:

$$N_{xx} = A_{11}\varepsilon_{xx}^0 + A_{12}\varepsilon_{yy}^0 + A_{16}\gamma_{xy}^0$$
$$+ B_{11}k_{xx} + B_{12}k_{yy} + B_{16}k_{xy}$$
$$N_{yy} = A_{12}\varepsilon_{xx}^0 + A_{22}\varepsilon_{yy}^0 + A_{26}\gamma_{xy}^0$$
$$+ B_{12}k_{xx} + B_{22}k_{yy} + B_{26}k_{xy}$$
$$N_{xy} = A_{16}\varepsilon_{xx}^0 + A_{26}\varepsilon_{yy}^0 + A_{66}\gamma_{xy}^0$$
$$+ B_{16}k_{xx} + B_{26}k_{yy} + B_{66}k_{xy}$$
$$M_{xx} = B_{11}\varepsilon_{xx}^0 + B_{12}\varepsilon_{yy}^0 + B_{16}\gamma_{xy}^0$$
$$+ D_{11}k_{xx} + D_{12}k_{yy} + D_{16}k_{xy}$$
$$M_{yy} = B_{12}\varepsilon_{xx}^0 + B_{22}\varepsilon_{yy}^0 + B_{26}\gamma_{xy}^0$$
$$+ D_{12}k_{xx} + D_{22}k_{yy} + D_{26}k_{xy}$$
$$M_{xy} = B_{16}\varepsilon_{xx}^0 + B_{26}\varepsilon_{yy}^0 + B_{66}\gamma_{xy}^0$$
$$+ D_{16}k_{xx} + D_{26}k_{yy} + D_{66}k_{xy} \qquad \text{(Eq 27a)}$$

In matrix notation,

$$\begin{bmatrix} N_{xx} \\ N_{yy} \\ N_{xy} \end{bmatrix} = [A] \begin{bmatrix} \varepsilon_{xx}^0 \\ \varepsilon_{yy}^0 \\ \gamma_{xy}^0 \end{bmatrix} + [B] \begin{bmatrix} k_{xx} \\ k_{yy} \\ k_{xy} \end{bmatrix} \qquad \text{(Eq 27b)}$$

and

$$\begin{bmatrix} M_{xx} \\ M_{yy} \\ M_{xy} \end{bmatrix} = [B] \begin{bmatrix} \varepsilon_{xx}^0 \\ \varepsilon_{yy}^0 \\ \gamma_{xy}^0 \end{bmatrix} + [D] \begin{bmatrix} k_{xx} \\ k_{yy} \\ k_{xy} \end{bmatrix} \qquad \text{(Eq 28)}$$

where N_{xx} = normal force resultant in the x-direction (per unit width), N_{yy} = normal force resultant in the y-direction (per unit width), N_{xy} = shear force resultant (per unit width), M_{xx} = bending moment resultant in the yz plane (per unit width), M_{yy} = bending moment resultant in the xz plane (per unit width), and M_{xy} = twisting moment resultant (per unit width).
$[A]$ = extensional stiffness matrix for the laminate (N/m or lb/in.):

$$[A] = \begin{bmatrix} A_{11} & A_{12} & A_{16} \\ A_{12} & A_{22} & A_{26} \\ A_{16} & A_{26} & A_{66} \end{bmatrix} \qquad \text{(Eq 29)}$$

$[B]$ = coupling stiffness matrix for the laminate (N or lb):

$$[B] = \begin{bmatrix} B_{11} & B_{12} & B_{16} \\ B_{12} & B_{22} & B_{26} \\ B_{16} & B_{26} & B_{66} \end{bmatrix} \qquad \text{(Eq 30)}$$

$[D]$ = bending stiffness matrix for the laminate (N-m or lb-in.):

$$[D] = \begin{bmatrix} D_{11} & D_{12} & D_{16} \\ D_{12} & D_{22} & D_{26} \\ D_{16} & D_{26} & D_{66} \end{bmatrix} \qquad \text{(Eq 31)}$$

The elements in $[A]$, $[B]$, and $[D]$ matrices are calculated from

$$A_{mn} = \sum_{j=1}^{N} (Q'_{mn})_j (h_j - h_{j-1}) \qquad \text{(Eq 32)}$$

$$B_{mn} = \frac{1}{2} \sum_{j=1}^{N} (Q'_{mn})_j (h_j^2 - h_{j-1}^2) \qquad \text{(Eq 33)}$$

$$D_{mn} = \frac{1}{3} \sum_{j=1}^{N} (Q'_{mn})_j (h_j^3 - h_{j-1}^3) \qquad \text{(Eq 34)}$$

where N = total number of laminas in the laminate, $(Q'_{mn})_j$ = element in the $[Q']$ matrix of the jth lamina, h_{j-1} = distance from the midplane to the top of the jth lamina, and h_j = distance from the midplane to the bottom of the jth lamina. It is important to note that, in the present coordinate system (Fig. 7), h is taken to be positive below the midplane and negative above the midplane.

The characteristics of the stiffness matrices $[A]$, $[B]$, and $[D]$ are summarized as follows [6]:

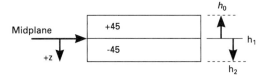

Fig. 9 Diagram for solving part (a) of Example 2

- $[A]$, $[B]$, and $[D]$ are functions of the elastic properties of each lamina and its location with respect to the midplane of the laminate.

- In the laminates where $[B]$ is a nonzero matrix, a normal force resultant (e.g., N_{xx}) will produce extension and shear deformations as well as bending-twisting curvatures. A bending moment (e.g., M_{xx}) will have the same result. Such "extension-bending coupling" is unique in laminated materials, regardless of whether the layers are isotropic or orthotropic. No extension-bending coupling exists for symmetric laminates, which are characterized by $[B] = 0$.

- For a balanced laminate, $A_{16} = A_{26} = 0$, and no normal stress/shear strain coupling exists. For a balanced and symmetric laminate, $[B] = A_{16} = A_{26} = 0$.

- For a special class of laminates, where for every lamina of $+\theta$ orientation above the midplane there is an identical lamina (in thickness and material) of $-\theta$ orientation at the same distance below the midplane, we have $D_{16} = D_{26} = 0$. No bending moment/twisting curvature coupling exists in these laminates. One simple example is $[0/+60/-60/+60/-60/0]$.

Example 2. Consider the same type of material as in Example 1, 60 vol% AS-4 fibers in a PEEK matrix. Determine $[A]$, $[B]$, and $[D]$ matrices for: (a) a $[+45/-45]$ angle-ply laminate and (b) a $[+45/-45]_s$ symmetric laminate. (Each lamina is 6 mm thick.)
Solution: From Example 1, we have obtained

$$[Q']_{45°} = \begin{bmatrix} 43.34 & 36.62 & 34.26 \\ 36.62 & 43.34 & 34.26 \\ 34.26 & 35.26 & 37.62 \end{bmatrix} \text{GPa}$$

$$[Q']_{-45°} = \begin{bmatrix} 43.34 & 36.63 & -34.26 \\ 36.62 & 43.34 & -34.26 \\ -34.26 & -34.26 & 37.62 \end{bmatrix} \text{GPa}$$

To solve part (a), draw a diagram (Fig. 9), where $h_0 = -6 \times 10^{-3}$ m, $h_1 = 0$, and $h_2 = +6 \times 10^{-3}$ m. In this laminate,

$$(Q'_{mn})_1 = (Q'_{mn})_{45°}$$
$$(Q'_{mn})_2 = (Q'_{mn})_{-45°}$$
$$A_{mn} = (Q'_{mn})_1 (h_1 - h_0) + (Q'_{mn})_2 (h_2 - h_1)$$
$$= [(Q'_{mn})_{+45°} + (Q'_{mn})_{-45°}] \cdot 6 \times 10^{-3}$$

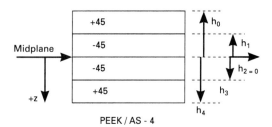

Fig. 10 Diagram for solving part (b) of Example 2

$$B_{mn} = \frac{1}{2}[(Q'_{mn})_{+45°} \cdot (h_1^2 - h_0^2)$$
$$+ (Q'_{mn})_{-45°} \cdot (h_2^2 - h_1^2)]$$
$$= 18 \times 10^{-6}[-(Q'_{mn})_{+45°} + (Q'_{mn})_{-45°}]$$
$$D_{mn} = 72 \times 10^{-9}[-(Q'_{mn})_{+45°} + (Q'_{mn})_{-45°}]$$

Substituting for various (Q'_{mn}) values, we obtain

$$A = \begin{bmatrix} 520.08 & 439.44 & 0 \\ 439.44 & 520.08 & 0 \\ 0 & 0 & 451.44 \end{bmatrix} \times 10^6 \text{ N/m}$$

$$B = \begin{bmatrix} 0 & 0 & -1233.4 \\ 0 & 0 & -1233.4 \\ -1233.4 & -1233.4 & 0 \end{bmatrix} \times 10^3 \text{N}$$

$$D = \begin{bmatrix} 6240.96 & 5273.28 & 0 \\ 5273.28 & 6240.96 & 0 \\ 0 & 0 & 5417.28 \end{bmatrix} \text{N-m}$$

The above data indicate that $A_{16} = A_{26} = 0$ for a [+45/−45] angle-ply laminate, because it is a balanced laminate. This angle-ply laminate also happens to exhibit $D_{12} = D_{26} = 0$.

To solve part (b), draw a diagram (Fig. 10), where $h_3 = -h_1 = 6 \times 10^{-3}$ m, $h_4 = -h_0 = 12 \times 10^{-3}$ m, and $h_2 = 0$. Also, $(Q'_{mn})_4 = (Q'_{mn})_1 = (Q'_{mn})_{+45°}$ and $(Q'_{mn})_3 = (Q'_{mn})_2 = (Q'_{mn})_{-450°}$. Therefore, after simple manipulations, we have

$$A_{mn} = 12 \times 10^{-3} \cdot [(Q'_{mn})_{+45°} + (Q'_{mn})_{-45°}]$$
$$B_{mn} = 0 \text{ (symmetric)}$$
$$D_{mn} = 1008 \times 10^{-9}(Q'_{mn})_{+45°} + 144 \times 10^{-9}(Q'_{mn})_{-45°}$$

Substituting for various Q'_{mn} values, we obtain

$$[A] = \begin{bmatrix} 1040.16 & 878.88 & 0 \\ 878.88 & 1040.16 & 0 \\ 0 & 0 & 902.88 \end{bmatrix} \times 10^6 \text{ N/m}$$

$$[B] = [0]$$

$$[D] = \begin{bmatrix} 49.93 & 42.19 & 29.60 \\ 42.19 & 49.93 & 29.60 \\ 29.60 & 29.60 & 43.34 \end{bmatrix} \times 10^3 \text{ N}$$

This is a symmetric, balanced laminate; therefore, $[B] = [0]$ and $A_{16} = A_{26} = 0$.

Earlier in this chapter we discussed the first step in applying the lamination theory. This step completed the calculation of stiffness matrices for the laminate. The elements of these matrices will be used in the subsequent steps. The next step is to calculate the midplane strains and curvatures for the laminate once the applied forces and moments are known. This can be accomplished by inverting Eq 27 and 28 to give:

$$\begin{bmatrix} \varepsilon_{xx}^0 \\ \varepsilon_{yy}^0 \\ \gamma_{xy}^0 \end{bmatrix} = [a]\begin{bmatrix} N_{xx} \\ N_{yy} \\ N_{xy} \end{bmatrix} + [b]\begin{bmatrix} M_{xx} \\ M_{yy} \\ M_{xy} \end{bmatrix} \qquad \text{(Eq 35)}$$

and

$$\begin{bmatrix} k_{xx} \\ k_{yy} \\ k_{xy} \end{bmatrix} = [c]\begin{bmatrix} N_{xx} \\ N_{yy} \\ N_{xy} \end{bmatrix} + [d]\begin{bmatrix} M_{xx} \\ M_{yy} \\ M_{xy} \end{bmatrix} \qquad \text{(Eq 36)}$$

where

$$[a] = [A^{-1}] + [A^{-1}][B][D^{*-1}][B][A^{-1}]$$
$$[b] = -[A^{-1}][B](D^*)^{-1}]$$
$$[c] = -[(D^*)^{-1}][B][A^{-1}] = [B']^T$$
$$[D^*] = [D] - [B][A^{-1}][B]$$
$$[d] = [(D^*)^{-1}] \qquad \text{(Eq 37)}$$

It is of interest to note that for a symmetric laminate, $[B] = [0]$; therefore, $[a] = [A^{-1}]$, $[b] = [c] = [0]$, and $[d] = [D^{-1}]$. In this case, equations for midplane strains and curvatures become

$$\begin{bmatrix} \varepsilon_{xx}^0 \\ \varepsilon_{yy}^0 \\ \gamma_{xy}^0 \end{bmatrix} = [A^{-1}]\begin{bmatrix} N_{xx} \\ N_{yy} \\ N_{xy} \end{bmatrix} \qquad \text{(Eq 38)}$$

and

$$\begin{bmatrix} k_{xx} \\ k_{yy} \\ k_{xy} \end{bmatrix} = [D^{-1}]\begin{bmatrix} M_{xx} \\ M_{yy} \\ M_{xy} \end{bmatrix} \qquad \text{(Eq 39)}$$

This shows that, for a symmetric laminate, in-plane forces cause only in-plane strains (no curvatures), and bending and twisting moments cause only curvatures (no in-plane strains) [6].

The third step is to use the value of the midplane strains and curvatures for the laminate to calculate the *strains* of the individual laminas. This is done on the basis of the linear relationship of strain in the thickness direction (assumption 3):

$$\begin{bmatrix} \varepsilon_{xx} \\ \varepsilon_{yy} \\ \gamma_{xy} \end{bmatrix}_j = \begin{bmatrix} \varepsilon_{xx}^0 \\ \varepsilon_{yy}^0 \\ \gamma_{xy}^0 \end{bmatrix} + z_j \begin{bmatrix} k_{xx} \\ k_{yy} \\ k_{xy} \end{bmatrix} \qquad \text{(Eq 40)}$$

where z_j = the distance from the laminate midplane to the midplane of the jth lamina.

The fourth step is to calculate stresses in the jth lamina using its stiffness matrix. Thus,

$$\begin{bmatrix} \sigma_{xx} \\ \sigma_{yy} \\ \tau_{xy} \end{bmatrix}_j = [Q'_{mn}]_j \begin{bmatrix} \varepsilon_{xx} \\ \varepsilon_{yy} \\ \gamma_{xy} \end{bmatrix}_j$$

$$= [Q'_{mn}]_j \begin{bmatrix} \varepsilon_{xx}^0 \\ \varepsilon_{yy}^0 \\ \gamma_{xy}^0 \end{bmatrix} + z_j [Q'_{mn}]_j \begin{bmatrix} k_{xx} \\ k_{yy} \\ k_{xy} \end{bmatrix} \qquad \text{(Eq 41)}$$

In some laminates, the physical significance of the extensional stiffness coefficients A_{ij} and the bending stiffness coefficients can become more transparent if we further carry out the fifth step. Consider a balanced symmetric laminate where $[B] = 0$ and $A_{16} = A_{26} = 0$. The extensional stiffness matrix is

$$[A] = \begin{bmatrix} A_{11} & A_{12} & 0 \\ A_{12} & A_{22} & 0 \\ 0 & 0 & A_{66} \end{bmatrix}$$

The inverse of the $[A]$ matrix is

$$[A^{-1}] = \frac{1}{A_{11}A_{22} - A_{12}^2} \begin{bmatrix} A_{22} & -A_{12} & 0 \\ -A_{12} & A_{11} & 0 \\ 0 & 0 & \dfrac{(A_{11}A_{22} - A_{12}^2)}{A_{66}} \end{bmatrix}$$

Therefore, Eq 38 gives

$$\begin{bmatrix} \varepsilon_{xx}^0 \\ \varepsilon_{yy}^0 \\ \gamma_{xy}^0 \end{bmatrix}$$

$$= \frac{1}{A_{11}A_{22} - A_{12}^2} \begin{bmatrix} A_{22} & -A_{12} & 0 \\ -A_{12} & A_{11} & 0 \\ 0 & 0 & \dfrac{(A_{11}A_{22} - A_{12}^2)}{A_{66}} \end{bmatrix} \begin{bmatrix} N_{xx} \\ N_{yy} \\ N_{xy} \end{bmatrix}$$

In the case of a uniaxial tensile load applied in the x-direction, $N_{xx} = h\sigma_{xx}$ (per unit width), $N_{yy} = 0$, and $N_{xy} = 0$. Thus we have

$$\varepsilon_{xx}^0 = \frac{A_{22}}{A_{11}A_{22} - A_{12}^2} h\sigma_{xx}$$

$$\varepsilon_{yy}^0 = -\frac{A_{12}}{A_{11}A_{22} - A_{12}^2} h\sigma_{xx}$$

$$\gamma_{xy}^0 = 0$$

which gives

$$E_{xx} = \frac{\sigma_{xx}}{\varepsilon_{xx}^0} = \frac{A_{11}A_{22} - A_{12}^2}{hA_{22}} \qquad \text{(Eq 42)}$$

$$\nu_{xy} = -\frac{\varepsilon_{yy}^0}{\varepsilon_{xx}^0} = \frac{A_{12}}{A_{22}} \qquad \text{(Eq 43)}$$

In turn, applying N_{xx} and N_{xy} separately, we can determine that

$$E_{xy} = \frac{A_{11}A_{22} - A_{12}^2}{hA_{11}} \qquad \text{(Eq 44)}$$

$$\nu_{yx} = \frac{A_{12}}{A_{11}} \qquad \text{(Eq 45)}$$

$$G_{xy} = \frac{A_{66}}{h} \qquad \text{(Eq 46)}$$

Equations 43 to 46 clearly suggest that the extensional stiffness coefficients, A_{ij}'s, represent nothing more than an alternative set of elastic constants of a laminate. Because the elastic constants E_{xx}, E_{xy}, ν_{xy}, and G_{xy} can be measured experimentally, these equations provide a way to check the accuracy of the classical lamination theory.

Now consider the bending behavior of a balanced symmetric laminate beam where $[B] = 0$. We have

$$[D] = \begin{bmatrix} D_{11} & D_{12} & D_{16} \\ D_{12} & D_{22} & D_{26} \\ D_{16} & D_{26} & D_{66} \end{bmatrix}$$

which gives

$$[D^{-1}] = \frac{1}{D^\circ} \begin{bmatrix} D_{11}^0 & D_{12}^0 & D_{16}^0 \\ D_{12}^0 & D_{22}^0 & D_{26}^0 \\ D_{16}^0 & D_{26}^0 & D_{66}^0 \end{bmatrix}$$

where

$$D_0 = D_{11}(D_{22}D_{66} - D_{26}^2) - D_{12}(D_{12}D_{66} - D_{16}D_{26})$$
$$+ D_{16}(D_{12}D_{26} - D_{22}D_{16})$$
$$D_{11}^0 = D_{11}(D_{22}D_{66} - D_{26}^2)$$
$$D_{12}^0 = -D_{12}(D_{12}D_{66} - D_{16}D_{26})$$
$$D_{16}^0 = D_{16}(D_{12}D_{26} - D_{22}D_{16})$$

$$D_{22}^0 = D_{22}(D_{11}D_{66} - D_{16}^2)$$
$$D_{26}^0 = -D_{26}(D_{11}D_{26} - D_{12}D_{16})$$
$$D_{66}^0 = D_{66}(D_{11}D_{22} - D_{12}^2)$$

If a bending moment is applied in the yz plane so that M_{xx} is present and $M_{yy} = M_{xy} = 0$, the specimen curvatures can be obtained from Eq 39:

$$k_{xx} = \frac{D_{11}^0}{D_0} M_{xx}$$

$$k_{xy} = \frac{D_{12}^0}{D_0} M_{xx}$$

$$k_{xy} = \frac{D_{16}^0}{D_0} M_{xx} \qquad \text{(Eq 47)}$$

The existence of K_{xy} suggests that, even though no twisting moment is applied, the specimen would tend to twist unless $D_{16}^0 = D_{16}(D_{12}D_{26} - D_{22}D_{16}) = 0$. This is possible only if the balanced symmetric laminate contains fibers in the 0° and 90° directions [6].

Discontinuous Fiber- and Whisker-Reinforced Polymer Composites

The above four sections have discussed the basic methods of estimating or predicting the elastic constants of various laminas and laminates containing continuous fibers as the reinforcement. However, many useful polymer composites are composed of discontinuous fibers (long or short) or whiskers dispersed in resin matrices. The design format for predicting stiffness for discontinuous fiber composites is well developed for two-dimensional structures [8]. By employing a combination of micromechanics (e.g., Halpin-Tsai equations) and macromechanics (e.g., lamination theory and the laminate analogy approach), one can calculate the in-plane stiffness and the Poisson ratio for any two-dimensional fiber orientation distribution [8-15].

Unidirectional Discontinuous Fiber 0° Lamina (Halpin-Tsai Equations)

The Halpin-Tsai equations are simple approximate forms of the generalized self-consistent micromechanics solutions developed by Hill [10]. These equations may only be used for calculating the in-plane stiffness of a single, transversely isotropic sheet having all the fibers unidirectionally aligned in the plane of the sheet. The stiffness values calculated by using these equations agree reasonably well with the experimental values for a variety of reinforcement geometries, including fibers, flakes, and ribbons.

The Halpin-Tsai equations are given collectively by

$$\frac{\overline{P}}{P_m} = \frac{1 + \zeta \eta V_f}{1 - \eta V_f}$$

$$v_{12} = v_f V_f + v_m V_m \text{ (in-plane Poisson ratio)} \quad \text{(Eq 48)}$$

where

$$\eta = \left(\frac{P_f}{P_m} - 1 \right) / \left(\frac{P_f}{P_m} + \zeta \right)$$

$V_f =$ fiber volume fraction
$\zeta_{E_{11}} = 2(l/d)$
$\zeta_{E_{22}} = 2(l/d) = 2$ for $l/d = 1.0$
$\zeta_{G_{12}} = 1$
$P = E_{11}, E_{22}, G_{12}$
$P_f = E_f, G_f$
$P_m = E_m, G_m$
$m =$ matrix
$f =$ fiber

E_{11} and E_{22} are the engineering tensile stiffnesses parallel and perpendicular to the fiber direction, and G_{12} is the in-plane shear stiffness. The fiber aspect ratio is l/d.

Implicit assumptions made in this and all other modulus theories are that:

- Both filler and matrix are homogeneous and linearly elastic.
- Filler and matrix are void-free.
- There is perfect contact between filler and matrix at the interface.
- The filler is perfectly dispersed.

Reliable estimates for the ζ factor are obtained by comparing the Halpin-Tsai equations with the numerical solutions of the micromechanics equations [9,14,15]. For example,

$$\zeta = 2\frac{l}{t} + 40v_f^{10} \qquad \text{for } E_{11}$$

$$\zeta = 2\frac{w}{t} + 40v_f^{10} \qquad \text{for } E_{22}$$

$$\zeta = \left(\frac{w}{t} \right)^{1.732} + 40v_f^{10} \qquad \text{for } G_{12}$$

where l, w, and t are the reinforcement length, width, and thickness, respectively. For a circular fiber, $l = l_f$ and $t = w = d_f$. For a spherical reinforcement, $l = t = w$. The term containing v_f in the expressions for ζ is relatively small up to $v_r = 0.7$; therefore, it can be neglected.

One of the advantages of the Halpin-Tsai equations is that they cover both the particulate reinforced case (fiber aspect ratio = unity; lower bound) as well as the continuous fiber case (fiber aspect ratio = infinity; upper bound). Indeed, one can mathematically express the equation limits as the rule of mixtures for continuous fibers and the modified Keener equation for spherical reinforcement

[8]. To account for the effect of maximum packing, modified Halpin-Tsai equations developed by Nielson [16] may be used.

Elastic Constant Prediction

Many short-fiber composites in reality have three-dimensional fiber orientations. Even thick sheets may have out-of-plane distribution components. No direct and efficient calculation format exists to treat the general three-dimensional problem. Nevertheless, some bounding cases have been examined.

Both the aggregate and laminate analogy approaches can be used to handle special cases of three-dimensional orientation. McGee [17] followed an approach that took second- and fourth-order tensors averaged under a more general three-dimensional orientation distribution. This formulation contains four orientation parameters. Both the two-dimensional planar and the three-dimensional axial (symmetry about one axis) orientation distributions are contained as special cases. Using the laminate analogy approach, Chang [18] has developed a single-parameter approach for two-dimensional orientations and a two-parameter method for three-dimensional axial orientation distributions. Halpin et al. [13] approached the orthogonal three-dimensional problem by modeling a plain, square-woven fabric pierced by a straight yarn perpendicular to the fabric plane. Their results showed that the moduli in the plane of the woven fabric in the three-dimensional case were about 5% lower than the comparable moduli for the two-dimensional material (plain square weave) at 50 vol% loading. This suggests that the three-dimensional weave overcomes the low shear strength between layers of a two-dimensional fabric laminate with only a small sacrifice in laminate in-plane stiffness. The effect of "stitching" on other mechanical properties of composites is discussed in Chapter 7.

Lavengood and Goettler [19] used an approximate averaging technique to generate an expression for the modulus of a structure having three-dimensional random fiber orientation:

$$E_{3D} = \frac{1}{5}E_{11} + \frac{4}{5}E_{22} \quad \text{(3-D random)} \qquad \text{(Eq 49)}$$

where E_{11} and E_{22} are the longitudinal and transverse engineering stiffness constants for a unidirectionally oriented ply (0°) of the same fiber aspect ratio and same fiber volume fraction as the randomly oriented discontinuous fiber composite. At 30 vol% loading, Eq 49 predicts a value about 20% lower than the in-plane stiffness for random two-dimensional orientation. An approximate expression for the latter is

$$E_{2D} = \frac{3}{8}E_{11} + \frac{5}{8}E_{22} \quad \text{(2-D random)} \qquad \text{(Eq 50)}$$

The corresponding shear modulus and the Poisson's ratio in the plane of the lamina are

$$G_{2D} = \frac{1}{8}E_{11} + \frac{1}{4}E_{22} \qquad \text{(Eq 51)}$$

$$v_{2D} = \frac{E_{2D}}{2G_{2D}} - 1 \qquad \text{(Eq 52)}$$

Example 3. Determine the elastic constants of a thin sheet of polycarbonate reinforced with short glass fibers:

$E_f = 82.7$ GPa	$E_m = 2.18$ GPa
$v_f = 0.22$	$v_m = 0.35$
$G_f = 27.6$ GPa	$G_m = 1.03$ GPa
$l/d = 300$	$V_m = 0.8$
$V_f = 0.2$	

Solution: The first step is to calculate Halpin-Tsai equation parameters:

$$\frac{E_f}{E_m} = \frac{82.7}{21.8} = 37.9$$

$$\frac{G_f}{G_m} = \frac{27.6}{1.03} = 26.8$$

Therefore,

$$\eta_L = \frac{(E_f / E_m) - 1}{(E_f / E_m) + 2(l / d)} = 0.058$$

$$\eta_T = \frac{(E_f / E_m) - 1}{(E_f / E_m) + 2} = 0.925$$

$$\eta_G = \frac{(G_f / G_m) - 1}{(G_f / G_m) + 1} = 0.928$$

The second step is to calculate E_{11}, E_{22}, and G_{12} for a unidirectional lamina (0°) containing discontinuous fibers:

$$E_{11} = \frac{1 + 2(l/d)\eta_L V_f}{1 - \eta_L V_f} E_m = 17.51 \text{ GPa}$$

$$E_{22} = \frac{1 + 2\eta_T V_f}{1 - \eta_L V_f} E_m = 3.72 \text{ GPa}$$

$$G_{12} = G_{21} = \frac{1 + \eta_G V_f}{1 - \eta_G V_f} G_m = 1.52 \text{ GPa}$$

The third step is to calculate E_{2D}, G_{2D}, and v_{2D} for random two-dimensional orientation:

$$E_{2D} = \frac{3}{8}E_{11} + \frac{5}{8}E_{22} = 8.89 \text{ GPa}$$

$$G_{2D} = \frac{1}{8}E_{11} + \frac{1}{4}E_{22} = 3.12 \text{ GPa}$$

$$v_{2D} = \frac{E_{2D}}{2 \cdot G_{2D}} - 1 = 0.42$$

Alternatively, we may use a quasi-isotropic laminate to simulate the two-dimensional random orientation sheet. Consider that each lamina in this laminate is composed of unidirectional discontinuous fibers (of the same aspect ratio and volume fraction as this sheet of two-dimensional random orientation) dispersed in the same matrix resin. It has been suggested [8] that only four angles (0°, ±45°, 90°) are sufficient to replicate random in-plane behavior. Therefore, one can consider a balanced, symmetric $[0/\pm45/90]_s$ laminate.

From Eq 10 and 25, we have

$$[Q]_0 = \begin{bmatrix} 17.93 & 1.24 & 0 \\ 1.24 & 3.79 & 0 \\ 0 & 0 & 1.52 \end{bmatrix}$$

$$[Q']_{45°} = \begin{bmatrix} 7.58 & 4.55 & -3.52 \\ 4.55 & 7.58 & -3.52 \\ -3.52 & -3.52 & 4.83 \end{bmatrix}$$

$$[Q']_{-45°} = \begin{bmatrix} 7.58 & 4.55 & 3.52 \\ 4.55 & 7.58 & 3.52 \\ 3.52 & 3.52 & 4.83 \end{bmatrix}$$

$$[Q']_{90°} = \begin{bmatrix} 3.79 & 1.24 & 0 \\ 1.24 & 17.93 & 0 \\ 0 & 0 & 1.52 \end{bmatrix}$$

From Eq 32, we obtain (per unit thickness):

$$[A] = \begin{bmatrix} 9.24 & 2.90 & 0 \\ 2.90 & 9.24 & 0 \\ 0 & 0 & 3.17 \end{bmatrix} \text{GPa}$$

For a balanced symmetric laminate we have, using Eq 42 to 44:

$$E_{xx} = \frac{A_{11}A_{22} - A_{12}^2}{A_{22}} = 8.34 \text{ GPa}$$

$$E_{yy} = \frac{A_{11}A_{22} - A_{12}^2}{A_{11}} = 8.34 \text{ GPa}$$

$$G_{xy} = A_{66} = 3.17 \text{ GPa}$$

$$\nu_{xy} = \frac{A_{12}}{A_{22}} = 0.31$$

These values are somewhat at variance with the values of E_{2D}, G_{2D}, and ν_{2D} obtained in step 3.

Another major class of reinforcement geometry has a rectangular cross section whose width-to-thickness ratio is large. In the continuous form these are commonly referred to as ribbons or tapes, while their discontinuous counterparts are classified as flakes. The elastic constants of the flake- and tape-reinforced composites can also be estimated using their constituent properties [8,15,20,23].

Woven, Braided, and Other Textile Structural Composites

Textile structural preforms are finding increased acceptance in advanced composites. For instance, woven fabric composites provide more balanced properties in the fabric plane than do unidirectional laminas; the bidirectional reinforcement in a single layer of fabric gives rise to excellent impact resistance [24]. Fabric-based materials are relatively easy to handle, and the fabrication cost for most of the two-dimensional preforms is relatively low. However, fabrication of multidirectional preforms, such as three-dimensional braided, three-dimensional woven, or orthogonal nonwoven, angle-interlock constructions, is still a tedious and expensive process.

Traditional two-dimensional laminated structures suffer from having weak interlaminar and thickness direction moduli and strength, which are practically resin-dominated properties. In using textiles as composite reinforcements, one can design the structure in such a way that the thickness-direction yarns add stiffness and strength to this third direction and act as crack arresters that impede interlaminar fracture [25]. This has also resulted in higher impact resistance and damage tolerance [25].

Three-dimensional constructions, in which the preform is woven or braided as one integral structure, are particularly suited to overcoming the inherent weakness of laminates [26]. In principle, braided structures can be manufactured to near-net shape. Braiding and weaving also permit the opportunity for novel means of matrix resin introduction, such as co-weaving, co-winding, and commingling.

New techniques in analysis and design, or modified versions of well-established ones, are required in order to fully use the potential of these new three-dimensional materials. In the next section, the methods that can be used to predict elastic constants of textile composites are reviewed. Prediction of elastic constants for hybrid composites is discussed in the following section, while the strength and failure behavior of these textile structural composites are discussed in Chapter 6.

Elastic Constants of Two-Dimensional Composites

A finite deformation theory has been developed by Kawabata et al. [27-29] to characterize the uniaxial, biaxial, and shear deformation behavior of plain-weave fabrics. This theory provides insight into the possible influence of a textile architecture on the composite behavior, even though this work pertains only to deformation of the fabric preform. However, this theory is semiempirical in that it depends on experiments to measure yarn compressibility and includes expressions for strain energy to take the compressibility into account.

Ishikawa [30] and Ishikawa and Chou [23,31-34] have developed several analytical models for the prediction of the stiffness and strength of woven fabrics. These micromechanical models are particularly suited for the analysis of laminas reinforced with plain- and satin-weave preforms. Basically, these models use lamination analogies to obtain the bounds for thermoelastic properties of a lamina, which may then be used in the classical lamination theory to calculate overall composite properties. The "mosaic model" was found to be effective in predicting the elastic properties of fabric composites, particularly satin-weave, cloth-based ones [23,30]. The "fiber undulation model" takes into account fiber continuity and undulation, and such a threadwise analysis is particularly suited for predicting elastic properties of plain-weave composites [23]. The "bridging model" was developed to simulate the load transfer among the interlaced regions in satin-weave composites. These modeling efforts [23,30-34] have included studies on linear and nonlinear behavior of hybrid and nonhybrid fabric composites, as well as on in-plane and bending thermal expansion coefficients. Ishikawa et al. [35,36] later extended the above work to include the effect of local coupling, geometrical, and material variables on the nonlinear behavior of fabric composites. They have also conducted experimental work to confirm the theory of elastic moduli of these composites [37].

By using geometric modeling and stress partitioning techniques, Dow et al. [38] have also developed a model for two-dimensional fabric composites. They used lamination analogies and strain energy expression to predict stiffness, assumed a linear strain through the thickness, and considered only plane stress conditions. The results showed a reduced upper bound of the elastic constants. Lower bounds can be reduced by assuming that the stresses in the matrix and transverse yarns are similar. Dow et al. also attempted to obtain strength predictions through stress partitioning, based on fiber volume fraction [38].

Summarized below are the key results of the efforts by Ishikawa and Chou in the analysis and modeling of two-dimensional composites [23,30-34,39]. We will begin with a brief introduction of the geometric patterns of two-dimensional fabrics. This will be followed by an outline of the modeling methodology that was proposed by Ishikawa and Chou. Then the three models developed by these workers will be discussed, with proper equations presented to allow for the prediction of the elastic constants of two-dimensional fabric composites. The advantages and limitations of each model will be assessed. We will limit our discussion to the cases where the fabrics are composed of two sets of mutually orthogonal yarns. The treatment on the elastic response of triaxially woven fabric composites can be found in Ref 40.

Geometric Characteristics of Fabrics. The woven fabrics under consideration are made by interlacing two yarn systems at a 90° angle. They typically exhibit good stability in the warp direction (along the fabric length) and in the fill, or weft, direction (width) [41]. The various types of orthogonal fabric can be identified by the pattern of repetition of the interlaced regions, as indicated in Fig. 11 [39]. These simple constructions represent the most commonly used woven fabrics in composites. Each of these fabric styles can be characterized by two basic geometrical parameters: n_{fg} denotes that a warp thread is interlaced with every n_{fg}^{th} fill thread, while n_{wg} denotes that a fill thread is interlaced with every n_{wg}^{th} warp thread. To simplify the discussion, only the fabrics with $n_{wg} = n_{fg} = n_g$ are considered here. Satin weaves are those fabrics with $n_g \geq 4$ in which the interlaced regions are not con-

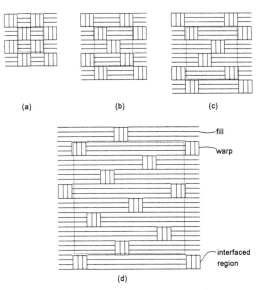

Fig. 11 Examples of woven fabrics. (a) Plain weave (n_g = 2). (b) Twill weave (n_g = 3). (c) Four-harness satin weave (n_g = 4). (d) Eight-harness satin weave (n_g = 8). Source: Ref 39

nected. Other orthogonal weaves are each given a name in accordance with their n_g values: plain weave ($n_g = 2$), twill weave ($n_g = 3$), four-harness satin ($n_g = 4$), and eight-harness satin ($n_g = 8$). The region confined within the dotted lines in each diagram of Fig. 11 defines the "unit cell," or basic repeating region, for each type of fabric.

Basic Approach of Analysis. Lamination analogies and the classical lamination theory form the framework of Chou and Ishikawa's approach. The constitutive equations (Eq 27, 28) can be expressed in a more compact form:

$$\begin{pmatrix} N_i \\ M_i \end{pmatrix} = \begin{bmatrix} A_{ij} & B_{ij} \\ B_{ij} & D_{ij} \end{bmatrix} \begin{pmatrix} \varepsilon_j^0 \\ k_j \end{pmatrix} \quad (i, j = xx, yy, \text{ and } xy) \quad \text{(Eq 53)}$$

The inverted form (from Eq 35, 36, and 37) can be written as

$$\begin{pmatrix} \varepsilon_i^0 \\ k_i \end{pmatrix} = \begin{bmatrix} a_{ij} & b_{ij} \\ b_{ij} & d_{ij} \end{bmatrix} \begin{pmatrix} N_j \\ M_j \end{pmatrix} \quad \text{(Eq 54)}$$

When the effect of temperature variation is taken into account, the constitutive equation becomes

$$\begin{pmatrix} N_i \\ M_i \end{pmatrix} = \begin{bmatrix} A_{ij} & B_{ij} \\ B_{ij} & D_{ij} \end{bmatrix} \begin{pmatrix} \varepsilon_j^0 \\ k_j \end{pmatrix} - \Delta T \begin{pmatrix} \tilde{A}_i \\ \tilde{B}_i \end{pmatrix} \quad (i, j = xx, yy, \text{ and } xy)$$

$$\text{(Eq 55)}$$

where

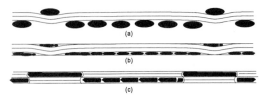

Fig. 12 Schematic of the mosaic model. (a) Cross-sectional view of a woven fabric before resin impregnation. (b) Woven fabric composite. (c) Idealization of the mosaic model. Source: Ref 39

$$(\widetilde{A}_i, \widetilde{B}_i) = \sum_{m=1}^{N} \int_{h_{m-1}}^{h_m} (1, z) q_i \, dz \qquad \text{(Eq 56)}$$

$$q_i = Q_{ij}\alpha_j \qquad \text{(Eq 57)}$$

Here, ΔT represents a small uniform temperature change, and α_j denotes the thermal expansion coefficient. The values of q_i are evaluated for each lamina located between h_m and h_{m-1} in the thickness direction. Upon inversion, Eq 55 becomes

$$\begin{pmatrix} \varepsilon_i^0 \\ k_i \end{pmatrix} = \begin{bmatrix} a_{ij} & b_{ij} \\ b_{ij} & d_{ij} \end{bmatrix} \begin{pmatrix} N_j \\ M_j \end{pmatrix} + \Delta T \begin{pmatrix} \widetilde{a}_j \\ \widetilde{b}_j \end{pmatrix} \qquad \text{(Eq 58)}$$

where

$$\begin{pmatrix} \widetilde{a}_j \\ \widetilde{b}_j \end{pmatrix} = \begin{bmatrix} a_{ij} & b_{ij} \\ b_{ij} & d_{ij} \end{bmatrix} \begin{pmatrix} \widetilde{A}_i \\ \widetilde{B}_i \end{pmatrix} \qquad \text{(Eq 59)}$$

Here, $\widetilde{a}_j$ and $\widetilde{b}_j$ represent the in-plane thermal expansion and thermal bending coefficients, respectively.

The above constitutive equations, in conjunction with the iso-stress and iso-strain assumptions, can be employed to provide the upper and lower bounds of the elastic constants of an orthogonal fabric composite [39]. The iso-stress assumption leads to the upper bounds of compliance constants and, by inversion, the lower bounds of stiffness constants. Conversely, the iso-strain assumption results in the upper bounds of stiffness constants and the lower bounds of compliance constants. The following three models were designed to simplify and simulate the basic fabric construction geometry with a more trackable, idealized laminate analogue so that the classical lamination theory can be applied directly.

The Mosaic Model [30,39]. In this model, a woven fabric (cross-sectional view shown in Fig. 12a) upon impregnation with a resin (Fig. 12b) can be idealized as a mosaic model (Fig. 12c). This mosaic model unit can be regarded as an assemblage of pieces of asymmetric cross-ply laminates. For instance, the unit cell of the mosaic model for an eight-harness satin composite (Fig. 13a) is constructed from several pieces of basic cross-ply laminates (Fig. 13b). This assumption ignores the fiber continuity and undulation (crimp) that exist in a real fabric.

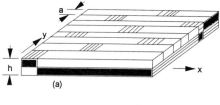

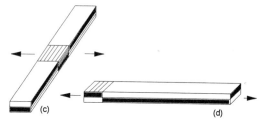

Fig. 13 Mosaic model for an eight-harness satin fabric composite. (a) A repeating region. (b) Idealization of a basic laminate. (c) The parallel model. (d) The series model. Source: Ref 39

Our first task is of course to derive the elastic constants of a cross-ply laminate, which serves as a building block for the entire composite. The laminate is composed of two unidirectional laminas of thickness $h/2$. Rewriting Eq 10, we obtain

$$Q_{ij} = \begin{bmatrix} E_1/D_v & v_{12}E_2/D_v & 0 \\ v_{21}E_1/D_v & E_2/D_v & 0 \\ 0 & 0 & G_{12} \end{bmatrix} \qquad \text{(Eq 60)}$$

where $D_v \equiv 1 - v_{12}v_{21}$. Again, if we position the xy plane to be the geometrical midplane of the laminate with a thickness h, and if we take $k = 1$ and $k = 2$ to define, respectively, laminas with fibers in the y- and x-directions, we have the following nonvanishing stiffness constants [39]:

$$A_{11} = A_{22} = (E_1 + E_2)h/2D_v$$
$$A_{12} = v_{12}E_2h/D_v$$
$$A_{66} = G_{12}h$$
$$B_{11} = -B_{22} = (E_1 - E_2)h^2/8D_v$$
$$D_{11} = D_{22} = (E_1 - E_2)h^3/24D_v$$
$$D_{12} = v_{12}E_2h^3/12D_v$$
$$D_{66} = G_{12}h^3/12 \qquad \text{(Eq 61)}$$

It is of interest to note that the bending-stretching coupling constants B_{11} and B_{22} are nonzero simply because

$E_1 \neq E_2$. This two-ply laminate is balanced (for nonhybrid fabric) but is not symmetrical. Yet, A_{ij}, B_{ij}, and D_{ij} are symmetrical constants.

Using Eq 61, we can rewrite Eq 53 in a more explicit form:

$$
\begin{bmatrix} N_{xx} \\ N_{yy} \\ N_{xy} \end{bmatrix} = \begin{bmatrix} A_{11} & A_{12} & 0 \\ A_{12} & A_{11} & 0 \\ 0 & 0 & A_{66} \end{bmatrix} \begin{bmatrix} \varepsilon_{xx}^0 \\ \varepsilon_{yy}^0 \\ \gamma_{xy}^0 \end{bmatrix} + \begin{bmatrix} B_{11} & 0 & 0 \\ 0 & -B_{11} & 0 \\ 0 & 0 & 0 \end{bmatrix} \begin{bmatrix} k_{xx} \\ k_{yy} \\ k_{xy} \end{bmatrix}
$$

$$
\begin{bmatrix} M_{xx} \\ M_{yy} \\ M_{xy} \end{bmatrix} = \begin{bmatrix} B_{11} & 0 & 0 \\ 0 & -B_{11} & 0 \\ 0 & 0 & 0 \end{bmatrix} \begin{bmatrix} \varepsilon_{xx}^0 \\ \varepsilon_{yy}^0 \\ \gamma_{xy}^0 \end{bmatrix} + \begin{bmatrix} D_{11} & D_{12} & 0 \\ D_{12} & D_{11} & 0 \\ 0 & 0 & D_{66} \end{bmatrix} \begin{bmatrix} k_{xx} \\ k_{yy} \\ k_{xy} \end{bmatrix}
$$

$$(Eq\ 62)$$

which can be inverted to give

$$
\begin{bmatrix} \varepsilon_{xx}^0 \\ \varepsilon_{yy}^0 \\ \gamma_{xy}^0 \end{bmatrix} = \begin{bmatrix} a_{11} & a_{12} & 0 \\ a_{12} & a_{11} & 0 \\ 0 & 0 & a_{66} \end{bmatrix} \begin{bmatrix} N_{xx} \\ N_{yy} \\ N_{xy} \end{bmatrix} + \begin{bmatrix} b_{11} & 0 & 0 \\ 0 & -b_{11} & 0 \\ 0 & 0 & 0 \end{bmatrix} \begin{bmatrix} M_{xx} \\ M_{yy} \\ M_{xy} \end{bmatrix}
$$

$$
\begin{bmatrix} k_{xx} \\ k_{yy} \\ k_{xy} \end{bmatrix} = \begin{bmatrix} b_{11} & 0 & 0 \\ 0 & -b_{11} & 0 \\ 0 & 0 & 0 \end{bmatrix} \begin{bmatrix} N_{xx} \\ N_{yy} \\ N_{xy} \end{bmatrix} + \begin{bmatrix} d_{11} & d_{12} & 0 \\ d_{12} & d_{11} & 0 \\ 0 & 0 & d_{66} \end{bmatrix} \begin{bmatrix} M_{xx} \\ M_{yy} \\ M_{xy} \end{bmatrix}
$$

$$(Eq\ 63)$$

Now we can consider two one-dimensional models where the pieces of cross-ply laminates are either in parallel (Fig. 13c) or in series (Fig. 13d). In these models, the two-dimensional extent of the plate is neglected, and the disturbance of stress and strain around the interface between the interlaced regions is also ignored.

In the parallel model, ε^0 and k are assumed to be uniform. For the one-dimensional repeating region of length $n_g w$, where w denotes the yarn width, an average membrane stress, $\overline{N}$, is defined as

$$
\begin{aligned}
\overline{N}_1 &= \frac{1}{n_g w} \int_0^{n_g w} N_{xx} dy \\
&= \frac{1}{n_g w} \left[\int_0^{w} (A_{11}\varepsilon_{xx}^0 + A_{12}\varepsilon_{yy}^0 + B_{11}k_{xx}) dy \right. \\
&\quad + \left. \int_w^{n_g w} (A_{11}\varepsilon_{xx}^0 + A_{12}\varepsilon_{yy}^0 + B_{11}k_{xx}) dy \right] \\
&= (A_{11}\varepsilon_{xx}^0 + A_{12}\varepsilon_{yy}^0) + \frac{1}{n_g w}[wB_{11}^T \\
&\quad + (n_g w - w)B_{11}^L]k_{xx} = A_{11}\varepsilon_{xx}^0 + A_{12}\varepsilon_{yy}^0 \\
&\quad + \left(1 - \frac{2}{n_g}\right) B_{11}^L k_{xx}
\end{aligned}
$$

$$(Eq\ 64)$$

The presence of the factor $(1 - 2/n_g)$ is due to the fact that the term B_{11} for the interlaced region (B_{11}^I) and noninter-

laced region (B_{11}^L) are of opposite signs, namely, $B_{11}^I = -B_{11}^L$. The term B_{11}^L was derived for a cross-ply with the same configuration as in Fig. 13(b), where the upper ply has its fibers aligned along the x-direction. If these two plies are exchanged in position, we have B_{11}^T. Other average stress resultant or moment results can be similarly obtained. For example:

$$
\overline{M}_1 = D_{11}k_{xx} + D_{12}k_{yy} + \left(1 - \frac{2}{n_g}\right) B_{11}^L \varepsilon_{xx}^0 \qquad (Eq\ 65)
$$

Therefore, if we use $\overline{A}_{ij}, \overline{B}_{ij}$, and $\overline{D}_{ij}$ to denote the stiffness constant matrices relating the average stress resultant $\overline{N}$ and moment resultant $\overline{M}$ with ε^0 and k, we come up with

$$
\overline{A}_{ij} = A_{ij}
$$

$$
\overline{B}_{ij} = \left(1 - \frac{2}{n}\right) B_{ij}^L
$$

$$
\overline{D}_{ij} = D_{ij} \qquad (Eq\ 66)
$$

These elements provide upper bounds for the stiffness constants of the two-dimensional orthogonal fabric composite, based on the one-dimensional mosaic model in parallel. Lower bounds of the elastic compliance constants can be obtained by inverting these stiffness constants. It should be noted that the stiffness constants A, B, and D are computed on the basic building block laminate, where the top lamina is composed of the fill threads (Fig. 13b).

If, instead, a series model is taken where the model is subjected to a uniform in-plane force N_1 in the longitudinal direction, the iso-stress assumption should lead to the definition of an average curvature. The average curvature $\overline{K}_{xx}$, for instance, can be obtained by

$$
\overline{K}_{xx} = \frac{1}{n_g w} \int_0^{n_g w} k_{xx} dx = \left(1 - \frac{2}{n_g}\right) b_{11}^L N_1 \qquad (Eq\ 67)
$$

where again the terms b_{11}^T for the interlaced region and b_{11}^L for the noninterlaced region are equal and opposite in sign. Now let $\overline{a}_{ij}, \overline{b}_{ij}$, and $\overline{d}_{ij}$ be the compliance matrices relating the average strain ε^0 and curvature K with the stress resultant N and moment resultant M. Thus, similar to Eq 66, we have

$$
\overline{a}_{ij} = a_{ij}
$$

$$
\overline{b}_{ij} = \left(1 - \frac{2}{n_g}\right) b_{ij}^L
$$

$$
\overline{d}_{ij} = d_{ij} \qquad (Eq\ 68)
$$

These equations provide the upper bounds for the compliance constants and, upon inversion, the lower bounds for the stiffness constants.

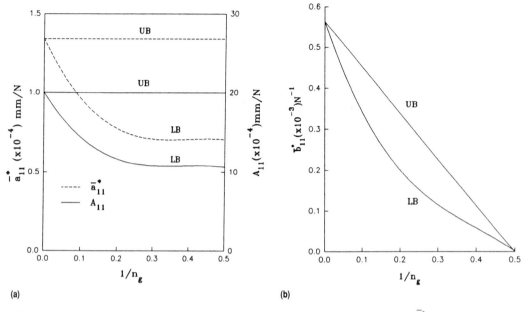

Fig. 14 Results of mosaic model prediction. UB, upper bound; LB, lower bound. (a) Variation of $\bar{a}_{11}^*$ and A_{11} with $1/n_g$. (b) Variation of the average coupling compliance with $1/n_g$. Source: Ref 39

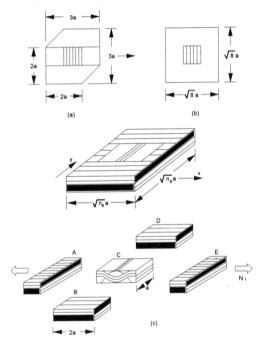

Fig. 15 Schematic of the fiber crimp model. Source: Ref 39

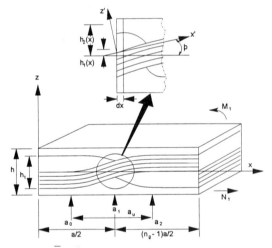

Fig. 16 $\bar{A}_{11}^c$ vs. $1/n_g$ for a graphite-epoxy composite ($V_f = 60\%$). Finite element results are indicated by ($\circ$) for the mosaic model and ($\bullet$) for the crimp model. UB, upper bound; LB, lower bound; CM, crimp model. Source: Ref 39

Ishikawa [30] and Ishikawa and Chou [23] have presented a numerical example to illustrate the relationship between bounds and $1/n_g$ for a graphite-epoxy (Fig. 14). Bidirectional fiber composites are represented by the limiting case of $1/n_g \to 0$ (or $n_g \to \infty$), and the upper and lower bounds of the elastic constants coincide with each other. The coupling effects for plain-weave composites ($1/n_g = 0.5$) disappear, as is implied in Eq 66 and 68. Whereas the upper and lower bounds of b_{ij} (B_{ij}) are identical (both = 0), the bounds of $\bar{A}_{ij}$ ($\bar{a}_{ij}$) do not coincide for plain-weave composites. Partly because of the omission of fiber continuity and undulation effect, the mosaic model tends to give a wide separation between the upper and lower bounds of an elastic constant. This deficiency is essentially removed in the fiber undulation (crimp) model and the related bridging model.

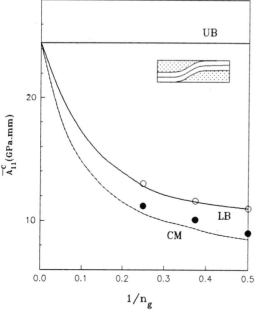

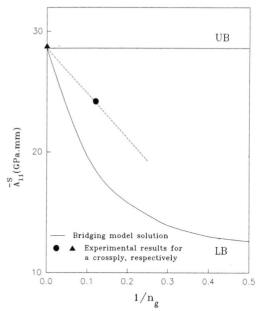

Fig. 17 The bridging model. (a) Shape of the repeating unit of an eight-harness satin. (b) Shape for the repeating unit. (c) Idealization of the bridging model. Source: Ref 39

Fig. 18 $\overline{A}_{11}^s$ vs. $1/n_g$ for a graphite/epoxy composite (V_f = 60%). UB, upper bound; LB, lower bound. Source: Ref 39

The Crimp (Fiber Undulation) Model [23,39]. Shown in Fig. 15 is the geometry of the crimp model, where the undulation shape is defined by the parameters $h_1(x)$, $h_2(x)$, and a_u. The parameter $a_0 = (a - a_u)/2$ and $a_2 = (a + a_u)/2$ are automatically determined by specifying au, which is geometrically arbitrary in the range from 0 to a. Ishikawa and Chou [23,39] have shown that the average in-plane compliance of the model under a uniformly applied in-plane stress resultant is given by

$$\overline{a}_{ij}^c = \left(1 - \frac{2a_u}{n_g a}\right) a_{ij} + \frac{2}{n_g a} \int_{a_0}^{a_2} a_{ij}(x)\,dx \qquad \text{(Eq 69)}$$

where a_{ij} in the first term on the right-hand side denotes the compliance of the straight portion of the yarns, namely a cross-ply laminate, and is independent of x. The other compliance coefficients, b_{ij}^c and $\overline{d}_{ij}^c$, are also given:

$$\overline{b}_{ij}^c = \left(1 - \frac{2}{n_g}\right) b_{ij} + \frac{2}{n_g a} \int_{a_0}^{a_2} b_{ij}(x)\,dx \qquad \text{(Eq 70)}$$

$$\overline{d}_{ij}^c = \left(1 - \frac{2a_u}{n_g a}\right) d_{ij} + \frac{2}{n_g a} \int_{a_0}^{a_2} d_{ij}(x)\,dx \qquad \text{(Eq 71)}$$

The average stiffness constants, $\overline{A}_{ij}^c$, $\overline{B}_{ij}^c$, and $\overline{D}_{ij}^c$, can be obtained by inversion of $\overline{a}_{ij}^c$, $\overline{b}_{ij}^c$, and $\overline{d}_{ij}^c$. The integrations in these equations are usually carried out numerically because of the complexity of the integrands. Numerical results were given to demonstrate the relationship be-

tween the in-plane stiffness $\overline{A}_{11}$ and $1/n_g$ (Fig. 16). It is quite clear that the fiber undulation model, based on a threadwise strip, provides better prediction than the mosaic model. The results of this analysis show that the in-plane elastic stiffness of fabric composites exhibits considerable softening as compared to that of cross-ply laminates. This approach is particularly suited for predicting elastic constants of plain-weave composites.

The Bridging Model [23,39]. The crimp model has provided a basis for the development of a bridging model for general satin fabric composites. The need for a separate model is justified by the observation that the interlaced regions in a satin weave are separated from one another. As shown in Fig. 17, the hexagonal shape of the repeating unit in a satin weave is modified to a square shape (Fig. 17b). A schematic view of the bridging model is also shown (Fig. 17c) for a repeating unit that consists of the interlaced region and its surrounding areas. The four regions labeled A, B, D, and E consist of straight fill yarns, and hence they can be regarded as pieces of cross-ply laminates. Region C has an interlaced structure with an undulated fill yarn. In this model, the undulation and continuity in the warp yarns are ignored, but their effect should be small because the load is applied in the fill direction.

The results in the crimp model have demonstrated that the in-plane stiffness in region C ($n_g = 2$) is much lower than that of a cross-ply laminate. This implies that regions B and D carry higher loads than region C and act as "bridges" for load transfer between region A and E. Regions B, C, and D are also assumed to have the same average midplane strain and curvature. With these assumptions, we can express the average stiffness constants for regions B, C, and D as

$$\bar{A}_{ij} = \frac{1}{\sqrt{n_g}}\left[(\sqrt{n_g} - 1)A_{ij} + \bar{A}_{ij}^c\right]$$

$$\bar{B}_{ij} = \frac{1}{\sqrt{n_g}}(\sqrt{n_g} - 1)B_{ij}$$

$$\bar{D}_{ij} = \frac{1}{\sqrt{n_g}}\left[(\sqrt{n_g} - 1)D_{ij} + \bar{D}_{ij}^c\right] \qquad \text{(Eq 72)}$$

where $\bar{A}_{ij}^c$ and $\bar{D}_{ij}^c$ for the undulated portion C are obtained from $\bar{a}_{ij}^c$ and $\bar{d}_{ij}^c$ of Eq 69 and 71. The coefficients A_{ij}, B_{ij}, and D_{ij} in Eq 72 for the cross-ply laminates of regions B and D are given in Eq 61.

If we further postulate that the total in-plane force carried by regions B, C, and D is equal to that carried by A or E, then the following average compliance constants are derived [39]:

$$\bar{a}_{ij}^s = \frac{1}{\sqrt{n_g}}[2\bar{a}_{ij} + (\sqrt{n_g} - 2)a_{ij}]$$

$$\bar{b}_{ij}^s = \frac{1}{\sqrt{n_g}}[2\bar{b}_{ij} + (\sqrt{n_g} - 2)b_{ij}]$$

$$\bar{d}_{ij}^s = \frac{1}{\sqrt{n_g}}[2\bar{d}_{ij} + (\sqrt{n_g} - 2)d_{ij}] \qquad \text{(Eq 73)}$$

where $\bar{a}_{ij}$, $\bar{b}_{ij}$, and $\bar{d}_{ij}$ are determined by inverting Eq 72. The quantities with a superscript "s" denote properties of the entire satin plane. Finally, the stiffness constants of the satin composite are obtained by inverting Eq 73. The theoretical predictions agree well with experimental data (Fig. 18).

Elastic Constants of Three-Dimensional Composites

A Review on Modeling of Three-Dimensional Composites. The structural geometry of three-dimensional fabrics can be classified according to the four basic methods of textile manufacturing: weaving, orthogonal nonweaving, knitting, and braiding. In addition, three-dimensional constructions can also be obtained by sewing or stitching layers of fabric together after the formation of the fabric. The three-dimensional woven fabrics are produced principally by the multiple-warp weaving method. The orthogonal nonwoven three-dimensional fabrics are fabricated by maintaining one stationary axis, by predeposition of either the yarn system or a spacer rod, which is later retracted and replaced by the axial yarn system. The placement of the planar yarn systems is conducted by inserting the yarns orthogonal to the axial yarn system in an alternating fashion [26]. Knitted three-dimensional fabrics are interlaced structures wherein the knitting loops are produced by the introduction of the knitting yarn, either in the cross-machine direction (weft knit) or along the machine direction (warp knit). The three-dimensional braided structures are constructed by the intertwining or orthogonal interlacing of two or more yarn systems to form integrated constructions [26].

Elastic constant design and prediction for three-dimensional composites requires a knowledge of the geometric arrangement of the component yarns in addition to the fiber and matrix properties. The geometry of the fabric preforms governs the integrity of the composite structure, fiber orientation, and fiber volume fraction. These in turn affect both the performance and processability of the composite system. A typical procedure for the design or analysis of three-dimensional fabric composites begins with the establishment of a unit cell geometry and definition of the geometric parameters. Given the fiber density and linear density, the yarn bundle size and yarn packing density are determined to calculate the fiber volume fraction for a required fiber orientation.

Wei et al. [42] investigated the structural mechanics of three-dimensional braided composites. They considered constraints involving fiber orientations, packing factors, and fiber volume fractions and established proper unit cell geometries. Elastic constants and strength data for the preform only were also reported [42]. Crane and Camponeshi [43,44] derived expressions relating certain parameters, such as braiding machine motions and numbers of rows and columns, to inclination angles within a rectangular three-dimensional braided preform. Extensive consolidation and experimental programs were carried out to study the mechanical properties of the resulting composite.

Pastore and Ko [45,46] developed a processing science model for three-dimensional braided preforms. Based on a mathematical approach involving permutations of yarn positions, they were able to calculate the geometry of an arbitrarily shaped braided domain. A fabric geometry model was developed [26,47] to predict elastic and strength properties of three-dimensional braided structures. This model presumably can be generalized to analyze other three-dimensional preform systems by a proper definition of the unit cell geometry and selection of failure criteria.

Halpin et al. [48-49] proposed the concept of lamination analogies on which much of the two-dimensional work was based. In this approach, spatially oriented discontinuous and continuous composites are mathematically modeled as laminates, of which the constituent laminas are oriented at off-axis angles representative of the angles found in the composite. Also, the relative thickness of each lamina is the same as the percentage of fibers oriented at the particular angle the lamina represents in the actual composite. This approach involves volume averaging of the stiffness constants to obtain in-plane properties. A series of Russian articles was subsequently published [50-56] that discussed further use of stiffness and volume averaging techniques to accomplish a variety of calculations: elastic constant prediction, elastoplastic and viscoelastic analyses, failure prediction, and analysis of stability problems.

A fiber inclination model was established by Yang et al. [57,58] for predicting the elastic properties of three-dimensional textile structural composites. This model represents an extension of the two-dimensional approach adopted by Ishikawa and Chou, discussed above. This

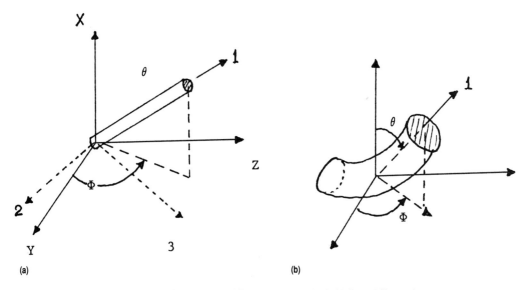

Fig. 19 (a) The coordinate system for the global stiffness average method. (b) Curved fiber path

model treats the unit cell of a composite as an assemblage of inclined unidirectional laminas. The analysis is again based on the classical lamination theory. Ma et al. [59] adopted an energy approach for analyzing the interactions among fiber yarns and took into account the strain energies associated with an interlock due to yarn axial extension, bending, and lateral compression. Closed-form expressions for elastic moduli as functions of fiber volume fraction and fiber spatial orientation were developed [59]. Lamination analogies have been further used by Whitney [60] to parametrically study the elastic properties of two-step braided and angle-interlock woven composites as functions of structural and geometric variables. The lamination analogy shows good agreement for tensile and flexural modulus and Poisson's ratio, less agreement for compressive modulus, and poor agreement for shear modulus.

Analytical and finite element modeling of three-dimensional orthogonally reinforced composites were conducted by Fowser and Wilson [61,62] with good success. By using the minimechanical analysis of a unit cell, Aboudi [63] developed a method to predict the effective elastic moduli and CTE's of orthogonal nonwoven three-dimensional composites. The variational bounding technique was used by Chatterjee and Kibler [64] to determine the upper and lower bounds on the elastic properties of the composite. A Voigt iso-strain model was used by Rosen et al. [65] to derive the overall elastic constants of tridirectionally reinforced composites. Finite element analysis was also conducted by Ross [66] and by Paramasivam et al. [67] to analyze three-dimensional composites.

Modeling efforts on three-dimensional carbon-carbon composites have been reviewed by Jortner [68]. Deineste and Perez [69] have successfully used global volume averaging techniques to arrive at a stiffness ma-

trix for a four-directional carbon-carbon finite element. In-depth treatment of spatially reinforced composites, including three-dimensional orthogonal structures, has been presented by Tarnopolskii et al. [70].

Global Volume Averaging Methods. A useful methodology for deriving elastic constant equations is based on the volumetric superposition of stiffness elements. Kregers and Melbardis [71] analyzed the composite by averaging the stiffness tensor of each individual unidirectional, reinforcing element in the system. Mathematically, the global element stiffness tensor, $C_{\alpha\beta\gamma\delta}$, can be expressed as:

$$C_{\alpha\beta\gamma\delta} = \frac{1}{V_f} \sum_{i=1}^{N} V_i C_{\alpha\beta\gamma\delta}(i)$$

$$V_f = \sum_{i=1}^{N} V_i \qquad \text{(Eq 74)}$$

where V_f = the total fiber volume fraction, V_i = the relative volume of the single unidirectional fiber element i, N = the number of discrete reinforcing directions, $C_{\alpha\beta\gamma\delta}(i)$ = the stiffness tensor of the unidirectional element (i), and $\alpha, \beta, \gamma, \delta \in \{x,y,z\}$.

By following the micromechanics approach developed by Fy [72], Kregers and Melbardis [71] obtained the following expressions for the elastic properties of the unidirectional element used in this model (Fig. 19a):

$$E_{11} = E_f V_f + E_m V_m$$
$$\nu_{12} = \nu_{13} = \nu_f V_f + \nu_m V_m$$
$$E_{22} = E_{33} = \frac{E_m}{f(1 - \nu_m^2) + q\nu_{12}(E_m / E_{11})}$$

$$v_{23} = (1 - v_{21} + v_{31})v_{13} \cdot \frac{E_{33}(1 + v_m)(1 - 2v_m)}{E_m}$$

$$+ \left[\frac{2V_f(1 - v_f^2)E_{33}}{E_m} \right]$$

$$\cdot \left[\frac{2 - 4v_m - (2 - 4v_f)G_m/G_f}{1 + V_m + V_f(3 - 4v_m) + (1 - V_f)(2 - 4v_f)G_m/G_f} \right]$$

$$G_{12} = G_{13} = \frac{G_m(1 + V_f + V_m G_m/G_f)}{V_m + (1 + V_f)G_m/G_f}$$

$$G_{23} = \frac{G_m(3 - 4v_m + V_f + V_m G_m/G_f)}{V_m(3 - 4v_m) + (1 + (3 - 4v_m)V_f)G_m/G_f}$$

$$\text{(Eq 75)}$$

where

$$f = \frac{2 + (2 - 4v_f)G_m/G_f}{Z} - \frac{2V_f(1 - G_m/G_f)}{3 - 4v_m + V_f + V_m G_m/G_f}$$

$$q = \frac{v_m - V_f(4 - 4v_m)(v_m - v_f)}{Z}$$

$$Z = 1 + V_m + (3 - 4v_m)V_f + (2 - 4v_f)V_m G_m/G_f$$

Here, E_{ij}, v_{ij}, and G_{ij} are the modulus of elasticity, Poisson's ratio, and the shear modulus of the unidirectional element, respectively (i = the applied stress direction and j = the induced strain direction).

The stiffness tensor components of the unidirectional element can be expressed by:

$$C_{1111} = \frac{(1 - v_{23}^2)E_{11}}{K}$$

$$C_{2222} = C_{3333} = \frac{(1 - v_{12}v_{21})E_{22}}{K}$$

$$C_{1122} = C_{1133} = \frac{(1 - v_{23})v_{21}E_{11}}{K}$$

$$C_{2233} = \frac{(v_{23} + v_{12}v_{21})E_{22}}{K}$$

$$C_{2323} = G_{23}$$

$$C_{1313} = C_{1212} = G_{12} \qquad \text{(Eq 76)}$$

where

$$k = 1 - 2v_{12}v_{21}(1 + v_{23}) - v_{32}^2 \qquad \text{(Eq 77)}$$

The local unidirectional stiffness elements C_{ijkl} (where $i, j, k, l \in \{1,2,3\}$) can be represented in the global coordinates x,y,z by using the tensor transformation technique. This is given by Eq 77:

$$C_{\alpha\beta\gamma\delta}(i) = C_{ijkl}(i)L_{i\alpha}L_{j\beta}L_{k\gamma}L_{l\delta} \qquad \text{(Eq 78)}$$

where α, β, γ, $\delta \in \{x,y,z\}$; $i, j, k, l \in \{1,2,3\}$; and $L_{i\alpha}$ is the cosine of the angle between axes i and α.

A composite can contain sterically curvilinear fibers. Kregers [73] extended the above analysis by considering

the curved part to be divided into N elements. The following modifications can be used to determine the stiffness element value for the whole fiber system (Fig. 19b):

$$C_{ijkl} = \sum_{n=1}^{N} \frac{V_i(n)}{V_i} \int_{L(n)} C_{ijk}(i) \frac{dl(n)}{L(n)} \qquad \text{(Eq 79)}$$

where $V_i(n)$ is the fiber relative volume fraction in the ith direction, V_i is the fiber total volume fraction in the ith direction, and $L(n)$ is the fiber arc length in the fiber trajectory.

In the present discrete model, the fibers and matrix are represented by a unidirectional rod in terms of their relative volume fractions. The stiffness matrix for each rod is then identified. Then each discrete stiffness is rotated in space to fit in place in the global stiffness matrix, using common transformation methods. In this approach, elasticity relations based on uniform strain assumption are used, and elasticity equations are incorporated in the solution. However, the approach does not take into consideration the possibilities of yarn crimp, yarn interlacing, matrix voids, and other effects, and varying fiber-matrix interfacial bond.

Hybrid Composites

Hybridization provides materials designers with an added degree of freedom in tailoring composites to achieve a better balance of stiffness and strength, increased failure strain, better damage tolerance, improved ability to absorb impact energy, and possibly a significant reduction in cost. This subject has been studied extensively over the past two decades [71-113] and has been reviewed by several researchers [e.g.,111-113]. Many hybrid configurations are possible, but only four are illustrated in Fig. 20.

Figure 20(a) shows two examples of intraply hybrids where each layer, either unidirectional (UD) or woven fabric-based (WF), is made up of two types of fibers. For composites containing hybrid fabrics where each WF layer is comprised of more than one type of yarn, Ishikawa and Chou [31,32,39] have developed a methodology to analyze the thermoelastic behavior. Although additional variables come into play due to the variation in material and geometry, the analysis is very similar to that for nonhybrid two-dimensional orthogonal fabric composites. Again, three models were developed for this purpose: mosaic, crimp, and bridging. Interested readers should consult the original references.

Hayashi [114] developed an expression for the elastic modulus of a hybrid laminate, composed of three kinds of UD fiber-reinforced materials. This is a more generalized rule-of-mixtures equation [114,115]:

$$E = E_{Af}V_{Af} + E_{Bf}V_{Bf} + E_{Cf}V_{Cf} + E_m V_m \qquad \text{(Eq 80)}$$

where E_{Af} and V_{Af} are, respectively, the axial modulus and volume fraction of type A fibers, and so on for type

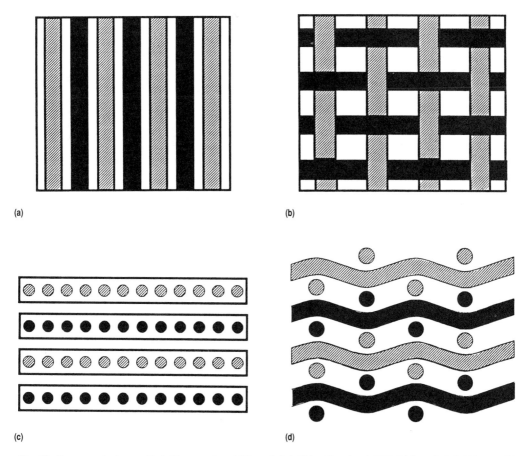

Fig. 20 Four generic classes of hybrid composites. (a) Intraply hybrid (unidirectional, UD). (b) Intraply hybrid (woven fabric, WF). (c) Interply hybrid (UD). (d) Interply hybrid (WF)

B and C fibers. This equation should provide a good estimation of the elastic modulus of hybrid composites at all levels of dispersion (e.g., for both intra- and interply UD hybrids).

An interlaminated hybrid composite can be constructed by laying up various sequences of two or more types of UD or WF laminas, $[A_m B_n C_1...]$. As shown in Fig. 20(b), each lamina or layer in these hybrid laminates is composed of only one type of fiber dispersed in a matrix. Each type of UD lamina will require a set of $[Q'_{ij}]$, $[A]$, $[B]$, and $[D]$ matrices. Different sets of matrices then can be combined, using stiffness and volume averaging techniques in the classical lamination theory to predict the overall elastic constants of the hybrid laminates.

A remarkable increase in the compressive modulus (from 14.0 to 35.0 GPa) of glass-fiber-reinforced polymer was reported to come from a small addition (4.7%) of carbon fibers [116]. The modulus of glass-rich hybrids was found [117] to exceed the value predicted by the rule-of-mixture law. The initial tensile modulus of glass-fiber-reinforced composites showed a marked increase with the addition of carbon fibers, but it exhibited a

progressively lower secondary modulus after failure of the carbon fibers [118]. Chamis et al. [97,98] reported reasonable agreement between measured and calculated flexural modulus data, using the properties of each of the k plies, given by

$$E_H = (1/3t) \sum_{i=1}^{k} (Z_{i+1}^3 - Z_i^3) E_i \qquad \text{(Eq 81)}$$

where E_H is the elastic flexural modulus of the hybrid, t is the thickness of the complete hybrid, Z_i is the distance to the bottom of the ith ply, and Z_{iL+1} is the distance to the top of the ith ply.

Analogous to the case of interlaminated UD hybrid laminates, each type of WF layer in interply or interlaminated woven hybrid composites can be treated separately by following Ishikawa and Chou's approach, described above. Their equivalent properties, obtained from either the crimp model or the bridging model, are then integrated by the same volume averaging and laminate analogy approach.

REFERENCES

1. R.M. Jones, *Mechanics of Composite Materials*, McGraw-Hill, 1975
2. B.C. Hoskin and A.A. Baker, Ed., *Composites for Aircraft Structures*, AIAA Educational Series, American Institute of Aeronautics and Astronautics, 1986
3. S.W. Tsai, *Mechanics of Composite Materials*, Technomic Publishing, 1984
4. R.A. Schapery, *J. Compos. Mater.*, Vol 2, 1968, p 380
5. S.W. Tsai and N.J. Pagano, *Composite Materials Workshop*, S.W. Tsai, J.C. Halpin, and N.J. Pagano, Ed., Technomic Publishing, 1968, p 233
6. P.K. Mallick, *Fiber-Reinforced Composites*, Marcel Dekker, 1988
7. S.W. Tsai, *Composites Design*, Think Composites Co., Dayton, OH, 1991
8. J.L. Kardos, *High Performance Polymers*, E. Baer and A. Moet, Ed., Hanser Publishing, 1991, p 199
9. J.C. Halpin, *Primer on Composite Materials: Analysis*, Technomic Publishing, 1984
10. J.C. Halpin and J.L. Kardos, *Polymer Eng. Sci.*, Vol 18, 1978, p 344, 496
11. B.D. Agarwal and L.J. Broutman, *Analysis and Performance of Fiber Composites*, Wiley-Interscience, 1980
12. J.L. Kardos, *Crit. Rev. Solid State Mater. Sci.*, Vol 3 (No. 4), 1973, p 419
13. J.C. Halpin, K. Jerina, and J.M. Whitney, *J. Compos. Mater.*, Vol 5, 1971, p 36
14. J.C. Halpin and S.W. Tsai, "Environmental Factors in Composite Materials Design," AFML-TR67-423, Air Force Materials Laboratory, June 1969
15. J.C. Halpin and R.L. Thomas, *J. Compos. Mater.*, Vol 2, 1968, p 488
16. L.E. Nielsen, *J. Appl. Phys.*, Vol 41, 1970, p 4626
17. S.H. McGee, "The Influence of Microstructure on the Elastic Properties of Composite Materials," Ph.D. dissertation, University of Delaware, 1982
18. S.L. Chang, "Prediction and Measurement of Orientation Distribution Effects on the Stiffness and Strength of Short Fiber and Molecular Composites," D.Sc. dissertation, Washington University, St. Louis, MO, 1981
19. R.E. Lavengood and L.A. Goettler, AD 886372, National Technical Information Service, U.S. Department of Commerce, Springfield, VA
20. R.L. Foye, Proc. 10th National SAMPE Symp. 9-11 Nov 1966 (San Diego, CA), Society for the Advancement of Material and Process Engineering
21. P.E. Chen and L.E. Nielsen, *Kolloid Z.Z. Polymer*, Vol 235, 1969, p 1174
22. R.A. Humphrey, Chapter 12, *Modern Composite Materials*, L.J. Broutman, R.H. Krock, Ed., Addison-Wesley, 1967
23. T. Ishikawa and T.W. Chou, Ed., *J. Mater. Sci.*, Vol 17, 1982, p 3211
24. T.W. Chou and F.K. Ko, *Textile Structural Composites*, Vol 3, *Composite Materials Series*, Elsevier, 1989
25. W.C. Chung, B.Z. Jang, T.C. Chang, L.R. Hwang, and R.C. Wilcox, *Mater. Sci. Eng. A*, Vol 112, 1989, p 157
26. F.K. Ko, Chapter 5, *Textile Structural Composites*, Vol 3, T.W. Chou and F.K. Ko, Ed., *Composite Materials Series*, Elsevier, 1989
27. S. Kawabata, M. Niwa, and H. Kawai, *J. Textile Inst.*, Vol 64, Feb 1973, p 20
28. S. Kawabata, M. Niwa, and H. Kawai, *J. Textile Inst.*, Vol 64, Feb 1973, p 47
29. S. Kawabata, M. Niwa, and H. Kawai, *J. Textile Inst.*, Vol 64, Feb 1973, p 62
30. T. Ishikawa, *Fiber Sci. Technol.*, Vol 15, 1981, p 127
31. T. Ishikawa and T.-W. Chou, *J. Compos. Mater.*, Vol 16, Jan 1982, p 2
32. T. Ishikawa and T.-W. Chou, *J. Compos. Mater.*, Vol 18, 1983, p 2260
33. T. Ishikawa and T.-W. Chou, *J. Compos. Mater.*, Vol 17, Mar 1983, p 92
34. T. Ishikawa and T.-W. Chou, *J. Compos. Mater.*, Vol 17, Sept 1983, p 399
35. T. Ishikawa, M. Matsushima, and Y. Hayashi, Proc. ICCM-V (San Diego, CA), International Committee on Composite Materials, 1985, p 1267-1281
36. T. Ishikawa, M. Matsushima, and Y. Hayashi, *AIAA J.*, Vol 25 (No. 1), Jan 1987, p 107
37. T. Ishikawa, M. Matsushima, and Y. Hayashi, *J. Compos. Mater.*, Vol 19, Sept 1985, p 443
38. N.F. Dow and V. Ramnath, *3-D Composites*, H.B. Dexter and E.T. Camponeshi, Ed., Conf. Pub. 2420, National Aeronautics and Space Administration, Sept 1986, p 5-30
39. T.W. Chou and T. Ishikawa, Analysis and Modeling of Two-Dimensional Fabric Composites, Chapter 7, *Textile Structural Composites*, Vol 3, T.W. Chou and F.K. Ko, Ed., *Composite Materials Series*, Elsevier, 1989
40. J.M. Yang and T.W. Chou, Thermo-Elastic Analysis of Triaxial Woven Fabric Composites, Chapter 8, *Textile Structural Composites*, Vol 3, T.W. Chou and F.K. Ko, Ed., *Composite Materials Series*, Elsevier, 1989
41. F. Scardino, An Introduction to Textile Structures and Their Behavior, Chapter 1, *Textile Structural Composites*, Vol 3, T.W. Chou and F.K. Ko, Ed., *Composite Materials Series*, Elsevier Science, 1989
42. W. Li, T.J. Kang, and A.E. Shiekh, *Materials—Pathway to the Futures, Society for the Advancement of Material and Process Engineers*, Proc. 33rd Int. SAMPE Symp. March 1988 (Anaheim, CA), Society for the Advancement of Material and Process Engineering
43. R.M. Crane and E.T. Camponeshi, *Exp. Mech.*, Vol 26 (No. 3), Sept 1986, p 259
44. E.T. Camponeshi and R.M. Crane, W. Taylor Naval Ship and Development Center, Annapolis, MD, Sept 1986
45. C.M. Pastore, and F.K. Ko, Proc. 1988 ASM/ESD Conf., ASM International, 1988
46. F.K. Ko and C.M. Pastore, *Recent Advances in Composites in the United States and Japan*, S.T.P. 864, J.R. Vinson and M. Taya, Ed., ASTM, 1985, p 428-439
47. D.W. Whyte, Ph.D. thesis, Drexel University, Philadelphia, June 1986
48. J.C. Halpin and N.J. Pagano, *J. Compos. Mater.*, Vol 3, Oct 1969, p 720
49. J.C. Halpin, K. Jerina, and J.M. Whitney, *J. Compos. Mater.*, Vol 5, Jan 1971, p 720
50. A.F. Kregers and Y.G. Melbardis, *Mekh. Polim.*, No. 1, Jan/Feb 1978, p 3-8
51. A.F. Kregers and G.A. Teters, Structural Model of Deformation of Anisotropic Three-Dimensionally Reinforced Composites, *Mekh. Kompoz. Mater.*, No. 1, Jan/Feb 1982, p 14-22
52. A.F. Zilauts, et al., *Mekh. Kompoz. Mater.*, No. 6, Nov/Dec 1981, p 987
53. A.F. Kregers and G.A. Teters, *Mekh. Kompoz. Mater.*, No. 1, Jan/Feb 1981, p 30
54. Y.G. Melbardis and A.F. Kregers, *Mekh. Kompoz. Mater.*, No. 3, May/June 1980, p 436
55. A.F. Kregers and G.A. Teters, *Mekh. Kompoz. Mater.*, No. 4, July/Aug 1979, p 617
56. A.F. Kregers and G.A. Teters, *Mekh. Kompoz. Mater.*, No. 1, Jan/Feb 1979, p 79

57. J.M. Yang, C.L. Ma, and T.W. Chou, *J. Compos. Mater.*, Vol 20, Sept 1986, p 472

58. J.M. Yang, "Modelling and Characterization of 2-D and 3-D Textile Structural Composites," Ph.D. thesis, University of Delaware, Dec 1986

59. C.L. Ma, J.M. Yang, and T.W. Chou, in *Composite Materials: Testing and Design*, STP 893, ASTM, 1986, p 404

60. T.J. Whitney, "Analytical Characterization of 3-D Textile Structural Composites Using Plane-Stress and 3-D Lamination Analogies," M.S. thesis, University of Delaware, Dec 1988

61. S.W. Fowser and D. Wilson, Analytical and Experimental Investigation of 3-D Orthogonal Graphite/Epoxy Composites, *3-D Composites*, H.B. Dexter and E.T. Camponeshi, Ed., Conf. Pub. 2420, National Aeronautics and Space Administration, Sept 1986, p 91-108

62. S.W. Fowser, "The Behavior of Orthogonal Fabric Composites," Master's thesis, University of Delaware, Aug 1986

63. J. Aboudi, *Fiber Sci. Technol.*, Vol 21, 1984, p 277

64. S.N. Chatterjee and J.J. Kibler, *Modern Developments in Composite Materials and Structures*, J.R. Vinson, Ed., ASME, 1979, p 269

65. B.W. Rosen, S.N. Chatterjee, and J.J. Kibler, in *Composite Materials: Testing and Design*, STP 617, ASTM, 1977, p 243

66. A.L. Ross, Paper No. 74-DE-25, Design Eng. Conf., April 1974 (Chicago), American Society of Mechanical Engineers, 1974

67. P. Paramasivam, J.I. Curiskis, and S. Vallappan, *Fiber Sci. Technol.*, Vol 20, 1984, p 99

68. J. Jortner, *3-D Composites*, H.B. Dexter and E.T. Camponeshi, Ed., Conf. Pub. 2420, National Aeronautics and Space Administration, Sept 1986, p 53-73

69. L. Deineste and B. Perez, *AIAA J.*, Vol 21 (No. 8), Aug 1983, p 1143

70. Y.M. Tarnopolskii, I.G. Zhegun, and V.A. Polyakov, *Prostranstveno-Armerovanigu Kompozitseoniye Materiali, Mashinostraeniye*, Moscow, 1987

71. D.S. Cordova, D.R. Coffin, J.A. Young, and H.H. Rowan, "Effects of Polyester Fiber Characteristics on the Properties of Hybrid Reinforced Thermosetting Composites," Proc. Ann. Conf. Reinforced Plastics/Composites Institute, Society of Plastics Industry, 1984

72. D.S. Cordova, H.H. Rowan, and L.C. Lin, "Computer Modeling of Hybrid Reinforced Thermoset Composites," Proc. Ann. Conf. Reinforced Plastics/Composites Institute, Society of Plastics Industry, 1986

73. A. Poursartip, G. Riahi, E. Teghtsoonian, N. Chinatambi, N. Mulford, Mechanical Properties of PE Fiber/Carbon Fiber Hybrid Laminates, Proc. ICCM-VI and ECCM-II, Vol 1, Elsevier Applied Science, 1987, p 209-220

74. J. Aveston and J.M. Sillwood, Synergistic Fiber Strengthening in Hybrid Composites, *J. Mater. Sci.*, Vol 11, 1976, p 1877-1883

75. M.G. Bader and P.W. Manders, The Strength of Hybrid Glass/Carbon Fiber Composites: Part I, Failure Strain Enhancement and Failure Mode, *J. Mater. Sci.*, Vol 16, 1981, p 2233-2245

76. M.G. Bader and P.W. Manders, Part II, A Statistical Model, *J. Mater. Sci.*, Vol 16, 1981, p 2246-2256

77. A.R. Bunsell and B. Harris, *Composites*, Vol 5, 1974, p 157-164

78. J. Aveston and A. Kelly, *Philos. Trans. R. Soc. (London) A*, Vol 294, 1980, p 519-534

79. J. Xing, G.C. Hsiao, and T.W. Chou, A Dynamic Explanation of the Hybrid Effects, *J. Compos. Mater.*, Vol 15, 1981, p 443-461

80. D.G. Harlow, Statistical Properties of Hybrid Composites, *Proc. R. Soc (London) A*, Vol 389, 1983, p 67-100

81. M.J. Pitkethly and M.G. Bader, Failure Modes of Hybrid Composites Consisting of Carbon Fiber Bundles Dispersed in a Glass Fiber Epoxy Resin Matrix, *J. Appl. Phys., Phys. D*, Vol 20, 1987, p 315-322

82. R.Y. Kim and S.R. Soni, "Suppression of Free-Edge Delamination by Hybridization," Proc. ICCM-V (San Diego, CA), International Committee on Composites Materials, 1985, p 1557-1572

83. Y.W. Mai, R. Andonian, and B. Cottrell, "On Polypropylene-Cellulose Fiber-Cement Hybrid Composites," SAMPE Symp. Proc., Society for the Advancement of Material and Process Engineering, 1981, p 1687-1699

84. B. Harris and A.R. Bunsell, Impact Properties of Glass Fiber/Carbon Fiber Hybrid Composites, *Composites*, Vol 6, 1975, p 197-201

85. M. Arrington and B. Harris, *Composites*, Vol 9, 1978, p 149-152

86. G. Marom, S. Fischer, F.R. Tuler, and H. Wagner, Hybrid Effects in Composites, *J. Mater. Sci.*, Vol 13, 1978, p 1419-1426

87. N.J. Parratt and K.D. Potter, "Mechanical Behavior of Intimately-Mixed Hybrid Composites," SAMPE Symp. Proc., Society for the Advancement of Material and Process Engineering, 1981, p 313-326

88. H.D. Wagner and G. Marom, On Composition Parameters for Hybrid Composite Materials, *Composites*, Vol 13, 1982, p 18

89. H.D. Wagner, "Elastic Properties and Fracture of Hybrid Composite Materials," Ph.D. thesis, The Hebrew University of Jerusalem, 1982

90. H.D. Wagner and G. Marom, "Delamination Failure in Hybrid Composites: The Effect of the Stacking Sequence," Session 12E, 38th Ann. Conf. Reinforced Plastic/Composite Institute, Society of Plastics Industry, Feb 7-11, 1983

91. D.F. Adams and R.S. Zimmerman, Static and Impact Performance of PE Fiber/Graphite Fiber Hybrid Composites, *SAMPE J.*, Nov/Dec 1986, p 10-16

92. C. Zweben, Tensile Strength of Hybrid Composites, *J. Mater. Sci.*, Vol 12, 1977, p 1325-1337

93. T.W. Chou and H. Fukuda, "Stiffness and Strength of Hybrid Composites," Proc. 1st Japan-U.S. Conf. Composite Materials (Tokyo), 1981

94. T.W. Chou and A. Kelly, Mechanical Properties of Composites, *Ann. Rev. Mater. Sci.*, Vol 10, 1980, p 229-259

95. N.M. Bhatia, Strength and Fracture Characteristics of Graphite-Glass Intraply Hybrid Composites, *Composite Materials: Testing and Design*, STP 787, I.M. Daniel, Ed., ASTM, 1982, p 183-199

96. K.S. Sivakumaran, "On the Vibration Characteristics of Interlaminated Hybrid Composite Plates," Proc. ICCM-V (San Diego, CA), International Committee on Composite Materials, 1985, p 1703-1710

97. C.C. Chamis and R.F. Lark, "Hybrid Composites State-of-the Art Review: Analysis, Design, Applications and Fabrication," NASA-TMX-73545, National Aeronautics and Space Administration, 1977

98. C.C. Chamis and J.H. Sinclair, "Prediction of Properties of Intraply Hybrid Composites," NASA-TM 79087, National Aeronautics and Space Administration, 1979

99. H. Fukuda and T.W. Chou, "A Statistical Approach to the Strength of Hybrid Composites," *Progress in Science and Engineering of Composites*, T. Hayashi, et al., Ed., Proc. ICCM-IV (Tokyo), International Committee on Composite Materials, 1982, p 821

100. H. Fukuda and T.W. Chou, Stress Concentrations around a Discontinuous Fiber in a Hybrid Composite Sheet, *Trans. Jpn. Soc. Compos. Mater.*, Vol 7, 1981, p 37-42

101. H. Fukunaga, T.W. Chou, and H. Fukuda, Strength of Intermingled Hybrid Composites, *J Reinf. Plast. Compos.*, Vol 3, 1984, p 145

102. X. Ji, On the Hybrid Effect and Fracture Mode of Interlaminated Hybrid Composites, *Progress in Science and Engineering of Composites*, T. Hayashi et al., Ed., Proc. ICCM IV (Tokyo), International Committee on Composite Materials, 1982, p 1137-1144

103. C.C. Chamis, M.P. Hanson, and T.T. Serafini, in *Composite Materials: Testing and Design*, STP 497, ASTM, 1972, p 324-349

104. P.W.R. Beaumont, P.G. Riewald, and C. Zweben, Method for Improving the Impact Resistance of Composite Materials, *Foreign Object Impact Damage to Composites*, STP 568, ASTM, 1974, p 134-158

105. G. Dorey, G.R. Sidey, and J. Hutching, Impact Properties of Carbon Fiber/Kevlar 49 Fiber Hybrid Composites, *Composites*, Vol 9, 1978, p 25-32

106. G. Dorey, "The Use of Hybrid to Improve Composite Reliability," Proc. Technical Symposium III on Design and Use of KEVLAR Aramid in Aircraft, E.I. Du Pont de Nemours & Co., Inc., Oct 1981

107. I.R. McColl and J.G. Morley, Crack Growth in Hybrid Fibrous Composites, *J. Mater. Sci.*, Vol 12, 1977, p 1165-1175

108. T.W. Chou, Hybrid Textile Structural Composites: An Overview, *Proc. Int. Symp. on Composite Materials and Structures* (Beijing, China), June 1986, T.T. Loo and C.T. Sun, Ed., p 181-186

109. J. Li, K. Mai, and H. Zeng, Studies on the Mechanical Properties and Fracture Morphology of Hybrid Carbon/Glass Reinforced Hybrid Matrix (PSF/ESF) Composites, *Proc. Int. Symp. on Composite Materials and Structures* (Beijing, China), June 1986, T.T. Loo and C.T. Sun, Ed., p 936-942

110. B.Z. Jang, L.C. Chen, L.R. Hwang, and R.H. Zee, The Response of Fibrous Composites to Impact Loading, *Polymer Compos.*, Vol 11, 1990, p 144-157

111. B.Z. Jang, L.C. Chen, C.Z. Wang, H.T. Lin, and R.H. Zee, *Compos. Sci. Technol.*, Vol 34, 1989, p 305-335

112. N.L. Hancox, Ed., *Fiber Composite Hybrids*, Applied Science, 1980

113. W.J. Renton, Ed., *Hybrid and Select Metal Matrix Composites*, AIAA, New York, 1977

114. T. Hayashi, *8th Reinforced Plastics Conference* (Brighton, UK), 1972, p 149-152

115. D. Short and J. Summerscales, *Fiber Composite Hybrids*, N.L. Hancox, Ed., Applied Science, 1980, p 69-117

116. R. Dukes and D.L. Griffiths, Paper 28, Int. Conf. Carbon Fibers, Plastics Institute, London, 1971

117. I.L. Karlin, in *Composite Materials: Testing and Design*, STP 497, 1973, p 551-563

118. T. Fuji and K. Tanaka, *J. Soc. Mater. Sci. Jpn.*, Vol 21, 1972, p 906-910

Chapter 6

Design and Prediction for Strength of Polymer Composites

Unidirectional Laminates

Longitudinal Strength and Failure Modes

Mechanics-of-Materials Approach to Tensile Strength. In order to appreciate the significance of fiber properties in dictating the tensile strength of a composite, we may choose to follow a simple analysis which assumes that all fibers break at the same stress level, at the same time, and in the same plane. We may further assume iso-strain condition, where the fiber and the matrix experience equal strain. In general, the fiber failure strain is lower than the matrix failure strain. This is exemplified by carbon fiber-epoxy systems. In response to a tensile load, fibers will fail first and the total load will be transferred to the matrix. At this moment, two composite failure modes are possible, depending on the fiber volume fraction V_f. At a high V_f, the matrix alone is not capable of carrying the full load. (Remember, we have shown in Chapter 5 that fibers generally bear a major portion of the load when V_f is not too low.) The matrix will thus fracture immediately after fiber fracture. The longitudinal tensile strength of a unidirectional continuous fiber composite can be estimated as

$$\sigma_{1u} = \sigma_{fu}V_f + \sigma'_m V_m \qquad \text{(Eq 1)}$$

where σ_{fu} is the fiber tensile strength and σ'_m is the stress carried by the matrix at the fiber failure strain.

At low V_f, there is sufficient matrix material to bear the full load after the fibers fracture. The matrix may also be able to carry additional load with increasing strain. Further, the fibers are generally assumed to carry no load at composite strains higher than the fiber fracture strain. The composite eventually fails when the matrix stress reaches the ultimate strength of the matrix. In this case, the composite tensile strength is given by

$$\sigma_{1u} = \sigma_{mu}V_m \qquad \text{(Eq 2)}$$

where σ_{mu} is the matrix ultimate tensile strength. This suggests the existence of a critical fiber volume fraction V'_f, above which the composite failure changes from a matrix-dominated mode to a fiber-dominated mode. This critical fiber volume fraction can be obtained by equating Eq 1 to Eq 2:

$$V'_f = \frac{\sigma_{mu} - \sigma'_{mu}}{\sigma_{fu} + \sigma_{mu} - \sigma'_m} \qquad \text{(Eq 3)}$$

Equation 2 predicts that the strength of the composite is always less than that of the unreinforced matrix. In contrast, Eq 1 predicts composite strength that can be higher or lower than the matrix strength, depending on the fiber volume fraction. It can also be shown that a minimum fiber volume fraction V_{min} exists, below which no fiber-reinforcing effect is expected (i.e., the composite strength is actually less than the strength of the matrix when alone). In order to have fiber-reinforcing or "strengthening" effect, one must have $\sigma_{1u} = \sigma_{fu}V_f + \sigma_m V_m \geq \sigma_{mu}$. Thus, this minimum fiber volume fraction is given by

$$V_{min} = \frac{\sigma_{mu} - \sigma'_m}{\sigma_{fu} - \sigma'_m} \qquad \text{(Eq 4)}$$

These equations demonstrate the importance of fiber failure strain and strength in dictating the composite failure behavior. If fibers of marginal strength are to be used to reinforce a strong matrix, large fiber volume fractions will be required to achieve a strengthening effect. V_{min} is a more important system parameter than V'_f. Note that σ'_m, V'_f, and V_{min} are generally quite small for composites containing high-strength, high-modulus fibers and low-modulus polymer matrices. For a glass

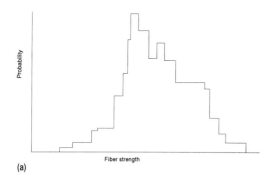

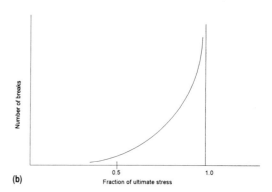

Fig. 1 Schematic of the statistical nature of fiber strength

fiber/epoxy composite (composite failure strain ≈ 4%; σ_{fu} = 2.8 GPa [406 ksi] and E_f = 70 GPa [10×10^6 psi]; σ_{mu} = 25 MPa [3.6 ksi] and σ_m' = 10 MPa [1.5 ksi]) we have V_f' ≈ 0.53% and V_{min} ≈ 0.54%. Similar treatments can be applied to systems where the matrix fails first, but the failure characteristics will be quite different. For nylon fiber-reinforced epoxy composites, for instance, a V_{min} of approximately 10% may be required to have an appreciable strengthening effect. The reader is encouraged to work out the analysis procedures in order to grasp this subject.

Statistical Aspects of Composite Strength. In deriving Eq 1 to 4 we have assumed that all fibers fail at the same strength in the same plane, and that when the fiber volume fraction is sufficiently high, the composite fails immediately after fiber failure. In reality, fiber strength is not a unique value; rather, it follows a statistical distribution. In a fiber bundle or in a composite, a few fibers are expected to break at low stress levels. The remaining fibers will carry higher stresses but may not fail at the same time. The exact behavior is related to the flaw sensitivity of the fibers and the fiber-matrix interaction. This is further explained as follows.

Brittle fibers generally contain surface flaws, and fiber breakages will likely occur at these imperfections at more or less random positions throughout the composite. Therefore, composite failure will not occur in a single plane. Further, fiber imperfections vary not only in position, but also in severity. Fiber fractures are therefore expected to occur throughout a range of stress levels up to ultimate composite failure (Fig. 1) [1].

The fiber-matrix interaction in the vicinity of a fiber fracture is another important feature of composite failure. When a fiber fractures, the normal stress at each of its broken ends becomes zero. Because of the shear stress transfer at the fiber-matrix interface, however, the stress builds back up to the average value over a distance of $l_c/2$ from each end, where l_c is the ineffective fiber length, to be defined later. Local stress redistribution results in the average normal stress in adjacent fibers. High stress concentrations are also expected to develop in the matrix near the fiber ends and at the void created by the broken fiber. As a consequence, partial or total debonding of the bro-

ken fiber from the surrounding matrix may occur due to high interfacial shear stresses at its ends. This considerably reduces the reinforcing effectiveness of fibers. Microcracks can initiate at the sites of stress magnification, and local plastic deformation mechanisms can operate when the matrix is ductile. The high stress concentrations may also trigger failure of fibers adjacent to a broken fiber.

The fracture of individual fibers in a composite normally starts at loads much smaller than the composite failure load. As the load is raised, more fibers break, and some fibers may break at many different cross sections. This phenomenon results in a statistical weakening of the cross sections of the composite. Further, fiber strength depends on fiber length. A longer fiber has a higher probability of having a weak link; thus longer fibers have smaller strengths. Composites typically contain bundles of fibers of nonuniform strength. This statistical nature of fibers in a composite led scientists to develop various cumulative-weakening failure models.

In one model, the tensile strength variation of a group of fibers and, to a lesser extent, that of a composite laminate are often represented by the Weibull distribution, which is more realistic than the normal distribution. Using two-parameter Weibull statistics, one can express the cumulative density function for the composite strength as

$$F(\sigma) = \exp\left[-(\sigma/\sigma_0)^\alpha\right] \qquad \text{(Eq 5)}$$

where $F(\sigma)$ = the probability of surviving a stress σ, α = a dimensionless "shape parameter," and σ_0 = a "location parameter." The mean tensile strength and variance of the composite are therefore given by [2]:

$$\sigma = \sigma_0 \, \Gamma((1 + \alpha)/\alpha)$$

$$S^2 = \sigma_0^2[\Gamma((2 + \alpha)/\alpha) - \Gamma^2(1 + \alpha)/\alpha)] \qquad \text{(Eq 6)}$$

where Γ is the gamma function. Note that the probability of surviving a higher stress is lower, as expected. A decreasing value of the shape parameter α would indicate greater scatter in the tensile strength data. This parameter

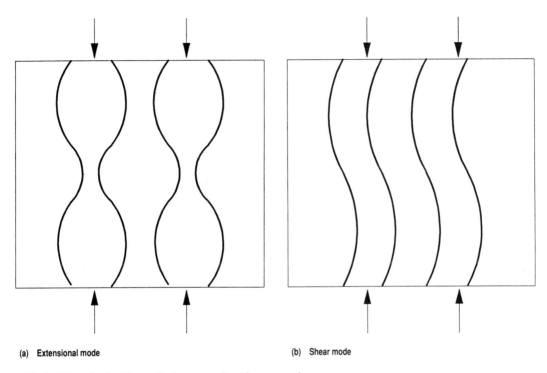

(a) **Extensional mode**　　　　　　　(b) **Shear mode**

Fig. 2　Fiber microbuckling modes in a composite under compression

is closely related to the fiber strength variation and the state of fiber-matrix interaction.

Mechanics-of-Materials Approach to Compressive Strength. Fibers under compression generally do not fail in a simple compression mode; rather, they fail by local buckling. Therefore, the above equations relating to tensile failure of composites do not apply to compressive strength. As shown in Fig. 2, two possible buckling modes can be envisaged: the extensional mode and the shear mode. The extensional mode involves stretch and compression of the matrix in an out-of-phase manner. It can be shown [3] that the compression strength of composite in an extensional mode is

$$\sigma_{cu} \approx 2V_f \cdot \left[\frac{V_f E_m E_f}{3(1 - V_f)} \right]^{1/2} \qquad \text{(Eq 7)}$$

In the shear mode the fibers buckle in phase and the matrix is sheared. The resulting buckling stress is [3]:

$$\sigma_{cu} \approx \frac{G_m}{(1 - V_f)} \qquad \text{(Eq 8)}$$

For advanced composites containing a reasonably high volume fraction of fibers, the extensional mode can be neglected. These equations are also found to considerably overestimate the compressive strength.

The shear mode of microbuckling is characterized by the formation of fiber kink bands in localized areas. In addition to fiber microbuckling, several other failure modes have been observed in composites under compression. These include compressive failure or yielding of fibers, longitudinal splitting in the matrix due to Poisson's effect, matrix yielding, fiber-matrix interfacial debonding, and fiber splitting or fibrillation. The latter is commonly found in polymer-based, high-strength fibers such as Kevlar-49® aramid and Spectra-1000® PE fibers. Equation 8 indicates that an increase in matrix shear modulus and fiber volume fraction will increase the composite compressive strength. Increases in fiber tensile modulus, fiber diameter, fiber-matrix interfacial strength, and matrix ductility would also tend to improve the compressive strength of composites. Fiber misalignment ("bowing") has been found to reduce the longitudinal compressive strength of a composite and therefore should be avoided during composite fabrication.

Transverse Strength and Failure Modes

Transverse Tensile Strength of a Composite. When a load is applied to the lamina at an angle of 90° with respect to the fibers, the fibers act as hard inclusions in the matrix and enhance the local strain and stress magnitude in the matrix near the fiber-matrix interface (Fig. 3). Near one such interface the radial stress was calculated to be tensile and more than 50% higher than the applied stress [2]. This intensified stress can cause cracking normal to the loading direction at the interface or in the nearby matrix. When the volume fraction of fiber becomes high, significant stress overlap will occur, leading to an even

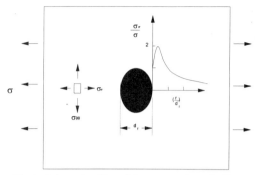

Fig. 3 Schematic of the radial stress at and near the fiber-matrix interface in a composite under transverse loading

higher stress concentration. The local stress increases when the (E_f / E_m) ratio is raised. Although the transverse modulus of the composite is raised with increasing V_f and (E_f / E_m), the transverse strength of the composite may be reduced [2].

By assuming that the lamina transverse strength is limited by the ultimate tensile strength of the matrix, Greszczuk[4] has derived a simple equation for predicting this strength:

$$\sigma_{2u} = \frac{\sigma_{mu}}{K} \qquad \text{(Eq 9)}$$

where

$$K = \frac{1 - V_f [1 - (E_m / E_f)]}{1 - (4V_f / \pi)^{1/2}[1 - (E_m / E_f)]}$$

In fact, K represents the maximum stress concentration in the matrix, assuming that fibers are arranged in a square array. If it is assumed that the transverse strength is limited by the ultimate strain of the matrix [5,6], then the transverse strength in terms of strain is calculated from

$$\varepsilon_{2u} = \frac{\varepsilon_{mu}}{K'} \qquad \text{(Eq 10)}$$

where

$$K' = \frac{1}{1 - (4V_f / \pi)^{1/2}[1 - (E_m / E_f)]}$$

represents the strain-magnification factor [5,6].

The above failure prediction was based on either a maximum stress or a maximum strain criterion. A more commonly used failure criterion is based on the maximum distortion energy. According to this criterion, the material failure occurs when the distortion energy at any point reaches a critical value. The corresponding strength reduction factor K'' can be written as

$$K'' = \frac{(U_{max})^{1/2}}{\sigma_c} \qquad \text{(Eq 11)}$$

where U_{max} is the maximum normalized distortional energy at any point in the matrix and σ_c is the applied stress on the composite. For a given σ_c, the distortional energy U_{max} is a function of several factors, including V_f, fiber packing, fiber-matrix interfacial adhesion, and constituent properties. This method is more accurate and rigorous and hence is anticipated to produce more reliable results [7].

Because of the stress (or strain) enhancement effect produced by fibers, the failure of unidirectional composites subjected to transverse tensile loads initiates as a result of matrix or interface tensile failure. In some cases they may fail by fiber transverse tensile failure, if the fibers are highly oriented and weak in the transverse direction. Therefore, the composite failure modes under transverse tensile loads may be characterized as matrix tensile failure or interfacial debonding/fiber splitting [7].

Transverse Compressive Strength. When subjected to a transverse compressive load, a unidirectional composite generally fails by shear failure of the matrix resin, possibly attended with interfacial debonding or fiber crushing [8]. As a first-order approximation, the transverse compressive strength can be taken to be the maximum shear strength of the matrix divided by a shear stress concentration factor, provided matrix shear failure is known to be the dominant mode of fracture.

In-Plane Shear Strength and Failure Modes

A unidirectional composite under in-plane shear loads generally fails by matrix shear fracture, interfacial debonding, or a combination of both. Harris [9] reported that the in-plane shear strength of a high-strength, carbon fiber-reinforced epoxy composite is higher than that of high-modulus carbon fiber composites. The higher degree of graphitization or higher level of crystallite orientation in higher-modulus carbon fibers leads to a smoother fiber surface that is deficient in the useful surface functions required to achieve good fiber-epoxy interfacial bonding. Contrarily, the higher concentration of crystallite edges and corners in high-strength carbon fibers (Chapter 4) provides active sites for improved interfacial bonding and hence a better in-plane shear strength for composites.

Failure Prediction for a Unidirectional Lamina

Because of the nonisotropic and nonhomogeneous nature of fiber-reinforced polymer composites, they are known to exhibit many unique characteristics. For example, principal stresses and principal strains may not be in the same direction, and coupling exists between extensional and shear deformations. Failure modes in polymer composites are generally noncatastrophic and may involve several localized damage mechanisms such as fiber breakage, matrix cracking, debonding, and fiber pull-out. Many failure modes can progress simultaneously and

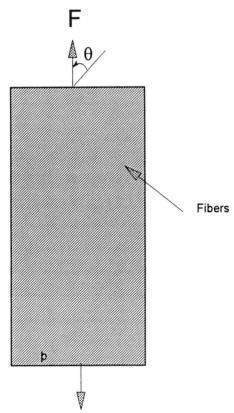

F

θ

Fibers

b

Fig. 4   Loading in a direction with an angle θ with respect to the fiber axis

interactively, making failure prediction for composites a very challenging task.

Failure prediction of a structure is generally performed by comparing stresses (or strains) due to applied loads with the allowable strength (or strain capacity) of the material. When biaxial or multiaxial stress fields are involved, an appropriate failure theory is used for this comparison. For isotropic materials that exhibit yielding, such as a low-carbon steel or an aluminum alloy, either the Tresca theory (maximum shear stress) or the von Mises yield criterion (the distortional energy theory) is commonly used for design against yielding. In general, fiber-reinforced thermoset composites are not isotropic and do not show gross yielding. Although the recently developed high-performance thermoplastic composites may undergo plastic deformation, they are by no means isotropic. One should not be surprised, therefore, to see that failure theories developed for metals are not directly applicable to these composites. Several failure theories have been proposed for composites, some of which are summarized in this section.

Consider a general orthotropic lamina under the plane stress condition (Fig. 4), where all the fibers are inclined at an angle θ with respect to the x-axis. We have learned in Chapter 5 that four independent elastic constants (E_{11}, E_{22}, G_{12}, and ν_{12}) are required to define the

elastic behavior of this lamina. The lamina strength properties are characterized by five independent strength constants:

- S_{Lt} = longitudinal tensile strength (or strain ε_{Lt})
- S_{Tt} = transverse tensile strength (or strain ε_{Tt})
- S_{Lc} = longitudinal compressive strength (or strain ε_{Lc})
- S_{Tc} = transverse compressive strength (or strain ε_{Tc})
- S_s = in-plane shear strength (or strain γ_s)

Again, the transformed stresses are denoted by σ_{11}, σ_{22}, and τ_{12} (in the material axes), while the applied stresses are denoted by σ_{xx}, σ_{yy}, and τ_{xy} (in the structural axes).

Maximum Stress Criterion. According to this theory, failure occurs when any stress in the principal material directions is equal to or greater than the corresponding allowable strength. This implies that, to avoid failure, we must have

$$-S_{Lc} < \sigma_{11} < S_{Lt}$$
$$-S_{Tc} < \sigma_{22} < S_{Tt}$$
$$-S_s < \tau_{12} < S_s \qquad \text{(Eq 12)}$$

Consider the case of simple tension, where only σ_{xx} is present and $\sigma_{yy} = \sigma_{xy} = 0$. Depending on the angle θ, the composite can fail in several different modes:

- failure normal to the fiber;
- failure parallel to the fibers by matrix rupture or fiber-matrix interface tensile failure (θ = 90°); or
- failure by shear of the matrix or fiber-matrix interface, or a combination of these modes.

It is a straightforward task to show that the tension stress parallel to fibers is $\sigma_{11} = \sigma_{xx} \cos^2\theta$, tensile stress perpendicular to fibers is $\sigma_{22} = \sigma_{xx} \sin^2\theta$, and shear stress parallel to fibers is $\tau_{12} = \sigma_{22} \sin\theta\cos\theta$. Failure will occur when $\sigma_{11} = S_{Lt}$, $\sigma_{22} = S_{Tt}$, or $\tau_{12} = S_s$. The critical failure stress for each mode is given by:

Mode (a):

$$\sigma_{xx} = \frac{S_{Lt}}{\cos^2\theta}$$

Mode (b):

$$\sigma_{xx} = \frac{S_{Tt}}{\sin^2\theta}$$

Mode (c):

$$\sigma_{xx} = \frac{S_s}{\sin\theta\cos\theta} \qquad \text{(Eq 13)}$$

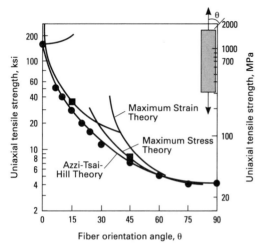

Fig. 5 Theoretical results based on three failure criteria: maximum stress, maximum strain, and Azzi-Tsai-Hill. Experimental data points were obtained from uniaxial strength measurements on a glass-epoxy composite. Source: Ref 10

These equations do not take into account the interaction of stresses and the occurrence of mixed-mode fracture. Figure 5 demonstrates that the composite strength falls rapidly with increasing θ. When the angle θ is high, the strength of the composite tends to be matrix- or interface-dominated. This statement is true not only of unidirectional composites but also of any lamina in a multidirectional laminate.

Maximum Strain Theory. This theory maintains that failure will occur if any of the strains in the principal material axes is equal to or greater than the corresponding allowable strain. Thus, to avoid failure, the following inequalities must be satisfied:

$$-\varepsilon_{Lc} < \varepsilon_{11} < \varepsilon_{Lt}$$
$$-\varepsilon_{Tc} < \varepsilon_{22} < \varepsilon_{Tt}$$
$$-\gamma_s < \gamma_{12} < \gamma_s \qquad \text{(Eq 14)}$$

For an orthotropic lamina subjected to a simple tension σ_{xx}, it can be easily shown that, in the principal material directions:

$$\varepsilon_{11} = \frac{\sigma_{11}}{E_{11}} - \nu_{21}\frac{\sigma_{22}}{E_{22}} = \frac{1}{E_{11}}(\cos^2\theta - \nu_{12}\sin^2\theta)\sigma_{xx}$$

$$\varepsilon_{22} = \frac{\sigma_{22}}{E_{22}} - \nu_{12}\frac{\sigma_{11}}{E_{11}} = \frac{1}{E_{22}}(\sin^2\theta - \nu_{21}\cos^2\theta)\sigma_{xx}$$

$$\gamma_{12} = \frac{\tau_{12}}{G_{12}} = \frac{1}{G_{12}}(\sin\theta \cos\theta)\sigma_{xx} \qquad \text{(Eq 15)}$$

Failure will occur when $\varepsilon_{11} \geq \varepsilon_{Lt}$, $\varepsilon_{22} \geq \varepsilon_{Tt}$, or $\gamma_{12} \geq \gamma_s$.

By assuming the material to be linearly elastic up to the failure strain, failure of the lamina can be predicted if the applied stress σ_{xx} exceeds the smallest of

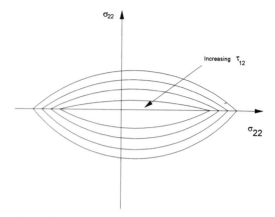

Fig. 6 Schematic of a failure envelope drawn according to the Azzi-Tsai-Hill criterion for a general orthotropic composite

$$\text{(a)} \quad \frac{E_{11}\varepsilon_{Lt}}{\cos^2\theta - \nu_{12}\sin^2\theta} = \frac{S_{Lt}}{\cos^2\theta - \nu_{12}\sin^2\theta}$$

$$\text{(b)} \quad \frac{E_{22}\varepsilon_{Tt}}{\sin^2\theta - \nu_{21}\cos^2\theta} = \frac{S_{Tt}}{\sin^2\theta - \nu_{21}\cos^2\theta}$$

$$\text{(c)} \quad \frac{G_{12}\gamma_s}{\sin\theta \cos\theta} = \frac{S_s}{\sin\theta \cos\theta} \qquad \text{(Eq 16)}$$

Figure 5 also shows that the maximum strain theory and the maximum stress theory give very similar results. This can be ascribed to the assumption of linear elasticity. The differences are a result of the Poisson's effect. Both theories are operationally simple. Again, no interaction between strengths in different directions is accounted for in the maximum strain theory.

Azzi-Tsai-Hill Theory (Maximum Work Theory). Following Hill's anisotropic yield criterion for metals [10], Azzi and Tsai [11] developed a maximum work theory for an orthotropic lamina. This theory states that in plane stress states, the failure initiates when the following inequality is violated:

$$\frac{\sigma_{11}^2}{S_{Lt}^2} - \frac{\sigma_{11}\sigma_{22}}{S_{Lt}^2} + \frac{\sigma_{22}^2}{S_{Tt}^2} + \frac{\tau_{12}^2}{S_s^2} < 1 \qquad \text{(Eq 17)}$$

where σ_{11} and σ_{22} are both tensile stresses. When normal stresses σ_{11} and/or σ_{22} are compressive, the corresponding compressive strengths are to be used in Eq 17.

Returning to the simple case of uniaxial tensile loading considered earlier, failure is predicted if

$$\sigma_{xx} \geq \frac{1}{\left[\dfrac{\cos^4\theta}{S_{Lt}^2} - \dfrac{\sin^2\theta \cos^2\theta}{S_{Lt}^2} - \dfrac{\sin^4\theta}{S_{Tt}^2} - \dfrac{\sin^2\theta \cos^2\theta}{S_s^2}\right]^{1/2}}$$

(Eq 18)

This equation, also plotted in Fig. 5, indicates a better agreement with experimental data than the maximum stress or the maximum strain criteria.

In order to illustrate the application of the Azzi-Tsai-Hill criterion, let us draw the failure envelope for a general orthotropic lamina (Fig. 6). A failure envelope is a graphic representation of the failure criterion in the stress coordinate system (in two or three dimensions) and forms a boundary between the safe and unsafe design spaces. In the present case of plane stress conditions, we select σ_{11} and σ_{22} as the coordinate axes. By rearranging Eq 17, we can represent this theory by the following equations:

In the first quadrant ($\sigma_{11}, \sigma_{22} > 0$):

$$\frac{\sigma_{11}^2}{S_{Lt}^2} - \frac{\sigma_{11}\sigma_{22}}{S_{Lt}^2} + \frac{\sigma_{22}^2}{S_{Tt}^2} = 1 - \frac{\tau_{12}^2}{S_s^2}$$

In the second quadrant ($\sigma_{11} < 0, \sigma_{22} > 0$):

$$\frac{\sigma_{11}^2}{S_{Lc}^2} + \frac{\sigma_{11}\sigma_{22}}{S_{Lc}^2} + \frac{\sigma_{22}^2}{S_{Tt}^2} = 1 - \frac{\tau_{12}^2}{S_s^2}$$

In the third quadrant ($\sigma_{11}, \sigma_{22} < 0$):

$$\frac{\sigma_{11}^2}{S_{Lc}^2} - \frac{\sigma_{11}\sigma_{22}}{S_{Lc}^2} + \frac{\sigma_{22}^2}{S_{Tc}^2} = 1 - \frac{\tau_{12}^2}{S_s^2}$$

In the fourth quadrant ($\sigma_{11} > 0, \sigma_{22} < 0$):

$$\frac{\sigma_{11}^2}{S_{Lt}^2} + \frac{\sigma_{11}\sigma_{22}}{S_{Lt}^2} + \frac{\sigma_{22}^2}{S_{Tc}^2} = 1 - \frac{\tau_{12}^2}{S_s^2}$$

Combining these equations, one can draw a failure envelope for a given τ_{12}/S_s ratio. As shown in Fig. 6, an increasing shear stress component will result in a shrinking failure envelope, making failure easier to initiate. The discontinuity in the failure envelope is a consequence of the anisotropic strength characteristics of a lamina. The Azzi-Tsai-Hill criterion provides a single function to predict the strength of a lamina. This criterion does take into consideration the interaction between strengths. Experimental support for this theory has been given by several investigators [12-15].

Tsai-Wu Failure Criterion [16]. According to this criterion, an orthotropic lamina when subjected to plane stress conditions will fail if and when the following equation is satisfied:

$$F_{11}\sigma_{11} + F_2\sigma_{22} + F_6\tau_{12} + F_{11}\sigma_{11}^2 + F_{22}\sigma_{22}^2$$
$$+ F_{66}\tau_{12}^2 + 2F_{12}\sigma_{11}\sigma_{22} = 1 \qquad \text{(Eq 19)}$$

where the strength coefficients are given by

$$F_1 = \frac{1}{S_{Lt}} - \frac{1}{S_{Lc}}$$

$$F_2 = \frac{1}{S_{Tt}} - \frac{1}{S_{Tc}}$$

$$F_6 = 0$$

$$F_{11} = \frac{1}{S_{Lt}S_{Lc}}$$

$$F_{22} = \frac{1}{S_{Tt}S_{Tc}}$$

$$F_{66} = \frac{1}{S_s^2}$$

Here, F_{12} is a strength interaction term between σ_{11} and σ_{22}. Although $F_1, F_2, F_{11}, F_{22},$ and F_{66} can be calculated using the tensile, compressive, and shear strength properties in the principal material directions, F_{12} cannot be determined without a suitable biaxial test [17]. For instance, let us consider an equal biaxial tension test ($\tau_{12} = 0$) in which $\sigma_{11} = \sigma_{22} = \sigma$ at failure. Under these conditions, Eq 19 is simplified to

$$(F_1 + F_2)\sigma + (F_{11} + F_{22} + 2F_{12})\sigma^2 = 1$$

which, upon simple substitution and rearrangement, gives

$$F_{12} = \frac{1}{2\sigma^2}\left[1 - \left(\frac{1}{S_{Lt}} - \frac{1}{S_{Lc}} + \frac{1}{S_{Tt}} - \frac{1}{S_{Tc}}\right)\sigma \right.$$
$$\left. - \left(\frac{1}{S_{Lt}S_{Lc}} + \frac{1}{S_{Tt}S_{Tc}}\right)\sigma^2\right]$$

Tsai and Hahn [18] recognized the difficulty in conducting reliable biaxial tests and hence recommended an approximate range of values for F_{12}:

$$-\tfrac{1}{2}(F_{11}F_{22})^{1/2} \leq F_{12} \leq 0 \qquad \text{(Eq 20)}$$

where the lower limit can be used as a first-order approximation when no biaxial test data are available. Experimental data [18] appear to provide the strongest support for this criterion as compared to the above three.

Failure and Strength Prediction for General Laminates

In order to predict failure and estimate the strength of a laminate, we must know the stresses and strains in each lamina. Procedures for calculating stresses and strains in individual laminas due to external loads on the laminate have been discussed in Chapter 5. Using the classical lamination theory, the individual lamina stresses or strains in the loading directions may be transformed into stresses or strains in the principal material directions for each lamina. These values in each lamina may be com-

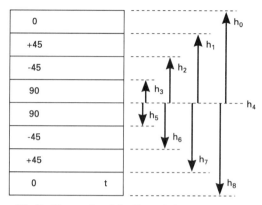

Fig. 7 Diagram for solving Example 1

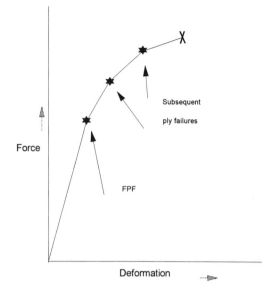

Fig. 8 Schematic of a tensile load-deformation curve of a laminate

pared with the corresponding allowable values to predict failure using any of the failure criteria discussed in the above section. Thus, for a given load, it may be easily determined whether any of the laminas in the laminate will fail. Conversely, the load at which first-ply failure (FPF) will occur can also be estimated.

After a lamina fails, the stresses and strains in the remaining laminas will increase and the laminate stiffness will be reduced. Because the strength of a lamina is a function of its fiber orientation, all laminas will not fail at the same load. Laminas will fail successively in the increasing order of strength in the loading direction. Because the transverse strength of unidirectional laminas is known to be much smaller than the longitudinal strength, the laminas with fibers perpendicular to the load are expected to fail first.

When FPF occurs, the laminate may not fail immediately, and the failed lamina may not cease sharing the load in all directions. Several approaches have been proposed to account for the effect of the failed lamina on the subsequent behavior of the laminate [2,3,19]. Examples include the total discount method, the limited discount method, and the residual property method. In the total discount method, zero stiffness and strength are assigned to the failed lamina in all directions. The limited discount method specifies that zero stiffness and strength are assigned to the failed lamina for the transverse and shear modes if the ply failure is in the matrix material. The total discount method is adopted, however, if the lamina fails by fiber rupture. In the residual property method, certain values of residual strength and stiffness are assigned to the failed lamina in the calculations of the subsequent properties of the damaged laminate.

Example 1. Determine the stiffness matrices before and after the FPF of a quasi-isotropic $[0/\pm45/90]_s$ laminate made from AS-4 carbon-fiber-reinforced PEEK that is subjected to an in-plane normal load N_{xx} per unit width. Use the same material properties as in Examples 1 and 2 in Chapter 5. Assume that each ply has a thickness t.

Solution. The values of various Q'_{mn} (in GPa) before FPF are:

Stiffness coefficient	0°	+45°	−45°	90°
Q'_{11}	146.11	43.34	43.34	9.08
Q'_{12}	2.36	36.62	36.62	2.36
Q'_{16}	0	34.26	−34.26	0
Q'_{22}	9.08	43.34	43.34	146.11
Q'_{26}	0	34.26	−34.26	0
Q'_{66}	3.36	37.62	37.62	3.36

Now draw a diagram to show the relative locations of each ply (Fig. 7). We can see from this diagram that $h_8 = -h_0 = 4t$, $h_7 = -h_1 = 3t$, $h_6 = -h_2 = 2t$, $h_5 = -h_3 = t$, and $h_4 = 0$. We also observe that

$$(Q'_{mn})_1 = (Q'_{mn})_8 = (Q'_{mn})_{0°}$$
$$(Q'_{mn})_2 = (Q'_{mn})_7 = (Q'_{mn})_{+45°}$$
$$(Q'_{mn})_3 = (Q'_{mn})_6 = (Q'_{mn})_{-45°}$$
$$(Q'_{mn})_4 = (Q'_{mn})_5 = (Q'_{mn})_{90°}$$

Because this is a symmetric and balanced laminate, we have $[B] = 0$ and $A_{16} = A_{26} = 0$. Now, with a uniform thickness t:

$$A_{mn} = \sum_{j=1}^{8} (Q'_{mn})_y (h_j - h_{j-1})$$
$$= 2t[(Q'_{mn})_{0°} + (Q'_{mn})_{45°} + (Q'_{mn})_{-45°} + (Q'_{mn})_{90°}]$$

Therefore:

$[A]$ (before FPF)

$$= t \cdot \begin{bmatrix} 483.74 & 155.92 & 0 \\ 155.92 & 483.74 & 0 \\ 0 & 0 & 163.92 \end{bmatrix} \text{GPa}$$

The 90° layers are expected to fail first. If the total discount method is used, then $Q'_{ij} = 0$ for the failed 90° layers. Therefore, the $[A]$ matrix is reduced to

$$[A] \text{ (after FPF)} = t \cdot \begin{bmatrix} 465.58 & 151.20 & 0 \\ 151.20 & 191.52 & 0 \\ 0 & 0 & 157.20 \end{bmatrix} \text{GPa}$$

Let us now consider the load-displacement curve of a hypothetical laminate (Fig. 8). In the beginning, when the applied load is small and no ply failure occurs, a linear curve is expected. As the external load is raised, the stresses in a lamina may become high enough to cause failure. After the FPF, the laminate response will deviate from this initial straight line, which is predicted by Eq 27 and 28 in Chapter 5. A discontinuity in the load-deformation curve may be observed. Further increases in load will cause more ply failures and correspondingly more discontinuities (knees) in the curve. Each segment between two discontinuities may be assumed to show a linear behavior up to fracture. Therefore, the load-strain relationship for each segment of the curve may be written for incremental load and incremental strain as follows [7]:

$$\begin{bmatrix} \Delta N_{xx} \\ \Delta N_{yy} \\ \Delta N_{xy} \end{bmatrix}_i = [\overline{A}]_i \begin{bmatrix} \Delta\varepsilon^0_{xx} \\ \Delta\varepsilon^0_{yy} \\ \Delta\gamma^0_{xy} \end{bmatrix}_i + [\overline{B}]_i \begin{bmatrix} \Delta K_{xx} \\ \Delta K_{yy} \\ \Delta K_{xy} \end{bmatrix}_i \qquad \text{(Eq 21)}$$

and

$$\begin{bmatrix} \Delta M_{xx} \\ \Delta M_{yy} \\ \Delta M_{xy} \end{bmatrix}_i = [\overline{B}]_i \begin{bmatrix} \Delta\varepsilon^0_{xx} \\ \Delta\varepsilon^0_{yy} \\ \Delta\gamma^0_{xy} \end{bmatrix}_i + [\overline{D}]_i \begin{bmatrix} \Delta K_{xx} \\ \Delta K_{yy} \\ \Delta K_{xy} \end{bmatrix}_i \qquad \text{(Eq 22)}$$

These equations were derived from Eq 27 and 28 in Chapter 5, where the new matrices $[\overline{A}]$, $[\overline{B}]$, and $[\overline{D}]$ have been modified to accommodate the fact that some of the plies have already failed. The incremental loads and strains determined from Eq 21 and 22 may now be added to the loads and strains at the previous ply failure, or simply those at the end of previous segment, to obtain their absolute values [7]:

$$\begin{bmatrix} N_{xx} \\ N_{yy} \\ N_{xy} \end{bmatrix} = \begin{bmatrix} N_x \\ N_y \\ N_{xy} \end{bmatrix}_{i-1} + \begin{bmatrix} \Delta N_x \\ \Delta N_y \\ \Delta N_{xy} \end{bmatrix}_i \qquad \text{(Eq 23)}$$

$$\begin{bmatrix} M_{xx} \\ M_{yy} \\ M_{xy} \end{bmatrix} = \begin{bmatrix} M_{xx} \\ M_{yy} \\ M_{xy} \end{bmatrix}_{i-1} + \begin{bmatrix} \Delta M_{xx} \\ \Delta M_{yy} \\ \Delta M_{xy} \end{bmatrix}_i \qquad \text{(Eq 24)}$$

$$\begin{bmatrix} \varepsilon^0_{xx} \\ \varepsilon^0_{yy} \\ \gamma^0_{xy} \end{bmatrix} = \begin{bmatrix} \varepsilon^0_{xx} \\ \varepsilon^0_{yy} \\ \gamma^0_{xy} \end{bmatrix}_{i-1} + \begin{bmatrix} \Delta\varepsilon^0_{xx} \\ \Delta\varepsilon^0_{yy} \\ \Delta\gamma^0_{xy} \end{bmatrix}_i \qquad \text{(Eq 25)}$$

$$\begin{bmatrix} K_{xx} \\ K_{yy} \\ K_{xy} \end{bmatrix} = \begin{bmatrix} K_{xx} \\ K_{yy} \\ K_{xy} \end{bmatrix}_{i-1} + \begin{bmatrix} \Delta K_{xx} \\ \Delta K_{yy} \\ \Delta K_{xy} \end{bmatrix}_i \qquad \text{(Eq 26)}$$

The strains and stresses in each lamina may now be calculated from Eq 40 and 41 in Chapter 5, respectively. Comparison of these stresses and strains with their allowable values can then be made to determine the load at which the next ply failure occurs. This stepwise procedure can be repeated until all plies have failed. The "ultimate strength," the load at which the final fracture of the laminate occurs, can also be calculated.

Whisker- and Short-Fiber-Reinforced Composites

In Chapter 5, we mentioned that a lamina can also be constructed from discontinuous (short) fibers dispersed in a matrix. The fibers can be arranged either in a unidirectional orientation, a planar (two-dimensional) orientation, or a three-dimensional random orientation. Discontinuous fiber-reinforced composites are known to have lower moduli than continuous fiber composites. However, with random orientation of fibers, it is possible to obtain more isotropic properties in all directions in the plane of the lamina. In this section, the effect of the discontinuous nature of fibers on the failure of composites is considered. The discussion begins with an introduction to the theories of stress transfer between the fiber and the matrix. This is followed by a discussion of the approaches that can be used to estimate the strength of a unidirectional, discontinuous fiber composite and that of a two-dimensional random composite.

Theories of Stress Transfer

The analyses for the mechanics of whisker- or short-fiber-reinforced composites have two major goals. The first is to establish the capability to predict the response of a given composite to specified loads so that these materials can be incorporated into the design of structures. The second goal is to develop an understanding of the influence of constituent properties on composite performance so that suitable materials can be designed or tailored for intended applications.

For discontinuous fiber composites, stress concentrations are built up in the vicinity of fiber ends or around fiber breaks during loading. Thus, the stress analysis of interest for the discussion of failure in composites is the one associated with the perturbed stress field in the vicinity of a fiber end or break.

The load applied to a discontinuous fiber composite is transferred to the fibers by a shearing mechanism between fibers and matrix. Because the matrix has a lower modulus, the longitudinal strain in the matrix is higher than that in adjacent fibers. If a perfect fiber-matrix interfacial bond is assumed, the difference in longitudinal strains between the two constituents creates a shear stress distribution across the interface. Ignoring the stress transfer at the fiber end cross sections and the interaction between neighboring fibers, the normal stress distribution in a short fiber can be calculated by a simple force equilibrium analysis. The variations of stresses along the length of a fiber were studied by Cox [20] and Outwater [21] in the 1950s. However, the most often quoted theory of stress transfer appears to be Rosen's shear-lag analysis [22], which is a modified version of an earlier analysis by Dow [23].

Consider the model shown in Fig. 9. The force balance equation for the infinitesimal length dx at a distance x from one of the fiber ends can be expressed as:

$$\left(\frac{\pi}{4}d_f^2\right)(\sigma_f + d\sigma_f) = \left(\frac{\pi}{4}d_f^2\right)\sigma_f + \pi d_f\, dx\tau \qquad \text{(Eq 27)}$$

where σ_f = longitudinal stress in the fiber at a distance x from one of its ends, τ = shear stress at the fiber-matrix interface, and d_f = fiber diameter.
Then

$$\frac{d\sigma_f}{dx} = \frac{4\tau}{d_f}$$

and

$$\sigma_f = \frac{4}{d_f}\int_0^x \tau\, dx \qquad \text{(Eq 28)}$$

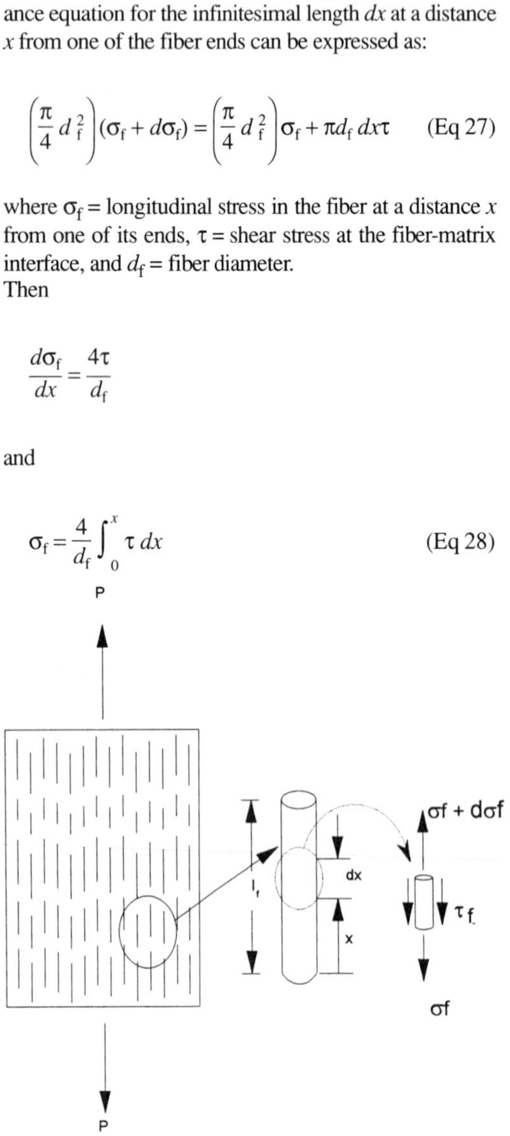

Fig. 9 Longitudinal tensile loading of a discontinuous parallel fiber lamina

The right-hand side of Eq 28 can be evaluated if the variation of shear stress along the fiber length is known. However, the shear stress is not known *a priori* and is actually determined as a part of the complete solution. To obtain analytical solutions, it is necessary to make assumptions regarding the deformation of material surrounding the fiber and the fiber end conditions. One frequently used assumption is that the matrix material surrounding the fiber is a rigid perfectly plastic material, the interface shear stress being constant (τ_i) along the fiber length and assumed to be equal to the matrix shear yield strength (τ_y). Then

$$\sigma_f = 4\tau_i x/d_f \qquad \text{(Eq 29)}$$

In a discontinuous fiber lamina, the stress along the fiber is not uniform. The stress is zero at the ends and builds up linearly to the maximum value at the center portion of the fiber. The maximum fiber stress that can be achieved at a given load is:

$$(\sigma_f)_{max} = 2\tau_i\frac{l_f}{d_f} \qquad \text{(Eq 30)}$$

However, it is important to note that the fiber stress has a limiting value, which is the stress that would be accepted by a fiber of continuous length at a given stress applied to the composite (σ_c):

$$(\sigma_f)_{max} = \frac{E_f}{E_c}\sigma_c \qquad \text{(Eq 31)}$$

where E_c is the composite modulus, which can be calculated from the rule-of-mixture law. The minimum fiber length, where the maximum fiber stress ($\sigma_f)_{max}$ can be achieved, may be defined as a load-transfer length, l_t ($l_t/2$ from each fiber end). It is over this length of the fiber that the load is transferred from the matrix to fiber. This is given by

$$\frac{l_t}{d_f} = \frac{(\sigma_f)_{max}}{2\tau_i} = \frac{\left(\dfrac{E_f}{E_c}\right)\sigma_c}{2\tau_i} \qquad \text{(Eq 32)}$$

Because $(\sigma_f)_{max}$ is a function of applied stress, so is l_t. A critical fiber length, l_c, independent of applied stress may be defined as the minimum fiber length where the fiber ultimate strength σ_{fu} can be achieved. The critical fiber length then can be calculated as:

$$l_c = \frac{\sigma_{fu}}{2\tau_i}d_f \qquad \text{(Eq 33)}$$

where σ_{fu} = ultimate fiber strength, l_c = the minimum fiber length required for the stress to equal the fiber ultimate strength at its midlength, and l_c = the maximum value of load transfer length, l_t.

The load-transfer length and critical length are sometimes referred to as the "ineffective length" because over this length the fiber supports a stress less than the maximum fiber stress. Shown in Fig. 10(a) are the fiber stress and interface shear stress distributions associated with fibers of different lengths. Figure 10(b) shows variations of fiber stress for increasing applied stress exerted on the composite containing a fiber longer than the critical length. A small length adjoining the fiber ends is stressed at less than the maximum fiber stress. Contrarily, when the fiber length is much greater than the load-transfer length, the composite performance approaches the performance of a corresponding continuous fiber composite.

Although normal stresses near the two fiber ends are lower than maximum fiber stress (i.e., at $x < \frac{1}{2}l_t$), their contributions to the total load-carrying capacity of the fiber cannot be completely ignored. With these end stress distributions included, an average fiber stress may be calculated as:

$$\bar{\sigma}_f = \frac{1}{l_f} \int_0^{l_f} \sigma_f \, dx = (\sigma_f)_{max} \left(1 - \frac{l_t}{2l_f}\right) \quad \text{(Eq 34)}$$

Note that the maximum load transfer length for $l_f < l_c$ is l_f, whereas that for $l_f > l_c$ is l_c.

For $l_f > l_c$, the longitudinal tensile strength of a unidirectional discontinuous fiber composite is calculated by assigning $(\sigma_f)_{max} = \sigma_{fu}$ and $l_t = l_c$. Thus:

$$\begin{aligned}\sigma_{Ltu} &= \bar{\sigma}_{fu}V_f + \sigma'_m (1 - V_f) \\ &= \sigma_{fu}\left(1 - \frac{l_c}{2l_f}\right)V_f + \sigma'_m (1 - V_f) \end{aligned} \quad \text{(Eq 35)}$$

This equation assumes that all fibers fail at the same strength level of σ_{fu}. It also shows that discontinuous fibers always strengthen a matrix to a lesser degree than continuous fibers. However, for $l_f > 5l_c$, more than 90% strengthening can be achieved, even with discontinuous fibers. For $l_f < l_c$, no fiber failure is expected; instead, two different failure modes are possible. The first is interfacial bond failure followed by fiber pull-out from the matrix. In this case, the longitudinal tensile strength of the composite is given by

$$\sigma_{Ltu} = 2\tau_i \frac{l_f}{d_f} V_f + \sigma'_m (1 - V_f) \quad \text{(Eq 36)}$$

where τ_i = fiber matrix interfacial shear strength and σ'_m = matrix stress at the instant of fiber pull-out. The second is matrix tensile failure, where the longitudinal tensile strength of the composite is given by

$$\sigma_{Ltu} = 2\tau_i \frac{l_f}{d_f} V_f + \sigma_{mu}(1 - V_f) \quad \text{(Eq 37)}$$

where σ_{mu} = tensile strength of the matrix material. The actual failure mode for $l_f < l_c$ will depend on the relative values of τ_i and σ_{mu}.

Strength Prediction for Discontinuous Fiber Composites

Considered in the above section has been the case of stress transfer to a single fiber. Kelly [24] has derived a multifiber shear lag expression for composites containing discontinuous, but perfectly aligned fibers. Kelly's analysis predicts that 95% of the strength obtainable with continuous fiber composites can be obtained with aligned discontinuous fibers. The properties are closely related to the critical fiber length l_c. For a yielding matrix such as metal, l_c is simply related to the fiber ultimate strength and the shear yield strength of the matrix [24]. For metal matrix systems where matrix yielding is predominant, Kelly's analysis is in reasonable agreement with the experimental data obtained from fiber pull-out measurements [25,26]. This approach should be applicable to select thermoplastic composites where the matrices are reasonably ductile. In systems where the matrix does not yield and stress concentrations about the fibers are significant (e.g., in carbon-fiber-reinforced tetrafunctional epoxy, polyimide, or BMI), the analysis may not be applicable.

Lees [27] modified Kelly's shear lag analysis to account for the effect of fiber lengths both below and above the critical load transfer length, as well as the residual shrinkage stress generated during composite fabrication. The agreement with the experimental results was rather poor. Following a micromechanics viewpoint, Chen and

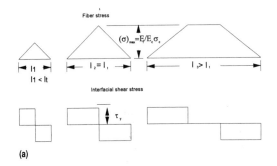

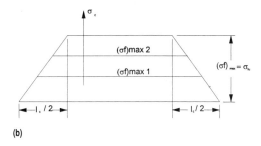

(a)

(b)

Fig. 10 (a) Fiber stress and interfacial shear stress on fibers of different lengths. (b) Influence of changing composite stress on fiber stress variation on a fiber longer than the critical length. Source: Ref 7

Lavengood [28] and Chen [29] used a finite element technique to calculate the composite strength. A much larger aspect ratio is required, as compared to that in the shear lag analysis prediction, for the discontinuous fiber composite to reach a given percentage of the continuous composite strength. Other workers [30-32] have also demonstrated an asymptotic strength limit for aligned discontinuous fiber composites as the aspect ratio increases. Dingle [33] presented experimental data to demonstrate that aligned short fibers could provide composites with strength values greater than 90% of the continuous fiber composite strength.

A semi-empirical approach developed by Halpin and Kardos [34,35] may be used to predict the strength of a uniaxially aligned discontinuous fiber lamina. This is based on a strength reduction factor (SRF), defined as the strength of a uniaxially aligned short fiber system divided by the strength of the corresponding unidirectional continuous lamina of the same fiber volume fraction. Thus

$$[\text{SRF}] = \frac{\overline{\sigma}_c}{\sigma_f V_f + \sigma_m V_m} \approx \frac{\overline{\sigma}_u}{\sigma_f V_f} \qquad \text{(Eq 38)}$$

As a lower bound, the [SRF] approaches that for a sphere-filled system where the aspect ratio approaches unity. This $[\text{SRF}]_0$ can be determined by using the results of Narkis et al. [36-38] for glass-bead-filled thermoplastic and thermoset composites:

$$[\text{SRF}]_0 = \frac{\sigma_{mu} E_c (1 - V_f^{1/2})}{\sigma_f V_f E_m} \qquad \text{(Eq 39)}$$

where σ_{mu} is the matrix strength. The upper bound of [SRF] for a very large aspect ratio, $[\text{SRF}]_\infty$, may be shifted vertically by using the (E_f / E_m) ratio, which practically controls the magnitude of the stress concentrations at the fiber ends [29-31]. The best fit for this shift produces

$$[\text{SRF}]_\infty = 0.5 + (E_f/E_m)^{-0.087} \quad \text{for } E_f/E_m > 5 \quad \text{(Eq 40)}$$

The horizontal shift factor (β) is related to the critical aspect ratio (l_c/d_f) and the ratio of fiber strength to matrix shear strength (or interface shear strength, whichever is lower):

$$\beta = (l_c/d_f)(\sigma_f/\tau_m) \qquad \text{(Eq 41)}$$

The final equation for the master curve may be normalized and expressed as follows [34,35]:

$$G = \frac{[\text{SRF}] - [\text{SRF}]_0}{[\text{SRF}]_\infty - [\text{SRF}]_0}$$
$$= 1 - 0.97 \exp[-0.42\beta] \qquad \text{(Eq 42)}$$

Equations 38 to 42, sometimes collectively referred to as the Halpin-Kardos equations, predict the lowering of continuous fiber strength due to the discontinuous nature of the fibers. Additional equations were developed [39,40] to estimate allowable failure strains for aligned short-fiber laminas.

In the Halpin-Kardos approach [39,40] to strength prediction, a short fiber-reinforced composite is envisioned as composed of several plies, each containing uniaxially aligned short fibers. The plies are oriented in the laminate to replicate the actual part fiber orientation distribution. For instance, a two-dimensional in-plane random orientation can be simulated by a quasi-isotropic laminate $[0°/\pm 45°/90°]_s$ comprising aligned short-fiber laminas of an identical fiber volume fraction. The stress-strain properties are then calculated by analyzing the individual ply response to the external applied load. The rest of the procedure is analogous to that used for the strength prediction of a general laminate, which requires repeated calculations of stiffness matrices, and thus incremental stresses and strains, at various stages of loading. A proper failure criterion is also needed to judge whether the stress, strain, or energy meets the failure requirement of each ply. The necessary equations and an example calculation for a two-dimensional planar random orientation are given in [40].

Woven and Braided Composites

Two-Dimensional Woven Composites

The crimp model and bridging model for the prediction of elastic constants in fabric composites were discussed in Chapter 5. These models may be extended for the study of the stress-strain behavior and hence for the strength prediction of woven fabric composites. At the outset of this analysis, it is important to recall that the breaking strain in the transverse laminas is much smaller than that of the longitudinal laminas in a cross-ply laminate. In both the crimp model and the bridging model, the fabric composites were considered to be composed of many repeating units, which were either cross-ply elements or interlaced elements. In the present analysis, taken after Chou and Ishikawa [41], only the failure of the transverse yarns (which occurs in the warp direction) is considered. Also, a maximum strain criterion is adopted for the strength prediction.

Consider the one-dimensional behavior of fabric composites under an applied N_1. By using the crimp model, Eq 54 in Chapter 5 reduces to

$$\varepsilon_1^0 = a_{11}N_1 + b_{11}M_1$$
$$K_1 = b_{11}N_1 + d_{11}M_1 \qquad \text{(Eq 43)}$$

where M_1 is the locally induced moment resultant due to the presence of N_1. By assuming that no bending deflection by the coupling effect is allowed along the x-axis, we have

$$K_1 = b_{11}N_1 + d_{11}M_1 = 0 \qquad \text{(Eq 44)}$$

and therefore

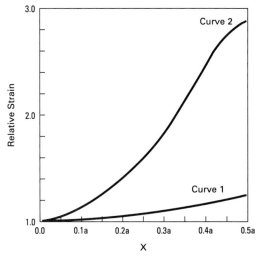

Fig. 11 Relative strain distribution along the *x*-axis in the fiber crimp model without bending. $a_u = a$. Curve 1, graphite-epoxy; curve 2, glass-polyester. Source: Ref 41

$$\varepsilon_1^0 = a_{11}^* N_1 \qquad\qquad\qquad\text{(Eq 45)}$$

where $a_{11}^* = a_{11} - b_{11}^2/d_{11}$ may be referred to as a modified in-plane compliance and is a function of *x*. Because N_1 is assumed to be uniform along the *x*-direction, a_{11}^* represents a strain distribution before the first internal failure. Given in Fig. 11 are two examples of the midplane strain distribution relative to that at the point $x = 0$ in Fig. 13 in Chapter 5 and for $a_u = a$. It is clear that the fiber undulation causes local softening and that the maximum strain appears at the center of undulation ($x = a/2$). Because no bending is allowed, the strain along the thickness direction at each section is uniform; this is a frequently made assumption in the classical lamination theory.

Suppose that the maximum allowable strain ε_2^b is exceeded by the highest strain in the region, leading to immediate failure of the adjacent area. The damaged area in the warp yarn will propagate when the load is raised. The effective elastic moduli of such a failed area in the yarn are lower than those of a sound area by a reducing factor *f*:

$$Q_{ij}^{'w} = \begin{bmatrix} Q_{11}^w/f & Q_{12}^w/f & 0 \\ Q_{12}^w/f & Q_{22}^w & 0 \\ 0 & 0 & Q_{66}^w/f \end{bmatrix} \qquad\text{(Eq 46)}$$

With the exception of Q_{22}^w, the stiffness constants of the warp yarns have been reduced by a factor *f*, say ≈ 100, to account for the weakening effect of transverse cracking. Such a successive failure process will be repeated until the lowest strain in the region reaches ε_2^b. At this point, all the warp regions will have failed completely. Beyond this point, the stress-strain curve becomes a straight line again, with a lower slope, until final failure of the fill

yarns occur. This procedure is rather similar to that used in the failure prediction of a general laminate, discussed in earlier sections. In the case where bending is allowed, modified stiffness matrices $[A_{ij}]$, $[B_{ij}]$, and $[D_{ij}]$ are to be calculated to obtain the strain values in both fill and warp regions for strength predictions [41].

The concept of successive warp yarn failure can also be followed in the bridging model analysis to predict the knee behavior or strength in satin composites. Again, if no bending is allowed, then [41]:

$$\overline{A}_{11}^* = (1/\sqrt{n_g})\overline{A}_{11}^{*c} + (1 - 1\sqrt{n_g})A_{11}^* \qquad\text{(Eq 47)}$$

where $\overline{A}_{11}^* = 1/\overline{a}_{11}^*$. Because N_1 is uniform along the *x*-direction, we have

$$\overline{a}_{11}^{*S} = (2/\sqrt{n_g})\overline{a}_{11}^* + (1 - 2\sqrt{n_g})a_{11}^*$$
$$\overline{A}_{11}^{*s} = 1/\overline{a}_{11}^{*s} \qquad\qquad\text{(Eq 48)}$$

where $\overline{a}_{11}^* = 1/\overline{A}_{11}^*$. The procedure for examining the knee phenomenon is quite similar to that of the plain-weave case using the crimp model.

The crimp model is particularly useful for examining the knee phenomenon of plain-weave composites. The predicted knee behavior of a glass-polyester composite without bending shows an excellent agreement with the stress-strain curve obtained by means of a finite element analysis [41]. The concept of successive failure of the warp yarns, in conjunction with the bridging model, is applicable for the prediction of knee behavior in satin composites. The theoretical results for an eight-harness satin glass/polyimide composite compare very well with the experimental results [41]. One important conclusion derived from these studies is that the bridging regions surrounding the interlaced regions are responsible for the higher stiffness and knee stress in satin composites than those in plain-weave composites.

An extensive experimental program has been conducted by Bishop [42] to study the strength and failure of woven carbon fabric-reinforced epoxy composites. It has been found that when carbon fiber laminates are made with woven fabric rather than with nonwoven material and are loaded parallel to the fibers, distortion of the fibers significantly reduces the tensile and compressive strengths, the stiffness, the toughness, and the fatigue performance. When woven fabric is oriented at 45° to the load direction, these properties compare well with those for nonwoven ($\pm 45°$) laminates. The high strength and stiffness of nonwoven composites made of unidirectional laminas may be maintained, but impact performance is improved if certain amounts of woven composite layers are incorporated in the laminate [42].

Three-Dimensional Composites

Whyte [43,44] has developed a fabric geometry model (FGM) to analyze the mechanical behavior of three-dimensional composites. This model requires a knowledge of the fiber properties, matrix properties, and fiber architecture. Proper geometric unit cells are identi-

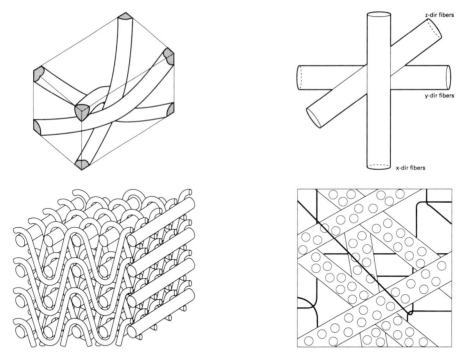

Fig. 12   Unit cells of four basic three-dimensional fabrics. Source: Ref 44

fied to represent the fiber architecture of each type of three-dimensional textile structures considered. The classical lamination theory is then modified to form the mechanics framework of the analysis.

Four types of three-dimensional fabrics (Fig. 12) were considered in Whyte's theory [43]. For a three-dimensional orthogonal nonwoven fabric there are three basic yarn components to the unit cell that are defined in terms of the yarn orientation: 0° (warp), 90° (weft), 45° (bias), –45° (bias), and the through-thickness stitching yarns. The fractional volume of fiber in each of the directions can be calculated geometrically on the basis of yarn size, yarn spacing, and stitch type [43]. The unit cell for the three-dimensional woven fabric depends on the type of weave construction, which is expected to be a combination of the orthogonal unit cell and crimp geometry. The unit cell for the three-dimensional braid may be represented by several yarns running parallel to the body diagonal of the cell. In some cases, however, yarns can run in the longitudinal (0°) and transverse (90°) directions of the fabric [43,44].

According to this FGM model, a generalized stiffness matrix may be formed for each system of yarns, using the stiffness matrix of a comparable unidirectional composite and transforming it by the proper fiber orientation. For each system of yarn we have:

$$[C_i] = [T_{\varepsilon,i}] [C] [T_{\varepsilon,i}]^{-1} \qquad \text{(Eq 49)}$$

where $[C_i]$ is the stiffness matrix for the ith system of yarns, $[T_{\varepsilon,i}]$ is the geometric strain transformation for the ith system, and $[C]$ is the stiffness matrix for a comparable unidirectional composite. Here $[C]$ may be given as

$$[C] = \begin{bmatrix} C_{11} & C_{12} & C_{13} & 0 & 0 & 0 \\ C_{12} & C_{22} & C_{23} & 0 & 0 & 0 \\ C_{13} & C_{23} & C_{33} & 0 & 0 & 0 \\ 0 & 0 & 0 & C_{44} & 0 & 0 \\ 0 & 0 & 0 & 0 & C_{55} & 0 \\ 0 & 0 & 0 & 0 & 0 & C_{66} \end{bmatrix} \qquad \text{(Eq 50)}$$

where $C_{11} = (1 - v_{23}^2)E_{11}/K$

$$C_{22} = C_{33} = (1 - v_{12}v_{21})E_{22}/K$$
$$C_{12} = C_{13} = (1 + v_{23})v_{21}E_{11}/K$$
$$C_{23} = (v_{23} + v_{12}v_{21})E_{22}/K$$
$$C_{44} = G_{23}$$
$$C_{55} = G_{13}$$
$$C_{66} = G_{12}$$
$$K = 1 - 2v_{12}v_{21}(1 + v_{23}) - v_{23}$$

The elastic constants of this equivalent unidirectional composite can be obtained from Eq 1 to 6 in Chapter 5 plus the following equations:

$$G_{12} = G_{13}$$
$$G_{23} = G_{m} / [1 - V_{f}^{1/2}(1 - G_{m} / G_{23f})]$$
$$v_{23} = v_{f}v_{23f} + V_{m}(2v_{m} - v_{21})$$

where G_{23f} = the shear modulus of the fiber and v_{23f} = the Poisson's ratio in the 2,3 plane.

Source: Ref 48

Table 1   Summary of properties of orthogonal nonwoven/phenolic resin composites

Material	In-plane	Through-thickness
Laminated quartz phenolic		
Tensile strength, MPa (ksi)	402 (58.3)	16 (2.32)
Young's modulus, GPa (10^6 psi)	30.6 (4.44)	11.5 (1.67)
Failure strain, %	1.78	0.18
3-D quartz phenolic		
Tensile strength, MPa (ksi)	307 (44.5)	414 (60.0)
Young's modulus, GPa (10^6 psi)	24.6 (3.57)	27.6 (4.0)
Failure strain, %	2.95	2.36
Laminated carbon phenolic		
Tensile strength, MPa (ksi)	130 (18.85)	21 (3.05)
Young's modulus, GPa (10^6 psi)	13.2 (1.91)	11.7 (1.70)
Failure strain, %	1.38	0.12
3-D carbon phenolic		
Tensile strength, MPa (ksi)	104 (15.1)	164 (23.8)
Young's modulus, GPa (10^6 psi)	9.9 (1.435)	12.3 (1.78)
Failure strain, %	1.78	2.02

The transformation matrix $[T_{\varepsilon,i}]$ is the Hamiltonian tensor transformation matrix, defined as:

$$[T_{\varepsilon,i}]$$

$$= \begin{pmatrix} l^2 & m^2 & n^2 & 2mn & 2ln & 2lm \\ l'^2 & m'^2 & n'^2 & 2m'n' & 2l'n' & 2l'm' \\ l''^2 & m''^2 & n''^2 & 2m''n'' & 2l''n'' & 2l''m'' \\ l'l'' & m'm'' & n'n'' & m'n''+m''n' & l'n''+l''n' & l'm''+l''m' \\ ll'' & mm'' & nn'' & mn''+m''n & ln''+l''n & lm''+l''m \\ l'l & m'm & n'n & mn'+m'n & ln'+l'n & lm'+l'm \end{pmatrix}$$

$$(\text{Eq } 51)$$

where $l = \cos\theta$, $m = 0$, $n = -\sin\theta$, $l' = \sin\theta\cos\beta$, $m' = \sin\beta$, $n' = \cos\theta\cos\beta$, $l'' = \sin\theta\sin\beta$, $m'' = -\cos\beta$, $n'' = \cos\theta\sin\beta$, and θ and β are the two angles defining the orientation of any fiber, as shown in Fig. 12. Specifically, θ is the orientation of the fiber with respect to the longitudinal axis of the structure and β is the azimuthal angle of the fiber. For example, for a three-dimensional braided construction, (θ,β) can be (0,0) for the longitudinal components, (90,0) for the transverse components, or (θ^*,β^*) for the braiding components, where $\theta^* = \tan^{-1}[\tan\theta' \cdot (1 + k^2)^{1/2}]$, $\beta^* = \tan^{-1}(1/k)$, θ' = the surface angle of the yarn, and k = the ratio of track movement to column movement [44].

In order to determine the stress-strain behavior of a three-dimensional fabric composite, the stiffness matrices for each system of yarns are first superimposed proportionately according to contributing volume to obtain the composite system stiffness:

Table 2   Effect of yarn bundle size on the mechanical properties of 3-D orthogonal nonwoven (ON) composites

Property	ON-3k	ON-6k	ON-12k	Laminate-3k
Yarn spacing, mm				
Z_x	0.7	1.0	1.4	...
Z_y	0.7	1.0	1.4	...
X_y	0.7	1.0	1.4	...
X_z	2.4	3.2	4.7	...
Y_x	0.7	1.0	1.4	...
Y_z	2.4	3.2	4.7	...
Fiber volume fraction, %	52.6	50.4	50.8	57.1
Specific gravity, g/cc (lb/in.^{3})	1.51 (0.0545)	1.49 (0.0538)	1.50 (0.0541)	1.53 (0.0552)
Tensile strength, MPa (ksi)	...	746 (108.2)	607 (88.0)	
Tensile modulus, GPa (10^6 psi)	61.8 (8.96)	56.1 (8.14)	52.0 (7.54)	62.9 (9.12)
Specific tensile modulus, GPa/(g/cc)	40.9	37.7	34.7	41.1
Flexural strength, MPa (ksi)	760 (110)	673 (97.6)	555 (80.5)	736 (106.7)
Specific flexural strength, MPa/(g/cc)	503	452	370	481
Flexural modulus, GPa (10^6 psi)	55.3 (8.0)	51.1 (7.41)	46.1 (6.69)	55.7 (8.08)
Specific flexural modulus, GPa/(g/cc)	36.6	34.3	30.7	36.4
Compressive strength, MPa (ksi)	477 (69.2)	415 (60.2)	327 (47.4)	459 (66.5)
Specific compressive strength, MPa/(g/cc)	316	272	218	300
Compressive modulus, GPa (10^6 psi)	...	52.9 (7.67)	52.2 (7.57)	...
Compressive modulus, GPa (10^6 psi)	61.4 (8.90)	48.5 (7.03)	45.6 (6.61)	62.7 (9.09)
Specific compressive modulus, GPa/(g/cc)	40.7	32.6	30.4	41.6
Shear strength, MPa (ksi)	>163 (23.6)	>161 (23.4)	>145 (21.0)	42.1 (6.10)
Poisson's ratio	...	0.130	0.139	...

Source: Ref 49

Table 3   Cut-edge effect on 3-D braided carbon/epoxy composites

Property	T-300 1×1 (uncut)	T-300 1×1 (cut)	T-300 3×1 (uncut)	T-300 3×1 (cut)	T-300 1×1×1/2F (uncut)	T-300 1×1×1/2F (cut)
Fiber volume fraction, %	68	68	68	68	68	68
Tensile strength, MPa (ksi)	665.6 (96.5)	228.7 (33.17)	970.5 (140.8)	363.7 (52.75)	790.6 (114.7)	405.7 (58.84)
Tensile modulus, GPa (10^6 psi)	97.8 (14.2)	50.5 (7.32)	126.4 (18.33)	76.4 (11.1)	117.4 (17.02)	82.4 (11.95)
Compressive strength, MPa (ksi)	...	179.5 (26.03)	...	226.4 (32.83)	...	385.4 (55.9)
Compressive modulus, GPa (10^6 psi)	...	38.7 (5.61)	...	56.6 (8.21)	...	80.8 (11.7)
Flexural strength, MPa (ksi)	813.5 (118.0)	465.2 (67.47)	647.2 (93.87)	508.1 (73.69)	816 (118)	632.7 (91.76)
Flexural modulus, GPa (10^6 psi)	77.5 (11.2)	34.1 (4.95)	85.4 (12.4)	54.9 (7.96)	86.4 (12.5)	60.8 (8.82)
Poisson's ratio	0.875	1.36	0.566	0.806	0.986	0.667
Apparent fiber angle, °	±20	±20	±12	±12	±15	±12

Source: Ref 50

Table 4   Effect of yarn bundle size on 3-D braided carbon/epoxy composites

Property	AS-4 3k 1×1	AS-4 6k 1×1	Celion 6k 1×1	AS-4 12k 1×1	Celion 12k 1×1	Thornel 300 30k 1×1	Thornel 300 8-harness satin weave
Fiber volume fraction, %	68	68	56	68	68	68	65
Tensile strength, MPa (ksi)	736.8 (106.9)	841.4 (122.0)	857.7 (124.4)	1067.2 (154.8)	1219.8 (176.9)	665.6 (96.5)	517.1 (75.0)
Short beam shear, MPa (ksi)	114.8 (76.65)	126 18.3)	71.4 (10.35)	121.4 (17.6)	71.4 (10.35)	...	69 (100)
Poisson's ratio	0.945	1.051	0.968	0.98	0.874	0.875	0.045
Flexural strength, MPa (ksi)	885.3 (128.4)	739.8 (107.3)	...	1063.3 (154.2)	...	813.5 (118.0)	689.5 (100)
Flexural modulus, GPa (10^6 psi)	84.5 (12.3)	95.2 (13.8)	...	1365.2 (198.0)	...	77.5 (11.2)	65.5 (9.5)
Apparent fiber angle, °	±19	±15	±15	±15	±17.5	±20	0

Source: Ref 50

$$[C_s] = \Sigma k_i [C_i] \qquad \text{(Eq 52)}$$

where k_i is the fractional volume of the ith system of yarns.

The potential knee behavior of a three-dimensional composite can be predicted by using a procedure similar to that used in the strength prediction of a general laminate. The stress, strain, strain energy, or distortional energy must be calculated in accordance with the failure criterion chosen. The system stiffness matrix must be calculated repeatedly at various stages of loading to account for the failure of certain systems of yarns. Thus, the incremental stress-strain behavior of the composite can be determined as

$$[\Delta\sigma] = [C'_s][\Delta\varepsilon] \qquad \text{(Eq 53)}$$

where $[\Delta\sigma]$ = the incremental stress vector (6×1), $[\Delta\varepsilon]$ = the incremental strain vector (6×1), and $[C'_s]$ = the reduced stiffness matrix at any stage. The stress vector at any stage of loading is then given by

$$[\sigma]_{i+1} = [\sigma]_i + [\Delta\sigma] \qquad \text{(Eq 54)}$$

where $[\sigma]_i$ is the instantaneous stress at the ith stage of loading. The entire stress-strain curve, including the final strength of the three-dimensional composite, can be predicted by this approach. The strain energy criterion was selected by Whyte and Ko [43,44] and was found to give predictions that agree well with the experimental results conducted on three-dimensional braided alumina/Al-Li composites.

As discussed in Chapter 5, several theories have been developed to predict the elastic constants of various three-dimensional composites. Most of these studies [e.g., 45-47] have yet to be extended to include prediction of composite strength. In principle, such an extension can be done by making some assumptions. For instance, in the models discussed in [45-47], the strategy might be to compute an x-y distribution of stiffness. Under a boundary force or displacement, the corresponding strains or stresses could be calculated at every point. Given a proper failure criterion based on strain, stress, or energy, it would be possible to predict the successive failure of various lamina analogues and the final strength of the three-dimensional composite.

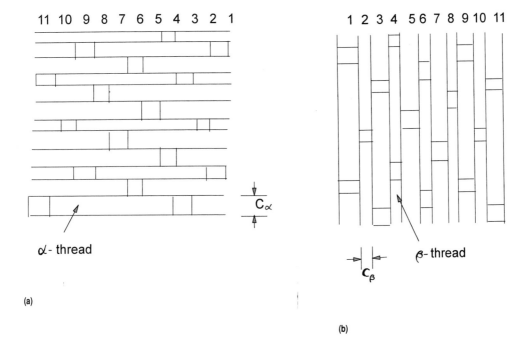

Fig. 13   An example of a hybrid fabric composite with $n_{fg} = n_{wg} = n_g = 8$. Source: Ref 59

Property studies on three-dimensional polymer matrix composites have been mainly for three-dimensional braided composites and, to a lesser extent, for orthogonal nonwoven (ON) structures. The data on the ON quartz- or carbon-reinforced phenolic resins, summarized in Table 1, indicate comparable strengths in the x- and y-directions but an order of magnitude higher strength for the ON composites as compared with the laminated composites [48]. Fukuta [49] has demonstrated that the overall properties of three-dimensional ON carbon fiber/epoxy composites are improved as the yarn bundle size decreases from 12k to 3k. With the yarn bundle size being 6k or 12k, the three-dimensional composites tend to exhibit lower strengths than the corresponding laminated composites (Table 2). However, the three-dimensional ON composites showed a four times greater shear strength than the laminated structures.

A study on three-dimensional braided carbon/epoxy composites has shown (Table 3 and 4) that the tensile, compressive, and flexural strengths of the test specimens are highly sensitive to the specimen cut edges [50]. By changing the braid construction, the strength of three-dimensional braid composites can be varied significantly. As also shown in Table 3, for instance, an increase in tensile strength from 665 to 970 MPa (96 to 140 ksi) is observed when a 1 × 1 construction is replaced with a 3 × 1 construction and the surface fiber orientation is re-

duced from 20° to 12° [50]. In the three-dimensional braid composites (Table 4) the tensile strength and modulus tend to increase as fiber bundle size increases [50]. This is probably related to the dependence of fiber orientation on yarn bundle size in a given braid construction. A larger yarn bundle size tends to produce lower crimp or lower fiber angles and hence higher strength and modulus [44].

Hybrid Composites

Dispersed Fiber Hybrids (Intraply Hybrids)

Intraply Hybrids, Unidirectional. Hayashi [51] addressed the problem of strength prediction for multifiber hybrid composites. Assume that the constituents have practically linear stress-strain curves and that their elastic moduli, ultimate strengths, and strains are represented by (E_A, E_B, E_C), $(\sigma'_A, \sigma'_B,$ and $\sigma'_C)$, and $(\varepsilon'_A, \varepsilon'_B, \varepsilon'_C)$, respectively, in a hybrid laminate composed of three kinds of unidirectionally fiber-reinforced materials. Note that each "constituent" in this case is a fiber-matrix composite by itself. According to Hayashi, the total modulus tensile strength and ultimate strength of the hybrid will be

$$E_i = E_A V_A + E_C V_C$$
$$\sigma_i = E_i \varepsilon$$
$$\sigma'_i = E_i \varepsilon'_A \qquad \text{(Eq 55)}$$

up to the failure of the low-elongation constituent (LEC). After this point of initial failure, the hybrid properties will be independent of the properties of the failed constituent:

$$E_{ii} = E_B V_B + E_C V_C$$
$$\sigma_{ii} = E_{ii} \varepsilon$$
$$\sigma'_{ii} = E_{ii} \varepsilon'_B \qquad \text{(Eq 56)}$$

until the medium-elongation constituent (MEC) fails. After the failure of the MEC and until the failure of the highest-elongation constituent (HEC):

$$E_{iii} = E_C V_C$$
$$\sigma_{iii} = E_{iii} \varepsilon$$
$$\sigma'_{iii} = E_{iii} \varepsilon'_C \qquad \text{(Eq 57)}$$

The total elastic strain energy per unit volume up to final failure will be given by

$$U = \frac{1}{2} E_A V_A \varepsilon'^2_A + \frac{1}{2} E_B V_B \varepsilon'^2_B + \frac{1}{2} E_C V_C \varepsilon'^2_C \quad \text{(Eq 58)}$$

The above equations were derived based on the assumption that all fibers of a single type failed at a definite strain and then no longer contributed to the load-carrying capability of the composite. This implies that the ultimate failure stress of a hybrid is determined by that of HEC multiplied by the volume fraction of HEC, provided that failure of the LEC and MEC causes no stress concentration on the HEC.

The real situation is much more complicated than the above simple multiple fracture mode. For instance, Kelly [52] pointed out that the failure strain of HEC may be reduced by the presence of the LEC. Further, failure of the LEC fibers will lead to an additional strain on the adjacent fibers. It is probable that the adjacent HEC fibers will bridge the crack rather than suffer consecutive failure [53]. In intimately mixed hybrids, there will be only a small distance from the failed fiber to the bridging fiber. Therefore, the full reinforcing strength will likely be redeveloped within the failed fiber within a short distance of the fracture surface of that fiber. As a result of the greater use of the fiber, the energy required to form multiple cracks will be increased.

Parratt and Potter [54] have extended the constant strain theory of hybrid tensile strength to show how the reinforcing strength of a set of fibers will assume higher values in finely mixed unidirectional hybrids than when the fiber types are segregated from each other. Statistical coordination solutions were developed that predicted, among other things, the tensile strength of the hybrid composite. A statistical approach to the strength of hybrid composites was developed by Fukuda and Chou [55,56], who first presented analysis of the state of local stress

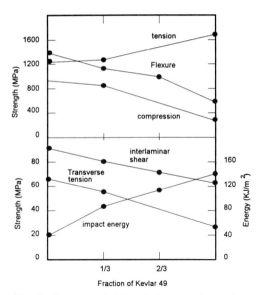

Fig. 14 Properties of carbon/Kevlar-epoxy interply hybrid composites (V_f = 60%). Source: Ref 64

redistribution at a fiber breakage. The results were then incorporated into a Monte Carlo simulation to model the strength and failure behavior of hybrid composites. Fariborz et al. [57] modified the basic chain-of-bundles probability model to analyze unidirectional intraply hybrid composites. The method of analysis was also a Monte Carlo simulation technique.

Intraply Hybrids, Woven Fabric. Ishikawa and Chou [41,58,59] have extended their earlier work on fabric composites to include the possibility of hybridizing the fibers in fabric composites. In addition to the geometric variations (n_{fg} and n_{wg} defining the geometric pattern), material arrangements are also considered in these studies. Here, the arrangement pattern of fiber types in the filling direction repeats for every n_{fm} warp threads, whereas that in the warp direction repeats for every n_{wm} filling threads. Figure 13 illustrates an example of a hybrid fabric composite with $n_{fg} = n_{wg} = n_g = 8$, where the front view is dominated by filling threads and the back side by warp thread. In this case, two kinds of fibers (α and β) are considered, although there is no restriction regarding the number of thread materials in a hybrid fabric. If n_g and n_m of a fabric are prime to each other in one or both directions, there is one unique interlacing pattern. Otherwise, there is more than one type of interlacing pattern [41].

Again, the analysis of the hybrid fabrics began with the use of a mosaic model, where the fabric composite was envisaged to be an assemblage of asymmetrical cross-ply laminates. Upper and lower bounds for elastic stiffness and compliance of the hybrid composites have been obtained by this approach using iso-strain and iso-stress assumptions, respectively. The influence of fabric parameters on elastic properties has been identified. For instance, the relative numbers of n_g and n_m, and thus the interlacing types, affect the magnitude of the coupling terms of B_{ij} and b^*_{ij}. The one-dimensional fiber undulation or crimp concept developed earlier has also been modified to treat the interlacing of two different types of fibers and incorporated into a general "bridging model" for predicting thermoelastic properties of hybrid fabric composites. This approach, in conjunction with the concept of successive failure of the warp yarns (or fill yarns), presumably can be combined to study the knee behavior in hybrid fabric composites and for the prediction of the final failure strength.

Hybrid Laminates (Interply or Interlaminated Hybrids)

Interply Hybrid, Unidirectional. For interlaminated (interply) hybrid composites comprising unidirectional laminas of different fiber types, the stiffness matrices $[A]$, $[B]$, and $[D]$ can be calculated from the classical lamination theory. The contribution of a separate $[Q'_{ij}]$ for any lamina to the resultant stiffness matrices is counted in direct proportion to the fractional volume of this lamina (volume averaging assumption). The stress-strain response of such a laminate subjected to a load represented by $[N_{xy}]$ and $[M_{xy}]$ can be predicted by following a procedure analogous to that described earlier for a gerneral laminate.

For example, consider a hybrid of boron, carbon, and glass fiber-reinforced composite: $[0_{2B}/(\pm45)_B/(\pm45)_C/90_C/90_G]_S$. The first step would be to calculate $[Q'_{ij}]_{0,B}$, $[Q'_{ij}]_{+45,B}$, $[Q'_{ij}]_{-45,B}$, $[Q'_{ij}]_{+45,C}$, $[Q'_{ij}]_{-45,C}$, and $[Q'_{ij}]_{90,G}$. The second step would be to use Eq 32 to 34 in Chapter 5 to calculate stiffness matrices $[A]$, $[B]$, and $[D]$. The midplane strains and curvatures, calculated from Eq 35 and 36 in Chapter 5, would then be used to compute the strains and stresses in individual layers using Eq 40 and 41 in Chapter 5, respectively. In order to determine the FPF, appropriate failure criteria must be selected, and the maximum allowable values must be given (in terms of strain, stress, or energy) for each type of fiber lamina. Successive ply failures can then be evaluated by using Eq 21 to 26.

The ultimate failure strength of four-layered hybrid laminated composites was addressed by Fukunaga et al. [60], who first applied a shear-lag model to obtain the stress redistributions after the breakage of layers. The results were then incorporated in a probabilistic approach to analyze the ultimate failure strength of the hybrids. This approach was an extension of an earlier method used by Fukunaga et al. [61] to predict the statistical strength of intraply composites. The approach was based on the basic chain-of-bundles probability model proposed by Harlow and Phoenix [62,63].

Hybridization provides an added degree of freedom in tailoring composite properties. The ability to engineer a balance in composite properties is usually achieved with some level of compromise, however. Dorey et al. [64] reported some experimental results to illustrate this compromise. As shown in Fig. 14, impact resistance and tensile strength are enhanced by the addition of Kevlar to the carbon fiber composite. Modulus of the carbon fiber composite is reduced because Kevlar fiber has a lower modulus. Compression and flexural strength are attenuated, possibly due to the lower intrinsic compressive strength and modulus of Kevlar fibers, which promotes buckling-type compressive failure of laminates.

Interply Hybrids, Woven Fabric. According to the models proposed by Ishikawa and Chou [41], each woven fabric layer can be simulated as an assemblage of cross-ply laminates. An interply woven hybrid can be visualized as a combination of different laminas to form a hybrid laminate. Therefore, the approaches for the interply hybrids discussed in the above section may be adopted to analyze the strength behavior of interply woven hybrids.

REFERENCES

1. B.C. Hoskin and A.A. Baker, Ed., *Composites Materials for Aircraft Structures*, AIAA Educational Series, New York, 1986
2. P.K. Mallick, *Fiber Reinforced Composites*, Marcel Dekker, 1988
3. R.M. Jones, *Mechanics of Composite Materials*, McGraw-Hill, 1975
4. L.B. Greszczuk, Proc. 21st Ann. Tech. Conf. Reinforced Plastics/Composites Institute, Society of Plastics Industry, 1966
5. J.A. Kies, "Maximum Strains in the Resin of Fiberglass Composites," NRL Report No. 5752, AD-274560
6. J.C. Schulz, 18th Ann. Conf. (Washington, D.C.), Society of Plastics Industry, 1963
7. B.D. Agarwal and L.J. Broutman, *Analysis and Performance of Fiber Composites*, 2nd ed., Wiley Interscience, 1990
8. T.A. Collings, *Composites*, Vol 5 (No. 3), 1974, p 108
9. B. Harris, *Engineering Composite Materials*, The Institute of Metals, London, 1986
10. R. Hill, *The Mathematical Theory of Plasticity*, Oxford University Press, 1950
11. V.D. Azzi and S.W. Tsai, *Exp. Mech.*, Vol 5, 1965, p 283
12. B.D. Agarwal and J.N. Narang, *Fiber Sci. Technol.*, Vol 10, 1977, p 37
13. B.H. Jones, in *Composite Materials: Testing and Design*, STP 460, ASTM, 1969, p 307
14. H.E. Brandmaier, *J. Compos. Mater.*, Vol 4, 1970, p 422
15. B. Harris, *Composites*, Vol 3, 1972, p 152
16. S.W. Tsai and E.M. Wu, *J. Compos. Mater.*, Vol 5, 1971, p 58
17. E.M. Wu, *J. Compos. Mater.*, Vol 6, 1972, p 472
18. S.W Tsai and H.T. Hahn, *Inelastic Behavior of Composite Materials*, C.T. Herakovich, Ed., American Society of Mechanical Engineers, 1975, p 73
19. P.H. Petit and M.E. Waddoups, *J. Compos. Mater.*, Vol 3, 1969, p 2
20. H.L. Cox, *Brit. Appl. Phys.*, Vol 3, 1952, p 72
21. J.O. Outwater, *Mod. Plast.*, Mar 1956, p 56
22. B.W. Rosen, Chapter 3, *Fiber Composite Materials*, American Society for Metals, 1964
23. N.F. Dow, Report No. TISR-63SD61, General Electric Co., Aug 1963
24. A. Kelly, Chapter 5, *Strong Solids*, Clarendon Press, 1966
25. A. Kelly and W.R. Tyson, *J. Mech. Phys. Solids*, Vol 13, 1965, p 329

26. A. Kelly and G.J. Davies, *Met. Rev.*, Vol 10, 1965, p 1
27. J.K. Lees, *Polym. Eng. Sci.*, Vol 8, 1968, p 195
28. P.E. Chen and R.E. Lavengood, *Proc. Int. Conf. Structure Sol. Mech. Eng. Des. Civil Eng. Mater.*, Part I, M. Te'eni, Ed., Wiley Interscience, 1971, p 75
29. P.E. Chen, *Polym. Eng. Sci.*, Vol 11, 1971, p 51
30. R.M. Barker and T.F. MacLaughlin, *J. Compos. Mater.*, Vol. 5, 1971, p 492
31. V.R. Riley, *J. Compos. Mater.*, Vol 2, 1968, p 436
32. F.J. Lockett, *Proc. Conf. Properties of Fibre Composites*, IPC Science and Technology Press, 1971
33. L.E. Dingle, Paper No. 11, Proc. Int. Conf. Carbon Fibers, Their Place in Modern Technol., The Plastics Institute, London, 1974
34. J.C. Halpin, J.L. Kardos, *Polym. Eng. Sci.*, Vol 18, 1978, p 496
35. J.L. Kardos, J.C. Halpin, and S.L. Chang, in *Rheology*, Vol 3, *Applications*, G. Astarita, G. Marrucci, L. Nicolais, Ed., Plenum Press, 1980, p 225
36. R.E. Lavengood, L. Nicolais, and M. Narkis, *J. Appl. Polym. Sci.*, Vol 17, 1973, p 1173
37. M. Narkis, *Polym. Eng. Sci.*, Vol 15, 1975, p 316
38. M. Narkis, *J. Appl. Polym. Sci.*, Vol 20, 1976, p 1597
39. E. Masoumy, L. Kacir, and J.L. Kardos, *Polym. Compos.*, Vol 4, 1983, p 64
40. J.L. Kardos, *High Performance Polymers*, E. Baer and A. Moet, Ed., Hanser, 1991, p 199
41. T.W. Chou and T. Ishikawa, Chapter 7, *Textile Structural Composites*, T.W. Chou and F.K. Ko, Ed., Elsevier, 1989
42. S.M. Bishop, Chapter 6, *Textile Structural Composites*, T.W. Chou and F.K. Ko, Ed., Elsevier, 1989
43. D.W. Whyte, Ph.D dissertation, Drexel University, Philadelphia, June 1986
44. F.K. Ko, Chapter 5, *Textile Structural Composites*, T.W. Chou and F.K. Ko, Ed., Elsevier, 1989
45. J.M. Yang, C.L. Ma, and T.W. Chou, *J. Compos. Mater.*, Sept 1986, p 472
46. C.L. Ma, J.M. Yang, and T.W. Chou, in *Composite Materials: Testing and Design*, STP 893, ASTM, 1986, p 404
47. T.J. Whitney, "Analytical Characterization of 3-D Textile Structural Composites Using Plane Stress and 3-D Lamination Analogies," M.S. thesis, University of Delaware, Dec 1988
48. E.M. Lenoe, R.M. Laurie, and G.N. Wassil, Proc. 10th National SAMPE Symp. 9-11 Nov 1966 (San Diego, CA), Society for Advancement of Material and Process Engineering
49. K. Fukuta, in *Proc. 2nd Textile Structural Compos. Symp.*, Drexel University, Philadelphia, 1987, p 64
50. A.B. Macander, R.M. Crane, and E.T. Camponeschi, in *Composite Materials: Testing and Design*, STP 893, ASTM, 1986, p 422
51. T. Hayashi, *8th Reinforced Plastics Conference* (Brighton, UK), 1972, p 149-152
52. A. Kelly, *Philos. Trans. R. Soc. A*, Vol 294, 1980, p 519-534
53. D. Short and J. Summerscales, *Fiber Composite Hybrids*, N.L. Hancox, Ed., Applied Science, London, 1980, p 69-117
54. N.J. Paratt and K.D. Potter, Mechanical Behavior of Intimately-Mixed Hybrid Composites, Proc. SAMPE Symp., Society for Advancement of Material and Process Engineering, 1981, p 313-326
55. H. Fukuda and T.W. Chou, Stress Concentrations around a Discontinuous Fiber in a Hybrid Composite Sheet, *Trans. Jpn. Soc. Compos. Mater.*, Vol 7, 1981, p 37-42
56. H. Fukuda and T.W. Chou, in *Progress in Science and Engineering of Composites*, T. Hayashi et al., Ed., Proc. ICCM-IV (Tokyo), International Committee on Composite Materials, 1982, p 1141-1150
57. S.J. Fariborz, C.L. Yang, and D.G. Harlow, *J. Compos. Mater.*, Vol 19, 1985, p 334-353
58. T. Ishikawa and T.W. Chou, *J. Mater. Sci.*, Vol 17, 1982, p 3211
59. T. Ishikawa and T.W. Chou, *J. Compos. Mater.*, Vol 17, 1983, p 94
60. H. Fukunaga, T.W. Chou, and H. Fukuda, *Compos. Sci. Technol.*, Vol 35, 1989, p 331-345
61. H. Fukunaga, T.W. Chou, and H. Fukuda, Strength of Intermingled Hybrid Composites, *J. Reinf. Plast. Compos.*, Vol 3, 1984, p 145
62. D.G. Harlow and S.L. Phoenex, *J. Compos. Mater.*, Vol 12, 1978, p 195, 314
63. D.G. Harlow, *Proc. R. Soc. (London)*, Vol 389, 1983, p 67-100
64. G. Dorey, G.R. Sidey, and J. Hutchings, *Composites*, Vol 9, 1978, p 25

Chapter 7

Design for Improved Fracture Resistance of Polymer Composites

There are two primary types of damage in fiber-reinforced composites: interlaminar cracks (delaminations) and through-the-thickness cracks. In order to design composites with improved fracture resistance, we must understand the causes of delamination, its effect on structural performance, and the analytical and experimental techniques that can be used to predict delamination. These subjects are therefore discussed in detail. Several examples of successful attempts to prevent or suppress delamination are given to illustrate the possible routes in improving the resistance to delamination. Also discussed are the models that have been proposed to predict the residual strength of delaminated composites. Then the experimental techniques for evaluating the interlaminar fracture toughness of composites are reviewed. The final section of this chapter is devoted to a discussion on the analytical and experimental methods that can be used to predict the notched strength of laminates.

Damage Characteristics of Polymer Composites

Damage in a polymer composite component can be caused during manufacture or service. Improper laminate fabrication or curing procedures can induce defects in the form of delamination, voids, debonds, wrinkles, inclusions, broken fibers, and fiber misalignment. The manufacturing defects may also be introduced by machining the components for fastener holes, design cutouts, and so on. [1,2]. The service damage in a composite structure (e.g., an aircraft) may result from the impact by runway debris, hailstones, bird strike, ground service vehicles, and so on. Impact damage can also be caused by dropping tools accidentally during maintenance or by ballistic projectiles in a combat environment. In many cases, especially when the incident energy is low, the damage created by such impacts may be barely visible or invisible on the surface but may considerably reduce the strength of

the structure. Fatigue loading, thermal cycling, or moisture absorption and desorption can also result in damage of various degrees.

Damage-tolerant design criteria are established for composite components to ensure that growth from invisible damage will not degrade the strength to an unacceptable level during the intended service life of a component. This requirement for the current graphite-epoxy composites is met by limiting the design strain to 0.4% [3,4] even though select carbon fibers have a breaking strain of about 1.8%. Such a stringent requirement inevitably results in an increase in structural weight (overdesign) and limits the exploitation of the full potential of composites. It has been shown that a weight saving of up to 12% can be achieved if the design strain is raised from 0.4 to 0.6% for an aircraft wing structure. A much greater weight saving is possible if the damage resistance of the composite can be further improved.

In response to a static or dynamic load, a composite laminate may exhibit initiation and propagation of damage that occurs in a series of cracking events, individually or interactively [5]. The anisotropic and heterogeneous character of composites naturally results in a large possibility of failure modes [5,6]. In order to make the problems more tractable, one may consider only the three principal failure modes: intraply cracking, interlaminar delamination, and fiber breakage. One may then seek to retard or suppress the initiation and propagation of these cracking modes. Other types of damage may be considered to simply alter the stress levels at which these three failure modes may occur [7].

The interlaminar mode of fracture has been a subject of extensive research over the past two decades [8-15]. The delamination failure mode is known to be the major life-limiting failure process in a composite laminate. Delamination can induce stiffness loss, local stress concentration, and a local instability that can cause buckling failure under compressive loading. Delamination may

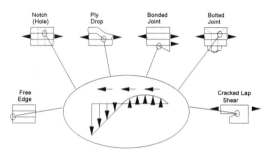

Fig. 1 Potential sources of out-of-plane loads from load path discontinuities. Source: Ref 8

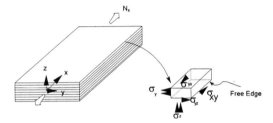

Fig. 2 Interlaminar stresses in a lamina of a symmetric balanced laminate under an axial load

also result in a redistribution of structural load paths, which can precipitate final structural failure. The presence of thickness-direction flaws such as sharp notches and holes may also significantly compromise the strength of the structure. The issue of notch sensitivity and residual strength prediction must also be understood in order to design the composite structural elements with improved notch strength.

Delamination

Causes of Delamination

The delamination in composites is caused by the interlaminar stresses produced by out-of-plane loading (e.g., impact), eccentricities in load paths, or discontinuities in the structure. O'Brien [8] has summarized in a diagram (Fig. 1) some of the design details that may cause the out-of-plane loads, resulting in interlaminar stresses. These potential sites of delamination initiation include straight or curved free edges (near holes), ply terminations or ply drop for tapering the thickness, bonded or co-cured joints, bolted joints, and cracked lap shear specimens. Even under an in-plane loading condition, the local loads near the discontinuities may be out of plane.

Interlaminar stresses in a laminate also may be created by variations in moisture content and temperature (the hygrothermal effect). Because of the differences in CTE between neighboring plies, residual thermal stresses may be developed when the laminate is cooled down from the cure temperature to the end-use temperature. Cure shrinkage of certain resins (e.g., unsaturated polyester) can induce additional residual stress. Moisture absorption or desorption can also lead to significant residual stresses. Differential CTE's between fibers and the matrix resin may also lead to residual thermal stresses at the fiber-matrix interface. This would modify the local stress threshold required for interfacial debonding, a potential precursor to delamination nucleation.

Free-Edge Delamination

Free-edge delamination in composite laminates under in-plane loading has been a subject of intensive research since the early 1970s [e.g., 9-19]. Several studies [e.g., 9,10] have indicated that free-edge delamination is attributable to interlaminar stresses that are highly local-

ized in the vicinity of a free edge under in-plane stress conditions. The magnitude and sign of the interlaminar stresses can be accurately calculated using the analytical model developed by Pagano and Soni [10,12]. These studies have demonstrated that the magnitude and distribution of the interlaminar stress components are widely varied, and that they depend on the laminate lay-up, stacking sequence, properties of the constituent materials, and the stress state.

As an example, a cross-plied laminate $(0°/90°)_s$ experiences less severe interlaminar stress components than a $(\pm30°/90°)_s$ laminate [14]. Consequently, the $(\pm30°/90°)_s$ laminate would exhibit a much more extensive delamination. The stacking sequence can also determine whether an interlaminar normal stress will be tensile or compressive at the free edges and control the magnitude of the interlaminar stress components. For instance, a $(\pm30°/90°)_s$ laminate experiences tensile interlaminar normal stress, while a $(90°/\pm30°)_s$ laminate produces compressive stress at the free edges under applied axial tension. As a consequence, the $(\pm30°/90°)_s$ laminate develops extensive delamination under uniaxial tension before final failure, but it does not easily delaminate under compression because of the compressive interlaminar normal stress produced [14].

Herakovich [20] has suggested that the interlaminar stress components originate from the mismatch in engineering properties between layers, for example a mismatch in Poisson's ratio (ν_{xy}) and in coefficient of mutual influence $(m_{xy,x})$. If an in-plane load N_x is applied along the x-direction (Fig. 2), the mismatch in ν_{xy} $(\Delta\nu_{xy})$ gives rise to interlaminar normal stress (σ_z) and shear stress (τ_{yz}), while the mismatch in $m_{xy,x}$ $(\Delta m_{xy,x})$ produces interlaminar shear stress (τ_{zx}) near the free edge of a laminate. The magnitudes of these stresses depend on the value of $|\Delta\nu_{xy}|$ and $|\Delta m_{xy,x}|$, elastic and shear moduli, stacking sequence, mode of loading, and environmental conditions [20]. The analysis by Herakovich has also indicated that the largest values of $|\Delta m_{xy,x}|$ and $|\Delta\nu_{xy}|$ occur for $(+11.5°/-11.5°)$ and $(22°/90°)$ lay-ups, respectively. These are in agreement with his test data on $(\pm\theta)_s$ laminates for $10° \leq \theta \leq 30°$, which have a great tendency to delaminate. High interlaminar shear stresses tend to develop if the angle plies $(\pm\theta)$ are adjacent to each other, so one should try to avoid placing them together in a laminate.

Herakovich has further confirmed the important influence of the stacking sequence on the sign and magnitude of interlaminar stresses. Under the same uniaxial tensile loading condition, $(\pm45°/0°/90°)_s$ and $(90°/45°/0°/-45°)_s$ laminates produce high tensile and compressive σ_z at midsurface, respectively. Thus, the latter laminate is not as prone to delamination as the former. Interspersing $\pm45°$ layers so as to create alternate positive and negative loads (σ) was found [20] to reduce σ_z and hence the chances of delamination.

Kim [14] has conducted a very extensive study on the strain levels at the onset of delamination. More than 40 laminates and stacking sequences were investigated by various authors, and the data were compiled by Kim for comparison. Stress analysis by these various authors indicates that in all the laminates studied, a tensile σ_z exists at the midplane of the laminate, where interlaminar shear components are zero. Kim concludes that σ_z plays a significant role in the onset of delamination for most of the laminates under in-plane axial loading. Different stacking sequences give rise to significantly varied threshold strain levels. A thickness effect on the delamination threshold is also noted by Kim. In general, the delamination threshold decreases with increasing number of layers.

Rodini and Eisenmann [21] observed that the delamination onset strain in the graphite-epoxy $(\pm45°_n/0°_n/90°_n)_s$ decreases with the value of $\sqrt{n}$. A systematic study on interply cracking was conducted by Crossman and Wang [22]. Interlaminar shear stress was also found to be a factor causing delamination, but failure under interlaminar shear is always affected by the presence of interlaminar normal stress. Such an interaction between different interlaminar stress components is probably very important in affecting the delamination, but this is not yet completely understood [14].

Residual thermal stresses produced during composite fabrication were also found to have considerable influence on interlaminar stresses and thus on the delamination tendency of composites [20,23-26]. Impact loading also produces delamination, but this subject will be discussed in Chapter 8. Matrix cracking in off-axis plies may generate interlaminar stresses to promote delamination. Reifsnider *et al.* [27,28] estimate that in quasi-isotropic laminates $(0°/90°/\pm45°)_s$, these interlaminar stresses occur near the tip of the matrix crack in the 90° ply. These interlaminar stresses tend to cause local delamination that grows along ply interfaces near matrix cracks.

Prediction of Interlaminar Stresses

To gain a better understanding of delamination behavior in composites, one must have a clear idea about the stresses in the boundary layer near the curved or straight free edges. Determination of the edge stresses is nonetheless a very difficult job, because one deals with an interfacial crack between two highly anisotropic composite laminates. Many efforts have been made to determine stress intensity factors at the tips of the interfacial cracks of two or more isotropic layers, for transversely isotropic half-planes [29-36], or for an edge delamination in an angle-ply laminate [37]. Because of the complex heterogeneous and anisotropic nature of composites, the analytical solutions to edge delamination in these materials have been rather limited. Therefore, the use of either finite difference methods [38-40] or finite element methods [41-61] has been emphasized. One advantage of using these numerical techniques is that the effect of moisture and temperature on the interlaminar stresses can be readily incorporated in the calculations. However, solving edge stress problems for thick structural laminates using such methods can be rather expensive.

Pagano [62] has developed a model to calculate the interlaminar normal stress distribution along the central plane of the free-edge laminate. The model consists of treating the quarter-section of the laminate as a plate fastened to a rigid foundation. The body is subjected to a constant axial strain and a constant temperature rise. The solutions of this elastic plate model have proven to be important in the subsequent formulation of the so-called "global-local model" discussed below, as well as in models developed by other authors [63,64]. This approximation method [9,62], although less accurate compared to other numerical methods, may provide some idea about the delamination tendency and the effect of the stacking sequence in laminates, based on the approximate σ_z distribution. This stress distribution, in conjunction with a statistical failure criterion, has been used to predict the onset of delamination in various composites [21].

By extending the Reissner variational principle, Pagano [65, 66] has developed a self-consistent model (local model) to define a complete stress field within an arbitrary composite laminate. This model guarantees satisfaction of "layer equilibrium" and permits the prescription of combinations of interfacial tractions and displacements, which allows for treatment of such conditions as interfacial continuity or cracks. However, this model met with great difficulty in attempting to analyze thick laminate free-edge problems.

Such a local model, in which each layer is represented as a homogeneous anisotropic continuum, becomes intractable as the number of layers becomes large. On the other hand, global models, which follow from an assumed displacement field and lead to the definition of effective laminate moduli, are not sufficiently accurate for the computation of stress fields. Therefore, Pagano and Soni [10,12] blended these concepts into a self-consistent model that can define detailed response functions in a region of interest (local), while representing the remainder of the domain by effective properties (global). Such a global-local variational model can be used to clearly define the interlaminar stresses in a very thick laminate. The effectiveness of this model appears to have been demonstrated by the use of numerical examples based on the free-edge class of boundary-value problems in laminate elasticity [10]. Additional results of using the global-local model to analyze interlaminar failures are given by Soni and Kim [67-69]. In these reports, the effect of interlaminar shear and tension on delamination are considered. This model has been quite useful in the investigation of the influence of material, geometric, and

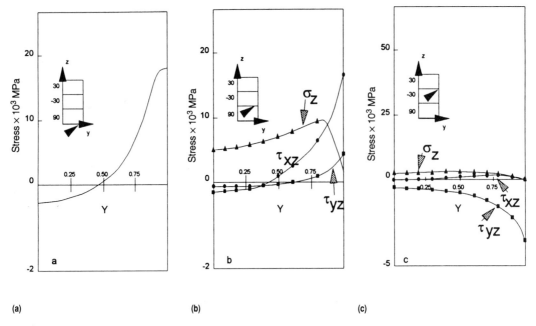

Fig. 3 Distribution of interlaminar stresses on various interfaces of a ($\pm$ 30°/90°) laminate of T-300 graphite/5280 epoxy under a uniaxial tensile load. Source: Ref 14

lamination parameters on the interlaminar response of thick laminates containing free edges [12].

Prediction of Delamination Initiation

The Fracture Mechanics Methods. In the fracture mechanics methods, the delamination is assumed to be an edge crack at the interface, and the prediction of delamination onset is based on the magnitudes of stress intensity factor K and the strain energy release rate G. Rybicki *et al.* [47] treated the delamination as an interfacial crack and analyzed its behavior by finite element methods. In this approach, an intrinsic flaw size equivalent to one ply thickness was assumed. S.S. Wang [37] has suggested that there exists a critical delamination length $a*$ for various composite laminates. For instance, $a* \approx 2h$ for angle ply laminates with $b/h = 16$, where b is the specimen width and h is the ply thickness. Wang indicated that any edge flaw $a_0 < a*$ will experience rapid, unstable growth as the load or G reaches a critical value. Contrarily, an initial crack greater than $a*$ will grow stably under a monotonically rising load. This suggests the existence of an inherent crack arrest mechanism for edge delamination. Other workers [38,41,42,46] have also observed this behavior of G remaining constant with delamination beyond a critical flaw size. Therefore, $a*$ may be an important quantity in the life prediction for delaminated composites [7].

A.S.D. Wang [13,70] has developed an energy method for free-edge delamination, which basically incorporates the method of ply elasticity and the concept of effective material flaws into the classical theory of fracture mechanics. In this theory, it is postulated that a distribution of effective interface flaws is present in each of the ply interfaces, and that, in particular, the effective flaws near the free edges will initiate delamination. The theory further assumes that, at some critical load, one of the most prevalent edge flaws, driven by the interlaminar stresses, will propagate into a delamination crack of macroscopic dimensions. This event is regarded as the onset of the free-edge delamination. The effective flaw size varies with the size of the laminate free-edge boundary-layer zone and ranges from one to a few ply thicknesses. Crossman *et al.* [71] went on to suggest that

$$a* = \frac{E^0 G_c}{\pi \, \sigma_c^2} \qquad \text{(Eq 1)}$$

where E^0 is the effective modulus and σ_c is the critical stress at which the pre-existing flaw in the material begins to grow. As a first-order approximation, E^0 may be taken to be the transverse modulus and $\sigma_c \approx$ transverse tensile strength. For a current dry epoxy-carbon fiber composite, $a*$ has been found to be approximately one ply thickness [7].

The strain-energy-release-rate approach to delamination prediction was further extended by O'Brien [43,72], who provided a simple method using only the laminate plate theory to analyze the onset of delamination. In this method, delamination-induced stiffness reduction was related to the strain energy release rate. The initiation of delamination in a laminate (thickness t) occurs at the critical strain ε_c, which is given by

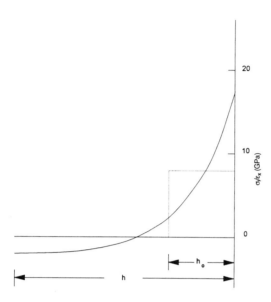

Fig. 4 Averaging the interlaminar stress over the distance h (one ply thickness) from the free edge. Source: Ref 14

$$\varepsilon_c = \left[\frac{2G_c}{t(E_1 - E^*)} \right]^{1/2} \qquad \text{(Eq 2)}$$

where G_c = the critical strain energy release rate; E_1 = the extensional stiffness of the laminate, computed from laminate plate theory; and E^* = the stiffness of the laminate completely delaminated along one or more interfaces, which is given by

$$E^* = \frac{\displaystyle\sum_{i=1}^{m} E_i t_i}{t} \qquad \text{(Eq 3)}$$

E_i = the stiffness of the ith sublaminate of thickness t_i.

Equation 3 has been used successfully to predict the onset of delamination in various laminates, and the prediction agrees well with the test data [72]. This equation was slightly modified by Whitney and Browning [73] to account for any discrepancies between the theoretical modulus and the experimentally determined modulus on the undamaged laminate.

The stress approach involves a detailed stress analysis near the free edge and the use of a failure criterion. Kim and Soni [14,44,68,69] have used the point stress and average stress criterion, originally developed by Nuismer and Whitney [74], to predict delamination initiation in various laminates. Kim and Soni [14,44] have used the global-local model discussed earlier to obtain the distributions of σ_z, τ_{yz} on the various interfaces of a laminate, for example ($\pm 30°/90°$)$_s$, under applied uniaxial tension (Fig. 3). A large gradient of interlaminar

stress components exists along the y-direction in the neighborhood of the free-edge region. The average value of each stress component, instead of the maximum value, is assumed to be the effective stress level that dictates the failure at the free edge. The effective stress is calculated by averaging the individual interlaminar stress components, over a fixed distance h_0, along the width of the laminate at its interface ahead of the free edge (Fig. 4). It is given by [14]:

$$\bar{\sigma}_i(z) = \frac{1}{h_0} \int_0^{h_0} \sigma_i(y,z) dy \qquad \text{(Eq 4)}$$

The fixed distance h_0, referred to as the critical length, was taken arbitrarily as one ply thickness in all cases investigated [14,44].

In order to predict the stress level at the onset of delamination, and its location, Soni and Kim [14,44] have applied the Tsai-Wu quadratic failure criterion (see Chapter 6). They incorporated the strength ratio R, defined as the ratio of the ultimate strength to the applied stress:

$$(F_{zz}\sigma_z^2 + F_{tt}\sigma_{xz}^2 + F_{uu}\sigma_{yz}^2)R^2 + (F_z\bar{\sigma}_z)R - 1 = 0 \quad \text{(Eq 5)}$$

where

$$F_{zz} = \frac{1}{zz'}$$

$$F_{tt} = \frac{1}{S_t S_t'}$$

$$F_{uu} = \frac{1}{S_u S_u'}$$

$$F_z = \frac{1}{z} - \frac{1}{z'}$$

z, z' = the interlaminar tensile and compressive strength
S_t, S_t' = the positive and negative shear strengths in the x and z planes ($S_t = S_t'$)
S_u, S_u' = the positive and negative shear strengths in the y and z planes ($S_u = S_u'$)

When R is greater than unity, stress can be applied safely. Failure occurs when R becomes unity. Delamination is expected to take place at the interface where the value of R is the smallest for different interfaces [14]. Reasonable agreement between the predicted behavior and the experimental results has been achieved for several laminates where delamination was predominantly caused by σ_z, τ_{xz}, or combined σ_z and τ_{xz} [14].

This approach has been used by other workers [e.g., 48] for the prediction of delamination onset in various laminates. For angle-ply ($\pm\theta$) laminates, $\theta = 15°$ is the most critical angle-ply laminate with the shear stress τ_{xz} dominating failure. As the fiber angle is increased above 15°, the failure mode of ($\pm\theta$)$_s$ laminates shifts to mixed shear (τ_{xz} and τ_{yz}), mixed shear and normal (τ_{xz}, τ_{yz}, and σ_z), and finally to transverse tension (σ_y). Contribution to failure is dominated by transverse tension (σ_y) for $\theta \geq$ 37.5°. The location of initial failure also shifts from the $\pm\theta$ interface to the midplane with increasing θ (when $\theta \geq$ 45°).

Prediction of Delamination Growth

The growth of a delamination crack depends on the crack tip stress state, which is governed by the mixed mode strain energy release rates (G_I, G_{II}, and G_{III}) or the stress intensity factors (K_I, K_{II}, and K_{III}). Wang [37] has shown that K_{II} is much lower than K_I and K_{III}, and that edge delamination is dominated by K_{III} (or the tearing mode) in angle-ply laminates ($\pm\theta$). However, mode III contribution is found to be negligible for the growth of an edge crack at the −30/90 interface of a [±30/±30/90/90]$_s$ laminate [43]. In general, it is not necessary that all three modes exist together; only one or two modes may dominate the growth of a delamination crack.

Many fracture mechanics equations have been developed to characterize the stable fracture process of edge delamination growth. O'Brien [43] has used the concept of crack growth resistance curves (R-curves) to describe such growth, whereby G_R versus delamination growth curves were determined. In this approach, the initial value of G_R represents the critical G_c, or the onset of delamination; G_c is considered a material property that is independent of ply orientation. However, Johannesson and Blikstad [75] have shown that G_c is strongly dependent on the ratio K_{III}/K_I for angle-ply laminates.

Many mixed-mode delamination growth laws have been proposed by different researchers. A simple law states that delamination growth occurs when the total strain energy release rate G_T reaches a critical value [43,47]:

$$G_T = G_I + G_{II} \rightarrow G_c \qquad \text{(Eq 6)}$$

This equation may describe mixed-mode delamination growth only for laminates where $G_{Ic} = G_{IIc}$. Unfortunately, $G_{Ic} \ll G_{IIc}$ for graphite-epoxy composites, so Eq 6 may not be appropriate for describing delamination growth in these materials.

Other, more generalized, equations characterize delamination growth [52,76-80]:

$$(G_I/G_{Ic})^m + (G_{II}/G_{IIc})^n = 1 \qquad \text{(Eq 7)}$$

and [81]

$$(1 - g)(G_I/G_{Ic})^{1/2} + g(G_I/G_{Ic}) + (G_{II}/G_{IIc}) = 1 \quad \text{(Eq 8)}$$

where $g = G_{Ic}/G_{IIc}$, and [82,83]:

$$G_c = -e^{-(C_1 M + C_2)} + C_3 \qquad \text{(Eq 9)}$$

where

$$M = [1 + (G_{II}/G_I)\sqrt{E_I}/\sqrt{E_T}]^{1/2}$$
$$C_1 = 0.25$$

Fig. 5 Through-width buckling of a composite plate. Source: Ref 14

$$C_2 = -6.31$$
$$C_3 = 503.3$$
$$E_L = \text{longitudinal modulus for the laminate}$$
$$E_T = \text{transverse modulus for the laminate}$$

The delamination growth for graphite-PEEK laminates was observed to follow the relation:

$$G_I/G_{Ic} + (G_{II}/G_{IIc})^{3/2} = 1 \qquad \text{(Eq 10)}$$

Finally, by extending Wu's tensor criterion [84] to plane strain conditions, Russel and Street [85] have demonstrated that mixed-mode delamination growth may be better represented by this modified polynomial criterion.

Effect of Delamination

Delamination has long been recognized as one of the most life-limiting damage modes in composite laminates. The initiation and growth of delamination during the manufacturing or service life cycles of composites may result in the progressive reduction of stiffness and strength of these materials. In some cases, the degradation process is sudden and may result in catastrophic failure of the laminates.

Reduction in Stiffness

Many workers [e.g., 27,43,80,86] have reported stiffness reduction in composites as the result of delamination initiation and growth. O'Brien [43] has developed a rule-of-mixture analysis to calculate the stiffness loss of a partially delaminated laminate:

$$E = (E^* - E_l)\frac{A}{A^*} + E_l \qquad \text{(Eq 11)}$$

where A is the delaminated area, A^* is the total interfacial area, and the other terms were defined in Eq 2 and 3. Equation 11 does not incorporate the effect of transverse cracking. An appropriate adjustment of E^* is necessary when the stiffness loss is significant due to transverse cracking, which usually precedes delamination in laminates under tension.

Strength Reduction

The delamination-induced strength reduction varies, depending on the delaminated area and type of loading for a given laminate. In general, the effect of delamination is much more severe under compressive loading than under tensile loading [78-80,83,86,87]. Buckling is the major failure mode of a delaminated composite under applied compression (Fig. 5). This phenomenon results in a high interlaminar stress concentration at the delamination crack tip, and under increased loads, the buckled area can increase to a critical size, possibly leading to a loss of material integrity or total collapse of the plate [78-80]. This usually occurs at load levels much lower than the undamaged compressive strength of the laminate or the stability of the plate.

Several analytical models [83,86,87,88] have been developed to predict the buckling growth of delamination and the effect of various parameters on the delamination growth. These models are based on the stress or on the strain energy release rate. Because of the complexity of the problem involved, these models have yet to be fully validated. Many factors must be taken into consideration if any of these models is to be successful: the geometric shape of the delaminated region; the thickness of the buckled laminate; selection of a proper failure criterion; the highly nonlinear nature of deformation; the type and stacking sequence of laminate; multiple delamination or crack interactions; and the stiffness and toughness of laminates [14].

Effect on Fatigue Behavior

The delaminated panel, when subjected to fatigue loading, may exhibit accelerated damage growth. This phenomenon may become particularly severe when the cyclic stresses include a compressive component. Fatigue loading therefore would further reduce the residual strength and stiffness. Such a behavior has been studied by several workers [89-90] where the delamination was caused either by impact or by a deliberately implanted flaw.

Experimental Evaluation of Delamination Resistance (Interlaminar Fracture Toughness)

Assessment of resistance to delamination is generally considered to be synonymous with characterization of interlaminar fracture toughness. The short beam shear test (ASTM D-2344) is commonly used to measure the interlaminar shear strength of composites. Based on observed failure modes, stress analysis, and fracture surface observations, considerable difficulty has been encountered in interpreting experimental data from the short beam shear test [15]. In particular, compression stresses in regions where high shear-stress components exist tend to induce vertical cracks, which appear to be inevitable precursors to horizontal interlaminar failures. For specimens without such initial damage, the failure mode appears to be compressive buckling or yielding in the upper portion of the beam under combined compression and shear [15]. This would make it extremely difficult to pinpoint the failure load that is used in the calculation of interlaminar shear strength.

Furthermore, in many cases, interlaminar normal stress was considered to be a dominant factor in causing delamination. Therefore, Wilkens et al. [41] suggested a double cantilever beam (DCB) test for mode I delamination. Other test methods include width-tapered DCB [91,92], the Arcan test [93,94] and free-edge delamination (FED) [43]. The FED specimen provides a simple means to determine delamination initiation, but it tends to give a higher G_{Ic} value than the DCB specimen does [73]. Also, the FED specimen does not necessarily produce a pure mode I delamination because of the complex stress state in the free-edge zone. DCB is therefore regarded as a more suitable test for determining G_{Ic}.

Mode II delamination behavior can be characterized by the crack lap shear (CLS) [41], Arcan [93,94] and end-notched flexure (ENF) tests [84,94]. ENF and Arcan are believed to produce a pure mode II (shearing) behavior, while a CLS specimen develops mixed modes, with mode II dominating [41]. The Arcan test requires very thick specimens, and the fixture is rather complicated. ENF uses the same test specimens and requires only a simple three-point bending test fixture. The split cantilever beam (SCB) test was suggested [95,96] as one viable method to determine mode III fracture toughness. The procedure is to simply change the loading mode of a DCB specimen to become one of "tearing." The off-axis center notch tensile specimen can be used to simulate interlaminar fracture between plies in a unidirectional composite. Although the loading is in-plane, rather than interlaminar, the failure occurs by propagating a crack between fibers, simulating matrix-dominated interlaminar fracture [97].

Briefly described below are a few of the more commonly used test methods. The common aim of these methods is to measure the critical strain energy release rate G_c during onset or extension of delamination. In the EDT test the increase in compliance at the onset of delamination is determined by the test. Other specimens are calibrated by means of a relation between strain energy release rate, G, and compliance, C [98]:

$$G = \frac{P^2}{2w} \frac{\Delta C}{\Delta a} \qquad (Eq\ 12)$$

where a is the crack length, w is the width of the specimen, P is the applied load, and C is the compliance, defined by the slope of the load (P) versus the displacement ($\dot{U}$) curve for the specimen:

$$C = \dot{U}/P \qquad (Eq\ 13)$$

Recently, Whitney [15] conducted an excellent review of the various approaches to characterizing interlaminar fracture toughness and presented analytical, numerical, and experimental data to indicate the uses for each technique. The reader may also find Ref 7 and 99 helpful.

DCB Test

The DCB test was originally developed for evaluating mode I fracture of adhesive bonded joints. Either a

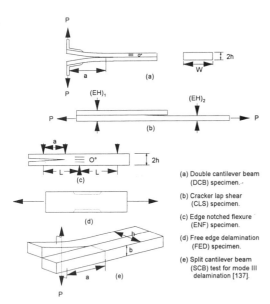

(a) Double cantilever beam (DCB) specimen.

(b) Cracker lap shear (CLS) specimen.

(c) Edge notched flexure (ENF) specimen.

(d) Free edge delamination (FED) specimen.

(e) Split cantilever beam (SCB) test for mode III delamination [137].

Fig. 6 Schematic of various interlaminar fracture test specimens. Source: Ref 7

tapered [92] or a straight-sided [41] DCB specimen can be used to characterize the interlaminar fracture toughness of composites. As shown in Fig. 6(a), a DCB test is conducted under displacement-controlled conditions, and P versus $\dot{U}$ curves are obtained for various crack lengths. These crack lengths are then used to calculate C and $\Delta C/\Delta a$ in Eq 12. The compliance (C) of the DCB specimen may be obtained from the elastic beam theory as

$$C = \frac{2a^3}{3EI} = \frac{8a^3}{WEh^3} \qquad \text{(Eq 14)}$$

where EI is the flexural rigidity of each beam of the specimen. Combining Eq 12 and 13, we have

$$G_I = \frac{P^2 a^2}{WEI} \qquad \text{(Eq 15)}$$

Critical conditions occur when $P = P_c$ and

$$G_{Ic} = \frac{P_c^2 a^2}{WEI} = \frac{3P_c^2 C}{2Wa} \qquad \text{(Eq 16)}$$

where the critical load (P_c) is determined at the point of delamination initiation. The elastic beam theory (Eq 14) will remain valid as long as the shear deformations are negligible and the deflections are small; otherwise some corrections must be made to Eq 14 and Eq 16 [100,101]. The reader may refer to Ref 15, 41, 100 for details of DCB data reduction.

Careful examination of the fracture surface is necessary before meaningful interpretation of the DCB data can be made. The matrix-dominated type of fracture

surface always represents a minimum value of G_{Ic}. Two points are noteworthy here. Firstly, in the case of multidirectional laminates, the crack tends to jump and move along ply interfaces above or below the laminate midplane, creating a non-self-similar crack growth pattern and mixed-mode response. This often produces apparent values of G_{Ic} that are considerably higher than those obtained with matrix-dominated 0° unidirectional specimens [41,102]. Secondly, even in the case of a unidirectional composite, the crack can move slightly off the centerline, creating "fiber bridging," where fibers are pulled from one side of the delamination plane to the other. This also results in a significantly increased apparent value of fracture energy. If fiber bridging actually occurs in a structural component, then the higher value of G_{Ic} associated with the phenomenon would be justified. These values will be very unconservative, because fiber bridging cannot be guaranteed in most applications [15].

CLS Test

The CLS specimen, used for measuring mixed-mode interlaminar fracture energy, is schematically shown in Fig. 6(b). The P-$\dot{U}$ curves may be obtained for various crack lengths, and $\delta c/\delta a$ may be determined. The substitution of P_c and $\delta c/\delta a$ in Eq 12 gives the value of G_c, which may also be obtained using

$$G = \frac{P_c^2}{2W^2} [1/Eh)_2 - 1/(Eh)_1] \qquad \text{(Eq 17)}$$

where the subscripts 1 and 2 refer to the sections indicated in Fig. 6(b). The CLS test represents a relatively straightforward method of determining the total energy consisting of mode I and II; specimen preparation is easy and the data reduction scheme is simple. The separation of individual components may require very sophisticated analysis [41]. Specimen design to obtain the desired percentage of mode I and II response also requires complex analysis.

ENF Test

The ENF specimen is identical to the DCB specimen, but ENF is loaded in three-point flexure, which results in almost pure in-plane shear delamination growth by mode II. Russel and Street [84] have expressed the determination of G_{IIc} using the beam theory as

$$G_{IIc} = \frac{9P_c^2 a^2 C}{2W(2C^3 + 3a^3)} \qquad \text{(Eq 18)}$$

where P_c is the critical load for delamination, W and a are defined in Fig. 6(c), and C is the compliance given by

$$C = (2L^3 + 3a^3)/(8EhW) \qquad \text{(Eq 19)}$$

where E is the flexural modulus, $2h$ is the thickness of the specimen, and L is defined in Fig. 6(c).

It may be noted that Eq 18 and 19 were derived [84] without considering the influence of interlaminar shear deformation and friction between the crack surfaces on

G_{II} and C. Carlsson *et al.* [94] modified the beam theory solution to include the influence of shear deformation and frictional effects. Gillespie *et al.* [103] later used the finite element method to refine the analysis of ENF specimens by including the effects of local shear deformation around the crack tip, which resulted in higher G_{II}. The inclusion of shear deformation in the derivation of G_{II} by beam theory appears to improve the results. However, the discrepancy between finite element and beam theory solutions could be as high as 20 to 40% for typical graphite-epoxy composites. An improved data reduction scheme, which retains the simplicity of the beam theory and the accuracy of the finite element method, was later suggested [104].

The ENF test provides a relatively straightforward method for measuring mode II interlaminar strain energy release rate. The method should be limited to 0° unidirectional composite specimens [15].

EDT Test

Developed by O'Brien [43], the EDT test provides a viable approach to determining delamination initiation. The test method makes use of tensile specimens (Fig. 6d) of the lay-ups that are delamination prone, such as $(\pm30/\pm30/90_n)_s$. The point at which the stress-versus-strain curves deviate from linearity is considered the point of onset of delamination, and thus G_c, the total strain energy release rate, may be obtained by Eq 2. The individual components G_{Ic} and G_{IIc} may be obtained by finite element analysis [43]. Though the EDT test is quite simple, one needs to know *a priori* the delamination interface in order to analyze the test data. In the EDT specimen with a midplane insert, initial crack propagation should be used in determining G_c. This is due to the asymmetric and nonuniform nature of the crack front, which makes the determination of the crack length difficult.

Johnson and Mangalgiri [76] compared the G_{Ic} and G_{IIc} obtained during delamination, using various specimens for T300/5208 composite materials, and found the DCB and ENF tests most suitable for characterizing modes I and II delamination behavior, respectively. They have also observed G_{IIc} to be about eight to ten times greater than G_{Ic}. Further, the EDT and CLS tests can only give total G_c for fracture, and it is rather difficult to obtain G_{Ic} or G_{IIc} from these specimens [76].

SCB Test

The SCB specimen (Fig. 6e) appears to provide a viable method of determining mode III delamination fracture toughness [96]. The specimen and method are similar to those of the DCB test in characterizing mode I delamination, but the loading mode is different. Using the SCB test, Donaldson [96] has determined that the G_{IIIc} values for AS4/3504 graphite-epoxy laminates are about 1.0 to 1.3 KJ/m^2.

Design against Delamination

Modifications of Thermoset Resins

It is generally believed that the polymer is likely to be tough if bulk, homogeneous yielding does occur. Even brittle crack propagation in polymers usually involves localized viscoelastic and plastic energy-dissipating processes taking place in the vicinity of the crack tip [105]. Such processes usually absorb a great amount of strain energy in the material. The two primary energy-absorbing mechanisms in glassy thermoplastics are shear yielding and crazing. Unfortunately, these mechanisms tend to be localized and produce inhomogeneous plastic deformation in the material. If they are confined to a very small volume compared to the size of the specimen, the total amount of plastic energy will be low. To maximize the toughness would require that a large volume of material be capable of undergoing these or other energy-dissipating mechanisms, while restricting the growth of voids to avoid premature crack initiation [105].

Lightly Cross-linked Thermosets. In an investigation of the strength of highly elastic materials, Lake and Thomas [106] identified an important material parameter, To, which is the energy per unit area required to produce new surfaces. It was shown that To can be calculated approximately by considering the energy required to rupture the polymer chains lying across the path of the crack. Because forces are transmitted primarily by the cross-links, to break a particular bond in a chain would require that all other bonds lying in the same chain be stretched virtually to the breaking point. Therefore, the energy required would be much greater than the dissociation energy of a single bond (although only one of these bonds will really be broken). An alternative approach was developed by King and Andrews [107], who made many similar assumptions. A major difference is that they considered the energy required to rupture a chain in terms of the bond dissociation energy, E_b, and not in terms of the energy needed to rupture a monomer unit. The energy necessary for chain rupture in this approach is approximately $n'E_b$, where n' is the number of bonds capable of storing the energy required to break the weakest bond, E_b. By considering the network to be a simple array of cross-links, the authors proceeded to show that

$$\text{To} = (1/2)\,N^{2/3}\,n'E_b \qquad \text{(Eq 20)}$$

The density of network chains, N, can be determined from the classical rubber elasticity theory [107]. In order to determine n', one must know how many monomer units there are per chain. This can be estimated from the calculated values of M_c (average molecular weight between two cross-link points) and the molecular weights of the resin monomer unit and the hardener. E_b can be estimated by considering the bond energy of the weakest link along a chain [107]. The extensibility defined in Eq 21 can also be related to N and n'.

Composite materials scientists aiming to improve the performance of resin matrices for, say, aerospace applications can either try to enhance the established matrix systems by modifying resin chemistry, or they can develop new systems based on different chemical structures. Both strategies have led to useful products. Improving damage tolerance and hot/wet properties without sacrificing other desirable properties, such as the glass transition temperature (T_g) and the modulus (E), was the goal of many efforts on resin modifications. Flexibilized epoxies generally possess a higher impact resistance but poor humidity and elevated temperature characteristics. The addition of a rubbery phase to the matrix resin, such as carboxyl terminated butadiene acrylonitrile elastomer (CTBN), has significantly improved composite damage resistance (discussed in the next section) at the expense of reduced T_g and E values.

Bravenec *et al.* [108] have investigated the concept of lightly cross-linked thermosetting (LXT) resins. Their approach was to build an LXT architecture that provides toughness properties without significantly compromising the end-use temperature. The cross-link density is reduced (to increase n') by incorporating predominantly difunctional constituents, while "stiffened" polymer backbone components are added to impart the required T_g's at reduced cross-link densities. These two effects seem to go in opposite directions, and therefore a compromise between toughness and T_g must be made. Nevertheless, a scope of material properties can be obtained by this approach. The work by Murakami *et al.* [109] appears to provide largely similar results. Both studies have provided qualitative support for Eq 20. Additional fundamental studies should be aimed at establishing basic quantitative relationships between toughness and the degree of cross-linking in a resin, and between resin properties and composite toughness.

Rubber-Toughening of Thermosets. In polymers such as high-impact polystyrene (HIPS) and ABS, crazing and/or shear yielding are made to grow from many sites, so that a much greater volume of the polymer is involved than solely that encompassed by the immediate crack tip. The rubber particles dispersed in these thermoplastic matrices are believed to play the role of triggering and controlling these microdeformation mechanisms [110,111].

Thermoset resins such as epoxy can also be toughened by adding a rubbery phase. Bucknall *et al.* [110,111] proposed that multiple crazing and shear yielding are the two energy-absorbing mechanisms in rubber-modified epoxies, but this issue remains controversial. A TEM micrograph was presented that showed a craze-like fibrillar structure in the epoxy [110], but no massive crazing was evident. Van den Boogaart [112] observed a transition from crazing to shear yielding on increasing the degree of cure for epoxy resins. Sultan, *et al.* [113,114] observed that small rubber particles (<0.1 μm) did not toughen epoxies, while large particles (1-22 μm) increased G_{Ic} by an order of magnitude. Stress whitening was observed on the fracture surface of the toughened epoxy. Although crazing was cited as being associated with large rubber particles, no direct evidence of crazes was presented in these reports. Stress whitening could

have been caused by microcavitation of the rubber particles beneath the blunt crack tip, where high hydrostatic stresses exist. Doubts have been expressed by many researchers [115-117] about the existence of crazes in thermoset resins. However, others [118-120] have felt strongly that they do form.

Based on a study concerning the toughening mechanism in ABS, Donald and Kramer [121] observed shear deformations that were promoted by rubber particle cavitation. Cavitation was considered the source of the observed stress whitening [121,122]. Further, individual or clusters of particles that can cavitate can help relieve the local build-up of hydrostatic tension created by the localized shear process [121,122]. The constraint conditions can be relieved, possibly soon after the development of some initial shear yielding, in such a way that even a thick specimen may behave as if the matrix were everywhere under plane-stress conditions. In order for such a mechanism to operate, the matrix must be readily able to shear. Matrices of this nature include rubber-modified polyvinyl chloride, polyethylene-modified polycarbonate [123,124], and possibly rubber-modified epoxies [106,116,125-128].

Bascom *et al.* [106,127] observed that the plastic zone sizes in rubber-modified epoxy adhesives are directly related to their toughness. Cavitation caused by the triaxial tension ahead of the crack tip can effectively increase the size of the plastic zone, permitting a greater degree of plastic flow of the epoxy matrix, presumably by the stress-relieving mechanisms discussed above. Additional evidence to support the cavitation-promoted shear yielding of the epoxy matrix came from Kinloch *et al.* [125,132]. Following the same line of reasoning, Yee and Pearson [116] proposed a very specific microfailure mechanism for rubber-modified epoxies. Using a crack in the opening mode as an example, they postulated that the rubber particles in a zone ahead of the crack tip cavitate in response to an increasing displacement and load. On further loading, cavities grow larger, with those at highest stress levels initiating shear bands. Optical microscopic examination of thin sections did reveal the presence of numerous shear bands, although only in somewhat lightly cross-linked epoxy systems.

Modification with a Particulate Phase. The dispersion of rigid particulate fillers in epoxy resins provokes significant increases in toughness. These include alumina, silica, silicon carbide, glass beads, and thermoplastic powder [129-131]. Such rigid particles also include tensile stress concentrations in the matrix, thereby enhancing shear yielding. Although both rubbery and rigid particles may promote shear yielding in the matrix, rubbery particles usually are more effective in toughness improvement. As suggested by Kinloch and Young [105], this may arise from differences in the post-yield behavior. The extent of post-yield deformation in rigid-particulate-filled polymers is often extremely limited; this would limit the energy absorbed by localized shear yield deformations and lead to more ready crack initiation and growth. Kinloch and Young speculated that the rubbery particles may cavitate to relieve the constraint, as discussed earlier, or stretch and stay well-bonded to sustain imposed loads. Contrarily, rigid fillers cannot deform sufficiently and may easily become debonded, losing their ability to bear the load. Those rigid particles therefore are not as effective in enhancing the post-yield deformation of the matrix.

Thermoplastic-Modified Thermosets. High-performance interpenetrating networks (IPNs) have been a subject of increasing research emphasis in aerospace, defense, electronic, and automotive materials communities. One objective of the IPN approach is to combine an easy-to-process but brittle thermosetting polymer with a tough but difficult-to-process thermoplastic polymer. This combination has been effective in developing numerous high-performance semi-IPNs that can be processed like a thermoset and possess good toughness, like a thermoplastic [132]. However, limited work has been done in correlating mechanical properties with the morphology of these resins. In a study of thermoplastic-modified epoxy resins, Recker et al. [133] observed a distinct maximum in resin toughness that was correlatable to a unique multiphase morphology. A spinodal, two-continuous-phased morphology was found to result in optimum toughness in poly (ether sulfone)-toughened epoxy [134]. No fundamental toughening mechanisms were discussed in these studies [133,134]. By manipulating the thermoplastic chemistry, the composition, and the cure conditions, a wide spectrum of phase morphology can be obtained. This provides a unique opportunity for studying plastic deformation/toughening mechanisms as a function of phase morphology.

Obviously the rigidity (and other types of mechanical behavior) of the dispersed phase in a high-performance IPN or semi-IPN lies somewhere between that of a highly rigid filler particulate and that of a flexible rubber particle. One would expect that in general, the toughening efficiency of the dispersed phase in IPNs or semi-IPNs would be at an intermediate level, given the same low second-phase fraction (<20%). However, one should not overlook the great opportunities of manipulating the phase morphology and interfacial adhesion in the interpenetrating network by controlling the thermodynamic driving force and the kinetic factors during phase separation. As opposed to rubber-modified resins, where only a limited proportion of a rubber component can be incorporated to toughen the matrix without compromising other properties, a much wider range of compositions is available for producing IPNs or semi-IPNs with balanced mechanical properties. This range usually encompasses the portion of a phase diagram where both spinodal decomposition (SD) and nucleation and growth (NG) are possible phase separation mechanisms. These different mechanisms, in combination with the coarsening and coalescence mechanisms at the later stage, can be used to produce a wide spectrum of morphologies and interfacial properties. This provides an added dimension in designing the materials for specific applications.

An urgent need exists to understand the possible plastic deformation and failure mechanisms in IPNs and semi-IPNs in relation to composition, phase morphology, and interfacial adhesion. Phase morphology studies must clarify the roles of size, shape, and dispersion of the second-phase particles in dictating the toughening mechanisms. When following the nucleation and growth route, the second phase tends to be dispersed as discrete spherical particles. It is of interest to know whether shear banding or diffuse shear yielding will occur in a penetrating network. If it does, how would each mechanism depend on the particle size distribution, spatial dispersion of particles (uniform? random? cluster of particles?), and interfacial bonding? Would the particles individually or collectively cavitate to promote shear bonding? Is there a critical particle size or a certain state of particle dispersion that particularly favors the development of shear bands or other stress relief mechanisms? These are all important questions, and no quick answers will be available without many experimental efforts.

Should phase separation follow the spinodal decomposition path, interconnected phases are expected. In such an interconnected phase morphology, the stress field will be very different from that in a conventional morphology, where discrete isolated particles are dispersed in a matrix. In the former, the deformation of one phase cannot take place without a cooperative deformation of the other. Both phases will make a significant contribution to the toughness, whereas in a discrete phase morphology, the dispersed particles largely serve only to trigger the plastic deformation in the matrix. It is the deformation of the matrix that is likely to make up the majority of the energy-absorbing capability. Contrarily, the SD morphology gives both phases a chance to contribute; propagation of a crack must involve stretching and tearing of both phases in a coordinated fashion. Whether such an interconnected phase morphology leads to a high energy-absorbing capacity remains to be determined.

Equation 20 can act as a theoretical framework to which modifications can be made to develop new theories for the intrinsic toughness of IPNs and semi-IPNs. The intrinsic toughness of a thermoset is expected to be controlled by To. Any deviation from To (in addition to the intrinsic toughness) must be accounted for by the shear yielding, crazing, or particle stretching and crack-bridging [115,124,135] mechanisms promoted by the presence of a second phase. This latter mechanism may turn out to be an effective energy-absorbing process in a co-continuous phase morphology or where the second phase volume is large.

Bauer et al. [136] and Stenzenberger et al. [137] have reviewed the current developments of BMI resins; in particular, various approaches to toughening BMI are discussed. The current BMI resins are complex blends of a basic BMI building block with comonomers such as Diels Alder or Ene-, which preferably are liquid at room temperature and which impart toughness to the network after copolymerization. Other formulation approaches are also possible. These include chain extension with diamine or aminobenzoic hydrazide and modifications with reactive elastomers and thermoplastics. The major emphasis in these developments was to improve toughness, prepreg handleability, moisture insensitivity, and processing.

Several studies [e.g., 136-138] deal chiefly with the modifications of BMI with thermoplastic resins. Interlaminar fracture toughness and damage tolerance can be significantly improved, but usually at the expense of reduced prepreg handleability. Again, the toughening mechanisms of thermosets with thermoplastics are not fully understood. How the backbone chemistry of the aromatic thermoplastic allows control over the morphology of the BMI/thermoplastic system and over the properties of composites deserves further studies.

PMR-15 polyimide (PI) resins are a group of high-temperature thermosetting resins that possess respectable strength and stiffness, but with poor toughness and inadequate microcracking resistance. Traditional approaches to toughening PI resins usually involved incorporating flexibilizing linking groups in the PI chain backbone. As a result, the toughness was improved with a reduced glass transition temperature. Pater [139,140] has suggested that simultaneous semi-IPN polyimide can be synthesized from an easy-to-process but brittle thermosetting PI and a tough but difficult-to-process thermoplastic PI. The products typically exhibit remarkable resin fracture toughness and composite microcracking resistance without compromising glass transition temperature and processability. The problem of insufficient information on the morphology-property relations must be overcome to permit the further improvement of mechanical performance. Deformation and failure mechanisms of both neat resins and fiber composites need to be investigated in relation to composition, phase morphology, and interphase adhesion.

Toughening Mechanisms. The above discussion has indicated that crazing in the thermoset remains to be a highly controversial issue. A reasonable speculation is that crazing can possibly occur in a lightly cross-linked thermoset but not in well-cured resins such as PI and epoxy. Also unresolved and to be determined is the question of whether rubber particles can really promote shear banding in a heavily cross-linked resin. Although shear banding was observed in the optical micrographs presented by Yee and Pearson [116] for lightly cross-linked model epoxy resins, this banding phenomenon became much less pronounced when the cross-link chain length, M_c, was decreased. There is no doubt that plastic deformation takes place in a thermoset, and possibly it can be provoked by a dispersed second phase. The question is whether shear yield deformation operates in the form of a micro-shear banding or a more diffuse shear yielding mode.

By increasing the degree of cross-linking, the chain network as a whole becomes more uniform. This should shift the tendency from a more localized shear banding mode to a more homogeneous diffuse shear yielding mode. More importantly, the reduced M_c with an increased level of cross-linking should severely limit the extensibility of the chains. This drawability is essential to formation of the characteristic fibrils in a shear band. This can be understood from the observation that a typical

shear strain in shear bands was about 2.5 (250%) [141]. This places a stringent demand on the chain extensibility of a network. Contrarily, the strain in the diffuse shear yield zones is relatively low [142,143]. Although in general a more homogeneous deformation is desirable, the extent of plastic deformation is critical. The problem associated with a highly cross-linked network is its inability to undergo plastic deformation to a large extent, thereby severely limiting the energy-absorbing capability.

Donald and Kramer [121,144,145] have conducted very detailed TEM studies showing that the microstructure of crazes, represented quantitatively by the parameter Δ_{cr}, is primarily governed by the entanglement network of the polymer glass. This network can be characterized in terms of the chain contour length between entanglements l_e, and the entanglement mesh size d, given as the root-mean-square end-to-end distance of a chain of molecular weight M_e, the entanglement molecular weight determined from melt elasticity measurements. A maximum theoretical extension ratio for the craze, Δ_{max}, is predicted by a model in which the entanglement points are treated as permanent cross-links and chain scission is neglected [144,145]. In this model,

$$\Delta_{max} = l_e/d \qquad (Eq\ 21)$$

Good agreement is obtained between Δ_{max} and experimental determination of Δ_{cr} for a wide range of homopolymers and copolymers [144] and for polymer blends [121].

Donald and Kramer [145-147] have observed that when polymers of high l_e, such as polystyrene, are stressed in air at room temperature, crazes form readily. However, polymers of low l_e, such as polycarbonate, show a marked tendency for diffuse shear deformation. The material within the shear deformation zone is also oriented, but the extension ratio Δ_{dz} is not as high as within a craze. Both Δ_{cr} and Δ_{dz} correlate strongly with $\Delta_{max} = l_e/d$. For the polymers studied, $\Delta_{cr} \approx 0.8\ \Delta_{max}$ (for materials of larger l_e), while $\Delta_{dz} \approx 0.6\ \Delta_{max}$ (for materials of lower l_e). The extension ratio within the shear zones in polycarbonate is found to be essentially constant throughout at $\Delta = 1.4$. This is in sharp contrast to Δ_{sb} (shear bands) = 2.0 to 2.5 [141,148] for polystyrene, which tends to develop localized plastic deformation processes (crazing in tension and shear banding in compression, for instance). For polymers with an intermediate l_e, such as high-molecular-weight blends of PS-PPO (polystyrene-polyphenylene oxide), shear deformation zone and crazes can coexist. Nevertheless, Δ_{cr} is always greater than Δ_{dz}, and Δ_{cr}/Δ_{dz} for PS-PPO have been found to be 3.3/2.3, 3.2/2.1, 2.8/2, 2.7/1.8, and 2.6/1.6 (an average of 1.5). These observations suggest that localized deformation modes (crazes and microshear bands) require a much greater draw ratio than does diffuse shear yielding. Further, crazing is greater than shear banding in the required extensibility.

Recently, the issue of craze-forming tendency in a cross-linked network has been addressed [149]. In the

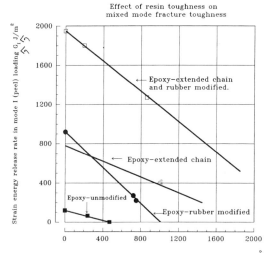

Effect of resin toughness on
mixed mode fracture toughness

Strain energy release rate in mode I (peel) loading G_I, J/m^2

Epoxy-extended chain
and rubber modified.

Epoxy-extended chain

Epoxy-unmodified

Epoxy-rubber modified

Strain energy release rate in mode II (shear), G_{II}, J/m^2

Fig. 7 The effects of extending the polymer chains and adding rubber particles on the fracture toughness of epoxy resins. Courtesy of NASA Langley Research Center

undeformed state, the cross-link mesh size d_x is assumed to be given by the root-mean-square end-to-end distance of a chain of molecular weight M_c. The cross-link molecular weight can be determined by the equilibrium dynamic shear modulus G' or from a theoretical calculation. By assuming free-rotating chains with fixed bond angles, one can calculate the mesh size d_x and the average chain contour length l_x between cross-links. The maximum extension ratio Δ_{max} for a cross-linked network is given by

$$\Delta_{max} = l_x/d_x \qquad (Eq\ 22)$$

The value of Δ_{max} for a highly cross-linked PI was found [149] to be approximately 1.6, suggesting the improbability that PI will craze. This recent study suggests that:

- The maximum extension ratio in a cross-linked network, $\Delta_{max} = l_x/d_x$, governs the mode of plastic deformation and controls the intrinsic toughness of a thermoset resin.

- For a given thermoset, there exists a threshold extension ratio Δ_{th} below which crazing is unlikely to occur (i.e., if $\Delta_{max} < \Delta_{th}$). This Δ_{th} can only be determined experimentally.

- With a low Δ_{max} (which is dictated by resin formulation and degree of curing), shear yielding will prevail where a combination of diffuse shear zone (DSZ) and microshear band (SB) can operate. The relative propensity of SB with respect to DSZ is determined by the magnitude of Δ_{max}, a lower Δ_{max} being in favor of DSZ.

- The "toughenability" concept, as proposed by Yee and Pearson [143], is a reflection of a proper Δ_{max}

above which desired plastic deformation modes such as shear banding in the matrix can be promoted by the dispersed phase (such as rubber particles) to impart toughness to the resin. With a low Δ_{max} where the extent of plastic deformation and the possibility of crazing or shear banding are low, the benefit of adding a second phase will be limited.

Shown in Fig. 7 are the interlaminar fracture toughness values, both G_{Ic} and G_{IIc}, of various epoxy-based carbon fiber composites. The toughness values of the unmodified baseline epoxy composites are very low. By either extending the chain length between cross-links (lightly cross-linked resin) or incorporating a rubbery phase in the epoxy resin, one can significantly increase the G_{Ic} and G_{IIc} of the composites. However, a dramatic improvement in both properties is observed when both approaches are applied to the same resin. This synergistic effect in toughness is probably due to the improved toughenability by a reduced cross-link density (ν_x), which makes it more likely that the rubber particles will promote shear yielding or even crazing in the epoxy matrix if the ν_x is sufficiently low.

Controlled Interlaminar Phase

The concept of controlled interlaminar phase (CIP) is suggested as a generic approach to toughening the interlaminar fracture resistance in continuous fiber-reinforced polymer composites. In this approach (Fig. 8), the interlaminar zone in an advanced composite is considered as a discrete *phase* and the microstructure and properties of this phase are properly designed and controlled. Three examples given in this section illustrate how the concept of CIP can be accomplished. These include the techniques of:

- Incorporating discrete layers of tough resins (interleaving)

- Adding either particulates, whiskers, or microfibers to the interlaminar zone (supplementary reinforcement)

- Introducing *z*-directional fibers (three-dimensional reinforcement)

In each case, the interlaminar fracture toughness and the damage tolerance of composites can be significantly improved. Also discussed in this section are the deformation behavior and fracture mechanisms of composites containing a CIP.

Interleaving. To account for the effect of possible impact damage, composite structures are usually designed with overly conservative allowables (e.g., a design strain of 0.004 in./in.) that are far less than those obtained with undamaged laminates. Such a conservative design would considerably reduce structural efficiency [150]. One approach taken to improve the damage tolerance of graphite composites without sacrificing hot/wet properties is to employ a high-strain "interleaf" resin between the plies of graphite-epoxy [150-163]. The conventional

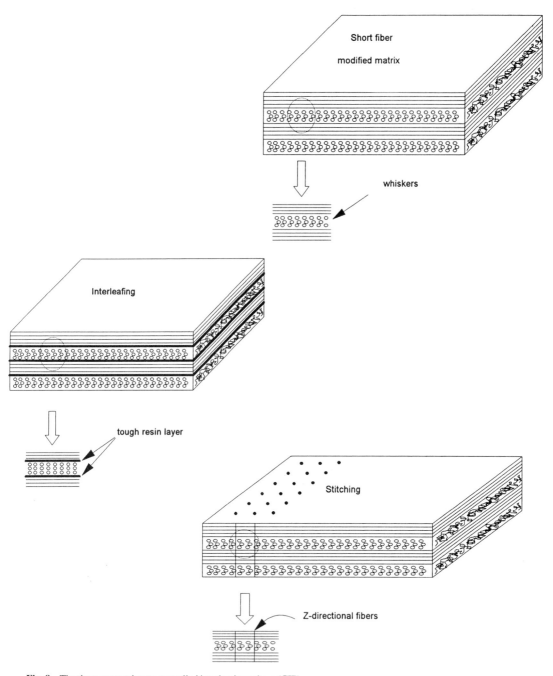

Fig. 8 The three approaches to controlled interlaminar phase (CIP)

wisdom behind this approach is that the matrix resin must provide both stiffness and toughness. A high resin modulus is required because the laminate compressive performance is believed to be directly proportional to the resin modulus [164]. Fiber kinking and fiber microbuckling, the predominant compressive failure modes in graphite-epoxy laminates, are controlled by the resin modulus [165]. A composite with a rubber-toughened

matrix will possess a reduced modulus. In the interleaf approach, the *matrix resin* within the lamina supports the fibers and provides the hot/wet stiffness. The *interleaf resin* has a high shear failure strain and provides toughness at the ply interfaces to prevent delamination [153,155, 156,159,163].

Interleaving has been shown to significantly improve the impact resistance of graphite-epoxy and graphite-bis-

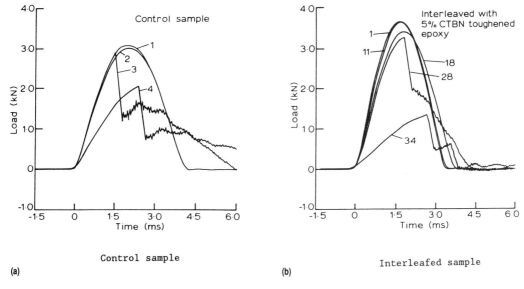

Fig. 9 Repeated impact load/time traces of carbon-epoxy laminates

Table 1 Experimental results of the end-notched flexure and double cantilever beam tests on interleaved and control composites

Graphite-epoxy laminates	G_{Ic} (KJ/m^2)	G_{IIc} (KJ/m^2)
Control composites	0.197	0.68
Epoxy 828	0.239	0.96
5% CTBN-toughened epoxy	...	1.56
10% CTBN-toughened epoxy	0.241	1.65
15% CTBN-toughened epoxy	...	1.45
20% CTBN-toughened epoxy	0.293	1.55
30% CTBN-toughened epoxy	0.280	...
100% CTBN, no epoxy	...	0.57
10% PU-toughened epoxy	...	1.34
15% PU-toughened epoxy	...	1.38
20% PU-toughened epoxy	...	1.24
30% PU-toughened epoxy	...	1.17
40% PU-toughened epoxy	...	1.06

Note: Data are average values of at least 9 tests, with a standard deviation of approximately 10%. CTBN, carboxyl-terminated butadiene acrylonitrile. PU, polyurethane

maleimide composites [154-156,159], as manifested by an increase in residual compression strength after impact. The results of interlaminar fracture toughness tests indicate a small increase in G_{Ic} and a large increase in G_{IIc} for composites with interleaving. In these studies [150,153-156,159,163], the shear failure strain of interleaf resin has been identified as a key parameter in the impact damage tolerance improvement. In a similar study [161], interlaminar adhesive layers have been found to be effective in suppressing edge delamination in the laminates subjected to in-plane tension. The adhesive layers in composite laminates also effectively suppress delamination up to very high impact velocities and help reduce the stress concentration effect [158]. These studies serve to demonstrate the advantages of using the interleaving concept to improve composite toughness. However, the material variables that dictate the composite toughness need to be defined and optimized. From the energy dissipation perspective, the "toughness" may be a better index for determining the best interleaf material. The composite interlaminar fracture properties (e.g., G_{IIc}) are essentially proportional to the interleaf resin toughness (G_{Ic}) [166].

Table 1 indicates that the G_{IIc} values of the interleaved composites are significantly higher than those of the control samples [166]. Although to a lesser extent, the G_{Ic} values are also improved. When a thin layer of uncured epoxy resin is applied as an interleaf phase between two prepreg layers, a 40% improvement in G_{IIc} is obtained. If the interleaf epoxy resin is modified with 5 wt% CTBN, the interleaved composites show a 230% improvement in the mode II critical strain energy release rate. The degree of improvement is observed to decrease when the CTBN content in the interleaf exceeds 15%. There appears to exist an optimum interleaf parameter where the composites will exhibit the maximum delamination resistance. However, the optimum conditions have yet to be identified.

Impact fatigue test represents one convenient way to assess the damage tolerance of a fiber composite [167]. In a typical impact fatigue test, a load-displacement curve obtained for a specimen in response to the first impact is recorded, which includes the numerical data of the maximum load ($P_{m°}$), the energy absorbed up to maximum load ($E_{m°}$), and the elastic modulus or stiffness ($S°$). This damaged specimen is then impact loaded for a second time with its load-displacement curve traced and Pm_1, Em_1, and S_1 calculated. The ratios $P_{m^1} / P_{m°}$, $E_{m^1} / E_{m°}$, and $S^1 / S°$ can be taken as measures of the degree of damage. However, we have found it very informative to overlay many curves together in the same diagram after n repeated impacts. Figure 9(a) shows such a diagram for

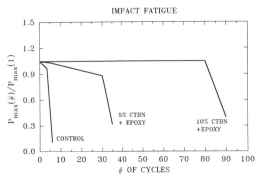

Fig. 10 Impact fatigue behavior of CTBN-toughened epoxy matrix composites. The ratio of maximum load (after n impacts)/maximum load (first impact) is plotted as a function of the number of repeated impacts.

Table 2 Energy absorbed and the ductility index of laminates

Laminates	Energy absorbed (J)	Ductility index
Control sample (not interleaved)	4.977 ± 0.330	1.171 ± 0.296
7.5% CTBN-828/Z	13.846 ± 2.482	4.172 ± 0.924
9.5% CTBN-828/Z	14.092 ± 1.101	4.788 ± 0.692

a graphite fiber-epoxy laminate (the control sample). As the number of repeated impacts increases, the maximum load tolerated by the material, an indication of the material strength, decreases. Also found to decrease was the slope of the load-time curve, which is proportional to the sample stiffness and modulus, given the same sample geometry. As a result of the stiffness reduction, the maximum deflection is found to increase. In general, the total energy absorption is also reduced as the degree of damage progresses.

Representative curves of impact fatigue tests on interleaved composites are shown in Fig. 9(b) and are further summarized in Fig. 10. The composites that contain a 10% CTBN-toughened epoxy resin as the interleaf material can sustain 89 events of repeated impacts and exhibit no significant reduction in the maximum load until just before fracture. However, the control sample can sustain only five events of repeated impacts before fracture. This observation demonstrates that an important beneficial effect of interleaving is improved resistance to transverse cracking and delamination, which are the two major failure modes leading to the reduced material integrity.

Figure 11(a) shows selected impact load-time curves of three thin plate samples subjected to impact penetration tests. The initial monotonic increase of the load is due to the front surface indentation by the impactor. The first major break on the load-time curve signals the onset of the failure process. Several humps after the maximum load, P_{max}, on the load-time curve suggests that there exists a multiple-step failure mode during the impact event. The load is distributed over a large portion of the

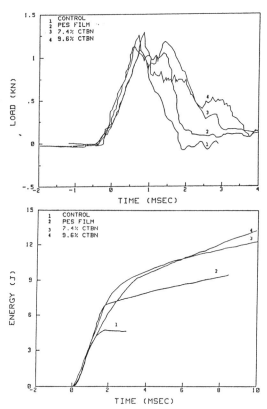

Fig. 11 Impact penetration test data of plate-type samples of control and interleaved composites. (a) Load vs. time curve. (b) Energy vs. time curve

material and creates more crack surface, reflecting a noncatastrophic failure mode. The incorporation of interleaf layers appears to increase the maximum load slightly. Because the penetration of a projectile through a composite laminate is a fiber-dominated phenomenon, the addition of thin interleaf layers was not expected to improve the impact strength to any significant extent. Surprisingly, the total energy absorbed, represented by the area under the load-time curve (Fig. 11b), is increased by three times in the interleaved specimens as compared to that of the control specimen (Table 2). The more gradual load drop with the interleaved specimens was probably the result of plastic deformation occurring in the interleaf layers. As shown in Fig. 11, the interleaved specimens also have higher strain-to-failure and higher propagation energy (the energy absorbed after maximum load), E_p. The impact toughness (total absorbed energy) of these interleaved laminates, although controlled by the through-the-thickness crack, did receive significant contributions from the interlaminar zones.

The ductility index (E_p/E_m) of the interleaved specimen is much higher than that of the control specimen, where E_m and E_p are the energy absorbed by the specimen before and after the maximum load, respectively (Table 2). This again indicates that a great amount of

Table 3 Impact strengths of composites, obtained from flex impact tests

Specimen	P_{max} (KN)	E_m (J)	E_p (J)	E_t (J)	Ductility index
Control	3.76 ± 0.29	7.85 ± 0.72	2.18 ± 0.20	10.03 ± 0.77	0.277
1% alumina	4.62 ± 0.31	8.43 ± 0.66	2.81 ± 0.38	11.50 ± 0.99	0.333
2.5% alumina	5.86 ± 0.45	9.93 ± 0.73	2.69 ± 0.26	12.62 ± 0.82	0.270
1% VME	4.14 ± 0.24	8.78 ± 0.36	3.11 ± 0.89	11.87 ± 0.80	0.354
2.5% VME	4.52 ± 0.32	8.76 ± 0.86	3.89 ± 0.27	12.57 ± 0.05	0.448
5% VME	5.75 ± 0.33	9.75 ± 0.87	3.54 ± 0.75	13.29 ± 0.30	0.363
1% VMD	4.38 ± 0.21	8.34 ± 0.46	4.51 ± 0.86	12.85 ± 0.49	0.540
2.5% VMD	5.78 ± 0.33	10.85 ± 0.86	4.54 ± 0.58	14.59 ± 0.98	0.451
5% VMD	5.25 ± 0.26	10.19 ± 1.21	4.34 ± 1.04	14.54 ± 1.03	0.425

Table 4 Falling-dart impact strengths of composites containing supplementary short fibers (plate-type specimens)

Specimen	P_{max} (KN)	E_m (J)	E_p (J)	E_t (J)	Ductility index
Control	2.48 ± 0.04	3.49 ± 0.28	13.18 ± 1.04	16.67 ± 1.18	3.776
1% VME	3.26 ± 0.01	4.92 ± 0.14	19.66 ± 2.11	24.57 ± 2.04	3.995
2.5% VME	3.41 ± 0.14	5.07 ± 0.26	19.96 ± 3.75	25.04 ± 3.51	3.936
5% VME	3.79 ± 0.09	6.00 ± 0.26	21.48 ± 1.15	27.49 ± 1.0	3.580
1% VMD	3.06 ± 0.13	4.84 ± 1.10	13.40 ± 1.24	18.20 ± 0.25	2.780
2.5% VMD	3.63 ± 0.16	5.28 ± 0.28	21.07 ± 1.63	26.35 ± 1.76	3.990
5% VMD	3.73 ± 0.16	5.79 ± 0.66	23.46 ± 0.84	29.24 ± 1.36	4.05
1% PE	2.67 ± 0.17	6.05 ± 0.06	15.39 ± 0.18	21.26 ± 0.40	2.543
2.5% PE	2.94 ± 0.10	7.20 ± 0.26	16.07 ± 0.58	23.28 ± 0.56	2.231
1% Kevlar	2.99 ± 0.12	11.77 ± 0.91	18.02 ± 1.65	29.80 ± 1.05	1.531

Source: Ref 48

energy is dissipated during the crack-propagation stage of an interleaved composite.

The mechanisms that led to these results became more apparent when the failure surfaces were examined. Scanning electron micrographs of the fracture surface of the baseline laminate showed very few "hackles," indicating a brittle matrix fracture mode. In contrast, hackles that were more extensive and larger (by an order of magnitude) were observed in the fracture surface of the 10% PU-828/Z interleaved composites. Similar fracture features were observed in the CTBN-828/Z interleaved composites during mode II shear loading [166].

Supplementary Reinforcement. Garcia *et al.* [168,169] have suggested a hybrid composite technique for improving the matrix-dominated properties of continuous fiber composites. In this technique, a supplementary reinforcement, such as particulates, whiskers, or microfibers, is added to the matrix prior to the resin impregnation. This technique does not require the development of new constituent material and is adaptable to the process and equipment currently used to fabricate composites. The transverse tensile strength and strains of

the graphite-epoxy-SiC hybrid composites containing 2 vol% SiC were significantly increased as a result of incorporating SiC whiskers [168]. Edge delamination resistance was also improved with 4 vol% SiC. However, fiber damage occurring during the whisker-resin impregnation process often resulted in significant reduction in the in-plane fiber-dominated properties [168]. Jang *et al.* [170,171] have found a significant improvement in impact energy of hybrid composites incorporating either particulates or ceramic whiskers. The relationships between the hybrid composite properties and the microstructure and morphology of the dispersed whiskers need to be established.

The results of the unnotched Charpy-type impact tests on the short fiber-modified composites are summarized in Table 3. The data indicate that the maximum load and the total fracture energy of the composites are significantly improved by modifying the resins with short fibers. When the volume fraction of short fibers increases, the load-carrying and energy-absorbing capabilities of the composites also increase. The data obtained from the plate-type impact samples (Table 4) also indicate that the contributions to the fracture energy from short fibers are remarkable. Both the maximum load (P_{max}) that can be tolerated by the composites and the total energy (E_t) absorbed are improved appreciably. A great amount of strain energy ($E_p = E_t - E_m$) is dissipated after P_{max}, where the composite is in the crack-propagating stage. The values of E_p are greater in the short fiber-modified composites than in the control composites.

The damage tolerance of a short fiber-modified composite is also found to be superior to that of an unmodified version. The impact fatigue results of a control sample (plate-type) and a sample containing 1% PE short fibers are shown in Fig. 12. While the maximum load of the control sample was reduced from 7 KN to 4 KN at the 15th impact, the load-carrying capability of the modified composite was reduced to the same extent after 30 repeated impacts. The results on the modified composites from the repeated Charpy-type falling dart tests also demonstrate a considerable improvement in damage tolerance. Short beam shear tests (ASTM D 2344) show that the interlaminar shear strength is also enhanced in the composites containing a small amount of short fibers [170].

The results of the DCB and ENF tests are summarized in Table 5. A distinct improvement in G_{Ic} for the specimens containing either 1% VME®, 2.5% VME®, or 1% VMD® short carbon fibers is observed. But G_{Ic} values decrease for the specimens containing 5% VMD carbon fibers, probably because of the presence of voids due to processing difficulty at high fiber loadings. A prominent improvement in G_{IIc} for the specimens containing either carbon or Kevlar short fibers is also seen from Table 5.

A SEM study [172] showed that the major fracture mechanisms involved in the DCB test for the control sample were interfacial debonding and matrix cracking with little plastic deformation. Obviously, the delamination resistance of the control sample was mainly dominated by the matrix properties. The same study also showed that in the fracture surfaces of samples containing both 1% VMD and 2.5% VME, fracture mechanisms such as interfacial debonding, matrix cracking, short-fiber pull-out and breakage, and short fiber/matrix debonding could be easily observed. The mechanical properties of the interlaminar region were apparently improved by these short carbon fibers.

The fracture surfaces of ENF test specimens were also examined by SEM [172]. A clean fracture surface with a few hackle structures around the continuous fibers could be seen on the micrograph of the control sample. However, the fracture surface micrographs for the specimen containing 1% VMD after short beam shear loading indicated the presence of a lot of hackles around the continuous fibers and short fibers. This implied that the matrix resin had been deformed to a great extent under shear loading. In the fracture surfaces of composites containing 2.5% VME and 5% VME, several large hackles around short fibers could be seen, probably because the available room for plastic deformation in the short-fiber-modified samples was larger than that in the control sample. The larger the area of hackle structure, the greater the amount of strain energy dissipated during crack propagation. In the composites containing PE short fibers, a clean fracture surface was observed with no hackle around the PE fibers. This was attributed to the poor

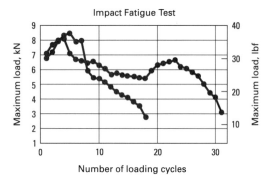

Fig. 12 Impact fatigue responses of a control carbon fiber/epoxy specimen and a carbon fiber/epoxy specimen containing 1% supplementary short polyethylene fibers

Table 5 Interlaminar fracture properties of composites containing supplementary short fibers

Specimen	ILSS (MPa)	G_{Ic} (KJ/m^2)	G_{IIc} (KJ/m^2)
Control	46.0	1.04 ± 0.09	1.25 ± 0.04
1% VME	48.0	1.37 ± 0.15	1.48 ± 0.07
2.5% VME	50.0	1.25 ± 0.04	1.66 ± 0.12
5% VME	50.0	1.05 ± 0.04	1.60 ± 0.04
1% VMD	47.0	1.21 ± 0.17	1.43 ± 0.05
2.5% VMD	43.7	1.04 ± 0.04	1.48 ± 0.06
5% VMD	41.5	0.99 ± 0.04	1.23 ± 0.01
1% PE	...	0.85 ± 0.01	0.55 ± 0.06
1% Kevlar	...	1.10 ± 0.02	2.08 ± 0.15

ILSS, interlaminar shear strength. Source: Ref 172

interfacial bonding between PE fibers and the matrix resin.

Stitched Three-Dimensioal Reinforcement. A third way to control the interlaminar properties is to introduce z-direction (thickness-direction) fibers by the technique of stitching [173-180]. The influence of the stitched z-direction fibers on the mechanical behavior of an otherwise two-dimensional composite has been investigated [176,179,180]. The three-dimensional composites have been found to possess greater damage resistance than two-dimensional laminates. The z-direction fibers effectively reduce the extent of delamination by increasing the interlaminar shear strength and toughness [179,180]. Stitching can not only help suppress delamination in laminated composites [173] but also serve as a "joint" in woven composite plates [177]. These studies [173-180] have demonstrated the great potential of using stitching in controlling the damage tolerance of composites. Further studies on the processing-structure-property relationships in stitched three-dimensional composites are essential to the proper use of this technique. The z-direction fibers will modify the interlaminar zones as well as the laminas. The technique of stitching, along with the other two approaches, can be conveniently considered under the same umbrella concept of CIP. Other techniques for introducing thickness-direction fibers, such as braiding and weaving, have been recently reviewed by Jang et al. [178,179].

The instrumented impact and four-point bending test data obtained from Kevlar and graphite composites have been published elsewhere [178]. In response to high-impact loading, three-dimensional Kevlar-epoxy composites demonstrated a higher level of load-carrying capability and a greater impact energy than their two-dimensional counterparts. The two-dimensional samples exhibited a stepwise yield drop (a few humps) in the load-displacement curve, characteristic of delamination crack propagation and arrest. The three-dimensional composites also showed some delamination, but to a much lesser extent, and they did not do so until a very

high load was reached. In the case of brittle graphite-epoxy composites, the three-dimensional materials tended to show a brittle impact response. This was believed to be caused by the high tendency for the graphite layers to suffer from fiber damage during stitching. However, three-dimensional graphite composites, just like three-dimensional Kevlar composites, exhibited a lower degree of delamination than their two-dimensional counterparts. When tested in a lower-rate flexural mode, the three-dimensional versions of both composite types demonstrated improved flexural strength and an increased amount of energy absorbed [178].

The results of impact fatigue tests on the stitched three-dimensional composites and the corresponding two-dimensional control material are summarized in Table 6. Each of these composites experienced a repeated constant impact load of approximately 4 J (3 ft · lbf) throughout all 25 impacts, the limit of our patience in this case. This table shows that the three-dimensional composites were able to maintain their strength throughout all 25 impacts, with a 17% strength reduction in the worst case. In contrast, the control sample showed a rapidly decreasing maximum load as the number of impacts increased. At the 25th impact, the control sample had lost 89% of its original strength. These data clearly demonstrate how applying stitching improves the damage tolerance of composites.

Design for Improved Damage Resistance of CIP Composites. Impact damage in composites will be briefly discussed here. A more detailed and extensive discussion is presented in Chapter 8. Many damage mechanisms are possible, and the extent to which any one damage mechanism predominates depends on the material properties, the structural geometry, and the stressing conditions.

In response to the impact of a low-velocity projectile, a laminate can deform by bending. No damage will occur to the composite if the projectile energy can be accommodated by the elastic strain energy in the material [181]. A critical condition is reached when a local stress exceeds a

Table 6 Impact fatigue results of stitched 3-D composites

Impact Number	Max load (KN)			
	Control 2D	Stitched with Kevlar	Stitched with graphite	Stitched with glass
01	3.26	3.78	3.85	4.00
03	...	3.55	3.61	4.20
05	2.10	3.82	3.27	3.94
07	...	3.92	3.18	4.17
09	...	3.90	3.35	4.09
10	2.60	...	...	...
11	...	3.87	3.35	4.00
13	...	3.81	3.25	3.79
15	...	3.79	3.26	3.89
19	...	3.92	...	3.74
21	...	3 83	...	3.71
23	...	3.84	...	3.69
25	0.35	3.77	3.40	3.64
Maximum reduction	89.3%	6%	17%	9%

local strength, or when a local strain energy density exceeds a local toughness in the composite. In a flexural beam with a small span-to-depth ratio, shear failure is more likely to occur in the form of delamination. Dorey [181] suggested that the elastic energy in a beam prior to shear failure would be:

$$U_s = \frac{2}{9} \frac{\tau^2}{E} \frac{wl^3}{t} \qquad \text{(Eq 23)}$$

where τ is the interlaminar shear strength, E is the flexural modulus, and w, l, and t are the width, span, and depth of the beam, respectively. For a beam with a greater span-to-depth ratio, flexural failure is more likely to occur near the top or the bottom surfaces of the beam (near the midspan). The threshold energy in this case would be [42]:

$$U_f = (1/18)\,(\sigma^2/E)\,(wtl) \qquad \text{(Eq 24)}$$

where σ is the flexural strength.

Since our major concern here is to improve the interlaminar fracture properties, we will discuss Eq. 23 in more detail. Keeping sample dimensions equal, one can see that the critical strain energy for crack initiation, U_s, is proportional to (τ^2/E). Our intention is to not decrease appreciably the magnitude of E, which would otherwise compromise the compression strength of composites. Introduction of a thin CIP layer was indeed found to have very little effect on the E of the resulting composite, as was evident from the slope of the load-displacement curve in either impact or three-point bending tests. A very small decrease, a small increase, and no change in E were found for the interleaved, the short-fiber-added, and the stitched composites, respectively.

For the interleaved composites, the interlaminar shear strength (τ), as measured by the maximum load in a short-beam flexural impact test and a short beam shear test, is only slightly better than that of a control composite. This implies that interleaving does not increase U_s to a great extent. The benefits of interleaving therefore do not lie in improving the resistance to crack initiation, but rather in enhancing the crack-propagation resistance. This was demonstrated by a remarkable increase in E_p and in the ductility index E_p/E_m, when the composite was interleaved with a small amount of tough resin (e.g., Fig. 11 and Table 2).

Tables 3 and 4 indicate a significant increase in P_{max}, a measure of τ, when a small amount of supplementary short fibers is incorporated in the laminates. The energy absorbed up to maximum load (E_m), being directly related to U_s, also exhibits a marked increase. A large increase in E_p is also noted. The resistance to both crack initiation and crack propagation clearly has been enhanced by supplementary reinforcement.

Four-point bending tests conducted on the stitched three-dimensional samples and their two-dimensional counterparts indicate that introducing some fibers in the z-direction could increase the critical load level below which no macroscopic crack would be initiated. A remarkable enhancement in the crack-propagation resis-

tance is observed from the much more gradual load-drop phenomenon and much greater area under the load-deflection curves in the stitched three-dimensional composites. Similar results are obtained from the impact penetration tests. However, in the case of stitching through graphite layers, stitching-induced fiber damage could result in a reduction in the in-plane strength of composites. A flexural failure, rather than shear failure, therefore ensues, and Eq 23 cannot be applied for calculating the resistance to failure initiation.

Mechanisms of Impact Damage of CIP Composites. In an investigation of low-velocity impact damage in laminated composites, Chang *et al.* [182] concluded that matrix cracking was the initial failure mode, caused predominantly by interlaminar shear stresses and in-plane tensile stresses. Delamination was initiated by the initial "critical" intralaminar matrix cracks, and delamination growth was dominated by the suddenly increased out-of-plane normal stress as a result of matrix cracking. This study has provided some important guidelines for designing damage-tolerant composites. Incorporating z-direction fibers, either short fibers or continuous fibers, obviously should increase the interlaminar shear strength and, therefore, the resistance to crack initiation. Some of the short fibers (in the case of supplementary fibers) and all of the long fibers (in the case of stitching) are oriented in the thickness direction (z-direction), so they should dramatically improve the ability of the composite to endure the out-of-plane normal stresses. The delamination resistance should therefore be enhanced drastically.

Krieger [157] demonstrated a possible tenfold reduction in the interlaminar shear stress concentration factor when an interleaf resin was added to the laminate. The presence of such a tough resin can also help dissipate the incident energy, by undergoing plastic deformation in response to in-plane tensile stresses and interlaminar shear stresses.

CIP composites are more resistant to crack initiation and delamination propagation than conventional laminates. They are able to absorb more impact energy per unit of delamination area created. Given the same incident energy, CIP composites would suffer much less damage and be more successful in maintaining their structural integrity. The damage tolerance in these composites therefore would be much better, as was demonstrated in the present study.

Interlaminar Fracture Toughness of CIP Composites. Several energy-absorbing mechanisms in a continuous fiber-reinforced resin composite during mode I and mode II loading have recently been considered by Friedrich *et al.* [183]: creation of the main-crack fracture surface; development of a damage zone around the crack; and crack bridging by fibers. The actual surface area produced during interlaminar cracking may be significantly larger than the projected area ($2 \times$ width $\times$ crack increment) used in the determination of G_c. The fracture surface is not planar; rather, it contains hackles, as can be clearly seen in the SEM micrographs. Both interleaving and short fiber modification have been found to result in the formation of more and larger hackles, and hence a much rougher surface. No appreciable variation in hackle pattern has been found with the composites when stitched. Because of the high complexity of the surface roughness encountered in these three CIP cases, no attempt was made to measure the profile length, as suggested by Crick *et al.* [184] and by Friedrich *et al.* [183].

Another important factor is the volume of the damage zone near the crack tip, where matrix plastic deformation and/or cracking and possibly fiber-matrix interfacial debonding take place. Bradley and Cohen [185] assumed that the process zone or damage zone for the composite had the same volume as in the bulk polymer, and that the toughness of the composite was derived from energy dissipation in the polymer matrix. They suggested that the composite interlaminar toughness (G_{co}) was given approximately by

$$G_{co} \approx G_{c,m} \cdot V_m \qquad \text{(Eq 25)}$$

where $G_{c,m}$ is the toughness of the bulk resin and V_m is the average volume fraction of matrix over the process zone. To account for the effect of the actual fracture surface area, Eq 25 was modified [183] to become

$$G_{co} \approx L_r / L_p V_m \cdot G_{c,m} \qquad \text{(Eq 26)}$$

where L_r is the profile length [184] and L_p is the projected crack length. With the introduction of an interleaf phase, which is capable of undergoing a great extent of plastic deformation, the effective values of L_r, V_m, and $G_{c,m}$ should increase. The effective magnitude of $G_{c,m}$ at the interlaminar zone clearly should increase when short fibers are added. In short-fiber-modified continuous fiber composites, the crack tip will interact not only with the misaligned continuous fibers but also with those dispersed short fibers. While the crack tip interacts with the short fibers, more energy will be consumed to break those short fibers. This will also induce fiber-matrix debonding, promote fiber pull-out, and alter the crack path, therefore producing more wrinkled fracture surfaces. We have observed previously that interfacial debonding between resin and *z*-directional fibers, as well as breakage and pull-out of these fibers, can be a major contributor to energy dissipation. We therefore can state that L_r and $G_{c,m}$, in a broader sense, can be raised effectively when thickness-direction fibers are introduced in an otherwise conventional two-dimensional laminate.

Additional energy absorption is possible when crack bridging by fibers occurs, resulting in fiber peeling and fractures [183]. In any attempt to improve the intrinsic fracture toughness of a laminate, these additional contributions to the interlaminar G_{Ic} and G_{IIc} can be misleading. In the commonly used interlaminar fracture toughness tests, this "fiber bridging" phenomenon may give a false indication of an otherwise matrix- and interface-dominated G_c value. In composites subjected to a low-velocity impact, transverse cracks and delamination may propagate through a naturally low-energy-density area, possibly avoiding the misaligned fibers. To discount these fiber bridging contributions to the interlaminar G_c

values would be a safer design practice as far as damage tolerance of composites is concerned.

Comparison of the Three CIP Techniques. Absolute rating of the effectiveness of the proposed three techniques (interleafing, supplementary reinforcement, and stitching) in improving the interlaminar fracture resistance of composites cannot be achieved without ambiguity. The technique of stitching appears to provide the highest interlaminar fracture resistance in terms of G_{Ic}, G_{IIc}, and interlaminar shear strength, but often at the expense of in-plane properties because of potential fiber damage. No significant increase is observed in the total impact energy absorbed by stitched composites. In addition, stitching through a thick stack of fabric or prepreg layers can be a difficult task in real practice.

Under comparable conditions, interleaving seems to be more effective than the supplementary reinforcement approach. This judgment is made on the basis of impact energy and interlaminar fracture toughness values. Improvements of 46%, 32%, and 66% in E_t, G_{Ic}, and G_{IIc}, respectively, can be achieved by introducing short fibers in the laminates. The corresponding values for interleaving are 180%, 53%, and 144%, respectively.

Conclusions above CIP. The following concluding remarks concerning CIP appear to be appropriate:

- The impact toughness and penetration resistance of a continuous fiber composite can be effectively improved by modifying the matrix resin with short fibers. Interleaving is also effective in improving these impact properties, probably because of the high plastic deformation capability of the interleaf resin. The stitched three-dimensional laminates generally demonstrate a higher level of load-carrying capability and impact resistance than their two-dimensional counterparts. However, the technique of stitching through prepreg layers could result in localized in-plane fiber damage and reduction in flexural strength.

- Short fiber modification of the matrix resin is also effective in improving the interlaminar shear strength (ILSS) of composites. The ILSS values usually increase with the volume fraction of short fibers. But too high a volume fraction of short fibers could create a high void content in composites, resulting in a reduced ILSS. Introduction of thickness-direction fibers via stitching also dramatically raises the ILSS of composites.

- With an adequate short fiber/matrix interfacial bonding, short fiber modification appears to be an effective method for improving the delamination resistance and damage tolerance of composites. The interlaminar fracture toughness (both G_{Ic} and G_{IIc}) can be improved by adding a small amount of carbon and Kevlar short fibers to the interlaminar phase. Through-the-thickness stitching has also proven to be effective in suppressing the free-edge delamination in two-dimensional laminates. The *z*-direction fibers tend to inhibit the propagation of delamination and allow energy-absorbing mechanisms to operate, such as interfacial debonding between z-direction fibers and matrix and pull-out of z-direction fibers. Interleaving also dramatically enhances the interlaminar fracture toughness of composites.

- Interleaving improves interlaminar crack propagation resistance, while both the crack initiation resistance and crack-propagation resistance can be enhanced by supplementary and z-directional reinforcements.

- The concept of CIP is a viable approach to improving the damage tolerance of composites. Additional work is needed to understand the correlation between CIP parameters and composite fracture behavior so that the delamination of a composite can be controlled without sacrificing other properties of composites.

Three-Dimensional Braided and Three-Dimensional Fabric Composites

In addition to stitching, three-dimensional braiding and weaving can be used to introduce thickness-direction fibers to possibly suppress delamination. In fact, in three-dimensional braided or woven structures, interlaminar boundaries are not well-defined. Perhaps it is more appropriate to discuss damage tolerance (postdamage strength retention) instead of delamination resistance.

Gause [186] has compared the effects of drill holes on the integrity of three-dimensional braided carbon fiber/epoxy composites and corresponding quasi-isotropic composites. The braided structures were found to retain over 90% of their original strength prior to drilling, as opposed to a 50% strength reduction in the case of quasi-isotropic composites. Gause has also observed a much smaller extent of impact damage in braided constructions as compared to that in conventional laminated composites. Similar observations were made by Ko et al. on glass-epoxy [187] and carbon-PEEK [188] composites. Significantly higher levels of energy were required to initiate and propagate damage in three-dimensional braided glass-epoxy composites than in laminated constructions when subjected to the same drop weight test. A much smaller impact damage area was observed with three-dimensional graphite-PEEK than with two-dimensional laminates.

As discussed in Chapter 6, the strength in the z-direction of an orthogonal nonwoven (ON) three-dimensional composite is an order of magnitude greater than that of a laminated composite [189]. Three-dimensional ON composites exhibit a shear strength four times higher than that of laminated composites [190]. These superior properties should lead to great damage resistance in three-dimensional ON and woven composites, but very few open-literature reports are available on this subject.

Stacking Sequence Design and Other Approaches

Other approaches to delamination control include stacking sequence design [19,65,191], free-edge region

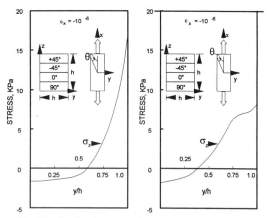

Fig. 13 Interlaminar normal stress distribution at the midplane for two different stacking sequences of a quasi-isotropic laminate of T-300 graphite/5208 epoxy

reinforcement by stitching or wrapping [192-194], and hybridization.

A proper stacking sequence can lead to considerable reduction in the interlaminar stress components in a given laminate [191]. Figure 13 shows the interlaminar stress components as a function of thickness location for two different stacking sequences of a quasi-isotropic laminate. A much higher interlaminar normal stress is present in the $(\pm 45°/90°/0°)_s$ laminate [14]. Other examples illustrating the effect of stacking sequence can be found in Ref 14, 19, 20, 65, and 191.

Free-edge delamination can be delayed or prevented by means of free-edge reinforcement to reduce the interlaminar stresses [193,194]. In one study, the first six reinforced laminates did not show any delamination until final failure. For the last three laminates, which did delaminate because they were subjected to a larger value of σ_z, the delamination threshold strain was raised by approximately 100%. Edge reinforcement was also shown to improve delamination resistance when the laminates were under fatigue loading [14].

Prediction of Notched Strength of Composites

The fracture properties of composites have been studied by many researchers in the last two decades. These investigations are concerned with topics such as initiation and growth of crack tip damage, critical damage zone dimension, notch sensitivity, and fracture crack arrest mechanisms [195]. Different composite systems appear to exhibit different failure modes and damage mechanisms. Furthermore, no consensus exists in regard to the proper set of failure criteria to follow. Consequently, many different analytical and experimental techniques have been developed to study the fracture mechanics of composites. In this section, the emphasis is placed on analytical models that predict the residual strength and fracture toughness of a notched laminate. A highly detailed review of the fracture models for composites has been recently conducted by Awerbuch and Madhukar [196]. Many methods are available to calculate the notched strength, but only a few important methods that are easy to operate as predictive tools will be presented here.

The Waddoups, Eisenmann, and Kaminski Model. Waddoups *et al.* [197], by embarking on a linear elastic fracture mechanics approach, have put forward the concept of a "damage zone." According to this model, a damage zone or region of intense energy exists adjacent to a stress concentration site, such as a notch or crack. This damage zone constitutes a characteristic volume of material that must be stressed to a critical level before fracture. This zone can be given a dimension, C_0, much similar to the plastic zone in a metallic material. Thus, a crack or notch of length $2C$ may be envisaged as having an effective length of $2(C + C_0)$ so that the stress intensity factor, K_{Ic}, at the crack tip may be given by

$$K_{Ic} = Y\sigma_N\sqrt{\pi(C + C_0)} \qquad \text{(Eq 27)}$$

where Y is the geometric correction factor and σ_N is the failure stress.

The failure stress σ_N will approach the unnotched failure strength σ_0 of the laminate when C approaches zero in Eq 27, and we obtain

$$K_{Ic} = \sigma_0\sqrt{\pi C_0} \qquad \text{(Eq 28)}$$

The notched strength may now be expressed as

$$\sigma_N = \frac{\sigma_0}{Y}\sqrt{\frac{C_0}{C + C_0}} \qquad \text{(Eq 29)}$$

where the two parameters σ_0 and C_0 may be determined by testing various notched specimens. Once the two parameters are known, one can proceed to calculate the notched strength of a laminate. Therefore, such a method is known as a two-parameter approach.

The Whitney-Nuismer Model. Another useful two-parameter method is due to Whitney and Nuismer [198,199], who have suggested two failure criteria known as point stress and average stress. The point stress criterion states that failure occurs when the normal stress at a characteristic distance, d_1, ahead of the crack reaches the unnotched strength σ_0 of the laminate. The average stress criterion states that failure occurs when the average value of the normal stress over some distance ahead of the notch tip first reaches σ_0. Two examples will be presented to illustrate these two criteria.

Example 1. Laminate Containing a Hole (Fig. 14). Consider an infinite orthotropic plate with a hole of radius R and subjected to a uniform stress σ, applied parallel to the x-axis at infinity. The normal stress σ_x along the y-axis in front of the hole can be approximated by [200]:

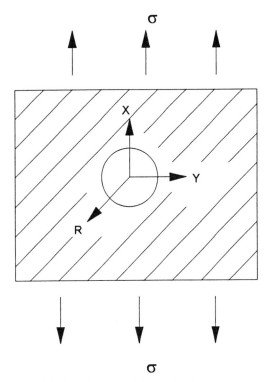

Fig. 14 A laminate containing a hole. The stress is applied in the x-direction.

$$\sigma_x(0,y) = \frac{\sigma}{2}\left\{2 + \left(\frac{R}{y}\right)^2 + 3\left(\frac{R}{y}\right)^4\right.$$
$$\left. - (K_T - 3)\left[5\left(\frac{R}{y}\right)^6 - 7\left(\frac{R}{y}\right)^8\right]\right\} \qquad \text{(Eq 30)}$$

where K_T is the orthotropic stress concentration factor for an infinite-width plate [201]:

$$K_T = 1 + \left[\frac{2}{A_{22}}\sqrt{A_{11}A_{22}} - A_{12} + \frac{A_{11}A_{22} - A_{12}^2}{2A_{66}}\right]^{1/2} \qquad \text{(Eq 31)}$$

and where A_{ij} are the in-plane laminate stiffness constants defined in Chapter 5. In terms of effective elastic moduli, K_T can be expressed as

$$K_T = 1 + \left[2\left(\sqrt{\frac{E_{xx}}{E_{yy}}} - \nu_{xy}\right) + \frac{E_{xx}}{G_{xy}}\right]^{1/2} \qquad \text{(Eq 32)}$$

Now, in the point stress criterion, failure is assumed to occur at $y = R + d_1$ where σ_x reaches σ_0:

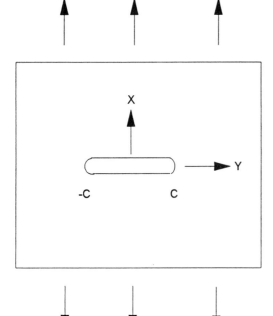

Fig. 15 A plate containing a crack of length 2C. The load is applied in the x-direction.

$$\sigma_x(0,R+d_1) = \sigma_0 \quad \text{(Eq 33)}$$

Combination of Eq 30 and 33 gives the ratio of notched to unnotched strength:

$$\frac{\sigma_N}{\sigma_0} = \frac{2}{2 + \lambda_1^2 + 3\lambda_1^4 - (K_T - 3)(5\lambda_1^6 - 7\lambda_1^8)} \quad \text{(Eq 34)}$$

where

$$\lambda_1 = \frac{R}{R + d_1} \quad \text{(Eq 35)}$$

Equation 34 reduces to the classical stress concentration result when the hole is very large; that is, $(\sigma_N/\sigma_0) = (1/K_T)$ when $\lambda_1 = 1$.

The average stress criterion assumes failure to occur when the average value of σ_T over a characteristic distance, d_2, ahead of the hole first reaches σ_0, that is, when

$$\frac{1}{d_2} \int_R^{R+d_2} \sigma_x(0,y)dy = \sigma_0 \quad \text{(Eq 36)}$$

Combining Eq 30 and 31, we have

$$\frac{\sigma_N}{\sigma_0} = \frac{2(1 - \lambda_2)}{2 - \lambda_2^2 - \lambda_2^4 + (K_T - 3)(\lambda_2^6 - \lambda_2^8)} \quad \text{(Eq 37)}$$

where

$$\lambda_2 = \frac{R}{R + d_2} \quad \text{(Eq 38)}$$

Using either Eq 34 or 37, we can compute the strength of a laminate containing a circular hole once the two parameters σ_0 and λ_1 or σ_0 and λ_2 are experimentally determined.

Example 2. A Plate Containing a Crack of Length 2C (Fig. 15). The exact elasticity solution for the normal stress in front of a crack of length 2C is given by [201]:

$$\sigma_x(0,y) = \frac{\sigma_y}{\sqrt{y^2 - C^2}} = \frac{K_1 y}{\sqrt{\pi C(y^2 - C^2)}}, \; y > C \quad \text{(Eq 39)}$$

where

$$K_1 = \sigma\sqrt{\pi C}$$

Embarking on the point stress criterion (Eq 33), we have

$$\sigma_x(0,C + d_1) = \frac{\sigma_N(C + d_1)}{\sqrt{(C + d_1)^2 - C^2}} = \sigma_0$$

or

$$\frac{\sigma_N}{\sigma_0} = \sqrt{1 - \lambda_3^2} \quad \text{(Eq 40)}$$

where

$$\lambda_3 = \frac{C}{C + d_1} \quad \text{(Eq 41)}$$

Similarly, if choosing the average stress criterion, we have

$$\frac{\sigma_N}{\sigma_0} = \frac{\sqrt{1 - \lambda_4}}{\sqrt{1 + \lambda_4}} \quad \text{(Eq 42)}$$

where

$$\lambda_4 = \frac{C}{C + d_2} \quad \text{(Eq 43)}$$

The corresponding fracture toughness K_Q may be given by

$$K_Q = Y\sigma_0\sqrt{\pi C(1 - \lambda_3^2)} \quad (point\; stress) \quad \text{(Eq 44)}$$

$$K_Q = Y\sigma_0\sqrt{\pi C(1 - \lambda_4)/(1 + \lambda_4)} \quad (average\; stress) \quad \text{(Eq 45)}$$

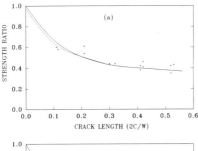

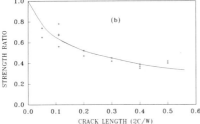

Fig. 16 Comparison between experimental results and predictions based on the Whitney-Nuismer fracture model for (a) a graphite-epoxy $[0/\pm45]_s$ laminate, and (b) a boron-aluminum $[0/90]_{2s}$ laminate containing a center crack. Solid line, point stress criterion, $d_0 = 0.42$ mm. Dashed line, average stress criterion, $a_0 = 1.81$ mm. $\sigma_0 = 684.7$ MPa; w = 25.4 mm. Source: Ref 196

For both cases, $K_Q \to 0$ when $C \to 0$ (vanishing small crack length). For large crack lengths, K_Q approaches a constant value for each case. These asymptotic values are

$$K_Q = \sigma_0\sqrt{2\pi d_1} \quad (point\ stress) \qquad (Eq\ 46)$$

$$K_Q = \sigma_0\sqrt{\pi d_2/2} \quad (average\ stress) \qquad (Eq\ 47)$$

Figure 16 compares the predicted values of σ_N/σ_0 and the experimental results obtained for graphite-epoxy and boron-aluminum laminates using the point stress and average stress criteria. The characteristic distances d_1 and d_2 were assumed to be independent of the hole size or the crack length. This assumption yields a very good agreement between experiments and predictions. Based on all experimental data reviewed by Awerbuch and Madhukar [196], the average stress criterion appears to yield a somewhat better fit with the experimental results.

The applicability of the Whitney-Nuismer fracture models would be greatly enhanced if d_1 and/or d_2 could be shown to have the same values for different notch shapes and for all laminate configurations and material systems. However, various studies have indicated that d_1 and d_2 do depend on the subject laminate, and that they must be determined *a priori* for each material system and laminate configuration independently. Once these characteristic distances are known, both criteria of Whitney-Nuismer models can be used to predict the trend for the notched strength and the critical stress intensity factor.

Karlak Fracture. Karlak [202] conducted fracture tests on quasi-isotropic graphite-epoxy laminates of dif-

ferent stacking sequences containing circular holes of various sizes and concluded that the notched strength does depend on the stacking sequence. From the test data available [202,203], Karlak concluded that the Whitney-Nuismer point-stress characteristic length, d_1, is not a material constant, but rather is related to the square root of the hole radius. Karlak therefore proposed a modified Whitney point stress criterion that incorporates the relationship between the hole radius and d_1 in the notched strength formula.

The Pipes, Wetherhold, and Gillespie Model. The work by Karlak [202], where d_1-R relationship has been incorporated, was further extended by Pipes, Wetherhold, and Gillespie. Two separate fracture models were proposed to predict the notched strength of a laminate containing circular holes [204,205] and straight cracks [206]. These models can be applied to any multidirectional orthotropic laminate for which $K_T \neq 3.0$. In these models, a more general exponential relationship between the characteristic dimensions and the size of the discontinuity (hole radius or crack length) was assumed. Furthermore, a notched strength-discontinuity size superposition principle was proposed that allowed superposition of all laminates and material systems into a single master curve. However, two additional parameters or shifting factors were introduced. These models, being a three-parameter approach, do provide a better fit with test data than the Whitney-Nuismer model, but they are much less convenient to operate.

The Lin Mar ML Model. Lin and Mar [207,208] proposed that the fracture of composite is governed by

$$\sigma_N = H_c(2C)^{-n} \qquad (Eq\ 48)$$

where H_c is the composite fracture toughness having the dimensions of (stress × length) to the nth power and is a property of the subject material. The exponent n is called "the order of singularity" of a crack with its tip at the interface of two different materials, and it is a function of the ratio of the shear moduli of the matrix and filament and of the two Poisson's ratios. A very good correlation between the experimental data and the Lin-Mar model has been established. However, the results for the various laminates have indicated that the exponent n depends on laminate lay-up, and therefore for this two-parameter model the two constants, n and H_c, must also be determined experimentally. The values of these two constants can be determined by testing two specimens containing discontinuities of two different sizes.

The Poe-Sova Model. The fracture models described in preceding paragraphs are suitable for notched laminates that are primarily fiber-dominated laminates. Such laminates exhibit essentially linear stress-strain behavior up to fracture. However, certain laminate lay-ups, such as $[\pm45]_s$ graphite-epoxy, may exhibit significant nonlinear behavior. This has led Poe and Sova [209,210] to propose a model, similar to the Waddoups model, to determine the notched strength of a laminate in terms of strain rather than stress to avoid the effect of nonlinearity. A general fracture toughness parameter, Q_C, is proposed

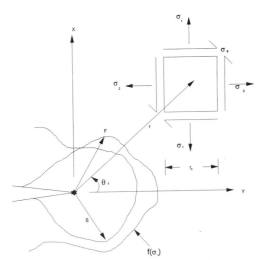

Fig. 17 The fracture direction as determined by the failure surface $f(\sigma_i)$ and the stress vector (**S**)

that is independent of laminate orientation. This parameter is derived on the basis of fiber failure in the principal load-carrying lamina and is proportional to the critical stress intensity factor, K_Q:

$$Q_C = \frac{K_{Q}\xi}{E_x} \qquad \text{(Eq 49)}$$

where ξ is a function depending on the orientation of the principal load-carrying fibers:

$$\xi = (1 - \nu_{xy}\sqrt{E_y/E_x})(\sqrt{E_x/E_y}\cos^2\theta = \sin^2\theta) \quad \text{(Eq 50)}$$

Here, θ is the fiber angle from the x-axis of the principal load-bearing lamina, while ν_{xy}, E_x, and E_y are laminate engineering elastic constants, E_x being along the 0° fiber direction.

Poe and Sova have suggested [209] that Q_C depends on the strain capability of the fibers in the principal load-carrying laminas, and that it is proportional to the ultimate tensile strain of the fibers, ε_f. Their studies have demonstrated that (Q_C/ε_f) almost remains constant for most of the laminates studied. Rewriting Eq 49, we have

$$K_Q = \frac{Q_C}{E_f} \cdot \frac{E_x E_f}{\xi} = \frac{Q_C}{E_f} \cdot \frac{\sigma_0}{\xi} \qquad \text{(Eq 51)}$$

where σ_0 is the tensile strength of the unnotched laminate. They have also determined that an average value of $Q_C/\varepsilon_f \approx 1.5 \sqrt{\text{mm}}$ can predict the toughness behavior of several laminates reasonably well, except for those laminates where splitting is predominant. In summary, based on the general fracture toughness parameter the critical stress and strain intensity factors, as well as the notched strength of individual laminates, can be calculated. The crack tip

damage zone size can also be predicted, based on the general fracture toughness parameter [196].

The Wu Fracture Model. In the fracture models discussed so far, crack growth is assumed to be self-similar, and thus the models cannot account for noncollinear crack growth or multiaxial loading conditions. Wu therefore proceeded to slightly modify the Tsai-Wu tensor polynomial failure criterion [211] to predict the fracture strength and the direction of crack propagation. According to Wu's model [212], the crack extension occurs when the elastic stress vector **S** at a radial distance r_c from the crack tip reaches or exceeds the strength vector, **F** [218]:

$$\mathbf{S} \geq \mathbf{F} \qquad \text{(Eq 52)}$$

Here, S is the stress vector acting external to the characteristic volume, defined in terms of unit vector $\mathbf{e}_i$ in the stress space:

$$\mathbf{S} = \sigma_i \mathbf{e}_i \quad i = 1, 2, ..., 6. \qquad \text{(Eq 53)}$$

The vector **F** represents the strength vector of the failure surface, $f(\sigma_i)$, which is defined as

$$f(\sigma_i) = \mathbf{F}_i\sigma_i + \mathbf{F}_y\sigma_i\sigma_j + ... \leq 1$$
$$(i, j = 1, 2, ..., 6) \qquad \text{(Eq 54)}$$

where $\mathbf{F}_i$ and $\mathbf{F}_{ij}$ are the second- and fourth-rank strength tensors, respectively. The direction of the fracture is decided by the intersection of the failure surface, $f(\sigma_i)$, and the stress vector, **S** (Fig. 17). Wu has demonstrated that, for the glass-epoxy laminates studied, the characteristic volume dimension, r_c, remains constant whether the loading is pure tensile, torsional, or a combination. Consequently, Wu postulated that fracture initiated from the failure of a constant volume r_c under complex states of stress, characterized by the tensor polynomial failure criterion represented by Eq 54. Lo *et al.* [213] have used this model to successfully predict the failure of notched composite laminates and the propagation direction of failure.

Concluding Remarks. Except for Wu's model, the above-discussed semiempirical fracture models proposed to predict the notched strength of composites have been critically reviewed by Awerbuch and Madhukar [196], who emphasized simplified, easy-to-operate predictive tools. Extensive experimental data were also reviewed and compared with the fracture models. The following concluding remarks can be gleaned from this review:

- The predictions from all fracture models reviewed agree well with the experimental notched strength data, provided that the parameters are properly determined. These fracture models are all semiempirical, and they can be applied when one or two notched strength data are known *a priori*, including the notched strength and lamina elastic properties.

- In these models, certain parameters are assumed to be material constants, but in reality these parameters depend on laminate configuration, material system, and a variety of intrinsic and extrinsic variables. Consequently, these parameters must be determined experimentally for each new material system or lay-up sequence.

- In these fracture models, the actual characteristics of the notch tip damage zone are bypassed by simulating the damage zone as some "effective" notch tip zone and assuming it to grow in a self-similar manner. When the characteristic dimensions are assumed to be independent of notch length, as in the Waddoups and Whitney-Nuismer models, model predictions appear to agree reasonably well with the experimental data. In most cases, these characteristic dimensions are practically independent of notch length or show only a slight increase with notch length. Although better agreement between experiments and prediction is observed when the characteristic dimensions are assumed to be dependent on notch length, as in the Karlak and Pipes models, this assumption demands determination of additional parameters.

- The general fracture-toughness parameter proposed in the Poe-Sova model can be applied to all laminates containing the same fiber, provided that crack tip damage is localized. However, when this parameter has been determined individually for each laminate, excellent agreement with experimental data has been established. Finally, all fracture models reviewed provided predictions that are identical for all practical purposes.

REFERENCES

1. A.K. Munjal, in *Advanced Composites*, Conf. Proc., 2-4 Dec 1985, American Society for Metals, 1985, p 309-321
2. D.E. Pettit, K.N. Lauratis, and J.M. Cox, AFWAL-TR-79-3095, Vol I, II, and III, Air Force Wright Aeronautical Laboratories, 1979
3. A.M. James and E. Williams, *Lockheed Horizon*, 1986, p 31-43
4. G. Dorey, AGARD-LS 124, 1982, p 6.1-6.11
5. K.L. Reifsnider, E.G. Hemmeke, W.W. Stinchcomb, and J.C. Duke, *Mechanics of Composite Materials: Recent Advances*, Z. Hashin and C.T. Herakovich, Ed., Pergamon Press, 1983, p 399-420
6. C.K.H. Dharan, *J. Eng. Mater. Technol.*, Vol 100, 1978, p 233-247
7. A.C. Gary, *J. Mech. Behav. Mater.*, Vol 1, 1988, p 103-210
8. T.K. O'Brien, NASA-TM-85768, National Aeronautics and Space Administration, 1984
9. N.J. Pagano and R.B. Pipes, *Int. J. Mech. Sci.*, Vol 15, 1973, p 679-688
10. N.J. Pagano and S.R. Soni, *Int. J. Solids Struct.*, Vol 19 (No. 3), 1983, p 207
11. N.J. Pagano, Ed., *Interlaminar Response of Composites Materials*, Vol 5, *Composite Materials Series*, Elsevier, 1989
12. N.J. Pagano and S.R. Soni, Models for Studying Free Edge Effects, *Interlaminar Response of Composite Materials*, N.J. Pagano, Ed., Vol 5, *Composite Materials Series*, Elsevier, 1989, p 1-68
13. A.S.D. Wang, Fracture Analysis of Interlaminar Cracking, *Interlaminar Response of Composite Materials*, N.J. Pagano, Ed., Vol 5, *Composite Materials Series*, Elsevier, 1989, p 69-110
14. R.Y. Kim, Experimental Observations of Free Edge Delamination, *Interlaminar Response of Composite Materials*, N.J. Pagano, Ed., Vol 5, *Composite Materials Series*, Elsevier, 1989, p 111-160
15. J.M. Whitney, Experimental Characterization of Delamination Fracture, *Interlaminar Response of Composite Materials*, N.J. Pagano, Ed., Vol 5, *Composite Materials Series*, Elsevier, 1989, p 161-250
16. R.B. Pipes and N.J. Pagano, *J. Compos. Mater.*, Vol 4, 1970, p 538-548
17. A.H. Puppo and H.A. Evensen, *J. Compos. Mater.*, Vol 4, 1970, p 204
18. J.M. Whitney and C.E. Browning, *J. Compos. Mater.*, Vol 6, 1973, p 300-303
19. N.J. Pagano and R.B. Pipes, *J. Compos. Mater.*, Vol 5, 1971, p 50-57
20. C.T. Herakovich, *J. Compos. Mater.*, Vol 15, 1981, p 336-348
21. B.T. Rodini and J.R. Eisenmann, *Fibrous Composites in Structural Design*, E.M. Lenoe, D.W. Oplinger, and J.J. Burke, Ed., Plenum Press, 1978, p 441-457
22. F.W. Crossman and A.S.D. Wang, *Damage in Composite Materials*, STP 775, K.L. Reifsnider, Ed., ASTM, 1982, p 118-139
23. C.T. Herakovich, *Int. J. Mech. Sci.*, Vol 18, 1976, p 129-134
24. F.W. Crossman and A.S.D. Wang, *J. Compos. Mater.*, Vol 12, 1978, p 2-18
25. P.W. Hsu and C.T. Herakovich, *J. Compos. Mater.*, Vol 11, 1977, p 422-428
26. C.T. Herakovich and D.M. Wong, *Exp. Mech.*, Vol 17, 1977, p 409-414
27. A.L. Highsmith, W.W. Stinchcomb, and K.L. Reifsnider, STP 836, ASTM, 1984, p 194-216
28. K.L. Reifsnider, E.G. Henneke, and W.W. Stinchcomb, "Defect Property Relationship in Composite Materials," AFML-TR-76-81, Air Force Materials Laboratory, 1979
29. F. Erdogan and G.D. Gupta, *Int. J. Solids Struct.*, Vol 7, 1971, p 39-61
30. F. Erdogan and G.D. Gupta, *Int. J. Solids Struct.*, Vol 7, 1971, p 1089-1107
31. A.H. England, *J. Appl. Mech.*, Vol 87, 1965, p 400-402
32. M. Comninou, *J. Appl. Mech.*, Vol 44, p 631-636
33. K.S. Parihar and A.C. Garg, *Int. J. Eng. Sci.*, Vol 14, 1976, p 831-843
34. A.C. Garg, *Int. J. Eng. Fract. Mech.*, Vol 7, 1976, p 751-759
35. K.S. Parihar and A.C. Garg, *Eng. Fract. Mech.*, Vol 7, p 751-759
36. J.R. Rice, *J. Appl. Mech.*, Vol 87, p 418-423
37. S.S. Wang, *J. Compos. Mater.*, Vol 17, 1983, p 210-223
38. E. Altus, A. Rotem, and M. Shmueli, *J. Compos. Mater.*, Vol 14, 1980, p 21-30
39. G.L. Farley and C.T. Herakovich, STP 658, ASTM, 1978, p 143-159
40. G.L. Farley and C.T. Herakovich, "Two Dimensional Hygrothermal Diffusion into a Finite Width Composite Laminate," VPI-E-77-20, 1977
41. D.J. Wilkins et al., *Damage in Composite Materials*, STP 775, K.L. Reifsnider, Ed., ASTM, 1982, p 168-183
42. T.K. O'Brien, NASA-TM 85728, National Aeronautics and Space Administration, 1984

43. T.K. O'Brien, *Damage in Composite Materials,* STP 775, K.L. Reifsnider, Ed., ASTM, 1982, p 140-167

44. R.Y. Kim and S.R. Soni, "Experimental and Analytical Studies on the Onset of Delamination in Laminated Composites," Report No. UDR-TR-83.40, University of Dayton Research Institute, 1983

45. T. Mohlin et al., FAA Rept. FAA-TN-183.57, The Aeronautical Research Institute of Sweden, 1983

46. J.D. Whitcomb, NASA-TM 86301, National Aeronautics and Space Administration, 1984

47. E.F. Rybicki, D.W. Schmueser, and J. Cox, *J. Compos. Mater.,* Vol 11, 1977, p 470-487

48. C.T. Herakovich, A. Nagarkar, and D.W. O'Brien, *Modern Developments in Composite Materials and Structures,* J.R. Vinson, Ed., American Society of Mechancial Engineers, 1981, p 53-66

49. T.K. O'Brien and I.S. Raju, 25th Structures, Structural Dynamics and Materials Conference May 1984 (Palm Springs), AIAA/ASME/ASCE/AHS, 1984, p 526-536

50. A.S.D. Wang and F.W. Crossman, *J. Compos. Mater.,* Vol 11, 1977, p 92-106

51. F.W. Crossman and A.S.D. Wang, *J. Compos. Mater.,* Vol 12, 1978, p 2-18

52. G.E. Law, STP 836, ASTM, 1984, p 143-160

53. S.S. Wang, STP 674, ASTM, 1979, p 642-663

54. A.S.D. Wang and F.W. Crossman, *J. Compos. Mater.,* Vol 11, 1977, p 300-312

55. I.S. Raju and J.H. Crews, Jr., *Comput. Struct.,* Vol 14, 1981, p 21-28

56. A.D. Reddy, L.W. Rehfield, and R.S. Haag, "Influence of Prescribed Delaminations on Stiffness Controlled Behavior of Composite Laminates," Symp. Effects of Defects in Composite Materials, 13-14 Dec 1982 (San Francisco), ASTM

57. S.R. Soni and N.J. Pagano, "Global Local Laminate Variational Model," AFWAL-TR-82-4028, Air Force Wright Aeronautical Laboratories, 1982

58. R.L. Spilker and S.C. Chou, *J. Compos. Mater.,* Vol 14, 1980, p 2-20

59. K.S. Kim and C.S. Hong, *J. Compos. Mater.,* Vol 20, 1986, p 423-438

60. C.G. Shah, Proc. Workshop-cum-Seminar on Delaminations in Composites, March 1987 (Banglaore), p 271-294

61. A.V. Krishnamurty and K. Vijaykumar, Proc. Workshop-cum-Seminar on Delaminations in Composites, March 1987 (Banglaore), p 271-294

62. N.J. Pagano, *J. Compos. Mater.,* Vol 8, 1974, p 65

63. J.M. Whitney and R.Y. Kim, in *Composite Materials: Testing and Design,* STP 617, ASTM, 1977, p 229-242

64. R.R. Valisetty and L.W. Rehfield, *AIAA J.,* Vol 23, 1985, p 1111-1117

65. N.J. Pagano, *Int. J. Solids Struct.,* Vol 14, 1978, p 385-400

66. N.J. Pagano, *Int. J. Solids Struct.,* Vol 14, 1978, p 401-406

67. S.R. Soni, *Progress in Science and Engineering of Composites,* T. Hayashi, et al., Ed., Proc. ICCM-IV (Tokyo), International Committee on Composite Materials, 1982, p 251

68. R.Y. Kim and S.R. Soni, *J. Compos. Mater.,* Vol 18, 1984, p 70

69. S.R. Soni and R.Y. Kim, in *Composite Materials: Testing and Design,* STP 893, ASTM, 1986, p 286-307

70. A.S.D. Wang, *Compos. Technol. Rev.,* Vol 6 (No. 2), 1984, p 45-62

71. F.W. Crossman, R.E. Mauri, and W.J. Warren, NASA-CR-3189, National Aeronautics and Space Administration, 1979

72. T.K. O'Brien, NASA-TM-85768, National Aeronautics and Space Administration, 1984

73. R.J. Nuismer and J.M. Whitney, STP 836, ASTM, 1984, p 104-124

74. R.J. Nuismer and J.M. Whitney, STP 593, ASTM 1975, p 117-142

75. T. Johannesson and. M. Blikstad, "A Fractographic Study of Delamination Process," Linkoping Institute of Technology, Linkoping, Sweden, 1984

76. W.S. Johnson and P.D. Mangalgiri, NASA-TM-87571, National Aeronautics and Space Administration, 1985

77. R.A. Jurf and R.B. Pipes, *J. Comp. Mater.,* Vol 16, 1982, p 386-394

78. J.D. Whitcomb, STP 836, ASTM, 1984, p 175-193

79. J.D. Whitcomb, *J. Compos. Mater.,* Vol 15, 1981, p 26-29

80. J.D. Whitcomb, *J. Compos. Mater.,* Vol 15, 1981, p 403-426

81. H.T. Hahn, *Compos. Technol. Rev.,* Vol 5, 1983, p 26-29

82. S.L. Donaldson, *Composites,* Vol 16, 1985, p 103-112

83. S.L. Donaldson, *Compos. Sci. Technol.,* 1987, p 33-44

84. E.M. Wu, Strength and Fracture of Composites, *Composite Materials,* Vol V, L.J. Broutman, Ed., Academic Press, 1974, p 191-247

85. A.J. Russel and K.N. Street, *Delamination and Debonding,* STP 876, W.S. Johnson, Ed., 1985, p 349-370

86. A.L. Highsmith, W.W. Stinchcomb, and K.L. Reifsnider, STP 836, ASTM, 1984, p 194-216

87. H. Chai, C.D. Babcock, W.G. Knauss, *Int. J. Solids Struct.,* Vol 17 (No. 11), 1981, p 1069-1083

88. D.Y. Konishi and W.R. Johnston, STP 674, ASTM, 1979, p 579-619

89. J.M. Hopper, E. Demuts, and G. Milizianto, Damage Tolerant Design Demonstration, 25th Structures, Structural Dynamics and Materials Conf., May 1984 (Palm Springs, CA), AIAA/ASME/ASCE/AHS, p 15-20

90. R.L. Ram Kumar, STP 813, ASTM, 1983, p 116-135

91. S.M. Lee, *J. Compos. Mater.,* Vol 20, 1986, p 185-196

92. W.D. Bascom et al., *Composites,* Vol 11, 1980, p 9-18

93. M. Arcan, Z. Hashin, and A. Voloshin, *Exp. Mech.,* Vol 18, 1978, p 141-146

94. L.A. Carlsson, J.W. Gillespie, and A.J. Smiley, *J. Compos. Mater.,* Vol 20, 1986, p 594-604

95. S.L. Donaldson, International Fracture due to Tearing (Mode III), *Conf. Compos. Mater.,* Vol 3, J. Morton, Ed., Proc. ICCM-VI and ECCM, Elsevier Applied Science, 1987, p 3.274-3.283

96. S.L. Donaldson, "Mode III for Laminar Fracture Characterization of Composite Materials," M.Sc. thesis, University of Dayton, 1987

97. S.L. Donaldson, *Composites,* Vol 16, 1985, p 103-112

98. J.F. Knott, *Fundamentals of Fracture Mechanics,* Butterworths, 1973

99. L.A. Carlsson and R.B. Pipes, *Experimental Characterization of Advanced Composites Materials,* Prentice-Hall, 1987

100. P.E. Keary et al., *J. Compos. Mater.,* Vol 19, 1985, p 154-177

101. J.M. Whitney, C.E. Browning, and W. Hoogsteden, *J. Reinf. Plast. Compos.,* Vol 1, 1982

102. D.J. Nicholls and J.P. Gallagher, *J. Reinf. Plast. Compos.,* Vol 2, 1983, p 2-17

103. J.M. Gillespie, Jr., L.A. Carlsson, and A.J. Smiley, *Compos. Sci. Technol.,* Vol 25, 1987, p 1-15

104. J.M. Gillespie, Jr., L.A. Carlsson, and R.B. Pipes, *Compos. Sci. Technol.,* Vol 24, 1986, p 177-197

105. A.J. Kinloch and R.J. Young, *Fracture Behavior of Polymers,* Applied Science, 1983

106. W.D. Bascom, R.Y. Ting, R.J. Moulton, C.K. Riew, and A.R. Siebert, *J. Mater. Sci.*, Vol 16, 1981, p 2657

107. N.E. King and E.H. Andrews, *J. Mater. Sci.*, Vol 13, 1978, p 1291-1302

108. L.D. Bravenec, A.G. Filippov, and K.C. Dewhirst, *34th Int. SAMPE Symp.*, 8-11 May 1989, Society for Advancement of Material and Process Engineering, p 714-725

109. S. Murakami, O. Watanobe, H. Inoue, M. Ochi, T. Shiraishi, and M. Shimbo, *34th Int. SAMPE Symp.*, 8-11 May 1989, Society for Advancement of Material and Process Engineering, p 2194-2205

110. C.B. Bucknall and T. Yoshii, *Br. Polym. J.*, Vol 10, 1978, p 53

111. C.B. Buckall, *Toughened Plastics*, Applied Science, 1977

112. A. Van den Boogaart, in *Physical Basis of Yield and Fracture*, Institute of Physics, London, 1966, p 167

113. J.N. Sultan, R.C. Laible, and F.J. McGarry, *Applied Polymers Symp.*, Vol 16, 1971, p 127

114. J.N. Sultan and F.J. McGarry, *J. Polymer Sci.*, Vol 13, 1973, p 29

115. S. Kunz-Douglass, P.W.R. Beaumont, and M.F. Ashby, *J. Mater. Sci.*, Vol 15, 1980, p 1109

116. A.F. Yee and R.A. Pearson, *Fractography and Failure Mechanisms of Polymers and Composites*, A.C. Roulin-Moloney, Ed., Elsevier Applied Science, 1989, p 291

117. A.J. Kinloch, S.J. Shaw, D.A. Tod, and D.L. Houston, *Polymer*, Vol 24, 1983, p 1341

118. R.J. Morgan, E.T. Mones, and W.J. Steele, *Polymer*, Vol 23, 1982, p 295-305

119. R.J. Morgan and J.E. O'Neal, *J. Mater. Sci.*, Vol 12, 1977, p 1966-1980

120. J. Lilley and D.G. Holloway, *Philos. Mag.*, Vol 28, 1973, p 215-220

121. A.M. Donald and E.J. Kramer, *J. Mater. Sci.*, Vol 17, 1982, p 1765

122. E.M. Hagerman, *J. Appl. Polym. Sci.*, Vol 17, 1973, p 2203

123. A.F. Yee, *J. Mater. Sci.*, Vol 12, 1977, p 757

124. A.F. Yee, W.V. Olszewski, and S. Miller, *Toughened and Brittleness of Plastics*, R.D. Deanin and A.M. Crugnola, Ed., American Chemical Society, Washington, D.C., 1976, p 97

125. A.J. Kinloch and G.J. Williams, *J. Mater. Sci.*, Vol 15, 1980, p 987

126. J.N. Sultan and F.J. McGarry, *Polym. Eng. Sci.*, Vol 13, 1973, p 29

127. W.D. Bascom and R.L. Cottington, *J. Adhes.*, Vol 7, 1976, p 333

128. J.R. Bitner, J.L. Rushford, W.S. Rose, D.L. Hunston, and C.K. Riew, *J. Adhes.*, Vol 13, 1982, p 3

129. B.Z. Jang et al. Thermoplastic Particulate and Whisker-Filled Epoxy, *J. Reinf. Plast. Compos.*, 1989

130. W.J. Cantwell and A.C. Roulin-Moloney, Ed., Elsevier Applied Science, 1989, p 233-290

131. A.J. Kinloch, D.A. Maxwell, and R.J. Young, *J. Mater. Sci. Lett.*, Vol 4, 1985, p 1276

132. R.H. Pater, in *Encyclopedia of Composites*, S. Lee, Ed., VCH Publishers, 1990

133. H.G. Recker et al., *SAMPE J.*, Vol 26 (No. 2), 1990, p 73-77

134. M.S. Sefton et al., *SAMPE Int. Tech. Conf.*, Vol 19, 1987, p 700

135. S. Kunz and P.W.R. Beaumont, *J. Mater. Sci.*, Vol 16, 1981, p 3141

136. R.S. Bauer, H.D. Stenzenberger, and W. Roomer, *34th Int. SAMPE Symp.*, 8-11 May 1989, Society for Advancement of Materials and Processes Engineering, p 899-909

137. H.D. Stenzenberger, and W. Roomer, P. Hergenrother, and B.J. Jensen, *34th Int. SAMPE Symp.*, 8-11 May 1989, Society for Advancement of Materials and Process Engineering, p 1877-1888, 2054-2068

138. C.R. Lin, W.L. Liu, and J.T. Hu, *34th Int. SAMPE Symp.*, 8-11, 1989, p 1803-1813

139. R.H. Pater, *ANTEC Proc.*, Society of Plastics Engineers, 1989, p 1434-1439

140. R.H. Pater and C.D. Morgan, LaRC-RP40, *SAMPE J.*, Vol 24 (No. 5), Sept/Oct 1988, p 25-32

141. P.B. Bowden, *Philos. Mag.*, Vol 22, 1970, p 455

142. E.J. Kramer, *J. Macromol. Sci.*, Vol B10, 1974, p 191

143. E.J. Kramer, *J. Polym. Sci. Polym. Phys. Ed.*, Vol 13, 1975, p 509

144. A.M. Donald and E.J. Kramer, Effect of Molecular Entanglements on Craze Microstructure in Glassy Polymers, *J. Polym. Sci., Polym. Phys. Ed.*, Vol 20, 1982, p 899-909

145. A.M. Donald and E.J. Kramer, The Competition between Shear Deformation and Crazing in Glassy Polymers, *J. Mater. Sci.*, Vol 17, 1982, p 1871-1879

146. A.C. Garg and Y.W. Mai, *Compos. Sci. Technol.*, Vol 31 (No. 3), 1988, p 179-223

147. A.M. Donald and E.J. Kramer, *Polymer*, Vol 23, 1982, p 1183-1188

148. A.S. Argon, R.D. Andrews, J.A. Godrick, and W. Whitney, *J. Appl. Phys.*, Vol 39, 1968, p 1899-1906

149. B.Z. Jang, R.H. Pater, M. Souchet, and J.A. Hinkly, *J. Polym. Sci., Polym. Phys. Ed.*, 1992

150. R.E. Evans and J.E. Masters, *Toughened Composites*, STP 937, N.J. Johnston, Ed., ASTM, 1987, p 413-436

151. C.M. Tung, T.T. Liao, and C.L. Leung, *33rd Int. SAMPE Symp.*, Society for Advancement of Materials and Process Engineering, 1988, p 503-511

152. B. Carroll, *33rd Int. SAMPE Symp.*, Society for Advancement of Materials and Process Engineering, 1988, p 78-90

153. R.B. Krieger, Jr., *29th Int. SAMPE Symp.*, Society for Advancement of Materials and Process Engineering, 1984, p 1570-1584

154. K.R. Hirshbuehler, *SAMPE Quart.*, Vol 17, 1985, p 46-49

155. G.L. Dolan and J.E. Masters, *20th Int. SAMPE Tech. Conf.*, Society for Advancement of Materials and Process Engineering, 1988, p 34-45

156. J.E. Masters, J.L. Courter, and R.E. Evans, *31st Int. SAMPE Symp.*, Society for Advancement of Materials and Process Engineering, 1986, p 844-858

157. R.B. Krieger, Jr., *SAMPE J.*, July/Aug 1987, p 30-32

158. C.T. Sun and S. Rechak, *Composite Materials: Testing and Design*, STP 972, J.D. Whitcomb, Ed., ASTM, 1988, p 97-123

159. J.E. Masters, *Int. Conf. Composite Materials*, Proc. ICCM-VI, International Committee on Composite Materials, 1987, p 3.96-3.107

160. A.K. Kaw and J.G. Goree, *Int. Conf. Composite Materials*, ICCM-VI, International Committee on Composite Materials, 1987, p 3.146-3.155

161. W.S. Chan, C. Rogers, and S. Aker, in *Composite Materials: Testing and Design*, STP 893, ASTM, 1986, p 266-285

162. C.T. Sun and P.M. Voit, *Compos. Technol. Rev.*, 1985, p 109-113

163. K.R. Hirschbuehler, in *Toughened Composites*, STP 937, ASTM, 1987, p 61-73

164. R.J. Palmer, Proc. 5th Conf. Fibrous Composites in Structural Design, NADC-81096-60, Department of De-

fense/National Aeronautics and Space Administration, Jan 1981

165. H.T. Hahn and J.G. Williams, Tech. Memo 85834, National Aeronautics and Space Administration, Aug 1984

166. S.F. Chen and B.Z. Jang, *Compos. Sci. Technol.*, Vol 41, 1991, p 77-97

167. B. Jang, L.R. Hwang, and R.C. Wilcox, *SAMPE Symp.*, May 1989, p 480-491

168. R. Garcia, R.E. Evans, and R.J. Palmer, *Toughened Composites*, STP 937, N.J. Johnston, Ed., ASTM, 1987, p 397-412

169. R. Garcia, NADC-83058-60, July 1983

170. T.L. Lin and B.Z. Jang, *Ann. Tech. Conf. (ANTEC)* (New York), May 1989

171. J.Y. Liau, B.Z. Jang, L.R. Hwang, and R.C. Wilcox, *Plast. Eng.*, Nov 1988

172. T.L. Lin and B.Z. Jang, *Polym. Compos.*, 1990

173. L.A. Mignery, T.M. Tan, and C.T. Sun, *Delamination and Debonding*, W.S. Johnson, Ed., STP 876, ASTM, 1985, p 371-385

174. J.G. Funk, H.B. Dexter, and S.J. Lubowinski, Conf. Proc. 2420, National Aeronautics and Space Administration, Nov 1985, p 185-205

175. X. Du, F. Xue, and Z. Gu, Int. Symp. Composite Materials and Structures (Beijing), 10-13 June 1986, Technomic, Lancaster, PA, 1986, p 912-918

176. B.Z. Jang and W.C. Chung, *Proc. 2nd Conf. Advanced Composites: The Latest Developments* (Dearborn, MI), 18-20 Nov 1986, American Society for Metals, 1986, p 183-191

177. D. Liu, K.G. Kim, and S. Hong, *Advanced Composites III: Expanding the Technology*, Proc. 3rd Ann. Conf. Advanced Composites, Sept 1987, ASM International, p 343-347

178. W.C. Chung, B.Z. Jang, T.C. Chung, L.R. Hwang, and R.C. Wilcox, *Mater. Sci. Eng. A*, Vol 112, 1989, p 157-173

179. B.Z. Jang, W.K. Shih, and W.C. Chung, *J. Reinf. Plast. Compos.*, Nov 1989, p 538-564

180. D. Liu, Proc. 2nd Tech. Conf. (Newark), Sept 1987, American Society for Composites

181. G. Dorey, *Structural Impact and Crashworthiness*, G.A.O. Davies, Ed., Vol 1, Elsevier Applied Science, 1984, p 155-191

182. F.K. Chang, H.Y. Choi, and S.T. Jeng, *34th Int. SAMPE Symp.*, 8-11 May 1989, Society for Advancement of Materials and Process Engineering, p 702-713

183. K. Friedrich, R. Walter, L.A. Carlsson, A.J. Smiley, and J.W. Gillespie, Jr., *J. Mater. Sci.*, Vol 24, 1989, p 3387-3398

184. R.A. Crick, D.C. Leach, P.J. Meakin, and D.R. Moore, *J. Mater. Sci.*, Vol 22, 1987, p 2094

185. W.L. Bradley and R.N. Cohen, *Delamination and Debonding of Materials*, STP 876, W.S. Johnson, Ed., ASTM, 1985, p 389

186. L.W. Gause, *Mech. Compos. Rev*, Air Force Materials Laboratory, Oct 1983

187. F.K. Ko and D. Hartman, *SAMPE J.*, Vol 22 (No. 4), 1986, p 26

188. F.K. Ko, H. Chou, and E. Ying, *Advanced Composites: The Latest Developments*, P. Beardmore and C.F. Johnson, Ed., Proc. 2nd Conf., ASM International, 1986, p 75-88

189. E.M. Lenoe, R.M. Laurie, and G.N. Wassil, *10th Nat. SAMPE Symp.*, 9-11 Nov 1966 (San Diego, CA), Society for Advancement of Materials and Process Engineering

190. K. Fukuta, *Proc. 2nd Textile Structural Composites Symp.*, Drexel University, Philadelphia, 1987, p 64-68

191. L.M. Lackman and N.J. Pagano, Proc. Structures Structural Dynamics and Materials Conf., April 1974 (Las Vegas), AIAA paper 74-355, 1974

192. R.B. Pipes and I.M. Daniel, *J. Compos. Mater.*, Vol 5, 1971, p 255

193. W.E. Howard, T. Gossard, Jr., and R.M. Jones, AIAA paper 86-0972, 1986

194. R.Y. Kim, *Materials and Processes—Continuing Innovations*, 28th Nat. SAMPE Symp., Society for Advancement of Materials and Processes Engineering, 1983

195. B.D. Agarwal and L.J. Broutman, Chapter 7, *Analysis and Performance of Fiber Composites*, 2nd ed., John Wiley & Sons, 1990

196. J. Awerbuch and M.S. Madhukar, *J. Reinf. Plast. Compos.*, Vol 4, 1985, p 3-159

197. M.E. Waddoups, J.R. Eisenmann, and B.E. Kaminski, *J. Compos. Mater.*, Vol 5, 1971, p 446-454

198. J.M. Whitney and R.J. Nuismer, *J. Compos. Mater.*, Vol 5, 1974, p 253-265

199. R.J. Nuismer and J.M. Whitney, STP 593, ASTM, 1975, p 117-142

200. H.J. Konish and J.M. Whitney, *J. Compos. Mater.*, Vol 9, 1975, p 157-166

201. S.G Lekknitskii, *Anisotropic Plates*, Gordon and Breach, 1968

202. R.F. Karlak, *Proc. Failure Modes in Composites*, IV, The Metallurgical Society of AIME, 1977, p 105-117

203. J.M. Whitney and R.Y. Kim, AFML-TR-765-177, Air Force Materials Laboratory, Nov 1971

204. R.B. Pipes, R.C. Wetherhold, and J.W. Gillespie, Jr., *J. Compos. Mater.*, Vol 12, 1979, p 148-160

205. R.B. Pipes, J.W. Gillespie, Jr., and R.C. Wetherhold, *Polym. Eng. Sci.*, Vol 19, 1979, p 1151-1155

206. R.B. Pipes, R.C. Wetherhold, and J.W. Gillespie, Jr., *Mater. Sci. Eng.*, Vol 45, 1980, p 247-253

207. K.Y. Lin, "Fracture of Filamentary Composite Materials," Ph.D. dissertation, Massachusetts Institute of Technology, Jan 1976

208. J.W. Mar and K.Y. Lin, *J. Aircraft*, Vol 14, 1977, p 703-704

209. C.C. Poe, Jr. and J.A. Sova, "Fracture Toughness of Boron/Aluminum Laminates with Various Proportions of 0° and 45° Plies," Tech. Paper 1707, National Aeronautics and Space Administration, Nov 1980

210. C.C. Poe, Jr., *Eng. Fract. Mech.*, Vol 17, p 153-171

211. E.M. Wu, *Composite Materials*, L.J. Broutman, Ed., Vol 5, Academic Press, 1974, p 191-247

212. S.W. Tsai and E.W. Wu, *J. Compos. Mater.*, Vol 5, 1971, p 58-80

213. K.H. Lo, E.M. Wu, and D.Y. Konishi, *J. Compos. Mater.*, Vol 17, 1983, p 384-398

Chapter 8

Design for Improved Impact Resistance of Polymer Composites

The present chapter highlights the structure-property relationships in fibrous composites subjected to various impact loading conditions. It emphasizes the effects of various materials variables on the impact resistance of composites. Topics include the impact response of polymer composites, more commonly used impact testing techniques, experimental approaches that can be used to characterize the damage state of polymer composites, the impact properties of various types of polymer composites, and modeling approaches that can be used to analyze the deformation and fracture of composites in response to both low- and high-velocity impacts. The ultimate goal of this chapter is to reveal some useful guidelines for designing against impact failure of composites.

Impact Response of Polymer Composites

Impact failure of fiber-reinforced polymer composites is a very complex phenomenon. This failure behavior is not a unique function of the constituent fibers and matrix; the fracture properties (e.g., impact energy) cannot be predicted by a simple rule-of-mixture law. The fiber-matrix interactions, the fiber orientations or stacking sequence, and the processing conditions are all important factors to consider. The stressing conditions and the environments to which the composites are subjected also play a key role in determining the impact failure processes of composites.

Fiber-reinforced polymer composites, particularly those containing thermoset matrices such as epoxy and polyimide, are known to be highly susceptible to internal damage caused by a low-velocity impact. The composites can be damaged beneath the surface with relatively light impact, while the surface appears to be undamaged upon visual inspection. As a result of this internal damage, the local strengths of the composites are significantly impaired and their service life is reduced. Research efforts on the low-velocity impact are usually aimed at reducing the degree of damage (in the form of transverse cracking and delamination) to improve the postimpact integrity or damage tolerance of composites. In the case of higher incident energy, a greater energy-absorbing capability for improved impact penetration resistance is the primary goal of using composites for ballistic applications.

The impact response of composites has become a critical subject of composites research [1-25]. However, our grasp of the impact response in composites is still quite limited, possibly because this complex phenomenon involves many different interactions and parameters [26-31]. Studies of impact damage in composites can be divided into three main areas: impact dynamics and damage mechanics [4,8,15-21,32], postimpact residual property characterization [6,7,9-12], and damage resistance improvements [5,14,33,34].

Various approaches have been used to improve the damage tolerance (low-velocity or low-energy impact) or penetration resistance (high-velocity or high-energy impact) of composite materials. These include control of fiber-matrix interfacial adhesion [35-37], matrix modifications (in particular, rubber toughening) [14,23,34], lamination design (e.g., selection of laminate stacking sequence) [38,39], introduction of through-the-thickness reinforcements (composites produced by braiding, three-dimensional weaving, and stitching) [40,41], insertion of interlaminar "interleaf" layers [42], fiber hybridization [43-45], and the use of high-strain fibers [33,43,47,48].

The last two approaches are particularly effective in improving the impact penetration resistance of composites under high-incident-energy conditions, although other approaches also work to some extent. The use of high-strain fibers appears to be very promising, because a major cause of impact penetration problems in high-performance composites is believed to be their low strain to failure. In general, it is the low-ductility fibers, rather than the relatively high-strain matrices, that limit the composite strain in a high-energy impact. The results of a few studies [e.g., 49,50] have indicated that both high fiber ductility and high fiber strength are required in order to achieve a great impact and penetration resistance of fiber-polymer composites. This chapter discusses the prevailing deformation and fracture mechanisms in several types of composite materials under various loading conditions and the major material parameters that dictate such mechanisms. The effects of fiber hybridization on the mechanical behavior of composites are also considered.

As discussed in Chapter 7, the techniques of stitching, weaving, and braiding can be selected to introduce the thickness-direction fibers in the hope that delamination can be reduced or suppressed in laminates. The high-rate mechanical properties of these three-dimensional composites are discussed in this chapter. The concept of controlled interlaminar bonding (CIB) is explored with two seemingly conflicting ideas evaluated. Perforated brittle polymer films (e.g., polyimide) can be inserted between fiber-epoxy prepreg layers to improve the impact toughness (energy-absorbing capability) without significantly sacrificing the static strength of composites. These CIB layers are added to promote delamination and to divert or blunt brittle cracks in composite. These layers are particularly useful when high energy-absorbing capability (e.g., for ballistic impact protection) is required. Conversely, relatively tough interleaved layers (e.g., thermoplastic or toughened thermoset adhesive films) can be inserted in composites to absorb energy from low-velocity impact and to provide some interlaminar ductility. These help to reduce the extent of delamination and possibly the extent of intraply cracking, and they improve the damage tolerance of composites.

Highsmith and Reifsnider [51-53] have studied extensively the damage mechanisms and the associated load redistribution effects in composites under conventional controlled-stress fatigue conditions. They have found that the approximate laminate analysis as suggested by Pagano [54] can be used to analyze microdamage patterns in composite laminates. They further noted that matrix cracks in adjacent plies of different orientation that crossed at the ply interface would create a region of significantly increased interlaminar normal and shear stress around the intersection point. This would conceivably lead to the formation of local delamination at the ply interface. Additional results on the cumulative damage mechanisms in composites were also reported by Masters and Reifsnider [55-58]. In this chapter, the response of composites to repeated impacts or "impact fatigue" will also be addressed.

It is difficult to precisely define the critical state of cracking (caused by impact, in particular) above which a reduction in material strength and stiffness will become appreciable. In the off-axis plies, transverse cracks generally tend to run across the width parallel to the fibers in these particular plies. When these minute cracks are still small in size (have yet to be extended across the width or through the thickness) and are isolated from one another, these plies should still be fully capable of carrying the load. Therefore, no strength or stiffness reduction is expected. When the transverse cracks run all the way across the width of a given off-axis ply, a change in the net load carried by that ply normal to the crack is expected. This can be understood from the results of a stress analysis based on either finite difference or shear-lag modeling [51,52,55]. When a transverse crack extends throughout the cross section of a ply, this ply would lose part or all of its load-carrying capability and the laminate would suffer from an overall degradation in integrity (a reduced in-plane strength and stiffness). At this stage, the considerably increased out-of-plane stress, with the assistance of a high interlaminar shear stress at the ply boundary, would readily initiate a delamination crack along this boundary. Therefore, the interlaminar shear strength would also be lowered.

Energy-based fracture mechanics approaches [59,60] were suggested to study the matrix-cracking behavior of laminates under an in-plane tensile loading condition. The observation of the crack pattern in the 90° ply of a [0/90]$_s$ glass-epoxy laminate after loading to 180 MPa (26 ksi) showed a moderate number of transverse cracks, yet no significant reduction in stiffness was observed at this stage of tensile loading, and no delamination was observed [60]. In response to repeated impacts, the epoxy-based laminates studied thus far usually did not exhibit such an extensive matrix cracking before one or two delamination cracks appeared. Such a failure mode difference is obviously due to the different stress fields between in-plane tensile loading [32] and the flexural impact loading condition. The latter tends to produce a large interlaminar shear stress and thereby promotes delamination.

Failure modes and damage accumulation in laminates with free edges were studied by Herakovich [61]. In tensile loading of quasi-isotropic laminates, a few transverse cracks were found when the strains were low (e.g., 0.4 to 0.55%). As the strain was further increased, the number of transverse cracks was increased and delaminations began to nucleate, usually at the interlaminar boundaries where the transverse cracks had some difficulty growing further. Thereafter, delaminations appeared to dominate the failure pattern, although the number of transverse cracks continued to increase on continued loading. Jang et al. [62,63] have observed very similar cracking modes in both quasi-isotropic and cross-plied laminates when impacts were repeated. Ply-cracking behavior was treated by Lim and Hong [175] using a shear-lag analysis and by Han and Hahn [260] following a resistance curve approach. The conditions for transverse cracking and property reductions were also analyzed in these reports.

Also discussed in this chapter are the various theoretical models that have been developed to analyze the impactor-target force, target laminate deflection and deformation, impact-induced internal stress field in composites, failure prediction, and the residual properties of composites after impact. This chapter also briefly reviews both destructive and nondestructive techniques that can

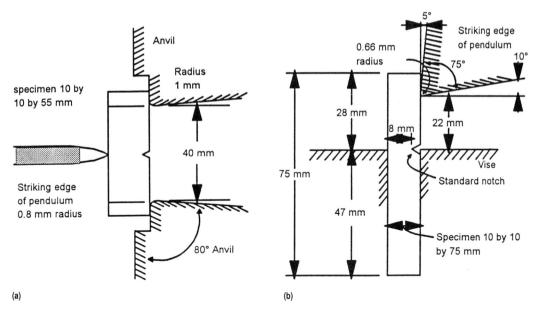

Fig. 1 Specimen arrangements for pendulum impact tests. (a) Charpy tests (top view). (b) Izod test (side view)

be used to assess the damage state and to observe the evolution of damage in composites.

Impact Testing of Composites

Izod and Charpy Impact Testing

An important type of dynamic loading involves the sudden application of the load, as from the impact of a moving mass. As the velocity and thus the kinetic energy of a striking mass are changed, a transfer of energy must occur; work is done on the parts receiving the blow. The energy of a blow may be absorbed in a number of ways: through elastic deformation of the specimen, through plastic deformation in the specimen, through hysteresis effects in the specimen, through friction action between the specimen and the test machine, and through the tested specimen's gaining certain kinetic energy. The effect of an impact load in producing stress depends on the extent to which the energy is expended in causing deformation. In the design of an impact-resistant structure, the aim is to provide for the absorption of as much energy as possible through elastic action and then to rely on some kind of damping to dissipate it. Therefore, the resilience or the elastic energy capacity of the material is a significant property to consider [64]. Also important are the various micro- and macromechanisms of energy absorption and dissipation in the material.

In an impact test, the load may be applied in flexure, tension, compression, or torsion. Flexural loading is the most common, tensile loading is less common, and compressive and torsional loading are used only in special instances [64]. The impact blow is normally delivered

through the use of a dropping weight, a swinging pendulum, a rotating flywheel, or a gas-gun-driven projectile.

The most commonly used impact tests for isotropic materials are the Charpy and the Izod tests, both of which employ the pendulum blow (Fig. 1). In the Charpy test, the specimen is supported as a simple beam, while in the Izod test it is supported as a cantilever (ASTM E 23 for metals and ASTM D 256 for plastics). For homogeneous isotropic materials, the tests are usually conducted on specimens with a notch on the tension side. The notch acts as a stress concentration site to minimize the energy required for initiation of fracture. Thus the total measured energy required for fracture is essentially the energy required for propagation of the fracture.

The Charpy and Izod impact tests are frequently used for comparing the impact response of isotropic materials that have different compositions or that are fabricated from different processing conditions. For this purpose of comparison, they are quite sufficient when applied to metals or polymers. When applied to polymer composites in which the fracture phenomenon is far more complex, these tests may not be adequate for providing data of basic physical significance. Several researchers [65-68] have observed that the Izod and Charpy test results depend on the dimensions of the composite test piece. More meaningful information can be obtained if these tests are instrumented to record the load history during the impact event [69,70].

Instrumented Falling Dart Test

To predict the impact performance of composites adequately, it is important to simulate the conditions under which the material will be loaded [3]. Compared to the traditional Izod and Charpy impact tests, instru-

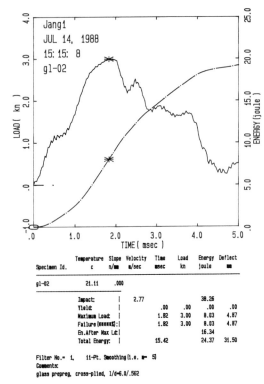

Specimen Id.	Temperature c	Slope n/m	Velocity m/sec	Time msec	Load kn	Energy joule	Deflect mm
g1-02	21.11	.000					
Impact:		2.77				38.26	
Yield:			.00	.00	.00	.00	.00
Maximum Load:			1.82	3.00	8.03	4.87
Failure (xxxxxX):			1.82	3.00	8.03	4.87
En.After Max Ld:						16.34	
Total Energy:			15.42		24.37	31.50

Filter No.= 1, 11-Pt. Smoothing (i.e. m= 5)
Comments:
glass prepreg, cross-plied, 1/d=6.0/.562

Fig. 2 Typical load-time and energy-time curves acquired by an instrumented impact tester

mented impact testing is believed to be a better method because it is capable of testing materials of almost any configuration at a wide range of velocities and in environments simulating service conditions [3]. This technique also gives a load-displacement curve and provides a time history of the energy dissipation process. Presented in this section are examples of how instrumented drop weight testing can be used to study the impact response of laminates.

The instrumented impact apparatus typically consists of a drop tower equipped with an instrumented indentor (tup or dart). This apparatus can have various specimen fixtures for performing Charpy- and Izod-type testing as well as for loading "plate" specimens over holes of various diameters. The annular hole on the specimen fixture used in a laboratory is typically 2.54 to 7.6 cm (1 to 3 in.) in diameter. The output of the load transducer can be recorded by an oscilloscope or be fed directly to a signal processing board installed in a personal computer. In a typical computerized system, both force and energy information can be acquired and shown on the monitor in a matter of seconds.

An example of a computer-generated data output is given in Fig. 2, where both the force and energy are plotted as functions of time during the impact event. Also presented are the basic numerical data, which include the incident impact energy, impact velocity, total energy absorbed (E_t), total deflection (l_t), maximum load (P_{max}),

energy (E_m) and deflection (l_m) at the maximum load, and energy ($E_p = E_t - E_m$) and deflection ($l_p = l_t - l_m$) after maximum load. It is occasionally possible to define a point of first damage (or point of incipient damage) that is the first significant deviation or break from the early portion of the load-time trace. The load and energy at this point are identified as P_i and E_i, respectively. In many cases, this incipient damage point coincides with the maximum load point.

It may be noted that both the falling dart and the hammer-type impact tester, the latter being used for the Charpy and Izod impact test methods, can be instrumented. The maximum energy E_0, obtainable by the hammer or instrumented tup assembly prior to impact with the specimen, is given by [114]:

$$E_0 = (\tfrac{1}{2})I \cdot V_0^2 \qquad \text{(Eq 1)}$$

where V_0 is the hammer velocity immediately before impact and I is the moment of inertia of the assembly defined as

$$I = \rho/g \qquad \text{(Eq 2)}$$

Here, ρ is the effective hammer weight and g is the gravitational acceleration. For falling dart testing, ρ is equivalent to the total weight of the hammer-tup assembly and E_0 is simply given by $E_0 = \rho h$. For pendulum impact testing [71]:

$$\rho \approx w_h + (\tfrac{1}{3})w_b \qquad \text{(Eq 3)}$$

where w_h is the hammer weight and w_b is the beam weight. If the hammer can be regarded as a free-falling body, $V_0 = \sqrt{2gh_0}$ where h_0 is the drop height. When the tup makes contact with a test specimen, the hammer energy is reduced by an amount ΔE_0:

$$\Delta E_0 = E_0 - E_f = (\tfrac{1}{2})I(V_0^2 - V_f^2) \qquad \text{(Eq 4)}$$

where E_f is the kinetic energy at time τ after initial contact between specimen and tup. Starting from the basic relationship between the impulse and momentum, one can easily show that

$$\int_0^\tau P\,dt = I(V_0 - V_f) \qquad \text{(Eq 5)}$$

which, in combination with Eq 4, yields [71-73]:

$$\Delta E_0 = E_a[1 - E_a/(4E_0)] \qquad \text{(Eq 6)}$$

where

$$E_a = V_0 \int_0^\tau P\,dt \qquad \text{(Eq 7)}$$

Equation 6 can be shown to be equivalent to [71]:

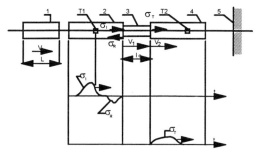

Fig. 3 Schematic of a split Hopkinson pressure bar technique. Source: Ref 75

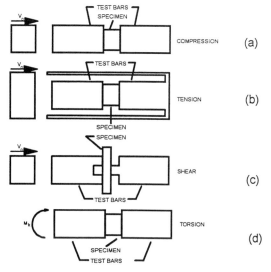

Fig. 4 Various modified versions of the split Hopkinson pressure bar technique

$$\Delta E_0 = \overline{V} \int_0^\tau P\,dt \qquad\qquad \text{(Eq 8)}$$

where, by definition,

$$\overline{V} = (\tfrac{1}{2})(V_0 + V_f) \qquad\qquad \text{(Eq 9)}$$

Note that E_a is the energy signal that would be recorded on the energy-time trace on the oscilloscope. Equation 6 simply provides the actual energy absorbed by the specimen, where a necessary correction has been made to account for the reduction in velocity of the dart during contact with the specimen.

Researchers disagree about the failure mechanisms involved in each stage of loading. The point P_i has been reported to be a result of either resin cracking or first fiber failure. Several failure mechanisms can also occur between P_i and P_{max} in a laminate. The point P_{max} was believed to represent the end of the crack initiation phase and the start of a catastrophic failure [11].

In order to identify the failure processes at various stages of impact loading and to define the physical significance of P_i, P_m, E_i, and E_t, one may conduct a corresponding static indentation test using the same impact indentor and the same specimen fixture. This "quasi-impact" event is then monitored continuously using acoustic emission, and/or it is observed intermittently using microscopy. This is based on the assumption that the impact and static responses of laminates are similar. This premise has been verified for epoxy laminates containing graphite, Kevlar-49, or glass fibers [49], which suggests that the viscoelastic nature of the matrix resin did not play a significant role in the range of impact velocity studied. For composites containing nylon, PET, or PE fibers, the impact and static specimens showed very similar load-displacement traces, although slight differences in moduli and maximum load were detected. Even so, the failure processes are likely to be similar between the static and the impact test specimens, provided that the impact loading rate is not excessively high (no effect of shock waves).

Hopkinson Bar Tests

A widely used technique for the impact testing of homogeneous isotropic materials under different simple stress conditions is based on the split Hopkinson pressure bar (SHPB). This technique has also been adapted to the testing of composite materials [74]. A simple version of the SHPB is designed for compression testing (Fig. 3). A specimen in the form of a cylinder with diameter a_0 and length l_0 is placed between two bars of circular cross section A (A slightly larger than a_0). A stress pulse $\sigma_I(t)$ is initiated at the end of the first bar (the punch bar). As a result of its interaction with the specimen, a stress pulse $\sigma_R(t)$ is reflected by the specimen, while a stress pulse $\sigma_T(t)$ is transmitted to the second bar. Strain gauges on the input bar record the incoming stress wave and the returning reflected wave from the specimen, while strain gauges on the output bar record the stress wave transmitted through the specimen. Both bars are made of a material that remains elastic during the loading process. From these three signals the dynamic stress-strain curve may be derived [74,75].

The SHPB can also be modified to provide shear, tension, and torsion tests (Fig. 4). In the punch-loading Hopkinson bar, a flat plate specimen is sandwiched between the loading bar and the output tube, and a circular hole is punched in the plate in a direction perpendicular to the plane of reinforcement (Fig. 4c). Although the punch loading configuration is simple, fundamental interpretation of the data is difficult. This test essentially measures the through-thickness shear strength. SHPB-based tensile impact tests have also been adapted to investigate the high loading rate behavior of composites [76,77]. The torsional version of SHPB has yet to be used for characterizing the impact response of fiber composites.

The punch load-displacement curves of fine-weave glass-epoxy composites obtained by a shear-type SHPB system are shown in Fig. 5, which indicates a very marked effect of loading rate on the impact response of these composites [74]. Both the maximum punch load and the energy absorbed increase greatly with an increased rate of loading. Fine-weave glass-epoxy composites also exhibit a significantly increased elastic modulus, fracture strain, and fracture strength when the tensile loading rate is raised (Fig. 6b). In contrast, over a strain-rate range of some seven orders of magnitude, no effect of strain rate could be detected with a unidirectional carbon-fiber-reinforced composite (Fig. 6a). Exactly why these two groups of materials respond so differently remains to be explained.

Gas-Gun Impact Test

There are two purposes to the design of a gas gun for impact testing of composites. The first is to provide a means to drive a lightweight mass to strike the target composite structure for low-velocity or low-incident-energy impact studies. The second is to propel a projectile to facilitate the investigation of ballistic impact penetration behavior of composites. Fiber-reinforced composites have become an important class of materials for construction of parasitic and structural armor. The ability to predict material behavior under impact loading is very important in the design and manufacture of ballistic protection products.

Gas-Gun for Ballistic Applications. The design of a gas gun for low-velocity impact studies can be found in many references [78-87]. An air gun based on helium propulsion was designed, constructed, and tested at Materials Engineering Laboratories of Auburn University (Fig. 7) [84]. A high-pressure helium storage tank (up to 6000 psi) was used to supply the helium gas required for propulsion. A high-pressure regulator and a high-pressure valve were used to transfer the high-pressure helium gas from the storage tank to two small cylinders (total volume of the cylinders was about 1000 cc). A pressure gage was used to ensure the proper pressure in the intermediate tanks. When the proper conditions were attained, a fast-acting valve (from fully closed to fully open condition in 50 ms) was activated and the pressure burst accelerated the projectile through the barrel to hypervelocities.

The gun itself consisted of two separate components. A liner (36 in. long) constituted the main barrel component, consisting of two halves for ease of machining (no boring was necessary). A long semicircular $\frac{1}{8}$ in. radius groove (36 in. long, the length of the liner) was machined on each half of the liner. When assembled, this liner

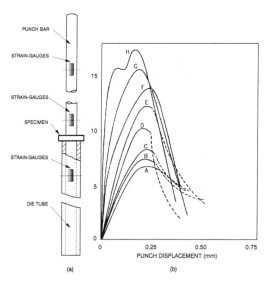

Fig. 5 High-speed punch loading of woven glass-epoxy composite. (a) Test assembly. (b) Punch load-displacement curves. Mean punch speeds (m/s): A—2.5×10^{-7}; B—2.5×10^{-6}; C—2.3×10^{-4}; D—2.3×10^{-2}. E—0.28; F—3.8; G—11.4; H—16. Source: Ref 77

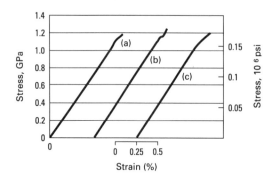

(a)

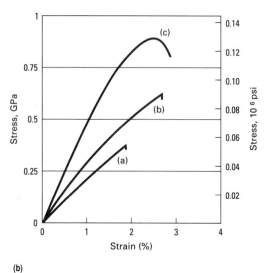

(b)

Fig. 6 Effect of strain rate on tensile stress-strain curves for (a) unidirectional and (b) fine-weave glass-epoxy specimens. Mean strain rate (S^{-1}): Curve (a): 10^{-4}; Curve (b): 10; Curve (c): 450 to 900. Source: Ref 77

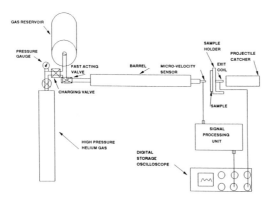

Fig. 7 Schematic of a helium gun system and the microvelocity sensor unit. Source: Ref 84

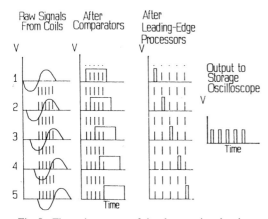

Fig. 8 The various stages of signal processing showing the principle of microvelocity sensor. Source: Ref 84

became a long barrel with a $\frac{1}{4}$ in. diameter hole through the center. This liner was clamped tightly between two clamping plates, using closely spaced bolts to ensure even distribution in pressure.

Prior to firing, the projectile was inserted from the open end using a long rod. During the flight of the projectile, it was accelerated to the desired velocity through the barrel, passed through the microvelocity sensor unit (to be described below), impacted the target, and then exited into the catcher at the end.

The projectiles used were typically 1.75 in. in length and $\frac{1}{4}$ in. in diameter. Different lengths and different materials can be used if required. A small magnet was inserted into the end of each projectile, for a purpose to be discussed in the next section. The magnets were cylindrical in shape and each measured $\frac{1}{8}$ in. in diameter and $\frac{1}{8}$ in. in length. Muzzle velocities were measured as a function of the helium gas pressure used.

An innovative microvelocity sensor unit capable of measuring the instantaneous velocity before, during, and after projectile-material interaction was also developed at Auburn [88,89]. The prefix "micro" refers to the superior spatial resolution of this device. In this system, a magnet is attached to the projectile rod. A detector made of a

series of coils on an electrically insulating tube is placed in front of the materials of interest. During the impact and the penetration of the projectile through the target, the magnet in the projectile triggers the coils in succession as a result of the rapidly changing magnetic flux the individual coils experience. The proper timing of the triggering therefore provides the needed information for real-time, real-space velocity mapping.

To determine the dynamic energy loss process, spatial resolution on the order of 0.1 in. is required. In the original detector design, there were separate raw signals in the form of a sine function from individual detector coils (five such signals are shown in Fig. 8), but in real situations, there will be 11 to 21 coils. The problem was that the signals were not separable when simply added together, due to overlapping. Therefore digital circuitry was designed, constructed, and tested. Each raw signal was fed individually into separate comparator circuits, and the output signal became a digitized 5V square wave. However, these signals were still too wide to be useful. They were then fed into individual lead edge detectors. The output of these signals have adjustable widths from 50 ns upward. These narrow signals can be summed to provide detailed information on the instantaneous positions of the projectile as a function of time. The final signals are displayed and stored on a digital storage oscilloscope. The pulses represent the passing of the magnet in the projectile through the eleven coils. In the case of a free-flying projectile through air, the constant distance between pulses signifies constant velocity without slowing down.

A secondary coil unit was designed and constructed to measure the exit velocity of the projectile. This unit consists of two coils spaced 2.5 in. apart and is placed behind the target. Due to the large distance between the coils (2.5 in.), the two raw signals are completely resolvable and therefore need not be processed. To further improve the spatial resolution of the sensor, a new coil was wound that has 0.05 in. intercoil spacings, 21 coils over an overall length of 1 in. From the signals obtained and processed, one can determine the velocity and hence the kinetic energy and the energy loss of the projectile almost continuously. This apparatus is of particular use in the investigation of energy dissipation processes in ballis-

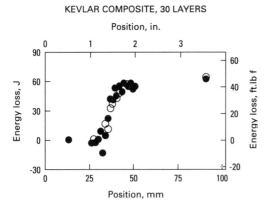

Fig. 9 Total energy loss vs. position curve for a 30-layer Kevlar-49/epoxy composite. Source: Ref 84

tic fabrics and composite armor materials. Shown in Fig. 9 is the projectile kinetic energy plotted as a function of time for a 30-layer Kevlar-49 fabric-reinforced epoxy composite [89].

Repeated Impacts or Impact Fatigue Testing

The increased use of fiber-reinforced polymer composites for primary structural applications has resulted in an increased need to characterize and predict the mechanical behavior of these materials. Strength and stiffness are usually the first two properties to consider when selecting a structural material. Nevertheless, the damage tolerance of brittle polymer composites, such as graphite fiber-epoxy in response to a low-energy impact, has been a major concern for both composite users and manufacturers [90-98]. The internal damage created by a light impact may not be visible from the surface, but it can result in a significant reduction in material integrity [94-98]. Reliable techniques for damage tolerance assessment of fiber composites are essential to the efforts to develop damage-tolerant materials.

The method formerly most widely used for determining damage tolerance was the compression-after-impact test [94-98]. A postimpact flexural test [97] was sometimes used as an alternative, possibly because of the difficulty involved in performing the compression test of continuous fiber-reinforced polymers. A special sample holding device was usually necessary to conduct an accurate compression test. In one example [94], a test fixture was needed to support the coupon at the sides and clamp it at the loaded edges. This technique of stabilizing the coupon allowed out-of-plane deflections associated with delamination growth.

Now a more convenient and reliable method for characterizing the damage tolerance of fiber composites has been developed [63]. This technique, impact fatigue or repeated impact testing, involves repeated instrumented impact testing of the same specimen using the same tester. During each impact event, load-time and energy-time traces can be recorded along with other significant numerical data. A wide range of incident ener-

gies can be applied while the material integrity, reflected by the stiffness (slope), strength (maximum load), and toughness (energy absorbed), is monitored as a function of the number of repeated impacts. This technique has been successfully applied to provide toughness measurements on unreinforced polymers [99-103].

In certain applications of fiber composites, the material can be subjected to repeated impacts. However, very limited work has been done in impact fatigue testing of composites. The residual tensile and compressive strengths of selected carbon-epoxy composites subjected to repeated impacts were measured by Wyrick and Adams [104]. Rotem [105] showed that when a low-energy impact was exerted on a laminate, invisible damage might occur as a result of the contact stresses between the impactor and the laminate. The damage zone grew and caused a certain amount of strength and modulus reduction. The variation of rigidity of fiberglass plastics during tensile impact fatigue testing was monitored by Tamuzs *et al.* [106,107]. A theoretical lifetime analysis was developed and microscopy was performed by Lhymn [108,109] on the impact fatigue specimens of short E-glass fiber/polyphenylene sulfide. But in these studies the possibility of applying impact fatigue testing for damage tolerance assessment was not discussed. Not much attention was directed to the failure mechanisms as repeated impacts were imposed on the composites.

The impact tester used for impact fatigue testing must be capable of plotting the impact load as a function of time (or force-deflection curve) and capable of plotting the energy absorbed by the specimen as a function of time (or absorbed energy-deflection). The sample is of Charpy-type and is normally tested using a span between supports of 50.8 mm (2 in.). The impactor tip is 12.7 mm (0.5 in.) in diameter. The sample is carefully placed in the same position after each test run. Impact energy depends on the mass and the drop height, and impact velocity depends on just the drop height. The mass may remain constant while the drop tower heights are changed slightly to obtain different impact energies and different impact velocity values. Alternatively, the mass may be changed and the drop tower heights kept constant to obtain different impact energy values at a constant impact speed.

A simplified elastic strain energy theory was proposed [62,63] to help to explain the condition of damage initiation in a composite subjected to a low-velocity impact. The critical incident impact energy values that were required to produce significant damage in several different composites were calculated to confirm this hypothesis. The macroscopic failure modes and the internal damage mechanisms as a consequence of repeated impacts were also studied. Parameters were identified that can be used as indices for composite damage tolerance.

The response of composites to repeated impacts was first investigated by testing unidirectional Cofab® fabric-epoxy composites fabricated from glass, aramid, and graphite fibers. Repeated impacts were also conducted on the epoxy composites prepared from unidirectional prepreg tapes of glass, aramid, and graphite fibers. Par-

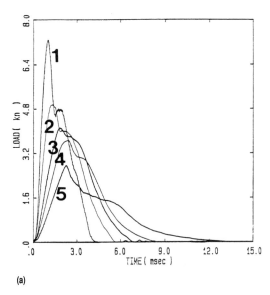 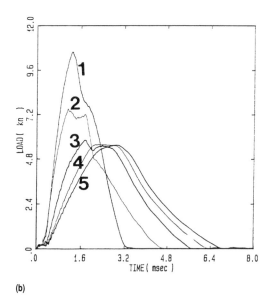

Fig. 10 Overlaid curves of repeated impact load-time response for (a) a Cofab® graphite-epoxy composite (E_{in} = 7.45 J) and (b) a Cofab® glass-epoxy composite (E_{in} = 9.94 J). Source: Ref 62

ticular attention was directed to observing the variations in load-displacement curves when the number of repeated impacts increased. Correspondingly macroscopic failure modes were examined by visual observations and by use of a light microscope. The following summarizes the more salient common features of the impact fatigue behavior of composites.

Load-Time and Load-Displacement Curves. The load-displacement curve obtained for a specimen in response to the first impact can be recorded, along with the numerical data of the maximum load ($P_m{}^0$), the energy absorbed up to maximum load ($E_m{}^0$), and the elastic modulus or stiffness (S^0). If the incident energy of the first impact is sufficiently high, the laminate will incur a damage. This damaged specimen is then impact loaded for a second time with its load-displacement curve traced, and $P_m{}^1$, $E_m{}^1$, and S^1 are calculated. The ratios $P_m{}^1 / P_m{}^0$, $E_m{}^1 / E_m{}^0$, and S^1 / S^0 can be taken as a measure for the damage tolerance of this composite. However, it is informative to overlay many curves together in the same diagram after N repeated impacts. Figures 10(a) and (b) show such a diagram for a Cofab® graphite- and a glass-fabric/epoxy laminate, respectively. As the number of repeated impacts increases, the maximum load that can be tolerated by the material (as an indication of the material strength) decreases. Also found to decrease is the slope of the load-time curve, which is proportional to the sample stiffness and modulus, given the same sample geometry. As a result of the stiffness reduction, the maximum deflection increases. The corresponding repeated impact energy-time curves for these two composites are overlaid in Fig. 11(a) and (b), respectively.

Curve 1 in Fig. 10(a) indicates that, in response to an incident energy of 7.45 J (5.5 ft · lbf) (during the first

impact), the graphite-epoxy composite experiences a maximum load at approximately 7.3 kN (1640 lbf). Beyond this maximum load comes a dramatic yield drop that indicates the formation of a delamination crack. The energy (E_m) absorbed up to this maximum load, as read off the first curve in Fig. 11(a), is approximately 4.2 J (3.1 ft · lbf). This value can be considered the minimum incident energy (E_c) to create delamination damage to the composite with given dimensions. Using the same method, we calculate the E_c value for the Cofab® glass-epoxy composite with comparable dimensions to be about 6.8 J (5 ft · lbf).

The Decay Diagrams. A wide scope of incident energies can be exerted to many different specimens of a sample by adjusting the drop weight and/or height. A large number of fiber composites subjected to impact fatigue testing with various incident energy values have been examined [62,63]. For a given material, there usually exists a critical incident energy (E_c) above which a significant delamination crack will initiate in response to just one impact. When impacts are repeated, the maximum load $P_m{}^N$ (at the Nth impact) would decrease. The normalized values, $\log_{10}(P_m{}^N / P_m{}^0)$, if plotted versus $\log_{10} N$ usually demonstrate a straight line (Fig. 12b). This trend has been observed for a majority of composites thus far studied (e.g., glass- and graphite-epoxy composites, both fabric- and prepreg-based).

The straight lines observed with most of the laminates suggest the existence of an exponential decay relationship, $P_m{}^N / P_m{}^0 = 10^{-Nb}$. A smaller value of the slope b would indicate a more damage-tolerant composite. Here, the slope in the curve for $\log P$ versus $\log N$ is a good indication of the impact fatigue durability or the damage tolerance of the material.

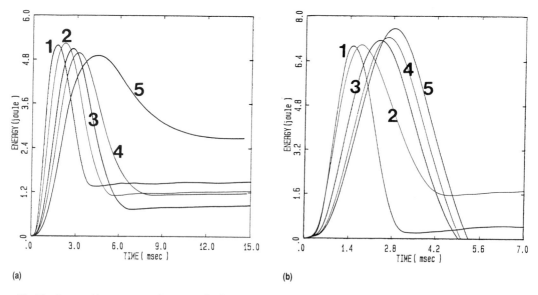

Fig. 11 Repeated impact energy-time curves for the same specimens as in Fig. 10. Source: Ref 62

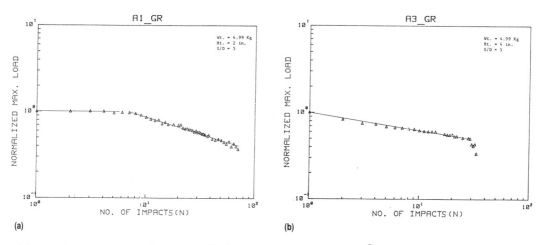

Fig. 12 Normalized maximum loads, $\log (P_m^N/P_m^0)$, vs. number of impacts for Cofab® graphite-epoxy composites. (a) $E_{in} = 2.48$ J and (b) $E_{in} = 4.97$ J

When the incident energy E_{in} is less than E_c, usually no appreciable reductions in the strength and stiffness are observed upon repeated impacts. However, when the number of cycles reaches a certain value (N_c), the maximum load and the curve slope begin to drop as N further increases (e.g., Fig. 1a for a graphite fabric-epoxy composite). This critical cycle number marks a state where an appreciable level of damage ensues, as usually characterized by the first observation of a significant delamination crack. This N_c can also be regarded as another indicator of the damage tolerance of a given composite. Prior to this incipient damage point, the curve for $\log P$ versus $\log N$ is essentially a horizontal line. Beyond this

critical damage state, the curve for the most part again exhibits a straight line except for the very late stage of impact fatigue, where additional one or two straight lines of different slopes or even nonlinear curves can be observed for a few cases. At this later stage, the data are no longer meaningful because the samples would have been severely damaged and distorted. For most epoxy-based composites studied, the following equations well describe the impact fatigue response:

$$P_m^N / P_m^0 = 10^{-Nb} \qquad \text{if } E_{in} \geq E_c$$
$$P_m^N / P_m^0 = N_c\, 10^{-b'(N-N_c)} \qquad \text{if } E_{in} < E_c$$

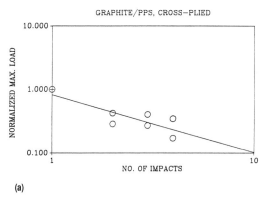

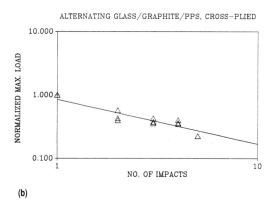

Fig. 13 Normalized maximum loads vs. number of impacts for (a) a graphite-PPS cross-plied laminate and (b) an alternating glass/graphite-PPS laminate

Impact fatigue data for carbon fiber-reinforced thermoplastic PPS composites and carbon/glass-PPS hybrids are shown in Fig. 13. These studies [62,63] on the impact fatigue response of continuous fiber reinforced resin composites have indicated the following impact damage evolution mechanisms:

- *Stage I.* If the incident impact energy exceeds a critical value (E_c), then the fiber-resin composite will suffer damage in the form of a delamination. In the cases where a strong interlaminar shear stress exists, this critical incident energy can be related to the critical elastic strain energy stored in the composite up to the point where this shear stress reaches its corresponding interlaminar shear strength (p_{ILSS}) to create a delamination crack.

- *Stage II*: In response to a subcritical incident energy impact, the composite will exhibit matrix microcracking and/or fiber-matrix interfacial debonding. On continued loading (e.g., repeated impact testing), these mechanisms will accumulate until one or more larger microcracks propagate to the interlaminar zone to form a delamination crack. Prior to the formation of such a critical crack, no significant loss in the material strength and stiffness is to be anticipated.

- *Stage III*: Once a delamination crack is initiated as a consequence of critical loading, it can grow or more delamination cracks can be nucleated in the composite subjected to continued loading or repeated impacts. This would lead to an ever-increasing loss in the strength and stiffness of this composite.

Wyrick and Adams [104] observed a continuing reduction in both the tensile strength and compressive strength of carbon-epoxy composites when an increasing number of repeated impacts were imposed on the materials prior to tensile or compressive testing. The impact-induced failure response of composites can also be found in many earlier publications [e.g., 50,78].

Experimental Characterization of Impact Damage

Methods for Impact Damage Assessment

Internal damage in composites subjected to impact loading usually consists of fiber breakage, matrix cracking, and delaminations. The damage state of a composite can be characterized by visual inspection [79-81,110-113], microscopic examination [82-85,104,110,113-128], ultrasound [86,125,132,133,136,144,145,149,151-160], radiography [79,85,86,104,114,115,117, 123,124, 129-131,146,147,261-264], shadow-moire technique [86,117,123,134-136], infrared thermography [84,137, 138], deplying method [82,139,140], acoustic emission [49,121,141], fiber optics [143], tapping [142], and vibration-based techniques [144,145]. Some of these techniques (e.g., deplying and microscopy) inevitably involve destruction or disturbance of the damaged specimens, while others are nondestructive. Some are considered to be whole-field methods (e.g., radiography, ultrasonic imaging, and infrared thermography) that can be used to cover larger areas, while others (e.g., microscopy and fiber optics) are more localized.

X-ray radiography requires the injection of an opaque penetrant into the damaged zone, but this technique can delineate overlapping delaminations. In contrast, conventional ultrasonic C-scan methods determine only the projected delamination area. Improved ultrasonic inspection procedures have been used by several researchers [115,120,129,146,147] to resolve the distribution and size of delaminations through the thickness of the specimen. A new acoustic backscattering technique appears to be capable of resolving interply delaminations, matrix cracks in each ply, and wavy or disaligned fibers [115]. Buynak et al. [146] reported a new ultrasonic method to nondestructively produce and display images of the damage on a ply-by-ply basis, with all of the data being collected during a single scan. This ultrasonic damage-mapping approach relies on digitizing of ultrasonic data with software processing. The method has the ability to resolve closely spaced delaminations that could not be separated using conventional methods. Liu *et al.* [145] have combined layer-sectioning and microscopic techniques to "reconstruct" the damage pattern along the thickness direction.

Thermography involves infrared sensing of the surface temperature distribution of a damaged laminate. A disturbance or temperature gradient could indicate the existence of delamination, but this technique does not yield accurate information as to the size and shape of delaminations, because the temperature difference at the delamination boundary is not sharp. Potet *et al.* [138] used vibrothermography to detect subsurface damage in composites. This technique involves the use of high-frequency, low-amplitude mechanical vibrations to induce local heating in a laminate. Local defects such as delaminations serve as heat sources by dissipating vibrational energy. The temperature distribution is then mapped out by means of an infrared camera. In conjunction with a two-dimensional heat equation, this technique can be used to determine a map of the heat generation rate per unit area, which is then used to locate the damage area.

The presence of internal defects can be detected by the "tapping" technique, in which the specimen is lightly tapped (e.g., hit by a small steel ball) while the change in radiated sound is measured and analyzed. Jang *et al.* [144,145] have developed a more general vibration-based technique to assess the damage state in composites. This technique takes advantage of the fast and powerful data acquisition and analysis capability of modern fast Fourier transform spectrum analyzers and the concept of experimental modal analysis. The dynamic properties of a laminate can be continually or intermittently monitored as the degree of damage progresses. This technique is potentially useful for real-time condition-monitoring of a large-scale composite structure. A fiber optic technique also provides the possibility of real-time monitoring of composite structural integrity [143].

A digital comparative holography technique has been developed for detecting flaws in composites and evaluating their severity. The method consists of creating two double-exposure holographic fringe patterns that indicate the displacement fields of flawed and unflawed components. Using digital image processing and numerical algorithms, the interference fringes are then "compared" to determine the net additional specimen displacement due to the presence of the flaw. A hybrid stress analysis computer model may be formulated and programmed to calculate the additional stress and strain fields due to the presence of the flaw. In this procedure, the experimental subtracted displacement data are used as input boundary and constraint conditions on a numerical plate solution, using the finite element method.

Impact Damage Development

Observation of Damage Evolution. Both low- and high-speed photography have been used to investigate the evolution of damage in composites. Moore and Prediger [149] studied crack propagation on the tension side of laminates during impact by using a 35 mm camera. The signal from a force transducer of the impact tup was used to trigger a flash with a preset time delay. One photo was taken in each impact run, but several identical specimens were tested and photographed at different delay times so that the evolution of damage could be reconstructed from these serial exposures.

High-speed photography was used by Takeda *et al.* [150] to measure the velocity of delamination crack propagation. Delamination crack speeds in a glass-epoxy laminate $[0°_5/90°_5/0°_5]$ can reach as high as 500 m/s (1640 ft/s). Delaminations were associated with an impact-generated flexural wave, while transverse cracks were caused by a tensile wave due to large deflections that traveled at almost the same speed as the flexural wave. Chai *et al.* [171] used high-speed photography in conjunction with the shadow-moire technique to investigate failure propagation phenomena in composite panels that were subjected simultaneously to compressive in-plane loading and low-velocity transverse impact. Although the global response of the panel was small with an impact speed of 0 to 75 m/s (245 ft/s), there was significant bulging of the surface as a result of the buckling of the delaminated area. High-speed photography was also used to study delamination growth patterns in glass and carbon fiber/epoxy composites [151,152] and the ballistic impact response of graphite-epoxy and boron-epoxy laminates [153].

An acoustic emission technique has been used to examine the damage evolution sequence during static and instrumented impact testing of laminates. The load-displacement curves of carbon-epoxy laminates obtained by a static test were very similar to those obtained by an impact test [49]. Static tests produced acoustic emission events analogous to those acquired during impact tests [121]. These suggest that the predominant failure modes may be the same for both static and very low-velocity impact tests.

Damage Evolution of Laminates. In an investigation of the single-impact response of laminated composites, Chang *et al.* [154] concluded that matrix cracking was the initial failure mode, which was caused predominantly by the interlaminar shear stresses and the in-plane tensile stresses. In response to the impact with a subcritical incident energy ($E_{in} < E_c$), the composite was not expected to delaminate. However, matrix cracking could occur in the laminate. These microfailure mechanisms preferentially evolve in the transverse plies or the plies with larger fiber inclinations (e.g., ≥45°) [63]. When the incident energy is sufficiently high, delamination initiation and propagation occur [63]. According to Chang *et al.* [154], a suddenly increased out-of-plane normal stress as a result of matrix cracking predominates the propagation stage of delamination during the impact event. Furthermore, once a large-scale delamination crack is formed, the neighboring off-axis plies are subjected to high normal stresses, leading to the formation of additional transverse cracks.

Wang [155] used X-ray radiography to study the low-velocity impact response of a 48-ply graphite-epoxy laminate of quasi-isotropic lay-up. The damage caused by

Table 1   Materials used in a study on impact response of fabric-epoxy composites

Material	Reinforcement style	Fiber characteristics (relative values)	Supplier
Polyethylene fibers	Plain weave	High strength and high strain	Spectra® 900 from Allied Signal
Graphite fibers	Plain weave	High strength and low strain	Hercules AW193P
Aramid fibers	Plain weave	High strength and medium strain	Kevlar®-49 fabrics from Burlington
Nylon fibers	Plain weave	Low strength and very high strain	Compet®-Nylon from Allied Signal
Polyester fibers	Plain weave	Low strength and very high strain	Compet®-PET from Allied Signal
Glass fibers	Plain weave	High strength and low strain	Unknown
Epoxy resin		Low curing temperature	Ciba-Geigy 507
Curing agent			Ciba-Geigy 956

Source: Ref 49

Table 2   Effects of fiber type on the impact properties of fabric composites

Fiber type	Composite thickness, mm (in.)	Max load, kN	E_m, J	E_p, J	E_t, J	Ductility index	Comments
Aramid	1.27 (0.050)	1.02	2.88	10.03	12.91	3.483	...
Graphite	1.09 (0.043)	1.01	2.78	6.48	9.26	2.331	...
Glass	1.9 (0.076)	1.52	5.65	8.39	14.04	1.485	...
Nylon	1.45 (0.057)	2.39	9.86	7.95	17.81	0.806	...
PET	1.6 (0.064)	4.26	58.50	25.99	84.49	0.444	...
PE	1.37 (0.054)	3.75	59.14	33.00	92.14	0.558	3 layers. No penetration.
PE	2.23 (0.088)	6.62	89.81	13.19	103.76	0.147	5 layers. No penetration.

E_m, energy at the maximum load. E_p, energy propagated. E_t, total energy. PET, polyethylene terephthalate. PE, polyethylene.
Source: Ref 49

the impact consisted primarily of matrix cracks in the various plies and the various ply interfaces (delaminations). Upon further loading, one major damage propagation mode was delamination growth. Although this study did not pinpoint the exact sequence of intraply cracking and delamination, the observations seemed to indicate that matrix cracking dominated the early stage while delamination growth prevailed at the later stage. Jang [62] and Buchar et al. [75] observed that in response to an impact of subcritical incident energy ($E_{in} < E_c$), graphite-, Kevlar-, and glass-epoxy laminates of various stacking sequences exhibited only intraply matrix cracking. When impacts were repeated, these intraply cracks grew in size and number, and many of them, when meeting the interply boundaries, acted to initiate delaminations. When a supercritical incident energy ($E_{in} > E_c$) was imposed upon the laminate, both intraply cracks and delaminations were observed, but it was almost always possible to trace the delamination initiation site back to an intraply matrix crack. Once delamination growth occurs, intraply and interply cracks tend to interact in a fashion that is not fully understood.

Cristescu et al. [185] investigated the failure mechanisms in five-layer glass-epoxy laminates impacted by blunt-ended penetrators. With sufficiently high impact velocities, the first phase of damage was the shearing out of a generator strip in the first layer. The strip was parallel to the fiber direction, and its width was equal to the diameter of the impactor. The impactor pushed down this strip, cut through the fibers in the first layer, and continued to load the second layer, causing delamination to initiate and grow. Delaminations were found to occur between two layers of different fiber orientations, but not between two layers with the same orientation. Clark [132] proposed a model to explain how delaminations occur at the interface between two plies during impact and how the distribution of delaminations through the thickness can be predicted. Many researchers have presented detailed maps of impact-induced delaminations for each interface of the laminate [79,81,87,110,113, 115,124,132,146]. The delamination shape is usually oblong or "peanut-shaped," with its major axis oriented in the direction of fiber orientation of the lower ply for that interply boundary.

The existence of a threshold incident energy (E_{in}) below which no detectable damage occurs is consistent with the observation that there is a threshold velocity above which significant damage, consisting of delaminations and matrix cracks, is introduced. As the impact velocity is further increased, surface damage due mainly to fiber breakage is introduced, in addition to the internal delaminations and matrix cracks. In the case of very high

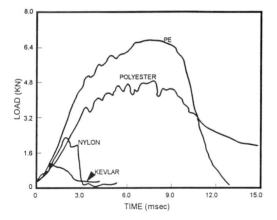

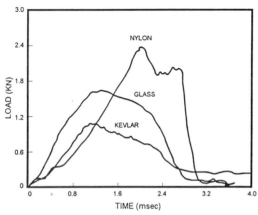

Fig. 14 Impact characteristics of several five-layer laminates. PE, polyethylene

speeds, the target would not have time to deform and therefore perforation occurs, leaving a clean hole in the target.

At subperforation impact velocities, the projected delamination area in graphite-epoxy and Kevlar-epoxy laminates, as observed on C-scan or radiographs, has been found to scale linearly with the kinetic energy of the impactor [80,111]. Correlation between the delamination area and the incident kinetic energy of the impactor has also been reported by many researchers [28,30,79,82,85,87,114,123,139,151,157,158].

Damage in Two-Dimensional Fabric and Braided Composites. The impact response of various woven fabric-reinforced composites was studied by Jang *et al.* [49] by using an instrumented drop weight impact tester. The laminates were clamped horizontally over an annular support of 10 cm (4 in.) inner diameter. The materials examined are listed in Table 1. The results of a series of instrumented impact tests are summarized in Table 2. Impact load versus displacement (time) curves of selected specimens are shown in Fig. 14. These data indicate that, under comparable conditions, the PE fiber-based com-

posites possess the greatest energy-absorbing capability. With an impact speed of 4.53 m/s (14.86 ft/s) and an incident energy of about 104 J (77 ft · lbf), practically 100% of the input energy was absorbed by the composite without creating perforation or penetration. Visual observations show that fibers in the bottom layer (the first layer of the back surface) were partially broken. A certain level of delamination can also be seen near the specimen edges. However, the most salient feature appears to be the observation that the specimen bulged out along the rim of the annular support to form a dome. The specimen has obviously undergone extensive plastic deformation, with the most severe deformation occurring at the center of the dome where the impact tip indented the specimen.

The polyester fiber-reinforced epoxy exhibits the second largest total energy (E_t) among this group of materials tested. Severe deformation and a certain level of delamination were also observed in the polyester fiber-based specimens. With few exceptions, no through penetration was found in these specimens. The impact toughness of the material is followed, in descending order, by nylon-, glass-, aramid-, and graphite-fiber composites. No obvious bulging or permanent deformation (except near the perforation) was visible in the last four materials. However, the specimens have been perforated and penetrated.

The data presented in Table 2 clearly show that a great maximum load leads to a high total energy absorption level. This implies that the impact toughness in these high-performance composites is controlled by the overall in-plane strength of these composites, the latter being dictated by the fiber properties. Except in the case of the aramid composite, the ductility index of each composite decreased with an increasing total impact energy. This suggests that, for tougher laminates, more energy will be dissipated before the maximum load is reached than afterwards. This observation again suggests that tough fibers are essential for achieving composite impact toughness.

The impact performance of woven carbon fiber-reinforced laminates with [0°,90°], [±45°], and mixed-woven [0, ± 45] lay-ups was evaluated by Bishop and Curtis [158,159], who utilized an incident energy of 1 to 9 J. The threshold energy for damage was similar for woven and nonwoven (unidirectional lamina tape-based) laminates, but the extent of damage in the woven material was much less. The damage was principally in the form of delamination between the layers, but at the point of impact there was evidence of tensile cracking toward the back surface and compressive buckling close to the front surface. The use of woven material restricted the extent of delamination, particularly the splitting along the fiber direction on the back surface [159].

Three-Dimensional Braided, Stitched, and Woven/Nonwoven Composites. The introduction of thickness-direction fibers through three-dimensional braiding, stitching, or weaving is very effective in restricting the growth of damage in composites. For instance, as compared with a quasi-isotropic laminate of carbon-epoxy, a braided three-dimensional composite exhibited a

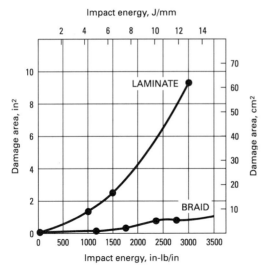

Fig. 15 Effect of impact energy on the damaged area of three-dimensional braided commingled and laminated carbon/PEEK composites. Source: Ref 41

much smaller degree of damage, even though the damage threshold energy values were very similar [160]. Ko *et al.* [41,161] made similar observations on glass-epoxy and carbon-PEEK composites. In a drop-weight impact test, the three-dimensional braided glass-epoxy composite required much higher levels of energy to initiate and propagate damage [161]. As shown in Fig. 15, the impact damage area in a three-dimensional braided carbon-PEEK composite was an order of magnitude lower than that in a corresponding laminated composite [41].

Three-dimensional weaver geometries can also be achieved through the interlocking of multilayer fabrics by a system of *z*-direction or through-the-thickness yarns. Majidi *et al.* [162] designed and fabricated two-dimensional and three-dimensional weave geometries using a combination of commingled and plies matrix graphite-PEEK yarn. Their impact response studies indicated that *z*-direction angle interlocking fibers significantly increased the energy-absorbing capacity of the composites and altered their damage characteristics. The damage zone of the three-dimensional angle interlock composites was less intense and more diffuse, while heavy damage occurred in a concentrated region of the two-dimensional composite. The residual stiffness and strength of three-dimensional composites therefore should be better than that of the two-dimensional versions.

The mechanical properties of stitched three-dimensional composites and their two-dimensional counterparts were studied by Jang *et al.* [40,163]. Failure mechanisms in these materials were also determined using *in situ* optical microscopy and SEM. Stitched multidimensional (three-dimensional) composites were found to possess greater damage resistance than conventional two-dimensional composite materials. The stitched third-direction Kevlar fibers have proven to be effective in arresting delamination propagation. A three-dimen-

sional composite had a smaller damage area (with little or no visible delamination) than a two-dimensional composite did. As observed in both flexural load-displacement curves (three-point and four-point bending), the plots of load versus time, and the plots of energy versus time (instrumented impact), the failure processes in three-dimensional composites usually proceeded gradually (retaining structural integrity for a longer period of time), while a two-dimensional composite usually tended to fail in a sudden catastrophic manner. This observation is also confirmed by microscopy.

The technique of stitching through prepreg layers could result in a localized in-plane fiber damage. The stitched three-dimensional composites, although having improved damage resistance, could exhibit a lower flexural strength when compared with two-dimensional composites. Composites fabricated from stitched fabrics were much less prone to such fiber damage than those from prepreg materials [195].

The failure mechanisms in both two-dimensional and stitched three-dimensional composites were strongly dependent on the composite types and the loading directions. In general, a two-dimensional composite loaded in the *z*-direction tended to have matrix-dominated fracture. Interlaminar delamination was usually found to be the dominant failure mode in two-dimensional composites. For a three-dimensional composite tested in the same direction, a fiber-dominated mode usually predominated. In this case, major fracture features were fiber buckling, fiber breakage, interfacial debonding, and fiber pull-outs. When loaded in the *y*-direction, both two-dimensional and three-dimensional composites fractured in the fiber-dominated mode. The three-dimensional composites usually experienced a smaller distortion (smashed) zone but a higher degree of fiber breakage and fiber pull-outs [163].

Hybrid Composites. Many researchers [e.g., 49,50,164-166] have observed that the incorporation of fibers with high failure strains (more precisely, high toughness) in carbon-epoxy composites significantly increases the impact energies of the materials. Variables that influence the impact response of hybrid composites include the volume fraction of each component fiber, lay-up sequence and orientation, relevant properties of fibers and resin, interlaminar strength between plies, and voids or other defects.

Both glass and Kevlar fibers have been selected for the purpose of hybridization, but recent studies have found high-strength PE fibers to be quite effective in improving the impact energy-absorbing capability of carbon-epoxy composites [44-46,49,50,167]. The energy-absorbing mechanisms of hybrid laminates in response to impact loading were investigated by Jang *et al.* [50]. The composites studied were prepared from epoxy resins reinforced with different combinations of various fabrics, including PET, nylon, Kevlar, gel-spun high-strength PE, glass, and carbon. Polyethylene, PET, and nylon fibers, when combined with epoxy resin, have been shown to absorb large amounts of energy prior to failure of composites. These fibers can be used in hybrids to improve the impact resistance of various composite materials. A large impact toughness of constituent fibers is essential to achieving improved impact resistance in hybrid laminates.

The impact energies of the interlaminated hybrids generally showed a negative hybrid effect (i.e., slightly lower than what would be predicted by the rule-of-mixture law). However, the maximum loads often showed a positive synergism. The technique of interlaminate hybridization was found to increase the tendency to delaminate in response to impact loading. Due to the considerable increase in the area of fracture surfaces, delamination was an effective energy-absorbing mechanism in hybrids. However, in designing for improved impact energy, one must also consider the potential influence of hybridization on other composite properties. The impact resistance of carbon-epoxy composites has been improved without compromising the maximum stress by hybridizing with Kevlar [166] and PE fibers [50].

The impact load-time traces confirmed the speculation that indentation of the front surface represents the very first stage of loading, which controls the initial laminate stiffness during impact. Perforation induced by indentation was found to be the commanding failure mechanism in the brittle laminates studied. The laminates containing tougher fibers were capable of responding to impact in flexure without perforation. This process of plastic deformation to form a dome helped to disperse a major portion of the strain energy.

Many different macroscopic failure mechanisms in hybrid laminates can occur during impact loading. Laminates with an alternating stacking sequence usually exhibited a combination of through penetration and delamination, the latter being in the dagger shape and visible from both sides. For asymmetric hybrid laminates containing two or three layers of PE or PET fabric, the impact energy depends on which side is facing the impact direction, but in general it is higher than its alternating-sequence counterpart. In most cases, failure in these materials involved perforation, delamination, and some tearing of the more brittle layers, in conjunction with deflection of the tougher layers, provided that the tougher side faced the impact direction. When the more rigid side was struck first, these stiff layers were perforated with a lesser degree of plastic deformation. With some exceptions, this process was followed by through penetration of the tougher layers, leading to an inferior energy-absorbing capability.

The ductility index was generally found to decrease with increasing energy-absorbing capability of a laminate, implying that for tougher composites, more energy will be dissipated to reach the maximum load than afterwards. This observation again suggests the significance of fiber properties in controlling the strength, and therefore the impact resistance, of composites.

Sandwich Structures. A composite sandwich panel comprises a lightweight foam, honeycomb, or corrugated core sandwiched between two composite facesheets. Such a combination offers exceptional specific stiffness and strength, buoyancy, dimensional stability, and thermal and acoustical insulation characteristics. The main applications of these composite sandwich panels are in building components, insulated structures, marine constructions, and aerospace vehicles.

The impact response of various composite sandwich panels was investigated in relation to the constituent material properties [148]. The impact energy absorbed by a composite sandwich panel containing single-layer facesheets is many times greater than that absorbed by the foam core when alone. Such a sandwich combination offers exceptional impact resistance and yet remains lightweight.

The impact response of composite sandwich panels is mainly controlled by the facesheets and is practically insensitive to the density of the polyvinyl chloride foam core, provided that the facesheet material is sufficiently tough (e.g., contains PET or high-strength gel-spun PE fibers). The impact failure mechanisms of composite sandwich panels made of less tough facesheet material, such as graphite fibers, tend to be foam-core-dominated, provided that the PVC foam core possesses a relatively high density.

The maximum load and the total absorbed energy of single-layer composite facesheets decline as the impact velocity increases. The energy absorbed by composite sandwich panels containing a low-density polyvinyl chloride foam core is about 15% to 100% greater than the sum of the energies separately absorbed by the two facesheets and the foam core. This percentage deviation from the rule-of-mixtures prediction increases as the density of the foam core increases. However, this deviation is insignificant if tougher facesheets, such as PE fibers, are used [148].

Sheet Molding Compounds (SMC). The impact behavior of SMC is of great interest to materials designers when considering intended applications such as automotive, appliance and furniture, and construction structural elements. Several researchers have investigated the impact damage of SMC [86,110,168-171]. Khetan and Chang [86] have indicated that the impact damage surface area increases nonlinearly with the incident energy of the impactor. Microscopic examinations of the impacted surfaces showed circular crack rings, corresponding to tensile precursor wave surfaces [110]. For a given impact condition, the energy-absorption capability is expected to be lower than that of a continuous fiber-reinforced composite.

Effects of Material Variables

The Effect of Fiber Properties. The impact failure mechanisms of composites containing various fibers with different strength and ductility were studied by a combination of a static indentation test, an instrumented falling-dart impact test, and scanning electron microscopy [49]. The composites containing fibers with both high strength and high ductility (e.g., PE fibers) demonstrated better impact resistance than those containing fibers with high strength (e.g., graphite) or high ductility (e.g., nylon fibers) but not both.

Upon impact loading with a high incident energy, the composites containing PE fibers usually exhibited a great degree of plastic deformation and some level of delamination. These permitted a greater volume of material to deform and crack, thereby dissipating a significant amount of strain energy, before the penetration phase proceeded. When the incident energy was less than 100 J (75 ft · lbf) and the sample-clamping ring was greater than 50 mm (2 in.), no through penetration was observed in any of the samples containing more than three layers of PE fabric except when loaded at relatively high rates and low temperatures. Although certain levels of delamination also took place in other composite systems, very little plastic deformation occurred, allowing ready penetration of the projectile. The penetration resistance of composites appeared to be dictated by the fiber toughness; a high modulus or a high ductility alone was not sufficient. Fiber toughness must be measured in a simulated high-rate condition. The importance of fiber toughness or work-to-fracture in dictating the impact behavior of composites was also demonstrated by Broutman and Mallick [164] and by Elber [23].

With a low incident energy and low impact velocity, tougher fibers also produce more impact-resistant composites. Given the same impact energy, higher fiber capacity to absorb energy results in less fiber breakage and a higher residual tensile strength of the composite. Secondary matrix cracking as a result of the initial fiber failure will also be reduced, so that the residual compressive strength of the composite is also increased.

The Effect of Resin Properties. The low-velocity impact resistance of a resin composite is to a great extent controlled by the resin toughness. A great ability of the resin to undergo massive plastic deformation during an impact event is essential to achieving improved damage resistance of the composite. Better resistance to delamination and matrix cracking achieved with a tougher resin also leads to improved residual strength after impact. This was discussed in Chapter 7, where various techniques that are capable of toughening the matrix resin were presented. Dorey [49,172] demonstrated that, as compared to more brittle carbon-epoxy composites, carbon-PEEK composites experienced significantly less extensive delaminations and that therefore their compressive strengths after impact were much higher.

In general, a tough matrix resin would also produce composites with a higher impact energy-absorbing capability under high-velocity impact conditions, but this effect occurs to a much lesser extent when ballistic penetration resistance is the desirable property. As pointed out by Elber [23], matrix properties govern the damage threshold and determine the extent of impact damage, while fiber properties control the penetration resistance.

The Effect of Fiber-Matrix Interfacial Adhesion. In a composite containing a notch (hole or other type of defect), there is a complex stress field near the notch tip. Local stresses near or ahead of the notch tip can exceed the low transverse strengths and cause fiber-matrix splitting parallel to the fibers. The splitting can effectively blunt the notch and possibly remove the stress concentra-

tion at the notch tip. However, the splitting parallel to the fiber is a relatively low-energy fracture process [172]. For carbon-epoxy composites, the initial splitting fracture energy is about 100 J/m² (similar to that for unreinforced epoxy resins) prior to the formation of a tied zone by misaligned fibers [172]. Fracture across fibers requires orders of magnitude higher energy, 50 to 100 KJ/m². Splitting parallel to the fibers within the individual plies, along with delamination between the plies, forms a damage zone at the crack tip. This damage zone absorbs energy, and the size of this zone and the amount of the absorbed energy depend on, among other factors, interface properties of the composite.

The toughness of a brittle fiber-reinforced brittle matrix composite largely originates from the fact that cracks get diverted along the fiber-matrix interface. If the interface bond is too weak, the material will not support loads in shear or compression. Contrarily, too strong an interface bond would not permit effective crack diverting, and this would make a brittle composite. A weaker interfacial bond would in general readily lead to intraply splitting and interply delamination and allow the composite to absorb a greater amount of impact energy, provided that the splitting and delamination are the dominant failure modes. However, when fiber tensile or compressive failure modes become significant, then a weaker interfacial bond can cause a weak flexural strength of the laminate; deteriorated integrity could mean low impact resistance.

Yeung and Broutman [173] investigated the Charpy impact energies of glass-polyester and glass-epoxy composites as functions of fiber surface treatments. As shown in Fig. 16, the initiation energy E_i increases with increasing shear strength (better interfacial bond) for both polyester and epoxy laminates. As the apparent shear strength (as determined from the short beam shear test) increases, the flexural strength of these laminates also increases because of greater interlaminar and intralaminar strengths. The curves for energy propagation, E_p, and total energy, E_t, of polyester laminates appear to have a minimum. Above a critical value of apparent shear strength, fiber tensile failure appears to be the dominant failure mode, and both E_p and E_t increase with increasing laminate shear strength. Below this critical value where delamination mode dominates, the impact energy decreases with increasing shear strength. For these polyester laminates, the greatest value of impact strength is achieved when the shear strength is lowest with a weakest interface bond. Here the initiation of failure requires less energy when the bond is weak, and a great amount of energy is absorbed during the delamination stage after initial failure. The specimen, although supporting less load during this stage, can sustain large deflections to permit absorption of more energy. The interfacial bonding in the epoxy laminates studied was believed to be too high to induce extensive delamination. Therefore, the laminate strength dictated the impact energies of these epoxy composites.

It should be noted that a high energy-absorbing capability is a desirable property for a composite armor. However, when damage tolerance or structural integrity is the

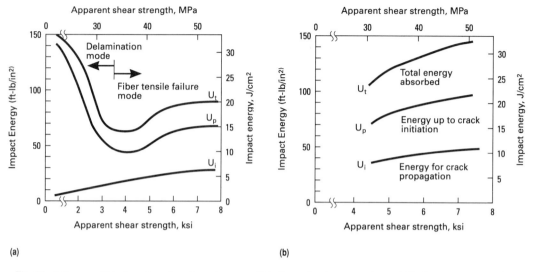

Fig. 16 Influence of interface strength on impact energy of (a) glass-polyester composites and (b) glass-epoxy composites. Source: Ref 173

primary design concern, delamination damage should be kept to a minimum and adequate interfacial bond is essential.

The Effect of Interlaminar Properties

Controlled Delamination. The concept of controlled interlaminar bonding (CIB) was proposed [174] as a means of improving the composite toughness (energy-absorbing capability, particularly under high-rate or ballistic conditions) without compromising the static strength of the composite. Failure mechanisms in several different types of composite laminates were investigated. Incorporation of an appropriate amount of perforated interlayers was found to increase the impact toughness of composite laminates without significantly decreasing the material strength. These interlayers appear to promote delamination that would aid in dissipating strain energy as well as diverting and blunting a propagating crack through the Cook/Gordon mechanism. The amount of delamination depends on the competition between the Griffith crack and the delaminating crack. The competition seems to be a function of the relative magnitudes of interlaminar strength and composite cohesive strength. The former is determined by the degree of perforation and lamina surface treatment. Failure mechanisms of composite laminates are found to be different when the loading directions of samples are varied in a three-point bending test or a Charpy impact test. The degree of delamination and the total absorbed energy are also varied correspondingly.

Interleaving. If improved damage tolerance of composites is a major concern, then delamination tendency must be reduced and therefore the above CIB approach should be avoided. Instead of deliberately weakening the interlaminar bond, tough and compatible interply layers may be introduced in the laminates. This approach of interleaving was discussed in Chapter 7. For instance,

Jang and Chen [176] reported that thin layers of epoxy resins modified with either liquid reactive rubber (CTBN) or polyurethane, when incorporated in the interlaminar zones in carbon fiber-reinforced epoxy composites, significantly improve their interlaminar shear strength, interlaminar fracture toughness, low-velocity impact damage tolerance, and impact penetration resistance.

Effects of Target Geometry and Testing Parameters

The Effect of Target Geometry. The impact response of a composite depends on the target bending stiffness, which is in turn a function of the dynamic modulus of the material and the target geometry ($D = EI$, where I is the moment of inertia). Using a rectangular beam as an example, $I = bh^3/12$ where b is the beam width and h is the thickness. It is obvious that the target is more rigid if E and/or b are larger. Being largely viscoelastic in nature, polymer composites will exhibit a higher dynamic modulus when submitted to a higher strain rate. Under high-rate conditions, an otherwise "flexible" target can behave like a rigid one.

In response to low-velocity impact loading, fiber-reinforced composites are capable of absorbing and dissipating large amounts of energy in a wide variety of elastic, plastic, and fracture processes. Up to the point of initial failure, the elastic response of the structure serves to absorb most of the projectile incident energy. This ability to absorb energy elastically depends on a large number of parameters, including fiber and resin properties, the interfacial adhesion, the projectile velocity, and the geometry of the structural component [82]. The first two parameters are discussed in the following paragraphs.

The dependence of the impact response of a composite on the projectile velocity is rather complex and not fully understood. Contwell [177] has shown that a large, slow-moving projectile tends to induce a global form of structural response, whereas light, fast-moving objects generate a much more localized form of target response. Consequently, in the former case the incident energy of the projectile tends to be absorbed by the whole structure, whereas in the latter case the energy is dissipated over a small zone immediate to the point of contact.

Under conditions of high-velocity impact loading, where the dynamic response of the target is very localized, geometrical effects are very small [139]. In a drop-weight impact test, where the contact time is greater and the target response is more significant, geometrical effects are likely to be more significant [82]. The thickness of a laminate plays an important role, for instance, in a high-velocity impact. A thick laminate does not bend much, due to its high bending stiffness ($I \propto h^3$), and therefore there is not much interlaminar failure. Fibers are fractured around the point of contact, starting in the impact ply to successive plies [82]. When subjected to an impact of a high incident energy, the laminate shows the formation of a shear plug that is ejected from the rear surface.

Contrarily, a thin laminate can deform to a great extent under the flexural wave generated by the high-velocity impact. Takeda et al. [178] studied wave propagation in the ballistically impacted panels through surface and embedded strain gauges. Initially, at a point slightly away from the point of impact, a weak tensile pulse arrives on both impact and rear surfaces. This is followed by a high magnitude flexural wave that induces the maximum delamination in the panel [150,178].

In the case of low-velocity impact, the damage evolution pattern is also rather complex. The geometrical effects in the low-velocity impact response of carbon-epoxy composites were investigated by Cantwell and Morton [82]. The variation of the critical energy to initiate first damage with target thickness for five (±45°) laminates is shown in Fig. 17. The curve indicates the existence of two distinct zones. Initially, the threshold damage energy increases with increasing target thickness and subsequently decreases. In the left-hand side of the curve, damage initiates at the lower surface of the flexible target, likely as a result of the flexural stress field in the bending target [139]. The high tensile stresses in the lower ply can cause matrix cracks, which then reach an interply boundary to form a delamination that in turn is deflected by matrix cracks in the layer above, and the process can repeat itself. The threshold incident energy increases with the target thickness, because according to Eq 18, the tolerable elastic strain energy sustained by a flexible bending beam scales with the beam thickness.

In thick laminates, damage initiates in the upper surface ply as a result of the locally high contact stress field [139]. In this case, increasing the target thickness reduces the damage threshold, presumably because of a reduced beam ability to bend (elastic strain energy contribution to the energy absorption is curtailed). At an intermediate thickness, damage could initiate simultaneously at the upper and lower surfaces of the target.

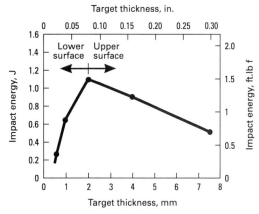

Fig. 17 Variation of the incident energy to initiate first damage with target thickness in the (±45°) laminates. Source: Ref 139

The Effect of Loading Rate. Matrix resins such as phenolic, polyester, and epoxy are known to be highly rate sensitive [179-182]. In general, both the initial modulus and maximum strength (as typically measured by the SHPB tests) increase with increasing strain rate. E-glass fibers have been found to be rate sensitive [183], but very little information is available on the rate dependence of the mechanical properties of other types of advanced reinforcement fibers. However, there are reasons to expect polymer-based fibers, such as aramid, PE, PBT, and so on to be loading rate dependent. One should in general anticipate some rate dependency of composites, although a direct correlation between the rate dependency of the composite and those of the constituent phases can be difficult or rather complicated [74].

Whereas unidirectional carbon fiber-reinforced plastic (CFRP) is relatively rate independent when loaded in the fiber direction, both woven and unidirectional glass fiber-reinforced plastic (GFRP) show a significant rate sensitivity [77,184,185]. Both modulus and strength increase as the test rates are raised. Shown as examples in Fig. 6 in Chapter 6 are the tensile impact stress-strain curves of two types of epoxy composites. The lack of a significant rate dependency of CFRP likely reflects the lack of rate dependence of the carbon fiber. It is important to note that a change in loading rate can result in a variation of failure modes. Furthermore, while at low strain rates the epoxy resin is more ductile than the composites, at high rates this may no longer be true. Under high strain rate conditions, the ductility or failure strain of a matrix resin could become a limiting factor for the composite strength.

Interactions between the fibers and the matrix are likely to be greater in the case of woven fabric composites as opposed to composites made up of unidirectional prepreg tapes. Localized strains in the matrix should increase as the fibers straighten under tensile loading or buckle under compressive loading [77]. A small rate dependence of the strength has been observed with woven-reinforced CFRP [184]. Effects associated with the fiber-matrix interface may introduce extra complications, but further studies are required in order to better understand this and other aspects of the rate sensitivity of fiber composites.

When subjected to an increasingly higher impact velocity, a laminate behaves like a more rigid beam or plate, less susceptible to bending. This shifts its behavior from that of a flexible beam (very low impact velocity), with failure preferentially initiated from the rear surface, to that of a rigid beam, with damage initiation occurring near the point of contact in the case of much higher impact velocity. At intermediate velocities, one should expect to see complex behavior of mixed fracture modes; this has yet to be verified by a systematic study. Bradshaw *et al.* [186] proposed the target span-to-depth ratio as a parameter for explaining geometrical effects under low-velocity impact conditions. Long thin structures fail in a bending mode, whereas short thick structures fail by interlaminar shear directly under the center of impact.

The Effect of Test Temperature. An extensive impact loading program was undertaken by Jang *et al.* [187] to study the fracture mechanisms in fiber-epoxy composites as a function of temperature and environment. It was found that, in general, the lower the test temperature, the higher the total impact energy absorbed. The specimens tested at a lower temperature are characterized by a greater level of microcracking and delamination. These phenomena are believed to be promoted by the higher residual thermal stresses. Slight exposure of composites to moisture or liquid nitrogen environment did not affect the impact response. A simple thermomechanical analysis was presented to estimate the residual thermal stresses on cooling from cure or crystallization temperature to an end-use temperature (23 and −196 °C, or 73 and −321 °F). Experimental efforts to measure the mechanical bond between a fiber and matrix appeared to yield results consistent with the prediction that the fabrication stresses would be higher with a larger temperature differential between the processing and the end-use temperature.

Theoretical Modeling and Predictions

Many investigators have developed various models to simulate the impact response of composite laminates: simplified analyses to highlight the main features of impact; indentation laws to facilitate the calculation of local stresses and the prediction of deformation in a target; prediction of damage initiation and propagation; and estimation of residual properties after impact. This subject has been recently reviewed by Abrate [188]. Presented below are a few select models that are useful in helping us understand the more important aspects of the impact response of composites.

Simplified Analyses

In spring-mass models, the structure is represented by an assemblage of springs and masses so that a discrete system with a few degrees of freedom is analyzed instead of a continuum. A single-degree-of-freedom system was

used by Caprino *et al.* [157], Rotem [189], and Chou and Flis [190], who also used a two-degree-of-freedom model to predict the impact response of beams. Two- or multiple-degree-of-freedom models were proposed by several other researchers [e.g., 191-193].

Rotem [189] considered the case of a simply supported composite beam that was loaded by an impactor of mass m and striking velocity V. The dynamic equation of motion is given by [189]:

$$-KX + mg = m\left(\frac{d^2X}{dt_2}\right) \qquad (Eq\ 10)$$

where g is gravitation and k is the spring constant of the beam:

$$k = 48\,E_yI/L^3 \qquad (Eq\ 11)$$

Here E_y is the apparent flexural modulus of the laminate in the beam (y) direction, I is the moment of inertia, and L is the span of the beam. The maximum deflection at midspan can be derived as:

$$X_{\max} = \frac{mg}{k}[1 + (1 + V^2k/mg^2)^{1/2}] \qquad (Eq\ 12)$$

while the maximum restoring force acting on the mass and the beam is given by:

$$F_{\max} = kX_{\max} \qquad (Eq\ 13)$$

The maximum strain ε_y at the outer layer is:

$$\varepsilon_y = \varepsilon_{y0} + X_{\max}\,kLX/4IE_y \qquad (Eq\ 14)$$

where x is the distance from the midplane and ε_{y0} is the strain of the midplane ($\varepsilon_{y0} = 0$ for a symmetric laminate). The stress at the outer layer is given by:

$$\sigma_y = \varepsilon_y(v_{yy} - V_y \cdot Q_{xy}) \qquad (Eq\ 15)$$

where Q_{yy} and Q_{xy} are the moduli of the outer lamina and V_y is the apparent Poisson's ratio of the laminate [189]. Equation 15 estimates the stress in the outer layer: tensile in the rear surface and compressive in the front surface facing the impactor. This stress can cause visible fracture of the outer layer. However, if an inner layer close to the outer one is weaker than the outer layer (e.g., a 90° ply), this layer may fail, causing an invisible damage. Such a damage can be localized or spread in the lamina [189]. Impact-induced bending of the beam would also produce interlaminar stresses to promote delamination. This type of damage is not addressed in the spring-mass models.

The spring-mass models are used to determine the entire history of the contact force and the displacements. When only the maximum contact force and the contact duration are needed, these data can be obtained more efficiently with energy-balance models [193-195]. In other simplified analyses, investigators either assumed a certain variation of the contact force and the form of its distribution over the contact area, or they considered only initial conditions. The deflection or deformation of the laminate (a beam, plate, shell) and possibly the internal stress distribution were then determined by using analytical or numerical methods [18,24,147,152,196-198]. Stress waves generated by transverse impact of laminated plates were analyzed using a modified version of Mindlin's plate theory, usually in conjunction with Laplace transform, Fourier transform methods, and a fast Fourier transform algorithm [199-203].

Indentation Laws

In addition to the stresses due to the bending, the stresses due to the contact between the impactor and the target can produce damage to the composite. This contact phenomenon is rather complex; the contact force and the pressure distribution over the contact zone are not known *a priori* and must be determined as part of the solution. In the case of flexible targets, the overall deflection of the structure as well as the local stresses near the contact region must be considered. Indentation of orthotropic half-space by a spherical or a flat-ended impactor has been studied by many workers [e.g., 204-208]. In a typical approach, for instance by Greszczuk [17,203], the analysis procedure consists of three steps: (1) determination of impactor-induced surface pressure and its distribution; (2) determination of internal stresses in the composite target caused by the surface pressure; and (3) determination of failure modes in the target caused by the internal stresses. The target and the impactor are assumed to be linear elastic and the impact duration is assumed to be long compared to stress wave transit times in the impactor. To obtain the magnitude and distribution of surface pressure in the target caused by impact, the dynamic solution to the problem of impact of solids is combined analytically with the static solution for the pressure between two bodies in contact.

Knowing the surface pressure, its distribution, and the area of contact, all as a function of impact velocity and time, one can proceed to determine the time-dependent internal triaxial stresses in various composite targets using finite-element codes, or in some instances closed form solutions [203]. Figure 18 compares the internal triaxial stress contours for σ_r, σ_z, and σ_θ, where the stresses have been normalized with respect to maximum surface pressure q_0 (at $r = 0$) while the dimensions have been normalized with respect to the radius of a, the area of contact. The maximum tensile, compressive, and shear stresses can be determined from these calculations, and their locations are noted in Fig. 19. These results suggest that, for brittle materials, the failure due to impact would be governed by tensile strength and would initiate at the periphery of the area of contact. For materials with low shear strength, subsurface shear failure will occur during impact loading.

The studies by Greszczuk [17,203] on a number of transversely isotropic materials indicated that:

- Resistance to damage increases as the fiber strength increases and the fiber modulus decreases (presumably as the fiber failure strain and work-to-fracture increase).
- Resistance to damage increases as the Young's modulus of the matrix decreases and the strength increases.
- Bidirectional lay-up is more efficient in resisting damage than tridirectional or unidirectional lay-up.
- In thick targets, impact causes local subsurface damage, whereas in thin flexible targets, the damage initiates on the backface.

Indentation of more flexible beams and plates requires consideration of local deformations analogous to those observed during the indentation of a half-space, as well as consideration of the overall deflection of the beam or plate. The state of stress in the beam (or the plate) results from the combination of the overall deformation of the beam (or plate) and local deformations in the contact area. Various experimental and numerical approaches have been taken to investigate the indentation laws during impact of a composite structure by a foreign object [17,85,203-224].

Low-Velocity Impact

Many of the studies on the indentation laws discussed earlier represent part of the efforts to understand the impact response of composites. Frequently, the results of indentation studies have been incorporated in global treatments of impact-induced deflection, deformations, internal stresses, and internal damage in composites [e.g., 203,213,215]. Additional impact response studies include impact on beams [213,225-227], impact on plates [219-224,226-232] and impact on shells [233,234]. This subject has been recently reviewed by Abrat [188] and will not be repeated here. Presented below is a simple strain energy-based approach to the prediction of the threshold incident energy required to create damage in a flexible beam. The results of this simple analysis will help to bring out some of the more salient features of the impact response of a beam.

Elastic Strain Energy of a Bending Beam. In order to explain the experimental observation that there exists a critical incident energy, E_c, above which damage will occur to a composite subjected to a low-energy impact, the concept of elastic strain energy in bending will be discussed. Assume that the composite beam can be treated as a quasi-isotropic body that is receiving an incident impact energy of E_{in}. If the E_{in} is small (and velocity is low), the laminate can respond by bending and no damage will occur, provided that E_{in} can be accommodated by the elastic strain energy in the laminate [235]. One can further assume that a critical condition is reached when a local stress exceeds a local strength. This can be further divided into two cases, according to the modes of failure. Consider the case where the beam is submitted to

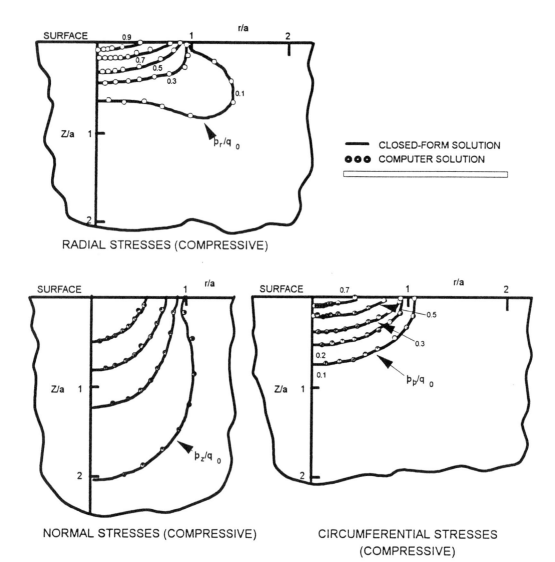

Fig. 18 Comparison of closed form and computer solutions for internal triaxial stresses in a solid subjected to surface pressure caused by impact. Source: Ref 203

a three-point bending load P. The work (elastic strain energy) for this beam is given by [236]:

$$W = \int_{\text{length}} \frac{M^2 dx}{2EI} \quad \text{(Eq 16)}$$

The bending moment in terms of the loads on the beam are

$$M_x = (P/2)x \qquad 0 \le x \le L/2$$
$$M_x = (P/2)x - P(x - L/2) \quad L/2 \le x \le L \quad \text{(Eq 17)}$$

Because E and I are constant along the length of the beam, we have

$$W = 1/(2EI) \int_0^{L/2} (Px/2)^2 dx + 1/(2EI) \int_0^{L/2} (Px/2$$

$$+ PL/2)^2 dx = \frac{P^2 L^3}{96EI} \quad \text{(Eq 18)}$$

The maximum normal stress in this case is given by

$$\sigma_{\max} = \frac{M z_{\max}}{I} = \frac{3PL}{2bh^2} \quad \text{(Eq 19)}$$

where the moment of inertia for a rectangular bar is $I = bh^3/12$. The maximum shear stress, existing at the centroidal axis, is given by

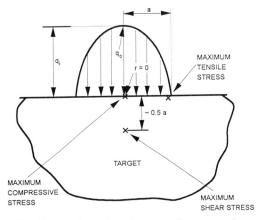

Fig. 19 Locations of maximum tensile, compressive, and shear stresses in an elastic half-space under surface loading. Source: Ref 203

$$\tau_{max} = (6V/bh^3)[h^2/4 - z^2]$$

With $V = dMx/dx = P/2$ and $z = 0$ this is reduced to

$$\tau_{max} = 3P / (4bh) \qquad \text{(Eq 20)}$$

Combining Eq 20 and 21, we have

$$\tau_{max}/\sigma_{max} = h/2L) \qquad \text{(Eq 21)}$$

Equation 21 suggests that the longer the beam, the greater is σ_{max} with respect to τ_{max}. Therefore, failure is expected to initiate near the top or bottom at the midspan, where maximum normal stress exists. This type of failure mode is referred to as "flexural failure," which occurs when σ_{max} reaches the flexural strength (σ_{fl}) of the beam:

$$\sigma_{max} = 3PL/(2bh^2) = \sigma_{fl} \qquad \text{(Eq 22)}$$

By substituting this equation back to Eq 18, one obtains

$$W_{fl} = P^2L^3/(96EI) = (Lbh/18)(\sigma_{fl}^2/E) \qquad \text{(Eq 23)}$$

When $E_{in.} \geq W_{fl}$, flexural damage will occur to the composite beam.

Contrarily, with a very short beam where $(h/2L)$ is large (e.g., greater than 1/10), τ_{max} will predominate. Shear failures will be more likely and will initiate when

$$\tau_{max} = 3P/4bh) = \tau_{ILSS} \qquad \text{(Eq 24)}$$

where τ_{ILSS} is the interlaminar shear strength of a composite. Substituting Eq 24 in Eq 18, we get

$$W_{ILS} = 2bL^3/9h)(\tau_{ILSS}^2) \qquad \text{(Eq 25)}$$

When the incident energy to a short beam exceeds this elastic strain energy (i.e., $E_{in} \geq W_{ILS}$), an interlaminar

shear failure will prevail. Equations similar to Eq 23 and 25 were presented and briefly discussed by Dorey [236]; no derivations were given. Slightly more complex equations have been derived by Jang [62] for orthotropic beams, primarily using expressions presented in Ref. 265. Corrections for the incident energy values may be needed if the relative movement between the impactor and the beam is taken into account [105].

Prediction of Impact Damage in Composites

Impact damage generally consists of delaminations, intraply matrix cracking, interfacial debonding, and fiber breakage, and these failure mechanisms can interact with each other. Furthermore, a large number of loading variables, target geometry factors, and environmental factors can influence the impact response of an orthotropic material. These make prediction of impact damage initiation and propagation in composites very difficult. Nevertheless, many investigators have reported various approaches to the prediction of impact damage in composites. Typically, the approach involves the determination of impact force and the induced stresses and strains, then the selection of a proper failure criterion to judge initial failure in composites.

Although beam and plate models (e.g., those cited in the last section) have proven effective in determining the overall target response and the contact force history, they are in general considered inadequate for analyzing the internal stresses caused by impact. More detailed modeling is required for the purpose of damage prediction in composites.

As pointed out by Greszczuk [17,203], in the case of semi-infinite isotropic solids subjected to a prescribed surface pressure, the internal stresses can be determined from various classical strength-of-material equations [e.g., 238,239]. However, the internal stresses in semi-infinite or flexible composite targets subjected to an impact can be best determined using finite-element methods [203].

Joshi and Sun [78] first used a two-dimensional finite-element model to analyze the impact response of three-layer crossply laminated beams. The contact force was represented by a half sine wave variation in time and was distributed over the contact area according to the classical Herzian theory. Near the impact point, flexure strains were shown not to vary linearly in the thickness direction of each ply, and transverse shear stresses were shown not to follow a parabolic distribution through the thickness of the laminate, as would be expected using lamination theory. Instead, a sharp shear stress concentration was observed close to the impacted surface, and a sharp increase in flexural stress was observed at the impacted surface. Later, Joshi and Sun [122] followed up with a study on quasi-isotropic laminates by conducting two-dimensional analyses for normal cross sections oriented in each one of the fiber orientations. This approach allowed them to interpret the major types of impact-induced damage.

Goning [240] used a special two-dimensional finite-element model to determine through-the-thickness transverse shear stress distributions. This model consisted of two translations and one rotation per node to account for several different layers of material inside each element. Results indicated that shear stress distributions followed the same trends as would be expected by the elasticity solution [241] to a static three-point bending problem.

Once detailed knowledge of transverse normal and shear stresses at each interface becomes available, one can then proceed to analyze delamination possibilities by considering it as a mixed-mode fracture process and by using a proper failure criterion. Various mixed-mode-fracture criteria were presented in Chapter 7, where some of the operationally easier-to-use criteria were identified. However, in view of the usually great variability in G_{IIc} values as measured by different methods, Jones *et al.* [242] suggested that perhaps simplified approaches, rather than the more complex polynomial fracture mechanics equations, should be used for damage prediction in composites.

Delaminations in a given plate were assumed to take place when the transverse shear force resultant exceeded a threshold value [228-230]. Liu [79] considered that delaminations were induced by bending stiffness mismatch between layers. A mismatch coefficient M was defined, and the shape of this M curve was found to correspond to that of observed delaminations. The orientation corresponding to the maximum value of M was the direction in which delamination developed. Wu and Springer [85] used eight-node brick elements to determine all six components of the stress tensors throughout the entire plate. They postulated that delamination was caused primarily by tensile stresses normal to the plate, and they selected the maximum value of interlaminar normal stress reached under the impact point as the critical parameter to consider. In-plane stiffness mismatch between two neighboring plies, similar to the M defined by Liu [79], was used as the basis for delamination prediction. Useful equations were developed to determine the size of delaminations between any two plies in the laminate. Sun and Grady [152] used a two-dimensional finite element model to analyze propagation of an existing delamination crack during impact of a composite beam by a rubber ball. The results indicated that the mode I contribution to the strain energy release rate was insignificant and that delamination propagation was dictated by G_{IIc}. Other studies on the initiation and propagation of delamination were reviewed in Chapter 5. Prediction of internal ply damage caused by impact has been studied by several investigators [243-247].

Prediction of Residual Properties after Impact

Broutman *et al.* [164] studied the residual tensile strength of laminates after impact as a function of the incident impact energy. Three distinct regions exist in a curve of residual strength plotted over a range of incident energy values (Fig. 20). Region I represents the condition where no appreciable damage occurs as a result of one impact with a subcritical incident energy ($E_{in} < E_c$). When the incident energy exceeds the threshold value ($E_{in} > E_c$), damage occurs and strength is rapidly reduced

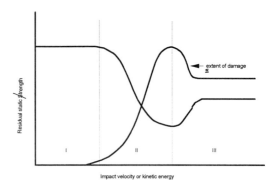

Fig. 20 Schematic of residual static strength of laminates after impact damage. Source: Ref 246

until a maximum damage size is reached. This region II is characterized by the development of numerous delaminations, intraply cracking, and possibly partial perforation. The extent of damage sustained by the laminate increases as the incident energy increases. For much higher incident energies (higher velocities), complete perforation occurs, typically leaving behind a "cleaner" hole with a diameter practically independent of impact velocity. Delamination damage in region III occurs to a much lesser extent than in region II. These factors are probably responsible for the observation that the residual strength is nearly constant in region III. The range of incident energies associated with each region is controlled by a number of parameters, including target thickness, size and shape of impactor, and the material properties.

When delamination is the dominant mode of failure in an impact-damaged composite, the residual strength and stiffness can be estimated by the methods discussed in the section "Effect of Delamination" in Chapter 7. When a clean perforation is incurred in a laminate, it is generally assumed that this perforation introduces the same strength and stiffness reduction as a hole or a crack of the same size [e.g., 248]. Then the various approaches that were developed to predict the notched strength of composites, reviewed in Chapter 7, can be used to predict the residual properties of composites.

The compression-after-impact test has been a popular technique in the aerospace industry to characterize the damage tolerance of a laminate. When loaded in compression, impact-damaged composites tend to exhibit severe strength reductions, mainly as a result of local instabilities. The influence of delamination damage on compressive strength has been a subject of many recent research efforts. Experimental and theoretical aspects of this subject have been recently reviewed by Baker *et al.* [249], Jones and Baker [250], and Abrate [188].

Energy-Absorbing Mechanisms in Composites

As alluded to earlier, the impact energy imposed upon a composite target can be absorbed by the material

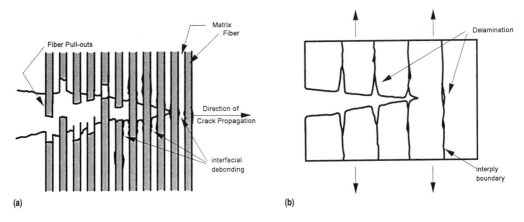

Fig. 21   Schematic of multiple failure mechanisms in (a) a unidirectional and (b) a multilayered composite

by two basic mechanisms: elastic and plastic deformation of the material and creation of new surfaces (through fibers, matrix, fiber-matrix interface, and interlaminar boundary zones). The material deforms first to absorb a portion of the incident energy E_i. If E_i is sufficiently high, a crack may initiate and propagate. The material deformation process continues ahead of the advancing crack. The energy-absorbing capability or toughness of a material can be enhanced by increasing either the total area of new surfaces created or the material deformation capability. For instance, several techniques suggested in Chapter 7 can be used to toughen the epoxy resin by promoting plastic deformation mechanisms such as shear yielding. These include reducing the cross-link density, adding a rubbery or thermoplastic phase, and forming an inter-penetrating network. Glass fibers (e.g., S-2) or PE fibers (e.g., Spectra®) may be introduced into carbon fiber composites to improve the composite toughness. However, the failure process of a fiber composite would normally involve a combination of several energy-absorbing mechanisms, which must be understood if one is to design composites with high toughness.

Presented in Fig. 21 are two simplified models that can be used to describe the fracture processes during the propagation of a crack in a unidirectional composite or a 0° layer in a laminate (Fig. 21a) and a general laminate (Fig. 21b), respectively. One may assume that failure in a fiber composite initiates from small defects such as broken fibers, matrix pores, and debonded interfaces. During the propagation of a crack, large transverse tensile stress and interfacial shear stress can cause fiber-matrix interfacial debonding (Fig. 21a) and interlaminar cracking or delamination (Fig. 21b) slightly ahead of the crack tip. In the high-stress region near the tip, some fibers are broken, although the breakage points are not necessarily on the crack plane. Matrix deformation and microcracking also occur near the crack tip region, forming a matrix plastic deformation zone. Immediately behind the crack tip the broken fibers can pull out of the matrix; fiber pull-out is an effective energy-absorbing mechanism.

The mechanisms of "fiber pull-out" [251,252], "debonding" [28], and "stress redistribution" [253,254] were proposed in the 1960s and early 1970s to explain the origins of toughness in brittle fiber/brittle matrix composites. Each separate mechanism was found to work reasonably well for certain fiber-matrix combinations. But none of them was adequate for boron fiber-epoxy composites, and this prompted Marston et al. [255] to develop a more generalized model. This analysis incorporates the energy absorbed in creating all new surfaces during the fracture of a composite material, in addition to the possible contributions from the mechanisms of fiber pull-out and stress redistribution. There are three types of newly formed surfaces: through the fiber cross section, matrix cross section (each with possible deviations from a planar fracture surface), and the surface created between fiber and matrix as a result of interfacial fractures. According to this theory, the total specific work of fracture (energy dissipated per unit area) is given by:

$$R_{total} = R_{surf} + R_{pull-out} + R_{redist} \qquad (Eq\ 26)$$

where R_{surf} = the specific work of fracture due to the new surfaces created, $R_{pull-out}$ = that due to fiber pull-out, and R_{redist} = that due to the mechanism of stress redistribution.

The first term, R_{surf}, consists of three components:

$$R_{surf} = V_f R_f + (1 - V_f)R_m + \frac{V_f l_c R_{if}}{d}$$
$$= V_f R_f + (1 - V_f)R_m + \frac{V_f \sigma_f R_{if}}{2\tau} \qquad (Eq\ 27)$$

where V_f = the volume fraction of fibers, d = fiber diameter, l_c = the ineffective length of fibers, σ_f = the fracture strength of fibers, τ = the interfacial shear strength, R_f = the specific work of fracture of fibers, R_m = the specific work of fracture of matrix, and R_{if} = the specific work of interface fracture or "interfacial fracture toughness."

Fiber Breakage. Brittle fibers such as carbon, especially high-modulus grades, have a low-fracture strain and hence possess a low energy-absorbing capability. For composites in tension, the energy required per unit area of the composite for fracture of fibers is

$$R_f = \frac{V_f \sigma_f^2 l}{6 E_f} \qquad \text{(Eq 28)}$$

where E_f is the fiber modulus and l is the fiber length. Fiber length l should be replaced by fiber critical length l_c in an impact loading case [256]. The magnitude of R_f for brittle fibers is usually negligible. However, the presence of fibers significantly influences the failure modes and thus the total impact energy absorbed by a composite.

Matrix Deformation and Cracking. Brittle thermoset resins, such as unmodified epoxy and polyester, can undergo only a limited extent of deformation prior to failure. However, the new generation of toughened thermoset resins and the advanced thermoplastic matrices may undergo extensive plastic deformation, which makes a much more significant contribution to the energy-absorbing capability than does cracking (creation of new surfaces).

According to Kelly [251], the work done associated with the material deformation is proportional to the work done in deforming the matrix to rupture per unit volume, U_m, times the volume of the matrix deformed per unit area of the crack surface. Following the equation derived by Cooper and Kelly [252] for the volume of matrix affected by fracture, the energy required for the matrix to deform to fracture per unit area of composite is given by Ref. 257:

$$R_m = \frac{(1 - V_f)^2}{V_f} \frac{\sigma_{mu} d}{4\tau} U_m \qquad \text{(Eq 29)}$$

where σ_{mu} is the tensile strength of the matrix. In the advanced thermoplastic composites, the contribution indicated by Eq 29 to the total absorbed impact energy may be quite significant. It should be noted that multiple matrix cracking, creating a considerable amount of new surfaces, may represent a significant source of toughness.

Interfacial Debonding. New surfaces can also be created when the fibers become separated from the matrix by cracks running parallel to the fibers. Propagation of these debonding cracks involves breaking of the primary and secondary bonds between the fibers and the resin. A debonding crack can run at the fiber-matrix interface or in the neighboring resin, depending on the respective strengths. In the absence of direct information for R_{if}, the toughness of the matrix, R_m, can be substituted. In assuming $R_{if} \approx R_m$, we realize that this provides only an upper bound for the work to generate debonded surfaces. Typically, one would expect $R_{if} < R_m$, especially in the thermoplastic composites where the interface is relatively weak and the debonding cracks should propagate at the interface.

Kelly [258] presented R_{if} values for a number of materials, and the values are typically of the order of interface shear strength (τ) times the failure strain of the resin (ε_{mu}). These values are typically ≤ 500 J/m^2. As pointed out by Kelly, the debonding energy R_{if} cannot be assumed to equal the elastic energy stored in the fibers after debonding, as originally suggested by Outwater and Murphy [259]. A deliberately weakened interface strength can promote extensive debonding, leading to a significant increase in the impact energy.

Fiber Pullout. The specific work of fracture due to fiber pull-out can be shown to assume the following expressions [251,256]:

$$
\begin{aligned}
R_{\text{pull-out}} &= \frac{V_f \tau l^2}{6d} \quad \text{when } l < l_c \\
&= \frac{V_f \tau l_c^3}{6dl} + V_f \tau (l - l_c)^3 \quad \text{when } l_c < l < 2l_c \\
&= \frac{V_f \tau l_c^3}{6dl} + V_f \tau (l - l_c) l_c^2 \quad \text{when } l > 2\,l_c \\
&= \frac{V_f \tau l_c^2}{6d} = \frac{V_f \sigma_f l_c}{12} \quad \text{when } l \to \infty \text{ (continuous)}
\end{aligned}
$$

The fiber pull-out phenomenon appears when brittle or discontinuous fibers are embedded in a tough matrix. The statistical nature of fiber strength suggests that the fibers can fracture early at their weak cross sections. These fiber breakage points do not necessarily lie in the plane of crack propagation. As the fracture proceeds, the broken fibers will likely be pulled out of the matrix, rather than allowed to fracture again at the crack plane.

Stress Redistribution. Another energy-dissipating mechanism is the redistribution of strain energy from the fiber to the matrix after a fiber breaks. Before fracture, matrix material adjacent to the impending fracture is assumed to be unstressed [253,254]. When rupture occurs, the severed ends must pick up load from the matrix, so that the matrix itself picks up load too. The strain energy lost in the fiber is considered to be the work of fracture and

$$R_{\text{redist}} = \frac{V_f \sigma_f^2 l_c}{3 E_f} = \frac{V_f \sigma_f^3 d}{6 E_f \tau} \qquad \text{(Eq 30)}$$

Concluding Remarks

Impact response of fiber-reinforced composites is a complex phenomenon; however, this is an important subject for practical applications. Many experimental methods have been developed to observe the evolution of impact damage or to simply assess the damage of composites after impact. Theoretical modeling approaches to analyzing the impact force, internal stresses, and target deformations have greatly improved our understanding of the impact behavior of composites, our ability to predict impact damage initiation and propagation, and our ability to estimate the residual properties of composites after impact. The material parameters and loading variables that may influence the impact properties of composites have been identified, along with the various impact energy-absorbing mechanisms in composites. Several approaches to improving the impact resistance in composites, under both low- and high-velocity impact conditions, have been discussed. Composite engineers now have a variety of effective tools to help in the design for improved impact performance of composites. The next decade will witness even more active research efforts toward a better understanding of this critical topic in composites research.

REFERENCES

1. *Foreign Object Impact Damage to Composites*, STP 568, ASTM, 1975
2. *Composite Materials: Testing and Design*, STP 497, ASTM, 1975
3. *Instrumented Impact Testing of Plastics and Composite Materials*, STP 936, ASTM, 1987
4. M.R. Kamal, Q. Samak, J. Provan, and V. Ahmad, Evaluation of a Variable-Speed Impact Tester for Analysis of Impact Behavior of Plastics and Composites, *Instrumented Impact Testing of Plastics and Composite Materials*, STP 936, ASTM, 1987, p 58-80
5. M.W. Wardle and G.E. Zahr, Instrumented Impact Testing of Aramid-Reinforced Composite, *Instrumented Impact Testing of Plastics and Composite Materials*, STP 936, ASTM, 1987, p 219-235
6. J.G. Avery and T.R. Porter, Comparison of Ballistic Impact Response of Metals and Composites for Military Aircraft Applications, *Foreign Object Impact Damage to Composites*, STP 568, ASTM, 1975, p 3-29
7. D.W. Oplinger and J.M. Slepetz, Impact Damage Tolerance of Graphite/Epoxy Sandwich Panels, *Foreign Object Impact Damage to Composites*, STP 568, ASTM, 1975, p 30-48
8. J.L. Preston, Jr. and T.S. Cook, Impact Response of Graphite-Epoxy Flat Laminates using Projectiles That Simulate Aircraft Engine Encounters, *Foreign Object Impact Damage to Composites*, STP 568, ASTM, 1975, p 49-71
9. J.A. Suarez and J.B. Whiteside, Comparison of Residual Strength of Composites and Metal Structures after Ballistic Damage, *Foreign Object Impact Damage to Composites*, STP 568, ASTM, 1975, p 72-91
10. G.E. Husman, J.M. Whitney, and J.C. Halpin, Residual Strength Characterization of Laminates Subjected to Impact, *Foreign Object Impact Damage to Composites*, STP 568, ASTM, 1975, p 92-113
11. L.J. Broutman and A. Rotem, Impact Strength and Toughness of Fiber Composite Materials, *Foreign Object Impact Damage to Composites*, STP 568, ASTM, 1975, p 114-133
12. R.C. Novak and M.A. DeCrescente, *Composite Materials: Testing and Design*, STP 497, ASTM, 1975, p 311-323
13. C.C. Chamis, M.P. Hanson, and T.T. Serafini, *Composite Materials: Testing and Design*, STP 497, ASTM, 1975, p 324-349
14. P.W.R. Beaumont, P.G. Riewald, and C. Zweben, Methods for Improving the Impact Resistance of Composite Materials, *Foreign Object Impact Damage to Composites*, STP 568, ASTM, 1975, p 134-158
15. N. Cristescu, L.E. Malvern, and R.L. Sierakowski, Failure Mechanisms in Composite Plates Impacted by Blunt-Ended Penetrators, *Foreign Object Impact Damage to Composites*, STP 568, ASTM, 1975, p 159-172
16. R.W. Mortimer, P.C. Chou, and J. Carleon, Behavior of Laminated Composite Plates Subjected to Impact, *Foreign Object Impact Damage to Composites*, STP 568, ASTM, 1975, p 173-182
17. L.B. Greszczuk, Response of Isotropic and Composite Materials to Particle Impact, *Foreign Object Impact Damage to Composites*, STP 568, ASTM, 1975, p 183-211
18. C.T. Sun and R.L. Sierakowski, Studies on the Impact Structural Damage of Composite Blades, *Foreign Object Impact Damage to Composites*, STP 568, ASTM, 1975, p 212-227
19. J.T. Kubo and R.B. Nelson, Analysis of Impact Stresses in Composite Plates, *Foreign Object Impact Damage to Composites*, STP 568, ASTM, 1975, p 228-244
20. T.M. Cordell, and P.O. Sjoblom, Low Velocity Impact Testing of Composites, *Proc. 1st Tech. Conf.* 7-9 Cot 1986 (Dayton, OH), American Society for Composites, p 297-312
21. J.C. Goering, Impact Response of Composite Laminates, *Proc. 1st Tech. Conf.* 7-9 Oct 1986 (Dayton, OH), American Society for Composites, p 326-345
22. H.T. Wu and G.S. Springer, Impact Damage of Composites, *Proc. 1st Tech. Conf.* 7-9 Oct 1986 (Dayton, OH), American Society for Composites, p 346-351
23. W. Elber, The Effect of Matrix and Fiber Properties on Impact Resistance, *Tough Composite Materials*, Conf. Pub. 2334, National Aeronautics and Space Administration, 1984, p 99-121
24. E.J. McQuillen and L.W. Gause, Low Velocity Transverse Normal Impact of Graphite Epoxy Composite Laminates, *J. Compos. Mater.*, Vol 10, 1976, p 79-81
25. N. Takeda, R.L. Sierakowski, and L.E. Malvern, Wave Propagation Experiments on Ballistically Impacted Composite Laminates, *J. Compos. Mater.*, Vol 15, 1981, p 157-174
26. M.W. Wardle, Impact Damage Tolerance of Composites Reinforced with Kevlar Aramid Fibers, *Progress in Science and Engineering of Composites*, T. Hayashi, Ed., Proc. ICCM-IV (Tokyo), International Committee on Composite Materials, 1982
27. G. Dorey, Relationships between the Impact Resistance and Fracture Toughness in Advanced Composite Materials, Conf. Proc., No. 288, AGARD, Aug 1980
28. G. Dorey, G.R. Sidney, and J. Hutchings, Impact Properties of Carbon Fibre/Kevlar 49 Fibre Hybrid Composites, *Composites*, Vol 9, 1978, p 25-32
29. J.G. Avery and T.R. Porter, Structural Integrity Requirements for Projectile Impact Damage—An Overview," Conf. Proc., No. 186, AGARD, Jan 1976
30. G. Dorey, Impact and Crashworthiness of Composite Structures, *Structural Impact and Crashworthiness*, Vol 1, G.A.O. Davies, Ed., Elsevier Applied Science, 1984, p 155
31. M.W. Wardle and E.W. Tokarsky, Drop Weight Impact Testing of Laminates Reinforced with Kevlar Aramid Fibers, E-Glass, and Graphite, *Compos. Technol. Rev.*, Vol 5 (No. 1), 1983, p 4-10
32. T.M. Tan and C.T. Sun, "Wave Propagation in Graphite Epoxy Laminates due to Impact," Report to NASA Lewis Research Center, Purdue University, Lafayette, IN, 1984

33. P.T. Curtis, "An Initial Evaluation of a High-Strain Carbon Fiber Reinforced Epoxy," TR 84004, Royal Aircraft Establishment, 1984

34. J.G. Williams and M.D. Rhodes, Effects of Resin on Impact Damage Tolerance of Graphite/Epoxy Laminates, *Composite Materials: Testing and Design*, STP 787, I.M. Daniel, Ed., ASTM, 1982, p 450

35. S.J. Bless, D.R. Hartman, and S.J. Hanchak, "Ballistic Performance of Thick S-2 Glass Composites," Symp. Composite Materials in Armament Applications, 20-22 Aug 1985, UDR-TR-85-88A, 1985

36. L.T. Drzal, M.J. Rich, and P.F. Lloyd, Adhesion of Graphite Fibers to Epoxy Matrices: I. The Role of Fiber Surface Treatment, *J. Adhes.*, Vol 16, 1982, p 1-30

37. P.S. Theocaris, The Mesophase and Its Influence on the Mechanical Behavior of Composites, *Character of Polymers in the Solid State (I)*, Adv. in Polymer Science Series, No. 66, H.H. Kausch and H.G. Zachmann, Ed., Springer-Verlag, 1985, p 150-187

38. R.W. Walter, R.W. Johnson, R.R. June, and J.E. McCarthy, Designing for Integrity in Long-Life Composite Aircraft Structures, STP 636, ASTM, 1977, p 228-247

39. C.S. Hong, Supperssion of Interlaminar Stresses of Thick Composite Laminates using Sublaminate Approach, *Advanced Mater. Technol. 87*, SAMPE Symp., Vol 32, Society for Advancement of Materials and Process Engineering, 1987, p 558-565

40. B.Z. Jang and W.C. Chung, Structure-Property Relationships in Three Dimensionally Reinforced Fibrous Composites, *Advanced Composites: The Latest Developments*, American Society for Metals, 1986, p 183-192

41. F.K. Ko, H.Chu, and E. Ying, Damage Tolerance of 3-D Braided Intermingled Carbon/PEEK Composites, *Advanced Composites: The Lateset Developments*, ASM International, 1986, p 75-88

42. R.B. Krieger, Jr., An Adhesive Interleaf to Reduce Stress Concentrations between Plys of Structural Composites, *Advanced Mater. Technol. 87*, SAMPE Symp., Vol 32, Society for Advancement of Materials and Process Engineering, 1987, p 279-286

43. J. Bradshaw, G. Dorey, and R. Sidey, "Impact Response of Carbon Fiber Reinforced Plastics," Tech Report 12240, Royal Aircraft Establishment, 1973

44. R.S. Zimmerman and D.F. Adams, "Impact Performance of Various Fiber Reinforced Composites as a Function of Temperature," Allied Fibers Technical Center, Petersburg, VA, 1987

45. D.S. Cordova and A. Bhatnagar, "High Performance Hybrid Reinforced Fiber Composites—Optimizing Properties with PE Fibers," Allied Fibers Technical Center, Petersburg, VA, 1987. Also in 32nd SAMPE Conf.,. 4-9 April 1987 (Anaheim, CA)

46. D.F. Adams, R.S. Zimmerman, and H.W. Chang, "Properties of a Polymer-Matrix Composite Incorporating Allied Spectra 900 PE Fibers," Proc. Nat. SAMPE Symp. 19 March 1985 (Anaheim, CA), Society for Advancement of Materials and Process Engineering

47. W.J. Cantwell, P.T. Curtis, and J. Morton, An Assessment of the Impact Performance of CFRP Reinforced with High-Strain Carbon Fibers, *Compos. Sci. Technol.*, Vol 25, 1986, p 133-148

48. W.J. Cantwell and J. Morton, "Ballistic Perforation of CFRP," Proc. Conf. Impact of Polymeric Materials (Guildford, Surrey), Sept 1985

49. B.Z. Jang, L.C. Chen, R.H. Zee, The Response of Fibrous Composites to Impact Loading, *Polym. Compos.*, Vol 11, June 1990, p 144-157

50. B.Z. Jang, L.C. Chen, C.Z. Wang, H.Y. Lin, and R.H. Zee, Impact Resistance and Energy Absorption Mechanisms in Hybrid Composites, *Compos. Sci. Technol.*, Vol 34, 1989, p 305-335

51. K.L. Reifsnider and A.L. Highsmith, "The Relationship of Stiffness Changes in Composite Laminates to Fracture-Related Damage Mechanisms," Proc. 2nd USA-USSR Symp. Fracture of Composite Materials 9-12 March 1981 (Bethlehem, PA)

52. K.L. Reifsnider and A.L. Highsmith, Stress Redistribution in Composite Laminates, *Compos. Technol. Rev.*, Vol 3 (No. 1), 1981, p 32-34

53. A.L. Highsmith and K.L. Reifsnider, Internal Load Distribution Effects during Fatigue Loading of Composite Laminates, *Instrumented Impact Testing*, STP 563, T.S. Desito, Ed., ASTM , 1987, p 233-251

54. N.J. Pagano, *Int. J. Solids Struct.*, Vol 14, 1978, p 385-400

55. A.L. Highsmith, "Stiffness Reduction Resulting from Transverse Cracking in Fibre-Reinforced Composite Laminates," M.S. thesis, Virginia Polytechnic Institute and State University, Blacksburg, VA, Aug 1981

56. J.E. Masters, "An Experimental Investigation of Cumulative Damage Development in Graphite Epoxy Laminates," Ph.D. dissertation, Virginia Polytechnic Institute, Blacksburg, VA, 1980

57. J.E. Masters and K.L. Reifsnider, An Investigation of Cumulative Damage Development in Quasi-Isotropic Graphite/Epoxy Laminates, *Foreign Object Impact Damage to Composites*, STP 568, ASTM, 1975, p 40-62

58. K.L. Reifsnider and J.R. Masters, "Investigation of Characteristic Damage States in Composite Laminates," Annual Winter Meeting 1978 (San Francisco), American Society of Mechanical Engineers

59. A.S.D. Wang, G.E. Law, Jr., and W.J. Warren, An Energy Method for Multiple Transverse Cracks in Graphite-Epoxy Laminates, *Modern Developments in Composite Materials and Structures*, American Society of Mechanical Engineers, Dec 1979, p 17-30

60. P. Lee-Sullivan and A. Poursartip, Matrix Cracking Behavior of Off-Axis Plies in Glass-Epoxy Laminates, *Int. Conf. Composite Materials*, Vol 1, ICCM-7, Nov 1989 (Goung-Dong, China), International Committee on Composite Materials, p 352-356

61. C.T. Herakovich, Failure Modes and Damage Accumulation in Laminated Composites with Free Edges, *Compos. Sci. Technol.*, Vol 36, 1989, p 105-119

62. B.P. Jang, "Impact Response and Testing of Fiber-Resin Composites," M.S. thesis, Auburn University, AL, June 1990

63. B.P. Jang, W. Kowbel, and B.Z. Jang, *Compos. Sci. Eng.*, 1992

64. H.E. Davis, G.E. Troxell, and G.F.W. Hauck, Chapter 13, *The Testing of Engineering Materials*, 4th ed., McGraw-Hill, 1982

65. N.L. Hancox, *Composites*, Vol 1, 1971, p 41

66. M.G. Bader and R.M. Ellis, *Composites*, Vol 4, 1974, p 253

67. C.D. Ellis and B.J. Harris, *J. Compos. Mater.*, Vol 7, 1973, p 76

68. B. Harris, P.W. Beaumont, and E. Montcunill De Ferran, *J. Mater. Sci.*, Vol 6, 1971, p 238

69. D.F. Adams and A.K. Miller, An Analysis of the Impact Behavior of Hybrid Composites, *Mater. Sci. Eng.*, Vol 19, 1975, p 245-260

70. D.F. Adam and J.L. Perry, *Fiber Sci. Technol.*, Vol 8, 1975, p 275

71. D.R. Ireland, *Instrumented Impact Testing*, STP 563, T.S. Desito, Ed., ASTM, 1987, p 3-29
72. B. Augland, *Br. Weld. J.*, Vol 9, 1962, p 434
73. M. Grumbach, M. Prudhomme, and G. Sanz, *Rev. Métall.*, April 1969, p 271
74. J. Harding, *Materials at High Strain Rates*, T.Z. Blazynski, Ed., Elsevier Applied Science, 1987, p 133
75. Z. Buchar, Z. Bilek, and F. Dusek, Chapter 3, *Mechanical Behavior of Metals at Extremely High Strain Rates*, Trans Tech, 1986
76. K. Kawata, A. Hondo, S. Hashimoto, N. Takeda, and H.L. Chung, *Proc. Japan-U.S. Conf. Composite Materials*, K. Kawata, and T. Akasaka, Ed., Japan Society Composite Materials, 1981, p 2
77. J. Harding and L.M. Welsh, *J. Mater. Sci.*, Vol 18, 1983, p 1810
78. S.P. Joshi and C.T. Sun, Impact Induced Fracture in a Laminated Composite, *J. Compos. Mater.*, Vol 6, 1985, p 51-66
79. D. Liu, *J. Compos. Mater.*, Vol 22, 1988, p 674-692
80. L.E. Malvern, C.T. Sun, and D. Liu, STP 1012, ASTM, 1989, p 387-405
81. N. Takeda, R.L. Sierakowski, and L.E. Malvern, *SAMPE Quart.*, Vol 12 (No. 2), 1981, p 9-17
82. W.J. Cantwell and J. Morton, *Comp. Struct.*, Vol 3, 1985, p 241-257; Vol 12, 1989, p 39-60
83. E. Dan-Jumbo, A.R. Leewood, and C.T. Sun, STP 1012, ASTM, 1985, p 221-237
84. C.Y. Hsieh, "Dynamic Study of Energy Loss Mechanisms during Ballistic Impact of Polymer Composites," Ph.D. dissertation, Auburn University, AL, March 1992
85. H.Y. Wu and G. S. Springer, *J. Compos. Mater.*, Vol 22, 1988, p 518-532
86. R.P. Khetan and D.C. Chang, *J. Compos. Mater.*, Vol 17, 1983, p 182-194
87. S. Hong and D. Liu, *Exp. Mech.*, Vol 29 (No. 2), 1989, p 115-120
88. R.H. Zee, B.Z. Jang, A. Mount, and C.J. Wang, *Rev. Sci. Instrum.*, Vol 60, 1989, p 3692-3697
89. R.H. Zee, C.J. Wang, A. Mount, and B.Z. Jang, *Polym. Compos.*, Vol 12, 1991, p 196-202
90. K.L. Reifsnider, Ed., *Damage in Composite Materials*, STP 775, ASTM, 1982
91. W.S. Johnson, Ed., *Delamination and Debonding of Materials*, STP 876, ASTM, 1985
92. N.J. Johnston, Ed., *Toughened Composites*, STP 937, ASTM, 1987
93. T.S. Desito, Ed., *Instrumented Impact Testing*, STP 563, ASTM, 1987
94. C.F. Griffin, Damage Tolerance of Toughened Resin Graphite Composites, *Toughened Composites*, STP 937, N.J. Johnston, Ed., ASTM, 1987, p 23-33
95. M.M. Sohi, H.T. Hahn, and J.G. Williams, The Effects of Resin Toughness and Modulus on Compressive Failure Modes of Quasi-Isotropic Graphite/Epoxy Laminates, *Toughened Composites*, STP 937, N.J. Johnston, Ed., ASTM, 1987, p 37-60
96. J.G. Williams, "Effects of Impact Damage and Open Holes on the Compressive Strength of Tough Resin/High Strain Fiber Toughness," TM-85756, National Aeronautics and Space Administration, 1984
97. K.R. Hirschbuehler, A Comparison of Several Mechanical Tests Used to Evaluate the Toughness of Composites, *Toughened Composites*, STP 937, N.J. Johnston, Ed., ASTM, 1987, p 61-73
98. B.A. Byers, "Behavior of Damaged Graphite/Epoxy Laminates under Compressive Loading," Contractor Report 159293, National Aeronautics and Space Administration, Aug 1980
99. G.C. Adams and T.K. Wu, Fatigue of Polymers by Instrumented Impact Testing, *Ann. Tech. Conf.* (ANTEC 1983), Society of Plastics Engineers, 1983, p 541-543
100. G.C. Adams, Impact Fatigue of Polymers Using an Instrumented Drop Tower Device, *Instrumented Impact Testing of Plastics and Composite Materials*, STP 936, ASTM, 1987, p 281-301
101. S.K. Bhateja, J.K. Rieke, and E.H. Andrews, Impact Fatigue Response of Ultra-High Molecular Weight Linear Polyethylene, *J. Mater. Sci.*, Vol 14, 1979, p 2103
102. F. Ohishi, S. Nakamure, B. Koyama, K. Minabe, and J. Fujisawa, Factors Affecting Dynamic Durability of Polycarbonate under Bending, Torsional, and Impacting Fatigue and in Cavitation Erosion and Dynamic Solvent Cracking Conditions, *J. Appl. Polym. Sci.*, Vol 20, 1976, p 79
103. M.T. Takemori, Biaxial Impact Fatigue of Polycarbonate, *J. Mater. Sci.*, Vol 17, 1982, p 164
104. D.A. Wywick and D.F. Adams, Residual Strength of a Carbon/Epoxy Composite Material Subjected to Repeated Impact, *J. Compos. Mater.*, Vol 22, 1988, p 749-765
105. A. Rotem, The Strength of Laminated Composite Materials under Repeated Impact Loading, *J. Compos. Technol. Res.*, Summer 1988, p 74-79
106. A.S. Prank and V.P. Tamuzs, Fiber-Glass Laminate Fatigue under the Effect of Repeated Shock Load, *Mekh. Polim.*, No. 4, 1969, p 195-199
107. V.S. Kuksenko and V.P. Tamuzs, Chapter 1, *Fracture Micromechanics of Polymer Materials*, M. Nijhoff, 1981
108. C. Lhymn, Impact Fatigue of PPS/Glass Composites—Theoretical Analysis, *J. Mater. Sci. Lett.*, Vol 4, 1985, p 1221-1224
109. C. Lhymn, Impact Fatigue of PPS/Glass Composites—Microscopy, *J. Mater. Sci. Lett.*, Vol 4, 1985, p 1429-1433
110. D. Liu and L.E. Malvern, *J. Compos. Mater.*, Vol 21, 1987, p 594-609
111. C.A. Ross, L.E. Malvern, and R.L. Sierakowsi, *Recent Advances in Composites in the United States and Japan*, STP 864, J.R. Vinson and M. Taya, Ed., ASTM, 1985, p 355-367
112. N. Takeda, R.L. Sierakowski, and L.E. Malvern, *J. Mater. Sci.*, Vol 16, 1981, p 2008-2011
113. N. Takeda, R.L. Sierakowski, and L.E. Malvern, *Compos. Tech. Rev.*, Vol 4, 1982, p 40-44
114. S.M. Bishop, *Comp. Struct.*, Vol 3, 1985, p 295-318
115. D.J. Boll, W.D. Bascom, J.C. Weidner, and W.J. Murri, *J. Mater. Sci.*, Vol 21, 1986, p 2667-2677
116. K.J. Bowles, STP 972, ASTM, 1988, p 124-142
117. H.Chai, W.G. Knauss, and C.D. Babcock, *Exp. Mech.*, Vol 23, 1983, p 329-337
118. G. Dorey, S. Bishop, and P. Curtis, *Compos. Sci. Technol.*, Vol 23, 1985, p 221-237
119. R.E. Evans and J.E. Masters, *Toughened Composites*, STP 937, N.J. Johnston, Ed., ASTM, 1987, p 413-436
120. B.G. Frock, M.T. Moran, K.D. Shimmin, and R.W. Martin, *Rev. Prog. Quant. NDE*, Vol 7B, 1988, p 1093-1099
121. D.S. Gardiner and L.H. Pearson, *Exp. Mech.*, Vol 9 (No. 11), 1985, p 22-28
122. S.P. Joshi and C.T. Sun, *J. Compos. Tech. Rev.*, Vol 9 (No. 2), 1987, p 40-46
123. D.C. Leach, D.C. Curtis, and D.R. Tamblin, *Toughened Composites*, STP 937, N.J. Johnston, Ed., ASTM, 1987, p 358-380

124. T.E. Preuss and G. Clark, *Composites*, Vol 19 (No. 2), 1988, p 145-148

125. P. Sjoblom, J. Hartness, and T. Cordell, *J. Compos. Mater.*, Vol 22 (No. 1), 1988, p 30-52

126. C.T. Sun and S. Rechack, *Composite Materials: Testing and* Design, STP 972, J.D. Whitcomb, Ed., ASTM, 1988, p 97-123

127. A. Vasudev and M.J. Meehlman, *SAMPE Quart.*, Vol 18 (No. 4), 1987, p 43-48

128. C.H. Yu and P.R. Kendrick, *Int. J. Impact Eng.*, Vol 5, 1987, p 729-738

129. C.F. Buynak and T.J. Moran, *Rev. Prog. Quantitative NDE*, Vol 6B, 1986, p 1203-1211

130. J.G. Williams and M.D. Rhodes, *Composite Materials: Testing and* Design, STP 787, I.M. Daniel, Ed., ASTM, 1982, p 450-480

131. R.T. Potter, *Comp. Struct.*, Vol 3, 1985, p 319-339

132. G. Clark, *Composites*, Vol 20 (No. 3), 1989, p 209-214

133. M. Clerico, G. Ruvinetti, F. Cipri, and M. Pelosi, *Int. J. Mater. Prod. Technol.*, Vol 4 (No. 1), 1989, p 61-70

134. R. Jones, J.F. Williams, and T.E. Tay, *Comp. Struct.*, Vol 8 (No. 1), 1987, p 1-12

135. R.F. Mousely, in *Structural Impact and Crashworthiness*, J. Morton, Ed., Elsevier, 1984, p 494-509

136. T.J. Van Blaricum, P. Bates, and R. Jones, *Eng. Fract. Mech.*, Vol 32 (No. 5), 1989, p 667-674

137. D.J. Hillman and R.L. Hillman, *Delamination and Debonding*, STP 876, W.S. Johnson, Ed., ASTM, 1985, p 481-493

138. P. Potet, C. Bathias, and B. Degrigny, *Mater. Eval.*, Vol 46, 1988, p 1050-1051

139. W.J. Cantwell, *Comp. Struct.*, Vol 10 (No. 3), 1988, p 247-265

140. S.M. Freeman, *Composite Materials: Testing and Design*, STP 787, I.M. Daniel, Ed., ASTM, 1982, p 50-62

141. S.I. Ochiai, K.Q. Lew, and J.E. Green, *J. Acoust. Emiss.*, Vol 1, 1982, p 191-192

142. V.H. Kenner, G.H. Staab, and H.S. Jing, *Delamination and Debonding*, STP 876, W.S. Johnson, Ed., ASTM, 1985, p 465-480

143. B. Hofer, *Composites*, Vol 18 (No. 4), 1987, p 309-316

144. M.D. Shelby, B.Z. Jang, H.B. Hsieh, and T.L. Lin, SAMPE National Tech. Conf. (Crystal City, VA), 1988

145. B.Z. Jang, M.D. Shelby, H.J. Tai, *Polym. Eng. Sci.*, Vol 31, 1991, p 47-55

146. C.F. Buynak, T.J. Moran, and S. Donaldson, *SAMPE J.*, Vol 24 (No. 2), 1988, p 35-39

147. R.L. Ramkumar and P.C. Chen, *AIAA J.*, Vol 21 (No. 10), 1983, p 1448-1452

148. W.C. Shih and B.Z. Jang, *J. Reinf. Plast. Compos.*, Vol 8, May 1989, p 270-298

149. D.Moore and S. Prediger, *Polym. Eng. Sci.*, Vol 28 (No. 9), 1988, p 626-633

150. N. Tekeda, R.L. Sierakowski, and L.E. Malvern, *Exp. Mech.*, Vol 22 (No. 1), 1982, p 19-25

151. L.E. Malvern, C.T. Sun, and D. Liu, *Recent Trends in Aeroelasticity Structures and Structural Dynamics*, P. Hajela, Ed., University of Florida Press, 1987, p 298-312

152. C.T. Sun and J.E. Grady, *Compos. Sci. Technol.*, Vol 31, 1988, p 55-72

153. E.F. Olster and P.A. Roy, STP 546, ASTM, 1974, p 583-603

154. F.K. Chang, H.Y. Choi, and S.T. Jeng, Characterization of Impact Damage in Laminated Composites, *34th Int. SAMPE Symp.*, May 1989, p 702-713

155. A.S.D. Wang, Chapter 2, *Interlaminar Response of Composite Materials*, N.J. Pagano, Ed., Composite Materials Series, Vol 5, Elsevier, 1989

156. N. Cristescu, L.E. Malvern, and R.L. Sierakowski, STP 568, ASTM, 1975, p 159-172

157. G. Caprino, I. Crivelli-Visconti, and A. Dillio, *Composites*, Vol 15 (No. 3), 1984, p 231-234

158. P.T. Curtis and S.M. Bishop, *Composites*, Vol 15 (No. 4), 1984, p 259-264

159. S.M. Bishop, Chapter 6, *Textile Structural Composites*, Vol 3, T.W. Chou and F.K. Ko, Ed., *Composite Materials Series*, Elsevier, 1989

160. L.W. Gause, "Mech. Composite Review," Air Force Materials Laboratory, Oct 1983

161. F.K. Ko and D. Hartman, *SAMPE J.*, Vol 22 (No. 4), 1986, p 26

162. A.P. Majidi, M.J. Rotermund, and L.E. Taske, *SAMPE J.*, Vol 24, Jan/Feb 1988, p 12-17

163. W.C. Chung, B.Z. Jang, T.C. Chang, L.R. Hwang, and R.C. Wilcox, *Mater. Sci. Eng. A*, Vol 112, 1989, p 157-173

164. L.J. Broutman and P.K. Mallick, "Impact Behavior of Hybrid Composites," AFOSR TR-75-0472, Nov 1974

165. D.F. Adams and A.K. Miller, *Mater. Sci. Eng.*, Vol 19, 1975, p 245-260

166. P.W.R. Beaumont, P.G. Reiwald, and C. Zweben, *Foreign Object Impact Behavior of Composites*, STP 568, ASTM, 1974, p 134-158

167. A. Poursartip, G. Riahi, E. Teghtsoonian, N. Chinatambi, N. Mulford, "Mechanical Properties of PE Fiber/Carbon Fiber Hybrid Laminates," ICCM-VI and ECCM-II, Vol 1, Elsevier Applied Science, 1987, p 209-220

168. S.K. Chaturvedi and R.L. Sierakowsi, *Comp. Struct.*, Vol 1 (No. 2), 1983, p 137-162

169. S.K. Chaturvedi, C.T. Sun, and R.L. Sierakowski, *Polym. Compos.*, Vol 4 (No. 3), 1983, p 167-171

170. S.K. Chaturvedi and R.L. Sierakowski, *J. Compos. Mater.*, Vol 19 (No. 2), 1985, p 100-110

171. H.T. Kau, *Proc. ANTEC '89*, Society of Plastics Engineers, 1989, p 1480-1485

172. G. Dorey, Chapter 11, *Advanced Composites*, I.K. Partridge, Ed., Elsevier Applied Science, 1989

173. P. Keung and L.J. Broutman, *Polym. Eng. Sci.*, Vol 18 (No. 2), 1978, p 62-72

174. B.Z. Jang, Y.K. Lieu, W.C. Chung, and L.R. Hwang, *Polym. Compos.*, Vol 8, 1987, p 94-102

175. S.G. Lim and C.S. Hong, Prediction of Transverse Cracking and Stiffness Reduction in Cross-Ply Laminated Composites, *J. Compos. Mater.*, Vol 23, 1989, p 695-713

176. S.F.Chen and B.Z. Jang, *Compos. Sci. Technol.*, Vol 41, 1991, p 77-97

177. W.J. Cantwell, "Impact Damage in Carbon Fiber Composites," Ph.D. thesis, University of London, 1985

178. N. Takeda, R.L. Cierakowski, and L.E. Malvern, *J. Compos. Mater.*, Vol 15, 1981, p 157-174

179. U.S. Lindholm, *J. Mech. Phys. Solids*, Vol 12, 1964, p 317

180. P.A.A. Back and J.D. Campbell, Session 6, Part 2, Proc. Conf. Properties of Materials at High Rates of Strain, Institute of Mechanical Engineering, London, 1957

181. E.W. Billington and C.J. Brissenden, *J. Phys. D*, Vol 4, 1971, p 272

182. L.M. Welsh and J. Harding, *Proc. 5th Int. Conf. Composite Materials*, ICCM-V, TMS-AIME, 1985, p 1517

183. A. Rotem, and J.M. Lifsjitz, Paper 10G, *Proc. 26th Ann. Tech. Conf.*, SPI Reinf. Plastics Composites Div., Society of Plastics Industry, 1971

184. L.M. Welsh and J. Harding, *J. de Physique*, Colloque C5, 1985, p 405

185. J. Harding and L.M. Welsh, *Proc. 4th Int. Conf. on Composite Materials*, ICCM-IV, Japan Society of Composite Materials, Tokyo, 1982, p 2
186. F.J. Bradshaw, G. Dorey, and G.R. Sidey, "Impact Resistance of Carbon Fiber Reinforced Plastics," RAETR 72240, 1972
187. B.Z. Jang, Y.K. Lieu, and Y.S. Chang, *Polym. Compos.*, Vol 8, 1987, p 188-198
188. S. Abrate, *Appl. Mech. Rev.*, Vol 44, 1991, p 155-190
189. A. Rotem, *SAMPE J.*, March/April 1988, p 19-25
190. P.C. Chou and W.J. Flis, *AIAA J.*, Vol 15 (No. 4), 1977, p 455-456
191. K.M. Lal, *J. Reinf. Plast. Compos.*, Vol 2, 1983, p 226-238
192. K.M. Lal, *J. Reinf. Plast. Compos.*, Vol 2, 1983, p 216-225
193. K.N. Shivakunar, *J. Appl. Mech.*, Vol 52, 1985, p 674-680
194. C.W. Bert and V. Birman, *Proc. 4th Japan-U.S. Conf. Composite Materials*, Washington, D.C., 1988, p 43-52
195. A.P. Christoforu, S.R. Swanson, and S.W. Beckwith, STP 1012, ASTM, 1989, p 376-386
196. T.S. Chow, *J. Compos. Mater.*, Vol 5, 1971, p 306-319
197. C.A. Ross, L.E. Malvern, R.L. Sierakowski, and N. Takeda, *Recent Advances in Composites in the United States and Japan*, STP 864, J.R. Vinson and M. Taya, Ed., ASTM, 1985, p 355-367
198. R.L. Ramkumar and Y.R. Thakar, *J. Eng. Mater. Technol.*, Vol 109, 1987, p 67-71
199. F.C. Moon, *J. Compos. Mater.*, Vol 6, 1972, p 62-79
200. F.C. Moon, *J. Appl. Mech.*, Vol 40, 1973, p 485-490
201. F.C. Moon, *Comp. Struct.*, Vol 3, 1973, p 1195-1204
202. B.S. Kim and F.C. Moon, *AIAA J.*, Vol 17, 1979, p 1126-1133
203. L.B. Greszczuk, *Impact Dynamics*, J.A. Zukas et al., Ed., John Wiley & Sons, 1982, p 55-94
204. H.D. Conway, *J. Appl. Math. Phys.*, Vol 6, 1955, p 402-405
205. H.D. Conway, K.A. Farnham, and T.C. Ku, *J. Appl. Mech.*, Vol 34 (No. 2), 1967, p 491-492
206. H.D. Conway, *J. Appl. Mech.*, Vol 34, 1967, p 1031-1032
207. M. Dahan and J. Zarka, *Int. J. Solids Struct.*, Vol 13, 1977, p 229-238
208. J.R. Willis, *J. Mech. Phys. Solids*, Vol 14, 1966, p 163-176
209. L.M. Keer and M.A.G. Silva, *Int. J. Mech. Sci.*, Vol 12, 1970, p 751-760
210. L.M. Keer and R. Ballarini, *AIAA J.*, Vol 21 (No. 7), 1983, p 1035-1042
211. L.M. Keer and G.R. Miller, *ASCE J. Eng. Mech.*, Vol 109 (No. 3), 1983, p 706-717
212. L.M. Keer and G.R. Miller, *Int. J. Eng. Sci.*, Vol 21 (No. 6), 1983, p 681-690
213. L.M. Keer and J.C. Lee, *Int. J. Eng. Sci.*, Vol 23 (No. 10), 1985, p 987-997
214. I.M. Keer and W.P. Schonberg, *Int. J. Solids Struct.*, Vol 22, 1986, p 87-103; 1033-1053
215. B.V. Sankar and C.T. Sun, *Int. J. Solids Struct.*, Vol 19 (No. 4), 1983, p 293-303
216. B.V. Sankar, *J. Appl. Mech.*, Vol 54, 1987, p 735
217. B.V. Sankar, *Int. J. Solids Struct.*, Vol 25 (No. 3), 1989, p 327-337
218. B.V. Sankar, *Compos. Sci. Technol.*, Vol 34, 1989, p 95-111
219. C.T. Sun and B.V. Sankar, *Int. J. Solids Struct.*, Vol 21, 1985, p 161
220. T.M. Tan and C.T. Sun, *J. Appl. Mech.*, Vol 52, 1985, p 6-12
221. J. K. Chen and C.T. Sun, *Comp. Struct.*, Vol 3, 1985, p 97-118; 59-73
222. J.K. Chen and C.T. Sun, *J. Compos. Mater.*, Vol 19 (No. 11), 1985, p 490-504
223. M.G. Koller, *J. Appl. Math. Phys.*, Vol 37, 1986, p 256-269
224. H.T. Wu and F.K. Chang, *Comp. Struct.*, Vol 31 (No. 3), 1989, p 453-466
225. C.T. Sun and S.N. Huang, *Comput. Struct.*, Vol 5, 1975, p 297-303
226. B.V. Sunkar and C.T. Sun, *Comput. Struct.*, Vol 20 (No. 6), 1985, p 1009-1012
227. B.V. Sunkar and C.T. Sun, *AIAA J.*, Vol 23 (No. 12), 1985, p 1962-1969
228. C.T. Sun and S. Chattopadhyay, *J. Appl. Mech.*, Vol 42, 1975, p 693-698
229. A.L. Dobyns, *AIAA J.*, Vol 19 (No. 5), 1980, p 642-650
230. A.L. Dobyns and T.R. Porter, *Polym. Eng. Sci.*, Vol 21 (No. 8), 1981, p 493-498
231. S. Chattopadhyay, *J. Acoust. Soc. Amer.*, Vol 82 (No. 2), 1987, p 493-497
232. B.V. Sunkar and C.T. Sun, *AIAA J.*, Vol 24 (No. 3), 1985, p 470-472
233. M.G. Koller and M. Busenhart, *Int. J. Impact Eng.*, Vol 4 (No. 1), 1986, p 11-21
234. W.E. Bachrach and R.S. Hansen, *AIAA J.*, Vol 27 (No. 5), 1989, p 632-638
235. G. Dorey Impact and Crashworthiness of Composite Structures, *Structural Impact and Crashworthiness*, Vol 1, G.A.O. Davies, Ed., Elsevier Applied Science, 1984
236. E.F. Byars, R.D. Snyder, and H.L. Plants, Chapters 7 and 12, *Engineering Mechanics of Deformable Bodies*, 4th ed., Harper & Row
237. J. Awerbuch and H.T. Hahn, *J. Compos. Mater.*, Vol 10, 1976, p 231
238. A.E.H. Love, *Philos. Trans. R. Soc. (London) A*, Vol 228, 1929
239. M.T. Huber, *Pisma*, Warsaw, 1956, p 2
240. J.C. Goering, 1st Tech. Conf., American Society of Composites, 1986, p 326-345
241. J.M. Whitney, *Compos. Sci. Technol.*, Vol 22, 1985, p 167-184
242. R. Jones, J. Paul, T.E. Tay, and J.F. Williams, *Comp. Struct.*, Vol 10 (No. 1), 1988, p 51-73
243. L.B. Greszczuk and H. Chao, in *Composite Materials: Testing and Design*, STP 617, ASTM, 1977, p 389-408
244. K. Shivakumar, W. Elber, and W. Illg, *AIAA J.*, Vol 23 (No. 3), 1985, p 442-449
245. Z.L. Gu and C.T. Sun, *Comp. Struct.*, Vol 7 (No. 3), 1987, p 179-190
246. W.E. Bachrach and R.S. Hansen, *AIAA J.*, Vol 27 (No. 5), 1989, p 632-638
247. M. Yener and E. Wolcott, *J. Pressure Vessel Technol.*, Vol 111, 1989, p 124-129
248. C.C. Chamis and C.A. Ginty, STP 1012, ASTM, 1989, p 338-355
249. A.A. Baker, R. Jones, and R.J. Callinan, *Comp. Struct.*, Vol 4, 1985, p 15-44
250. R. Jones and A.A. Baker, *Composite Structures*, Vol 3, I.H. Marshall, Ed., Elsevier, 1985, p 402-414
251. A. Kelly, *Strong Solids*, Oxford University Press, 1966
252. G. Cooper and A. Kelly, STP 452, ASTM, 1969
253. M. Poggott, *J. Mater. Sci.*, Vol 5, 1970, p 669
254. P.W.R. Beaumont, J. Fitz-Randolph, D.C. Phillips, and A.S. Tetelman, *J. Compos. Mater.*, Vol 5, 1971, p 542; *J. Mater. Sci.*, Vol 7, 1972, p 289

255. T.U. Marston, A.G. Atkins, and D.K. Felbeck, *J. Mater. Sci.*, Vol 9, 1974, p 447-455

256. P.W.R. Beaumont, *J. Adhes.*, Vol 6, 1974, p 107-137

257. B.D. Agarwal and L.J. Broutman, Chapter 8, *Analysis and Performance of Fiber Composites*, 2nd ed., Wiley, 1990

258. A. Kelly, *Proc. Roy. Soc. A*, Vol 319, 1970, p 95

259. J.O. Outwater and M.C. Murphy, Paper IIc, 24th Ann. Tech. Conf. Compos. Div., Society of Plastics Industry, 1969

260. Y.M. Han and H.T. Hahn, Ply Cracking and Property Degradations of Symmetric Balanced Laminates Under General In-Plane Loading, *Compos. Sci Technol.*, Vol 35, 1989, p 377-397

261. G.V. Gandhe and O.H. Friffin, *SAMPE Q.*, Vol 20 (No. 4), 1989, p 55-58

262. D.C. Leach and D.R. Moore, *Compos. Sci. Technol.*, Vol 23, 1985, p 131-161

263. R.L. Ramkumar, STP 813, ASTM, 1983, p 116-135

264. B.V. Sanka and C.T. Sun, *AIAA J.*, Vol 24 (No. 3), 1986, p 470-472

265. J.M. Whitney, *Structural Analysis of Laminated Anisotropic Plates*, Technomic Pub. Co., 1987

266. A.H. Cottrell, *Proc. Roy Soc.*, Vol A-282, 1964, p 2

Chapter 9

Design for Improved Fatigue Resistance of Composites

Composite Fatigue Versus Metal Fatigue

Fatigue is the process whereby mechanical damage, caused by repetitive or fluctuating stresses, results in a material failure at lower stress levels than would be required under static loading. Fatigue damage in isotropic metals typically occurs from the initiation and growth of a single crack, and thus damage is localized. Fatigue of composite materials differs from that of more isotropic metals in that the damage mechanisms are more complex, and may exhibit a synergistic damage effect from several micro-failure mechanisms.

These multiple-damage mechanisms can blunt the stress concentration effect of a notch or crack on the load bearing fibers, resulting in good fatigue behavior of composites. Since metals undergo fatigue by the localized growth of a crack from a stress concentration, perpendicular to the local tensile stress, they are more sensitive to tensile loading than compressive loading. The more widespread damage in composites leads to a reduced support for the fibers and hence a loss in compressive strength, so composites are more sensitive to compressive loading than tensile loading.

The damage that progresses results in a reduction of laminate stiffness and load carrying capability frequently observed in fatigued laminates. Most materials exhibit a fatigue limit for stress that, if kept below, no fatigue failure of the material will occur. Because of their unique microstructure, composite materials may or may not have a fatigue limit depending on the material combination chosen. However, composites do exhibit a gradual softening due to microscopic damage long before visual damage is detected. Fiber reinforced composite laminates can sustain fatigue damage by matrix cracking, fiber-matrix debonding, ply cracking, delamination and fiber breakage. The extent of fatigue damage is typically measured as crack length in metals and crack densities in composites.

Fatigue of all materials depends upon a large number of important external variables. Three of these variables are most critical to the extent of fatigue damage. They are a maximum tensile or compressive stress (or strain) of sufficient magnitude, a variation or fluctuation in the applied stress (or strain), and a sufficiently large number of stress (or strain) cycles. Other variables which play a role in the extent of fatigue damage include stress concentrations, temperature, environment, stress fluctuation frequency, overloading stresses, the material composition and structure, residual stresses and combined stresses.

Fatigue damage in continuous and discontinuous fiber composites is different. In discontinuous fiber composites, matrix crack initiation can occur at the fiber ends as a result of stress concentrations in these areas. Cracking and fiber-matrix debonding can also occur at low stress levels if the discontinuous fibers are not of adequate length to support the applied load. Fiber distribution and orientation are also important factors in fatigue damage resistance. The fatigue life of discontinuous fiber composites exhibits slower crack growth initially, followed by rapid growth near failure.

This chapter will concentrate on the fatigue process for continuous fiber-reinforced composites. Types of stress loading, control mode, and test methods are presented first, followed by a description of the fatigue performance, damage mechanisms, techniques for damage assessment and consequence of fatigue damage. Testing parameters and material variables that influence the fatigue response of composites will then be discussed. Various methods that have been developed for fatigue life prediction will be briefly explained. A discussion is included on design approaches for improved fatigue resistance of composites.

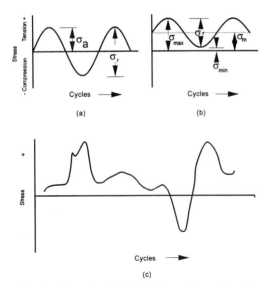

Fig. 1 Schematic of a typical fatigue stress cycles. (a) Reverse stress. (b) Repeated stress. (c) Irregular or random stress cycles. Source: Ref 1

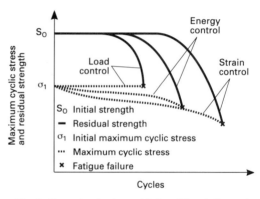

Fig. 2 Strength reduction and fatigue life as influenced by three test control parameters. Source: Ref 2

Material Type:	Polymer matrix
	Metal matrix
	Ceramic matrix

Stress State:	Axial-bending
	Axial-aff-axis
	Axial-shear

Test Geometry:	Plates - tubes
	Notched - unnotched
	Thick - thin

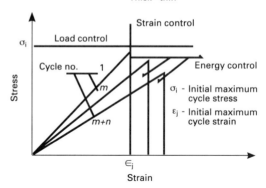

Fig. 3 Parameters that affect fatigue life of composite laminates. Source: Ref 2

Fatigue Testing and Fatigue Performance of Composites

Types of Stressing

Figure 1 shows different types of fatigue stressing commonly applied. The stress states include tension-tension, tension-compression, compression-compression, and flexure. The tension-tension state is the most frequently used for composites. The compression states are not as common, primarily because of the difficulty in gripping the specimens to prevent buckling failure of thin laminates. A complete tension-compression state can be achieved with the flexure fatigue test method.

These stress states are further broken down into how the stress is applied in each of these states. The stress can be applied as completely reversed, repeated in either tension or compression, or in a random fashion. The way the stress is applied is imperative to the understanding and prediction of damage that will occur under loading.

Some common indicators used in the description of fatigue loading are: the stress range ($\Delta\sigma$ = the difference between the maximum, σ_{max}, and minimum, σ_{min}, applied stress levels); the mean stress (σ_m = the algebraic average stress applied); the R-value (the ratio of minimum stress to maximum stress); and the A-value (the ratio of alternating stress σ_a to the mean stress).

Control Mode

When considering fatigue testing or fatigue data comparison, it is also essential to know how the test was performed. Three methods that can be used for fatigue testing are load control, energy control, and strain control [2]. Each type of test control exhibits certain unique stress

loading and will, therefore, affect the type and extent of the damage to the laminate. Figure 2 shows the S-N relation for these different test controls.

Load control fatigue testing maintains a constant magnitude cyclic stress on the tested specimen. As damage in the specimen occurs, the strain increases to accommodate the constant stress on the damaged laminate. This increases the strain-energy per cycle applied as the test progresses. Load-controlled fatigue life is typically the shortest of the three control methods. Figure 3 better illustrates the variations between load, energy, and strain-controlled fatigue testing.

Energy Control. In energy control fatigue testing, the magnitude of the stress applied will decrease as the specimen is damaged. However, the strain energy per cycle remains constant throughout the test. This is

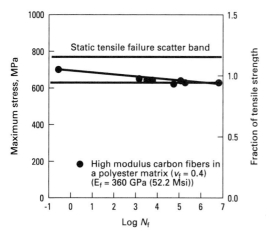

Fig. 4 Tension-tension *S-N* curves for a 0° high-modulus carbon fiber/epoxy composites. Source: Ref 3

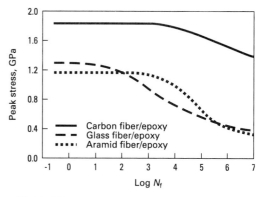

Fig. 5 *S-N* curves for three unidirectional composite materials. Source: Ref 6

achieved by increasing the strain to keep the area under the cyclic stress-strain curve constant. Fatigue life is greater than with load control tests but less than with strain control tests. Energy control fatigue tests may better represent the fatigue life of a structure in service than do the other test methods [2].

Strain control is a constant displacement fatigue test where the magnitude of applied stress decreases as the laminate is damaged. The strain energy per cycle decreases as the test proceeds. This method of testing generally exhibits longer fatigue lives than load or energy control methods.

Frequency of application of the cyclic stress or strain can influence the fatigue properties of a polymer matrix composite. Because polymer matrices are viscoelastic in nature, a phase lag between the applied stress and resulting strain exists. This viscoelastic effect results in the generation of heat in the matrix. Because of the usually poor thermal conductivities in polymers, local areas of heat buildup can exist within the composite. These local hot spots affect the matrix material properties and therefore, the fatigue properties of the composite. The magnitude of local hot spots increases with increasing cyclic frequency. Researchers recommend keeping the cyclic frequency below 30 Hz where possible.

Fatigue Performance of Composites

Tension-Tension Fatigue. Owen and Morris [3] have reported data on the tension-tension and flexure fatigue of high-modulus carbon fiber-reinforced epoxy and polyester composites. The tension-tension *S-N* curves for unidirectional composites are almost horizontal and fall within the scatter band of the static tensile strength (Fig. 4). The fatigue effect is said to be slightly greater for relatively low modulus carbon fiber-reinforced composites [4]. Unidirectional 0° boron and Kevlar-49® fiber composites also exhibit very good fatigue strength in tension-tension [5]. Additional fatigue data of unidirectional carbon, aramid and glass fiber

composites (Fig. 5) indicate that the ratio of fatigue stress at high life to static strength is smallest for the glass fiber material and largest for the carbon fiber-reinforced material [6]. Laminates of other stacking sequences are found to show similar *S-N* curves [7], but the actual fatigue effect depends on stacking sequence, the proportion of fibers aligned with the loading axis [6], and the mode of cycling. For instance, Ramani and Williams [8] demonstrated that a tension-compression cycling ($R = -1.6$) produced a steeper *S-N* curve than the tension-tension curve ($R = 0.1$) (Fig. 6).

Flexural Fatigue. In general, the flexural fatigue resistance of fiber composites is inferior to the corresponding tension-tension response. Figure 7 demonstrates that the slope of the flexural *S-N* curve is higher than that of the tension-tension *S-N* curve for high-modulus carbon fiber composites [3]. The lower flexural fatigue strength is generally ascribed to the weakness of laminates on the compression side [4]. In a three-point bending type specimen, the span-to-depth ratio (s/d) can play a critical role in dictating the fatigue failure behavior. A short beam (small s/d) would tend to exhibit shear failure modes while a longer beam would have combined compression-tension modes. A four-point bending type specimen consisting of a thick core material (e.g., honeycomb) sandwiched between two composite facing layers can provide a convenient way of testing the tensile and compressive fatigue response of a composite simultaneously.

Interlaminar Shear Fatigue. Short beam shear specimens have been used [9-11] to investigate the interlaminar shear τ_{xz} mode under cyclic stress conditions. The interlaminar shear fatigue strength of a unidirectional 0° carbon/epoxy composite was reduced after 10^6 cycles to less than 55% of its static interlaminar shear strength (ILSS), whereas its corresponding tension-tension strength was nearly 80% of its static tensile strength (Fig. 8). A similar phenomenon was observed for a unidirectional 0° boron/epoxy system. However, a reverse trend was observed for a glass fiber/epoxy system [9]. This could be attributed to a better interfacial adhesion between glass fiber (presumably coated with a good cou-

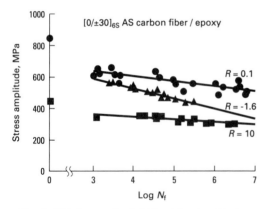

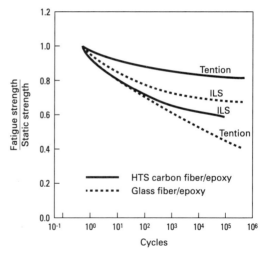

Fig. 6 S-N diagrams for [0/±30]₆ₛ AS-carbon fiber/epoxy laminates at various fatigue stress ratios. Frequency, 10 Hz. Source: Ref 8

Fig. 8 Interlaminar shear S-N curves for 0° carbon and glass fiber/epoxy laminates. ILS, interlaminar shear. Source: Ref 9

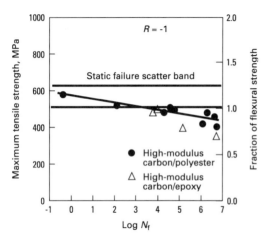

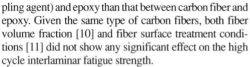

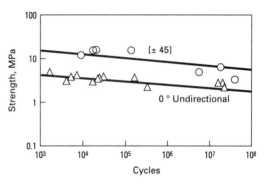

Fig. 7 Flexural S-N diagrams for a 0° carbon fiber/epoxy laminate

Fig. 9 Torsional S-N diagrams for a 0° and [±45]ₛ high-strength carbon fiber/epoxy composites. Source: Ref 15

pling agent) and epoxy than that between carbon fiber and epoxy. Given the same type of carbon fibers, both fiber volume fraction [10] and fiber surface treatment conditions [11] did not show any significant effect on the high cycle interlaminar fatigue strength.

Interlaminar fracture behavior of composite materials under cyclic loading conditions has been studied using a double cantilever beam (DCB) [12] and a width-tapered DCB specimen [13]. For a unidirectional carbon/PEEK and a carbon/epoxy composite, the cyclic delamination growth behavior at an elevated temperature (93 °C or 200 °F) was found to be significantly different than the response at room temperature under load-controlled conditions [13]. For both composites, the high-temperature response was dominated by the formation and growth of a process zone ahead of the delamination crack [13]. The interlaminar fatigue crack growth behavior of a glass/epoxy composite was studied using a test that involves a constant range of cyclic strain energy release rates [14].

The well-known Paris power law was modified to interpret fatigue crack growth data under the influence of fiber bridging [14]. DCB-type specimens provide a convenient way to characterize the delamination resistance of a composite under cyclic loading conditions.

Torsional Fatigue. Limited torsional fatigue data is available in open literature. Fujczak [15] investigated the torsional fatigue behavior of carbon/epoxy thin tubes. On a log-log scale, the S-N plots in alternating torsional fatigue ($R = -1$) for both [0°] and [±45°]ₛ materials exhibit a linear response (Fig. 9). At an equivalent number of cycles, the fatigue strength of [±45°]ₛ specimens is approximately 3.7 times higher than that of the [0°] specimens. Whereas the [0°] specimens failed by a few longitudinal cracks, the [±45°]ₛ specimens failed by cracking along the ±45° lines accompanied by extensive delamination. The [0°] specimens retained a much higher post-fatigue static torsional strength, even though they showed a lower torsional fatigue strength.

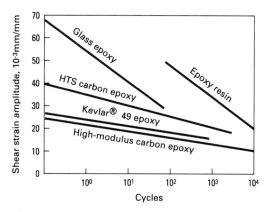

Fig. 10 Torsional shear strain-cycle diagram for various 0° fiber reinforced composites. Source: Ref 16

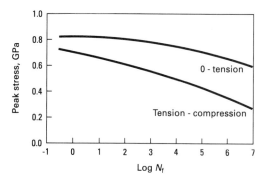

Fig. 11 A comparison of the tensile and reversed axial fatigue behavior of [±45/0]$_s$ CFRP materials. Source: Ref 6

Phillips and Scott [16] have conducted torsional fatigue tests on a neat epoxy resin, and on epoxy-based carbon (both high-strength and high-modulus), glass, and Kevlar®-49 composites. The experiments were performed by shear strain cycling of solid rod specimens. All the unidirectional composites fail at approximately 10^3 cycles at a strain level of approximately half the static shear failure strain (Fig. 10).

Compressive Fatigue. In compression, the fibers are the principal load-bearing constituents, but they must be supported from becoming locally unstable, undergoing buckling failure. This support is provided by the matrix and the fiber-matrix interface, which both play a more important role in compressive loading than in tensile loading. Local resin and interfacial damage lead to fiber instability in compressive loading. The need to support the specimen from undergoing macro-buckling, combined with the limitations imposed on specimen geometry by the anisotropic nature of the materials, has made it very difficult to conduct compressive tests (both static and cyclic). Therefore, little information is available on the compressive fatigue of composites [17].

The worst fatigue loading condition for composites is tension-compression. Figure 11 demonstrates the inferior behavior of composites in reversed axial loading compared with tensile loading [6]. Many laminas, without fibers in the test direction, develop intraply damage which causes local ply delamination at relatively short lifetimes. Tensile-induced damage of this type can lead to local ply instability and buckling when the laminate is in compression during cycling. Thus, fatigue lives of composites in reversed axial loading are usually shorter than in zero-compression or zero-tension loading [6].

Fatigue Damage in Composites

Fatigue Damage Mechanisms

Micro-failure Mechanisms in Fatigued Composites. Fatigue in composite materials exhibits both microscopic and macroscopic damage mechanisms at various stages in the fatigue process. Generally, matrix cracking occurs early in the weakest ply. Crack multiplication then dominates most of the fatigue life with cracks initiating in the progressively stronger plies. As cracks intersect each other, areas of delamination can occur. At some later stage, the fibers break in the loading direction causing the laminate to fail.

The damage mechanisms most commonly attributed to fatigue are matrix cracking, fiber-matrix debonding, delamination, and fiber breakage. Figure 12 shows damage as a function of applied cycles for a typical unidirectional graphite/epoxy laminate. Both SEM and optical micrographs may be taken periodically to detect which damage mechanism is dominant.

Matrix cracking is characterized by the microscopic cracks that form predominantly in the matrix areas of a laminate under loading. Their orientation may be in any direction depending on the applied stress. In a multidirectional laminate, matrix cracks will appear first in the weakest ply layer. As the laminate is further stressed, cracking will subsequently occur from the weakest to the strongest layers. Matrix cracks are initiated early in the fatigue process. The majority of the fatigue life of a laminate is spent in the crack multiplication stage where the density of cracks increases.

As the crack density increases, cracks begin to grow into each other, forming larger cracks. When the laminate is still further loaded, the crack density will be increased to a limiting value where stress redistribution would limit the initiation of new cracks. At this point, matrix cracking becomes a macroscopic form of damage that could dictate the initiation of other damage mechanisms. The subsequent damage is usually delamination, and then fiber breakage. The stacking sequence of the laminate affects the magnitude of crack damage during fatigue, because stress distribution and differences in Poisson's ratios and in coefficients of mutual influence between ply layers vary.

Fiber-Matrix Debonding. Fatigue damage by this mechanism is dependent on the interfacial bond strength between the fiber and the matrix. High interfacial bond strengths will permit little or no fiber-matrix interfacial

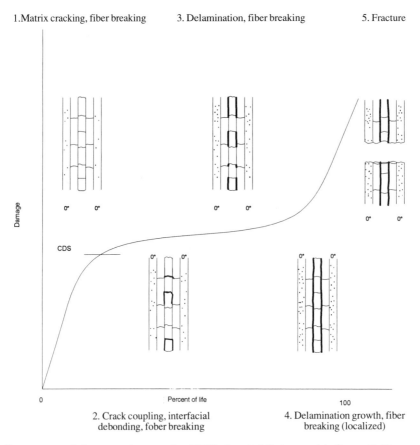

Fig. 12 Damage accumulation process in composites. CDS is characteristic damage state. Source: Ref 2

debonding. In contrast, laminates with low bond strength may exhibit large areas of interfacial debonding that either combines with or intensifies other damage mechanisms to speed-up laminate failure.

Delamination is the separation of plies in a laminate, usually induced by high interlaminar stresses. These local stresses occur near free edges of a composite. Their magnitude depends upon the materials used, stacking sequence, laminate shape and type of stress loading. During fatigue loading, delamination begins early in the life and progresses as the fatigue cycles increase. Other damage mechanisms may also influence the extent and propagation of delamination in a fatigued laminate. For example, growing matrix or fiber breakage cracks may initiate delamination.

Fiber Breakage. Fiber breaks occur either by overloading or extensive strain. Because polymer matrix composites generally have greater strain-to-failure values than the fibers used, this damage mechanism is seen quite frequently. Stress concentrations from matrix cracks and fiber-matrix debonding can also provoke fiber breakage.

Unidirectional Fiber Composites. Fatigue damage mechanisms in unidirectional composites depend on the loading mode, loading direction, stress level, fiber type and matrix fatigue properties [14,18-20]. Only the dam-

age mechanisms in tension-tension fatigue will be discussed here. At very high stress levels, the fibers may experience a stress exceeding the lower limit of the fiber strength scatter band. The weakest fibers will break on the first application of the maximum stress. High stress concentrations at the broken fiber ends can initiate more fiber breakages in nearby areas. However, the strong shear concentration at the interface close to the tip of the broken fiber can also result in fiber-matrix debonding. The debonded area acts as a stress concentration site for the longitudinal tensile stress. The intensified tensile stress may exceed the fracture stress of the matrix, resulting in the formation of a matrix crack. Fiber breakage is believed to dominate at high stress levels, leading to catastrophic failure, possibly in a few hundred cycles.

At lower fatigue stress levels, the fiber stress may be lower than the lower limit of the fiber strength scatter band, but the matrix strain may exceed the cyclic strain limit of the matrix. Since the constrained matrix is subjected to strain-controlled fatigue, a high matrix strain will lead to failure initiation by matrix microcracking instead of fiber failure. High stress concentrations at the matrix cracks may cause interfacial debonding and occasional fiber failure. The propagation of matrix macrocracks can be retarded by interfacial debonding.

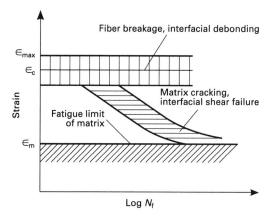

Fig. 13 Fatigue-life diagram for unidirectional composites under loading parallel to fibers. Source: Ref 19

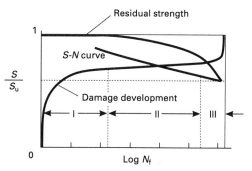

Fig. 14 Three stages of fatigue damage development in multidirectional laminates. Source: Ref 23

The fatigue failure process is progressive and, depending on stress level, may span over 10^6 cycles [4,20]

Talreja [19] illustrated the relationship between fatigue life and damage mechanisms (Fig. 13). The coordinates are the maximum strain, and the logarithm of the number of cycles to failure. Strain instead of stress was chosen as an independent variable, as both the fiber and the matrix would be subjected to the same strain while stresses in the two constituents would differ, depending on the constituent volume fractions and elastic moduli [19]. Figure 13 indicates a horizontal band centered about the composite failure strain (ε_c), corresponding to principally fiber breakage and the resulting interfacial debonding mechanisms. A second region, corresponding to matrix cracking and interfacial shear failure, is shown as a sloping band between the lower bound of the fiber breakage scatter band and the horizontal line representing the fatigue strain limit of the matrix, ε_m. However, the fracture strain of the composite ε_c will depend on the fiber stiffness. The fracture strain ε_c is much higher than ε_m for composites containing low-stiffness fibers. For high-stiffness composites, the difference is not as great; ε_c could even be less than ε_m.

Other continuous fiber composites. For off-axis laminas with an angle between 0 and 90°, the tip of a crack initiated in the matrix will be subjected to two displacement components: an opening displacement normal to the fibers and a sliding one parallel to the fibers. A mixed mode crack should grow parallel to the fibers. The limiting value of the crack-tip displacement, under which no growth can occur, will depend on the off-axis angle. The opening mode of crack growth, which is the more critical of the two modes, will become more important with increasing off-axis angle. Therefore, the fatigue strain limit will decrease with increasing off-axis angle [19]. The basic damage mechanisms in angle-plied laminates are those observed in off-axis fatigue of unidirectional composites (or of the off-axis laminas), with the added mode of delamination between plies. Fatigue life data for symmetric angle-plied laminates of glass-epoxy [21] indicate that the limiting fatigue strains vary between

an upper and a lower bound. The upper bound is close to the fatigue limit of the unreinforced epoxy, whereas the lower bound is the minimum strain for transverse fiber debonding.

In cross-plied laminates, fatigue damage was shown [22] to begin with the debonding of the transverse fibers. Interfacial debonding is expected to represent the first phase of intralaminar crack initiation for composites that contain relatively weak fiber-matrix interface bonds (e.g., carbon fiber-reinforced thermoplastic composites). Strong interface bonds tend to result in composites exhibiting matrix microcracking. In either case, the intralaminar cracks grow toward the interlaminar boundaries as fatigue proceeds. Some of the microcracks are then deflected parallel to the 0° layers, causing delaminations between 0° and 90° layers. The sequence of damage mechanisms in other laminates containing predominately 0°, ±45°, ±30°, ±60° and 90° plies normally begins with the failure of the 90° plies by transverse fiber debonding followed by delaminations of the off-axis plies, leading to overstressing of 0° plies, and eventually final failure [19].

Reifsnider *et al.* [23] divided the sequence of fatigue damage in a general laminate in three regimes (Fig.14). Regime I is characterized by the initiation and growth of matrix microcracks through the thickness of off-axis or 90° laminas. These transverse cracks also extend along the fiber direction. They normally would develop quite early in the life cycle, but quickly stabilize to a nearly uniform pattern with a constant spacing. The crack pattern in Regime II is called the characteristic damage state (CDS), which depends on the properties of individual laminas, their thicknesses, and the stacking sequence. Surprisingly, the CDS is found to be independent of load history, environment, residual stresses, or stresses due to moisture absorption.

In Regime II, coupling and growth of matrix microcracks take place, ultimately leading to interfacial debonding and delaminations between laminas. Both modes occur as the consequence of high normal and shear stresses produced at the tips of matrix macrocracks. Edge delaminations may also develop in laminates where high interlaminar stresses exist. Delaminations force the 0° layers to carry additional load, causing fiber breakages, accelerating the fatigue failure process.

Regime III involves the fiber fracture mechanism in the 0° laminas, followed by longitudinal splitting at fiber-matrix interfaces in these laminas. These fiber fractures usually develop in local areas adjacent to the matrix microcracks in off-axis or 90° plies. The rate of fiber fracture in Regime III is much greater than that in I and II. Fiber fractures quickly lead to laminate failure.

Consequence of Fatigue Damage and Methods of Damage Assessment

Fatigue damage influences the mechanical properties of composite laminates. Reductions in both stiffness and strength increase as fatigue damage increases. The crack initiation and multiplication mechanisms that begin early in the fatigue process lead to a definite reduction in modulus [24-26] and a variation in Poisson's ratio [27] proportional to the crack density. The strength reduction of composites during fatigue varies with the type and stacking sequence of the laminate. Unidirectional laminates do not show much of a change in strength until immediately before final failure. Cross-ply laminates show a gradual reduction in strength as the fatigue process continues. The difference in the variation of strength between these two stacking sequences is in the mutual influence that one ply exerts on neighboring plies.

The variations in dynamic modulus and damping coefficient can be used to monitor the development of fatigue damage in a composite [28]. Since these two parameters can be measured by vibration-based and acousto-ultrasonic techniques, the fatigue damage state or the integrity of a composite can be continuously monitored in a non-destructive fashion. Ultrasonic scanning can be used to detect interlaminar damage such as edge cracks and delamination as fatigue progresses. However, the ultrasonic beam normally requires a transfer medium (usually water) and therefore immersion of the specimen in a water tank is inevitable. X-ray inspection of composites during fatigue loading relies upon the introduction of a dye penetrant into the damage that is opaque to X-rays. Optical microscopy, possibly in conjunction with an edge replication technique, can be used to observe the damage on the specimen edge intermittently. High-resolution infrared thermography provides an uninterrupted and non-interfering method to assess the damage of a composite during fatigue testing.

Factors That Influence Composite Fatigue

Material Variables

S-N curves for three unidirectional composites (epoxy resin reinforced with carbon fiber, glass fiber, and aramid fiber) are shown in Fig. 5. In unidirectional composites under tensile loading, the fibers carry virtually all the load. Carbon fibers, possessing the highest elastic modulus, provide the composites with the highest peak stress given the same fatigue life cycle. However, some experimental evidence has shown that the slopes of the S-N curves are determined principally by the strain in the

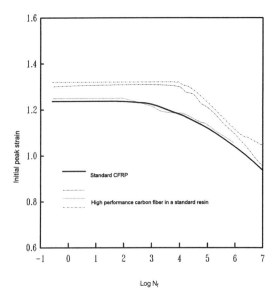

Fig. 15 Fatigue strain-life data for unidirectional carbon fiber composites with different fiber types in the same basic resin. Source: Ref 6

matrix [6,19,20,29]. For intermediate stress levels, matrix cracking and possibly interfacial shear failure are the dominant fatigue failure mechanisms, both being closely related to the matrix properties. The use of very stiff fibers (such as carbon fibers, typically with moduli of 200 to 700 GPa, or 29 to 101 Msi, and failure strains of 0.6 to 1.8%) will impose upon the matrix a relatively low strain, permitting a delay in matrix cracking and leading to a shallow S-N curve. Contrarily, less stiff fibers (such as glass, typically with moduli of 70 to 80 GPa, or 10 to 12 Msi, and failure strains of 2.5 to 4.0%) allow the composite to deform to a greater extent, under a stress-controlled condition, forcing the matrix to experience a greater strain and resulting in steeper S-N curves.

Several high-strain carbon fibers have recently been developed. Curtis [6,29] reported that using higher performance carbon fibers in the same standard epoxy resin matrices did not lead to great improvement in fatigue behavior. As shown in Fig. 15, only small changes in the fatigue behavior are apparent, usually just a slight shift along the ordinate in keeping with changes in the static tensile strength.

Very little work has been done in systematically investigating the effects of resin type and coupling agents on the fatigue behavior of fiber composites. Early experiments conducted by Boller [30] on balanced E-glass fabric-reinforced laminates have demonstrated the superiority of epoxies over polyesters and other thermosetting resins. Owen and Rose [31] suggested that the principal effect of flexibilizing the resin is to delay the onset of fatigue damage and the long-term fatigue lives are not affected by the resin flexibility. However, Curtis [6,29] observed that toughened epoxy resins exhibit poor fatigue behavior with steeper S-N curves (Fig. 16), even though their static strengths are better than those of the

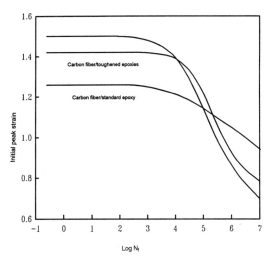

Fig. 16 Fatigue strain-life data for unidirectional carbon fiber composites with the same high-strain fiber in different epoxy matrices. Source: Ref 6

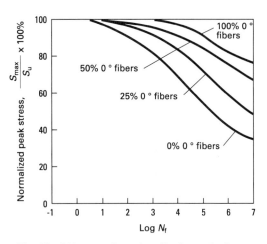

Fig. 17 S-N curves for carbon fiber/epoxy laminates with varying percentages of 0° fibers. Normalized with respect to static strength. Source: Ref 6

standard epoxy resins. The reason for this behavior is not clear.

The percentage of off-axis laminas in a laminate also influences fatigue behavior. In general, increasing the proportion of off-axis laminas reduces the static tensile strength and stiffness, since fewer fibers are available to carry the load (Fig. 17). Since off-axis plies tend to develop matrix cracks and interfacial shear failure at a much lower stress level, S-N curves will have a deeper slope if the laminate contains more off-axis laminas. However, a small percentage of some off-axis plies may act to prevent the tendency of splitting in an otherwise unidirectional laminate, whose low transverse strengths can lead to the development of cracks running parallel to fibers [30,32].

Woven fabric composites offer significant advantages in handling and fabrication over conventional non-woven materials (e.g., prepregs). Impact performance is generally better with fabric composites. However, the laminate stiffness and strength of fabric composites are normally lower than can be achieved with prepreg based composites, mainly because of the inevitable yarn distortion in the weave and a lower achievable fiber volume fraction. The stresses at short lifetimes in a fatigue test are similarly reduced in the woven composites. Furthermore, the fiber cross-over points in the fabric act as stress concentration sites to promote resin cracking and interfacial debonding, resulting in steeper S-N curves [6,33].

Short fibers provide poor static stiffness and strengths to the resulting composites, due to the less efficient utilization of the fibers and the stress concentration effect of the fiber ends. Additional damage initiates at fiber ends, resulting in failures at shorter lifetimes than in continuous fiber composites. In addition, specimens tested transverse to the direction of preferred fiber orientation show poorer fatigue response than those tested along the fiber orientation direction [6].

Hybridization has been a good way to achieve improved toughness or stiffness matching in composites. In hybrid composites, the fatigue behavior is usually dominated by the lower-strain phase. A lower-strain fiber is usually stiffer and carries a greater portion of the load. The other fiber and the matrix are thus usually subjected to strains below normal and show a smaller fatigue degradation effect than when tested alone [6]. Figure 18 shows a series of S-N curves for carbon/Kevlar® fiber hybrid composites with varying ratios of fiber type. On increasing the Kevlar® proportion of the hybrid, there is a gradual variation from the flat S-N behavior of the carbon fiber composite to slightly steeper S-N plots. This change is not directly proportional to the large difference in the fatigue behavior observed between the carbon and Kevlar® materials when tested alone. This observation suggests that the behavior of the hybrids with up to 50% Kevlar® is still dominated by the lower-strain carbon fibers [34].

Loading Parameters and Environment

Tensile mean stress has been shown [35,36] to have an important effect on the fatigue properties of fiber reinforced composites. For 0° and ±15° E-glass/epoxy laminates, the stress amplitude at a constant life tends to decrease with increasing tensile mean stress [35]. In fatigue testing of composites, the test frequency should be chosen so as to minimize the hysteresis heating of the material, which could originate from the viscoelastic nature of the resin and perhaps fiber-matrix interfacial friction. The effects of test frequency have been studied by several workers [37-40], who produced conflicting results. For instance, Dally and Broutman [37] have found a modest decrease in fatigue life with increasing frequency up to 40 Hz for cross-plied as well as quasi-isotropic E-glass/epoxy composites. In contrast, Mandell and Meier [38] observed a decrease in the fatigue life of a cross-ply E-glass/epoxy laminate as the cyclic frequency was reduced from 1 to 0.1 Hz. Sum and Chan [39] have found a moderate increase in fatigue life to a peak value between 1 and 30 Hz. The frequency at which the peak life is observed shifts toward a higher value as the stress level is reduced. However, the effect of frequency is considered negligible for most unidirectional continuous fiber composites tested in the fiber direction.

Fatigue Life and Residual Strength Prediction

Damage mechanics is the study of a physical mechanism by which damage may initiate or propagate. Kachanov [41] proposed the first damage mechanics approach to the prediction of creep rupture of metals at high temperatures. This approach was recently extended for lifetime prediction under complex loading, static failure, crack propagation and bifurcation. Sidoroff [42] later developed a concept of continuum damage mechanics for treating the failure behavior of a material containing a wide spectrum of microdefects. In this approach, damage mechanics and fracture mechanics are considered to be complementary. The final stage of damage mechanics is viewed as the initial stage of fracture mechanics.

By assuming a relationship between fatigue damage and variations in moduli, Poursartip, Ashby and Beaumont [43] developed a fatigue damage mechanics approach for composite laminates. The damage growth rate is assumed to depend on the cyclic stress amplitude $\Delta\delta$ and on the current damage state D:

$$dD/dN = f(\Delta\delta) \qquad (Eq\ 1)$$

which was later extended to include the effect of the load ratio R [44]:

$$dD/dN = f(\Delta\delta, R, D) \qquad (Eq\ 2)$$

For both situations, the relationship between the fatigue damage state D and the variation of moduli is expressed as:

$$E = E_0\, g(D) \text{ or} \qquad (Eq\ 3)$$

$$D = g^{-1}(E/E_0) \qquad (Eq\ 4)$$

Integration of Eq 1 and 2 gives an estimate of life N_f once the damage accumulating function $f(\Delta\delta, R, D)$ has been determined. This function can be obtained experimentally by using [44]:

$$f(\Delta\delta, R, D) = 1 \,/\, [g'g^{-1}(E/E_0)]\,(1/E_0)\,(dE/dN) \quad (Eq\ 5)$$

By varying $\Delta\delta$, R and D in turn and keeping the other variables constant, the results may be plotted out and the function $f(\Delta\delta, R, D)$ determined. This approach was further extended [45] to provide a means for predicting fatigue life of a composite under variable load cycling.

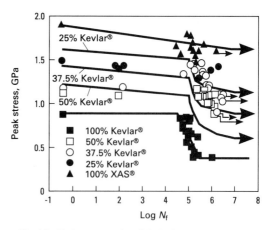

Fig. 18 Fatigue properties of Kevlar/carbon hybrid composites. $R = 0.1$. Source: Ref 34

Waddoups *et al.* [46] evaluated the effect of stress concentration in polymer laminates under both static and cyclic loading conditions. Whitney and Nuismer [47] investigated the macroscopic fracture mechanics approach to the strength prediction of advanced composites containing small holes or through-thickness narrow slits. This approach in general requires heavy testing and data collection.

A quantitative methodology addressing the effect of notches on composite laminate performance in tension was later developed by Kortschot and Beaumont [48] and Spearing *et al.* [49]. For laminates $[90/0]_s$ and $[0/\pm45/90]_s$, predictions of fatigue-induced damage extent and of the resulting residual strength correlated with experimental observations reasonably well. Soutis and Fleck [50] studied compression failure and damage mechanisms in polymer composites with and without holes. They demonstrated that microbuckling of fibers in the 0° layers were the dominant failure mechanism based on both theory and experiments. These efforts [43-45,49-51] constitute a quantitative damage mechanics approach to the failure prediction of notched and unnotched laminates subjected to both tension and compression static and fatigue loading.

Fong [52] proposed a conceptual definition of fatigue damage to assist in the selection of measurement techniques and parameters for correlating damage with fatigue life of composites. A review was conducted [52] of some typical damage parameters for composites, and a survey was conducted on new techniques for fatigue damage monitoring. Hashin [53] formulated the cumulative damage problem in terms of a damage function which must satisfy certain conditions. Such a damage function is shown to unify the residual life and the residual strength methods. Various residual life theories given in the literature can be simply reproduced merely by assuming a general functional dependence of the damage function on fatigue variables.

A large amount of research is currently in progress on developing models to predict the fatigue life of composite

Table 1 Constants in *S-N* representation of composite laminates [4](a)

Material	Lay-up, °	R	m	b	Ref
E-glass/ductile epoxy	0	0.1	−0.1573	1.3743	[20]
T-300 carbon/ductile epoxy	0	0.1	−0.0542	1.0420	[20]
E-glass/brittle epoxy	0	0.1	−0.1110	1.0935	[20]
T-300 carbon/brittle epoxy	0	0.1	−0.0873	1.2103	[20]
E-glass/epoxy	[0/±45/90]$_s$	0.1	−0.1201	1.1156	[22]
E-glass/epoxy	[0/90]$_s$	0.05	−0.0815	0.934	[23]

(a) See Eq 6

laminates. Many of the models are for well-defined fatigue processes, where the types of damage and amount of damage can be predicted using the characteristic damage states (CDS) for given laminates. In actual structural applications, the loading history may be irregular or have overload situations that make prediction of damage and fatigue life less accurate. Furthermore, as discussed earlier, the damage mechanisms of fatigued laminates are more complex than isotropic materials. Variations in the manufacturing process and in-service conditions can also affect the fatigue behavior.

A few simple methods for the prediction of fatigue life and residual strength of composites are briefly described here. Detailed derivations of methods and equations can be found in several references [4,54-64]. Techniques for fatigue life and residual strength predictions of composites can be divided into two categories: empirical models [59,60] and wear-out models [61-64]. The former rely on the assumption that if a set of coupons could be tested both statically and in fatigue, their strengths in both tests would rank in the same order. These models thus require large amounts of data, including static strength, fatigue life, and residual strength distributions, in order to predict residual strengths at any number of cycles [6]. Useful statistical means of manipulating data have been developed in these approaches. The wear-out models are all based on the degradation of a select property, usually stiffness or strength, and an ability to predict the decay of this property during fatigue. Unfortunately, the parameter chosen to describe the wear-out is inevitably a function of stress. This would require testing at many stress levels to describe its behavior.

Simple approaches have been developed to give an analytical representation of the *S-N* curve for the fatigue of composite materials. Many of these approaches are based on linear extrapolations of fatigue data. One such relationship is [4]:

$$S = \sigma_u \, (m \log N_f + b) \qquad \text{(Eq 6)}$$

where S is the maximum fatigue stress, N_f is the number of cycles to failure, σ_u is the average static strength, and m and b are constants. Table 1 lists fatigue data constants for several material combinations for use in the above fatigue life prediction method.

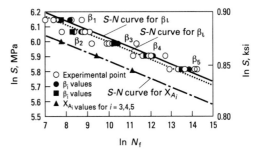

Fig. 19 *S-N* curves for T300 carbon fiber/5208 epoxy composites. Source: Ref 55

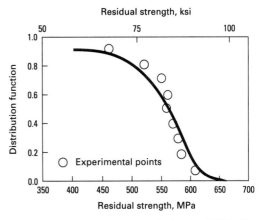

Fig. 20 Residual strength distribution for [0/90/±45] graphite epoxy laminate after 10,000 cycles at 414 MPa (60 ksi). Source: Ref 55

Another method used in the prediction of fatigue life is based on a strength degradation model [55]. The residual strength is monitored and used as an indication of fatigue damage. This model uses a two-parameter Weibull distribution that yields an expression for residual strength (Eq 2) and fatigue life (Eq 3) distributions as follows [65]:

$$R(x) = \exp[-\{(x/\beta_0)^c + n/(1/KS^b)\}\alpha_0/c] \qquad \text{(Eq 7)}$$

$$R(n) = \exp[-(n/N)_f^\alpha] \qquad \text{(Eq 8)}$$

where K and b are data parameters, α_0 is a Weibull shape parameter, β_0 is a Weibull scale parameter, S is the stress level applied, N is the number of fatigue cycles, and c is a constant. Figure 19 shows the S-N relation for a particular laminate with varying Weibull parameters [55]. Experimental values are denoted by open circles. The theoretical distribution of residual strength values are compared to the experimental distribution for the same laminate in Fig. 20.

Still another method uses the characteristic damage states (CDS) estimations at various stages of fatigue damage and performs an estimation of strength reduction based on the estimated damage [56]. This model is primarily used for cross-ply laminates. The model predicts the life during the matrix crack initiation and multiplication stages of fatigue up to the crack saturation point. After the crack saturation point is reached, a standard model to predict fatigue life for unidirectional laminates is used.

References

1. G.E. Dieter, *Mechanical Metallurgy*, 2nd ed., McGraw-Hill, 1987, p 403-447
2. W.W. Stinchcomb and K.L. Reifsnider, The Life-Limiting Process in Composite Laminates, in *Basic Questions in Fatigue Design*, Vol II, STP 924, R.P. Wei and R.P. Gangloff, Ed., ASTM, Philadelphia, PA, 1988, p 294-303
3. M.J. Owen and S. Morris, in *Carbon Fibers: Their Composites and Applications*, Plastics Institute, London, 1971, p 292
4. P.K. Mallick, *Fiber-Reinforced Composites*, Marcel Dekker, Inc., Basel and New York, 1988, p 214-248
5. L.H. Miner, R.A. Wolfe, and C. Zweben, in *Composite Reliability*, STP 580, ASTM, Philadelphia, PA, 1975
6. P.T. Curtis, in *Advanced Composites*, I.K. Partridge, Ed., Elsevier Applied Science, 1989, p 331
7. R.C. Donat, *J. Compos. Mater.*, Vol 4, 1970, p 124
8. S.V. Ramain and D.P. Williams, *Failure Modes in Composites III*, AIME, 1976, p 115
9. R.B. Pipes, in *Composite Materials: Testing and Design* (Third Conference), STP 546, ASTM, Philadelphia, PA, 1974, p 419
10. C.K.H. Dharan, *J. Mater. Sci.*, Vol 13, 1978, p 1243
11. M.J. Owen and S. Morris, *Plastics and Polymers*, Vol 4, 1972, p 209
12. G.M. Newax and S. Mall, *J. Compos. Mater.*, Vol 23, 1989, p 133
13. W. Hwang and K.S. Han, *J. Compos. Mater.*, Vol 23, 1989, p 396
14. C.K.H. Dharan, in *Fatigue of Composite Materials*, STP 569, ASTM, Philadelphia, PA, 1975, p 171
15. B.R. Fujczak, U.S. Army Armament Command, Report No. WVT-TR-74006, March 1974
16. D.C. Phillips and J.M. Scott, *Composites*, Vol 8, 1977, p 233
17. J.D. Donners, J.F. Mandell, and F.J. McGarry, *Proc. 34th Annual Tech. Conf.*, SPI, 1979
18. H.C. Kim and L.J. Ebert, *J. Compos. Mater.*, Vol 12, 1978, p 139
19. R. Talreja, *Proc. Royal Soc. London, A*, Vol 378, 1981, p 461
20. R. Talreja, *Fatigue of Composite Materials*, Technomic Publishing Co., Lancaster, PA, 1987
21. K.E. Hofer, Jr., J.W. Stander, and L.C. Bennett, *Polymer Eng. Sci.*, Vol 18, 1978, p 120
22. L.T. Broutman and S. Sahu, *Proc. 24th Annual Tech. Conf.* SPI, 1969, p 1
23. K. Reifsnider, K. Schultz, and J.C. Duke, in *Long-Term Behavior of Composites*, STP 813, ASTM, Philadelphia, PA, 1983, p 136
24. I.M. Daniel, J.W. Lee, and G. Yanic, in *Proc. 6th Int. Conf. Comp. Materials (ICCM-6) and 2nd European Conf. Comp. Material (ECCM-2)*, F.L. Matthews *et al.* Ed., Elsevier Applied Science Publishing, Ltd., 1987, p 129
25. A.L. Highsmith and K.L. Reifsnider, in *Damage in Composite Materials*, STP 775, Philadelphia, PA, 1982, p 103
26. K.H. Leong and J.E. King, joint FEFG/Int. Conf. on Fracture of Eng. Materials and Structures, Singapore, Aug 1991
27. P.A. Smith and J.R. Wood, *Compos. Sci. Technol.*, Vol 38, 1990, p 85
28. M.D. Shelby, H.B. Hsieh, and B.Z. Jang, *Polymer Compos.*, 1991
29. P.T. Curtis, "An Investigation of the Mechanical Properties of Improved Carbon Fiber Composite Materials," RAE TR 86021, 1986
30. K.H. Boller, *Modern Plastics*, Vol 41, 1964, p 145
31. M.J. Owen and R.G. Rose, *Modern Plastics*, Vol 47, 1970, p 130
32. J.W. Davis, J.A. McCarthy, and J.N. Schrub, *Mater. Design Eng.*, 1964, p 87
33. P.T. Curtis and B.B. Moore, RAE TR 85059, 1985
34. T. Adam, R.F. Dickson, C. Fernando, B. Harris, and H. Reiter, *Proc. Intl. Conf. Fatigue of Engineering Materials and Structures*, Sheffield, 1986, Inst. of Mech. Eng., London, 1986, p 329
35. K.H. Boller, U.S. Air Force Materials Lab., Report No. ML-TDR-64-86, June 1964
36. M.J. Owen and S. Morris, *Carbon Fibers: Their Composites and Applications*, Plastics Institute, London, 1971, p 292
37. J.W. Dally and L.J. Broutman, *J. Compos. Mater.*, Vol 1, 1967, p 424
38. J.F. Mandell and U. Meier, in *Long-Term Behavior of Composites*, STP 813, ASTM, 1983, p 55
39. C.T. Sun and W.S. Chan, in *Composite Materials: Testing and Design* (Fifth Conf.), STP 674, ASTM, Philadelphia, PA, 1979, p 418
40. C.R. Saff, in *Long-Term Behavior of Composites*, STP 813, ASTM, Philadelphia, PA, 1983, p 78
41. L.M. Kachanov, Time to Failure under Creep Conditions, *Ak. Nauk. SSR. Otd. Tekh. Nauk.*, No.8, 1958, p 26-31
42. F. Sidoroff, Damage Mechanics and its Application to Composite Materials, in *Mechanical Characterization of Load Bearing Fibre Composite Laminates*, Cardon and Verchery, Ed., Elsevier, New York, NY, 1985
43. A. Poursartip, M.F. Ashby, and P.W.R. Beaumont, Damage Accumulation During Fatigue of Composites, *Scripta Metall.*, Vol 16, 1982, p 601-606
44. A. Poursartip, M.F. Ashby, and P.W.R. Beaumont, The Fatigue Damage Mechanics of a Carbon Fibre Composite Laminate: I-Development of the Model, *Compos. Sci. Technol.*, Vol.25, 1986, p 193-217
45. A. Poursartip and P.W.R Beaumont, The Fatigue Damage Mechanics of a Carbon Fibre Composite Laminate: II-

Life Prediction, *Compos. Sci. Technol.*, Vol 25, 1986, p 283-299

46. M.E. Waddoups, J.R. Eisenmann, and B.E. Kaminski, Macroscopic Fracture Mechanics of Advanced Composite Materials, *J. Compos. Mater.*, Vol 5, 1971, p 446-454

47. J.M. Whitney and R.J. Nuismer, Stress Fracture Criteria for Laminated Composites Containing Stress Concentrations, *J. Compos. Mater.*, Vol 8, 1974, p 253-265

48. M.T. Kortschot and P.W.R. Beaumont, "Damage-Based Notched Strength Modeling: A Summary," presented at ASTM Conf. on Fracture and Fatigue of Composites," Orlando, FL, 1989

49. S.M. Spearing, P.W.R. Beaumont, and M.F. Ashby, "Fatigue Damage Mechanics of Notched Graphite/Epoxy Laminates," Presented at ASTM Conf. on Fracture and Fatigue of Composites, Orlando, FL, 1989

50. C. Soutis and N.A. Fleck, Static Compression Failure of Carbon-Fibre T800/924C, Composite Plate with a Single Hole, *J. Compos. Mater.*, Vol 24, 1989, p 536-558

51. K.T. Kedward and P.W.R. Beaumont, "The Treatment of Fatigue and Damage Accumulation in Composite Design," Cambridge Univ. Eng. Dept. Report, CUED/C-MATS/TR.185, June 1991

52. J.T. Fong, in *Damage in Composite Materials*, STP 775, K.L. Reifsnider, Ed., ASTM, Philadelphia, PA, 1982, p 243-266

53. Z. Hashin, *Compos. Sci. Technol.*, Vol 23, 1985, p 1-19

54. C. Zweben, H.T. Hahn, and T. Chou, Mechanical Behavior And Properties of Composite Materials, in Delaware Composites Design Encyclopedia, Vol 1, Technomic Publishing Co., p 73-127

55. R.Y. Kim, in *Engineered Materials Handbook*, Vol 1, *Composites*, ASM International, Metals Park, OH, 1988, p 436-444

56. J.W. Lee, I.M. Danial, and G. Yaniv, Fatigue Life Prediction of Cross-Ply Composite Laminates, in *Composite Materials: Fatigue and Fracture*, Vol 2, STP 1012, P.A. Lagace, Ed., ASTM, Philadelphia, PA, 1989, p 19-28

57. W. Hwang and K.S. Han, Fatigue of Composite Materials—Damage Model and Life Prediction, in *Composite Materials: Fatigue and Fracture*, Vol 2, STP 1012, P.A. Lagace, Ed., ASTM, Philadelphia, PA, 1989, p 87-102

58. R. Talreja, Estimation of Weibull Parameters for Composite Material Strength and Fatigue Life Data, in *Fatigue of Fibrous Composite Materials*, STP 723, ASTM, Philadelphia, PA, 1981, p 291-311

59. H.T. Hahn, USAF Wright Patterson Labs Report, AFML-TR-78-45, 1978

60. J.N. Yang and S. Du, *J. Compos. Mater.*, Vol 17, 1983, p 511

61. J.N. Yang, D.L. Jones, S.H. Yang, and A. Meskini, *J. Compos. Mater.*, Vol 24, 1990, p 753

62. Lin Ye, *Compos. Sci. Technol.*, Vol 36, 1989, p 339

63. R.B. Pipes, S.V. Kulkarni, and P.V. McLaughlin, *Mater. Sci. Eng.*, Vol 30, 1977, p 113

64. A. Poursatip, M.F. Ashby, and P.W.R. Beaumant, in *Proc. Intl Conf. Composite Materials IV (ICCM-4)*, Tokyo, 1982

65. R.Y. Kim, Fatigue Behavior, in *Composites Design*, S.W. Tsai, Ed., Think Composites, 1989, p 19-1 to 19-22

Part III

Polymer Composites for Special Applications

Chapter 10

Vibration Testing and Damping Analysis in Composites

Introduction

Vibrational and damping characteristics of composites are important in many applications, including ground-based and air-borne vehicles, space structures, and sporting goods. In response to a transient or dynamic loading, structures can experience excessive vibrations that create high noise levels, stress fatigue failure, premature wear, operator discomfort, and unsafe operating conditions. An improved understanding of various theoretical and experimental aspects to vibration in composites is essential to the efforts to avoid or eliminate these potential problems.

The most widely used vibration properties are damping and dynamic modulus. Consider the stress-strain curve for a load cycle applied repeatedly to a material (Fig. 1). The process of "friction" causes a loss of energy per cycle which is equal to the area, ΔW, contained within the hysteresis loop. The modulus is defined by the average slope of the stress-strain loop. For most materials the loop almost degenerates into a straight line, which provides a unique definition of modulus. The energy dissipation under steady-state vibration is usually defined in terms of specific damping capacity ϕ, which is the ratio of $\Delta W/W$, where $W = \sigma^2/2E$ is the maximum strain energy per unit volume. Damping can be considered as the transformation of vibrational energy into some other form of irrecoverable energy.

The bandwidth of half-power points in steady-state sinusoidal excitation can be specified in terms of the dimensionless damping ratio ε, which can be calculated from $\varepsilon = (\omega_2 - \omega_1)/(2\omega_n)$, where ω_2 and ω_1 are the two nearby frequencies at which the amplitude drops to $1/\sqrt{2}$ of that at the resonant frequency ω_n. The quality factor Q is defined as $1/(2\varepsilon)$. By using the complex modulus approach, we have $E^* = E' + iE'' = E'(1 + i\eta)$, where E^* is the complex modulus E' and E'' are the respective real (storage) and imaginary (dissipation)

components of E^*, and η is called the loss factor or loss tangent, since $\tan \theta = E''/E' = \eta$. A measure of the decay of free vibration is the logarithmic decrement, defined as $\delta = \ln(a_i/a_{i+1})$. The above measures of damping can be interrelated through the following equation:

$$\eta = 2\varepsilon = 1/Q = \phi/2\pi = \delta/\pi = (\omega_2 - \omega_1)/\omega_n$$
$$= E''/E' = \tan \theta \qquad \text{(Eq 1)}$$

Analysis of Vibration in Composites

Macromechanical Modeling

Both macromechanical and micromechanical modeling of vibration in composites have been reviewed [1-3]. These review articles discuss the analytical and modeling techniques that have been developed for predicting vibra-

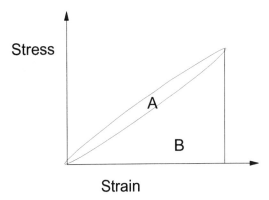

A = ΔW = energy dissipated
W = A + B = energy input
$\Delta W/W$ = specific damping capacity

Fig. 1 Specific damping capacity in a mechanical system

tion properties of laminated composites. Sun *et al.* [4] used a force-balance approach in conjunction with classical lamination theory to predict modulus and damping properties in composites. The classical lamination theory requires eighteen material constants in the constitutive equations of laminated composites. These include six extensional stiffness [A], six coupling stiffness [B], and six flexural stiffness constants [D]. The damping corresponding to [A], [B], and [D] were obtained by first expressing [A], [B], and [D] in terms of the stiffness matrix $[Q]_k$ and η_k of each lamina and then using Q_{ij} in relation to the four basic engineering constants E_L, E_T, G_{LT}, and ν_{LT}. In light of the elastic and viscoelastic correspondence principle, these constants were then respectively replaced by E_L^*, G_{LT}^*, ν_{LT} and $[A^*]$, $[B^*]$, and $[D^*]$. By separately equating the real parts and the imaginary parts, A_{ij}', A_{ij}'', B_{ij}', B_{ij}'', and D_{ij}', D_{ij}'' can be expressed in terms of the material damping η_L^k and η_T^k of each lamina. The results of this study are summarized as follows:

$$A_{ij}^* = A_{ij}' + i\ A_{ij}'' = \Sigma(Q_{ij}' + Q_{ij}'')_k\,(h_k - h_{k-1})$$
$$B_{ij}^* = B_{ij}' + i\ B_{ij}'' = 1/2\Sigma(Q_{ij}' + Q_{ij}'')_k\,(h_k^2 - h_{k-1}{}^2)$$
$$D_{ij}^* = D_{ij}' + i\ D_{ij}'' = 1/3\Sigma(Q_{ij}' + Q_{ij}'')_k\,(h_k^3 - h_{k-1}{}^3)$$
$$(Eq\ 2)$$

where $i = \sqrt{-1}$ and both Q_{ij}' and Q_{ij}'' are functions of E_L', E_T', G_{LT}', E_L'', E_T'', and G_{LT}''. The expressions for E_L', E_L'', etc. have been derived by Sun, Gibson and co-workers [5,6].

The material damping in laminated composites is therefore expressed as [4]:

$$\eta_{ij} = (\text{in–plane}) = A_{ij}''\ /\ A_{ij}'$$
$$\eta_{ij} = (\text{coupled}) = B_{ij}''\ /\ B_{ij}'$$
$$\eta_{ij} = (\text{flexural}) = D_{ij}''\ /\ D_{ij}'$$

As a second example, Adams and Bacon [7] have studied the effect of fiber orientation and laminate geometry on the dynamic properties of composites. In their theoretical treatment, both the strain energy and the energy dissipated in any element of each layer were separated into three components, corresponding to longitudinal tension/compression, transverse tension/compression, and longitudinal shear loading. The total energy dissipation ΔW was obtained by integrating the sum of the three energy dissipation components over the whole plate section. The maximum strain energy W stored by the plate can also be calculated by the lamination theory. The specific damping capacity of the laminate is then given by $\phi = \Delta W/W$. In this approach, three lamina-specific damping capacity parameters used in the calculation have to be measured experimentally. The solution to the equations in this method is usually only feasible using computerized numerical techniques. An improved version of this method has been reported [8]. The variational principles have been utilized by Alan and Asnani [9,10] to derive the governing equations of mo-

tion for a laminated plate. Other closed form solutions for vibration properties of laminated beams [11-14], laminated shells [15-19], and composite membranes [20] have also been reported. Finite element methods have also been employed to analyze vibration of composite plates [21-27].

Micromechanical Modeling

Considerable progress has been made in the recent research on micromechanical models for improvement and optimization of damping in composites [4,6,28-34]. The loss factor of unidirectional continuous fiber composites can be derived from a simple rule-of-mixture equation in conjunction with the elastic-viscoelastic correspondence principle [35]. Assuming that the composite is loaded along the fiber direction and that all the energy dissipation occurs in the matrix:

$$\eta_c = (1 - V_f)\,\eta_m\,E_m/E_c \qquad (Eq\ 4)$$

where η_m is the loss factor of the matrix under direct loads and the elastic modulus E_c of the composite in the fiber direction is given by the rule-of-mixtures equation, $E_c = E_f V_f + E_m(1 - V_f)$. The subscripts c, m, and f represent composite, matrix, and fiber, respectively. Equation 4 underestimates considerably the measured value of η_c. This discrepancy is likely due to the following contributing factors [1]:

- The smaller the fiber diameter, the larger the surface area per unit volume. A consistent increase in η_c with reduction in fiber diameter was observed [36] with glass fiber-polyester system. This implies that fiber contribution cannot be neglected.
- The effect of fiber misalignment was not considered in the equation.
- Structural defects, such as voids and interfacial debonding cracks, can lead to interfacial rubbing and hence additional energy loss.

A theoretical analysis of internal damping and dynamic stiffness for aligned discontinuous fiber composite was developed based on the micromechanics models for the complex moduli. In this approach, the expression for the elastic stiffness of the discontinuous composite was derived first. Then the elastic-viscoelastic correspondence principle was used to obtain the expression for the complex modulus $E_c^* = E_c' + E_c''$ [5,6]:

$$E_c' = E_f'\,V_f\left(\frac{1 - \tanh\beta l\,/\,2}{\beta l\,/\,2}\right)$$

$$-\ \frac{E_f''\,V_f}{2}(\eta_{G_m} - \eta_{rf})\left(\frac{\tanh\beta l\,/\,2}{\beta l\,/\,2} - \frac{1}{\cosh^2\beta l\,/\,2}\right)$$

$$+\ E_m'\,V_m \qquad (Eq\ 5)$$

$$E_c'' = E_f'' V_f \left(\frac{1 - \tanh \beta l / 2}{\beta l / 2} \right)$$

$$- \frac{E_f' V_f}{2} (\eta_{G_m} - \eta_f) \left(\frac{\tanh \beta l / 2}{\beta l / 2} - \frac{1}{\cosh^2 \beta l / 2} \right)$$

$$+ E_m'' V_m \qquad \text{(Eq 6)}$$

$$\eta_c = E_c'' / E_c' \qquad \text{(Eq 7)}$$

$$\beta^2 = (G_m' / E_f') [2 / (r_0^2 \ln(R/r_0))] \qquad \text{(Eq 8)}$$

The symbols η_c and η_f are the extensional loss factor of composite and fiber, respectively, while η_{G_m} is the shear loss factor of matrix. G_m and E_m are the shear storage modulus and extensional loss modulus of matrix. The fiber length is l while the ratio (R/r_0) is related to the dimensions, distribution, and volume fraction of fibers [5].

This method predicts an optimum fiber aspect ratio for maximum damping. The complex moduli of composites subjected to off-axis loading can also be obtained using a similar approach [29,32,33,37,38]. When the anisotropic nature of the fibers is taken into account, the micromechanics equations can be modified to give expressions for the complex transverse modulus and the complex in-plane shear modulus of a composite [4,31]. This approach was used [4,31] to study the influence of fiber length and fiber orientation on damping and stiffness of polymer composites. These results show that very low fiber aspect ratios are required to produce significant improvements in damping. Predictions and measurements also demonstrate that the control of lamina orientation in a continuous fiber composite may be a better approach to the improvement of damping than the control of the fiber aspect ratio.

For longitudinal shear loading of a viscoelastic composite, Hashin [39] has shown that

$$\eta_{LT} = \frac{\eta_m (1 - V_f)[(G + 1)^2 + V_f (G - 1)^2]}{[G(1 + V_f) + 1 - V_f][G(1 - V_f) + 1 + V_f]}$$
$$\text{(Eq 9)}$$

where $G = G_f / G_m$, the ratio of shear moduli between the fiber and the matrix. No reliable theories exist for predicting η_T, the loss factor of a unidirectional composite in transverse flexure. However, Willway and White [40] used the mechanics of materials approach to derive the following equation, assuming all energy dissipation to occur in the resin:

$$\eta_T = \eta_m E_T (1 - V_f)/E_m \qquad \text{(Eq 10)}$$

where E_T is the elastic modulus of the unidirectional lamina in transverse flexure, which is given by

$$E_T = E_m/[(1 - V_f) + V_f(E_m/E_{Tf})] \qquad \text{(Eq 11)}$$

Here, E_{Tf} is the transverse modulus of elasticity of the fibers. Both Eq 10 and 11 are used for both continuous and discontinuous unidirectional composites. The same approach was adopted [41] to derive an equation for predicting the transverse shear loss factor of a unidirectional composite:

$$\eta_s = (1 - V_f) \eta_m G_c/G_m \qquad \text{(Eq 12)}$$

where G_c is the transverse shear modulus of a unidirectional composite given by

$$G_c = (G_m G_f)/[(1 - V_f)G_f + V_f G_m] \qquad \text{(Eq 13)}$$

Alternative micromechanical models for the prediction of damping properties of continuous fiber reinforced laminas from constituent material properties and volume fractions have been proposed [42,43]. Research also has been conducted to analyze the vibration response of randomly oriented short fiber-reinforced laminae [30]. A combined micromechanical strain energy/finite element technique has been developed [44,45] for modeling damping in discontinuous aligned fiber composites. This technique has the capability of predicting fiber interaction and fiber interface effects on composite damping. The numerical data and experimental measurements show that fiber interaction does affect the damping of discontinuous fiber composites, and that damping can be improved by increasing the fiber end gap size or by decreasing the fiber aspect ratio. The finite element implementation of the strain energy approach is a powerful method for predicting damping in composites [44].

Experimental Techniques

The macroscopic, anisotropic nature of fiber-reinforced composites with respect to damping and stiffness makes the determination of the vibration characteristics considerably more complicated than for homogeneous, isotropic materials. However, the basic techniques used in these two cases are very similar [46]. Modal analysis is the process of determining the dynamic properties of structures and is considered to be the most effective method for investigating the vibration problems of composites. Modal analysis determines a mathematical model of a structure based on its natural frequencies, mode shapes, and damping factors [47]. Both storage modulus and loss modulus can be obtained from modal testing data.

Measurement of Vibration Properties in Composite Structures

Vibration measurement techniques for composite structures can be divided into two categories: time domain methods and frequency domain methods. Time domain methods include the logarithmic decrement technique and the reverberation technique. Frequency domain methods include the peak amplitude technique, the

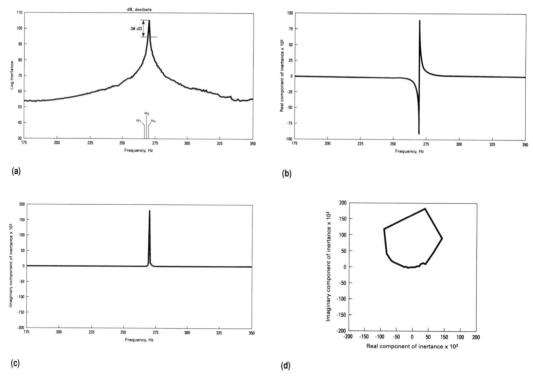

Fig. 2 Frequency response function. Inertance is acceleration divided by force, where acceleration is expressed in g's (gravitational acceleration) and force is in lbf. (a) Frequency response function (FRF), in the frequency domain, of a dynamic system. (b) The real part of a FRF (The CO-plot). (c) The imaginary part of a FRF (The QUAD-plot). (d) The Argand plane (the Vector plot, or Nyquist plot)

CO-and-QUAD Plots and the Kennedy and Pancu Method. See [3] for a recent review of these methods.

Consider the structural system to be a multi-degree-of-freedom linear system governed by

$$\{m\}\{\ddot{U}\} + \{c\}\{\dot{U}\} + [k]\{U\} = \{F\} \qquad \text{(Eq 14)}$$

where $\{U\}$ is the displacement vector, $\{m\}$ is the system mass matrix, $\{c\}$ is the system damping matrix, and $[k]$ is the system stiffness matrix.

If the system is assumed to be proportionally damped, then the transformation

$$\{U\} = [\phi]\{q\} \qquad \text{(Eq 15)}$$

permits the system to be described by:

$$M_i(d^2 q_i / dt^2) + c_i(dq_i / dt) + K_i q_i = F_i \qquad \text{(Eq 16)}$$

where q_i is the ith modal coordinate, M_i is the ith modal mass ($= \{\phi_i\}^T[m]\{\phi_i\}$), c_i is the ith modal damping constant ($= \{\phi_i\}^T[c]\{\phi_i\}$), and K_i is the ith modal stiffness ($= \{\phi_i\}^T[k]\{\phi_i\}$). The damping matrix $[c]$ can be measured using system identification techniques [48,49]. However,

in the modal testing method, only the modal damping ratio ε_i is measured, where $\varepsilon_i = c_i / (2K_iM_i)$.

Recent advances in high-speed data processing techniques have permitted accurate and efficient measurements of the modal damping ratio for each mode. The introduction of the fast Fourier transform (FFT) algorithm and the development of software programs for processing modal testing data have made modal testing a powerful technique for structural dynamic analysis. In the following sections, select time domain and frequency domain techniques will be briefly discussed.

Time Domain Methods. The logarithmic decrement technique is one of the time domain methods of estimating damping coefficients. The method involves exciting the structure at one of its natural modes of vibration and then suddenly disconnecting the shakers from the structure. The time-history of the response (acceleration) measured at the driving points can be used to measure the damping coefficient associated with that particular mode:

$$\varepsilon = \delta/(2\pi) = (1 / 2\pi n) \ln(a_1/a_n) \qquad \text{(Eq 17)}$$

where a_1 is the response amplitude at the first cycle of decay curve and a_n is the amplitude after n cycles. This technique can be used for a lightly damped system. A curve fitting technique can also be used for time domain applications, where high modal density and modal interference exist. However, this technique involves extensive data reduction [50]. The reverberation time method uses the short time-averaged squared response, a^2, and decibel measurements. The displacement level L_a is defined as

$$L_a = (10 \text{ dB}) \log (a^2/a_{ref}^2) \qquad \text{(Eq 18)}$$

where a is the displacement and a_{ref} is a reference displacement. The vibration reverberation time, T_R, the time for the displacement level to be reduced by 60 dB, is given by

$$T_R = 6.90/\varepsilon\omega_n \qquad \text{(Eq 19)}$$

Using the FFT algorithm, the time history of test data can be transformed into frequency response data. The techniques used in the frequency domains are generally more versatile.

Frequency Domain and Modal Testing Methods. The vibrational motion (response) can be measured in terms of displacement, velocity or acceleration. The corresponding ratios of motion to force are defined as follows [51]:

- Receptance (compliance or admittance) = (displacement)/(force)
- Mobility = (velocity)/(force)
- Inertance (or accelerance) = (acceleration)/(force)

These terms are further classified as "point," if the excitation and measurement points coincide, or "transfer," if they are different; for example, "point receptance" and "transfer receptance". These ratios are broadly termed "frequency response functions" (FRF).

Peak Amplitude Method. In this technique, the structure is excited by a sinusoidal force from a shaker and the frequency response curves of the total amplitude, obtained at several points on the structure, are recorded as a function of frequency. The frequencies where peaks are attained are usually identified as the natural frequencies, although some reservation exists on the accuracy of this definition for natural frequencies [51,52]. These frequencies, in many cases, can be used to calculate the dynamic storage modulus of a structure [53]. The amount of damping in a particular mode is determined from the sharpness of the peak (Fig. 2a):

$$\eta = 2\varepsilon = (\omega_2 - \omega_1)/\omega_n \qquad \text{(Eq 20)}$$

where ω_n is the natural frequency and ω_1 and ω_2 are the frequencies on either side of ω_n where the peak amplitude is reduced by a factor of $\sqrt{2}$.

Methods Using CO and QUAD Plots. The frequency response of a linear system can be decomposed into two parts: that in phase with the input (or the real component); and that 90° out of phase with the input. When plotted versus frequency, they are referred to as the CO-plot and QUAD-plot, respectively [50]. From the peaks of the CO-plot (Fig. 2b), the damping ratio can be computed from:

$$\varepsilon = [(\omega_2 / \omega_1)^2 - 1] / 2[(\omega_2 / \omega_1)^2 + 1] \qquad \text{(Eq 21)}$$

where ω_1 and ω_2 are the frequencies on either side of ω_n where the CO response reaches a maximum [50]. From the QUAD-plot (Fig. 2c), the damping ratio can be computed from:

$$\varepsilon = (\omega_2 - \omega_1) / (2\omega_n) \qquad \text{(Eq 22)}$$

where ω_1 and ω_2 are frequencies where the amplitude of the QUAD response is equal to half of the value of the response at resonance.

Kennedy and Pancu Method. Instead of plotting the amplitude and phase response as a function of frequency, the complex admittance can be plotted in the Argand plane (Fig. 2d). In such a diagram (vector plot, polar plot, CO-QUAD plot, or Nyquist plot), the real component is plotted against the imaginary component of the admittance. Both damping ratio and natural frequency can be obtained from such plots [51,54].

For more complex structures, multiple exciter techniques are usually required. Using the automatic force appropriation techniques (e.g., Asher's methods [55]), the force ratios necessary for tuning the principal modes of vibration can be calculated from the complex admittance matrix, the columns of which are measured experimentally using sinusoidal excitation. By plotting around the resonance frequencies using the force ratios calculated, the modal parameters can be extracted [51].

The newly developed multichannel real-time FFT analyzers have made it possible to measure response to wide band excitation signals. Using these analyzers the FRF of a structure can be measured at a single point due to impulse excitation at various points on the structure. The structure can also be excited at a single point, using various forms of wide band random signals, and the FRF measured at several points. The modal parameters are then extracted by analytically curve-fitting the measured data in both the time and frequency domains [51]. The results of applying this approach to determine the vibration response of a simple graphite/epoxy composite plate is shown in Fig. 3, where a two-dimensional view of a three-dimensional animated structure for the first three modes is given.

Measurement of Dynamic Mechanical Properties of Composite Materials

The general techniques used to measure the dynamic mechanical properties of composite materials are the same as those used in measuring the vibration properties of composite structures. However, in measuring the dynamic mechanical properties of materials, one is more interested in knowing the complex moduli than just the natural frequency and mode shapes. As with isotropic materials, the complex moduli of composite materials can

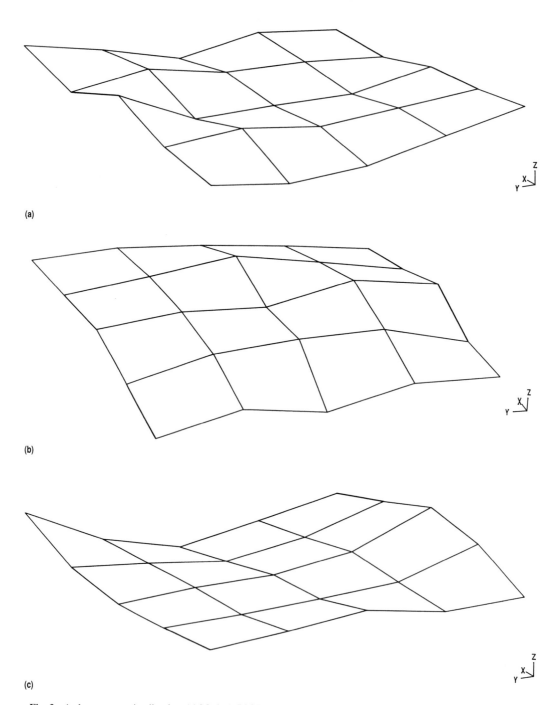

Fig. 3 A plate structure in vibration. (a) Mode 1, 76.25 Hz, damping 0.17%. (b) Mode 2, 139.02 Hz, damping 0.60%. (c) Mode 3, 191.76 Hz, damping 0.39%

be measured by the resonance method, non-resonant forced vibration method, free vibration method, and wave propagation method [56-61]. The type of motion, specimen configuration and fiber orientation determine the kind of complex moduli measured. Although it is conceptually possible to determine all of the complex moduli of a composite, this is very difficult to achieve experimentally because of such complicating factors as shear-nor-

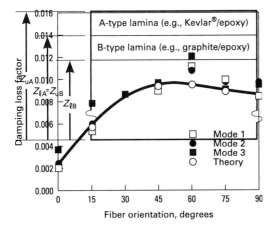

Fig. 4 Comparison between model prediction and experimental results on the damping loss factor of a glass fiber/epoxy laminate as a function of fiber orientation

mal coupling, torsion-flexure coupling, and large number of orthotropic complex constants to be determined [46].

Effects of Material Variables and Structural Parameters

Four primary mechanisms have been suggested to contribute to damping in composites: viscoelastic response of the constituents, friction at the fiber-matrix interface, thermoelastic damping due to cyclic heat flow, and damage initiation and growth. The damping ratio of a composite is dictated primarily by the viscoelastic or microplastic phenomena in the matrix and relative slipping at the fiber-matrix interface.

In an experimental investigation of material damping in laminated resin composites, Yim *et al.* [62] have found damping to be highly sensitive to stacking sequence and the type of matrix used in the composites studied. These composites include off-axis, cross-ply and quasi-isotropic continuous fiber laminates constructed from glass/epoxy, carbon/epoxy and carbon/PEEK. In this study [62], damping loss factor for the 32 ply glass/epoxy composite as a function of ply orientation and frequency is shown in Fig. 4. Damping is found to be lowest in the fiber direction, increase to a maximum value in the range of 45 to 60° and decrease slightly as the orientation approaches 90°. An increase in damping by a factor of 4 to 5 in this material for the 45-60° orientations relative to 0° is observed. This increase is attributed to the viscoelastic response excited by the off-axis stacking sequence where transverse normal and shear stresses are present [62]. A comparison of the unidirectional, bi-directional and quasi-isotropic data for the 32 ply carbon/epoxy laminates indicated that increasing the proportion of 0° plies in the laminates reduced damping [62]. This observation further confirms the viscoelastic response of the

matrix being the primary damping mechanism in the case of carbon fiber-reinforced composites. The same authors [62] also concluded that cyclic heat flow was not an important mechanism of energy dissipation, since no significant dependence of damping on specimen thickness was noted.

The contribution of fiber *per se* to the total internal damping of composite is usually negligible. However, an emerging class of high performance thermoplastic fibers may possess an unusually high damping coefficient. Kevlar® aramid fiber has a loss factor of 0.018, which is more than ten times that of graphite or glass fibers, and in the same range as the damping of the resins [63]. Epoxy has a loss factor of approximately 0.026, while vinyl ester and polyester have loss factors of about 0.045.

Except for composites containing these fibers and excluding the contribution from any cracks and other defects, the internal damping of a composite is determined by the following variables [1]:

- Properties and relative proportions of the matrix and the reinforcement
- Dimensions of the inclusions
- Orientation of the reinforcement with respect to the loading axis
- Surface treatments of the reinforcement
- Void content

The dynamic modulus of a composite is also a function of these factors. Testing parameters and environmental factors such as amplitude, frequency, and temperature will also affect both the dynamic modulus and the damping values of a composite. Detailed discussion on this subject can be found in [1,2,46].

Chottiner *et al.* [64] investigated the effect of fiber type and matrix modifiers on the damping properties of laminates. Fibers evaluated included Kevlar, graphite, glass and polyester. The epoxy resin matrices were modified by incorporating reactive diluents, carboxy-terminated polybutadiene (CTBN) and a second phase or interpenetrating polymer network of a thermoplastic polymer (semi-IPN). Two- to three-fold increases in the damping factor of composite laminates have been achieved by the substitution of Kevlar® and Dacron® (polyester) reinforcements for glass fibers. Additional enhancement of the damping properties has been achieved with thermoplastic additives capable of forming semi-IPNs. Five-fold improvements have been obtained with combinations of Kevlar and IPN modifiers. Using a thermoplastic component to produce semi-IPN is particularly attractive, since it can also serve to toughen the otherwise brittle epoxy matrix.

Damping of Interply Hybrid Composites

Laminas of different materials can be combined to make a hybrid laminate. In order to tailor the properties of a hybrid laminate to design specifications, it is necessary to develop predictive rules for property calculation. Hoa and Ouellette [65] have derived the following equation for the prediction of the loss factor of hybrid laminates using combined rule-of-mixture law and strain-energy approach:

$$\eta_h = 2 / 3\, d_{11}\, [\Sigma \eta_A E_A (Z_u^3 - Z_l^3)_A \\ + \Sigma \eta_B E_B (Z_u^3 - Z_l^3)_B] \qquad \text{(Eq 23)}$$

where (see Fig. 5) Z_l and Z_u represent the lower and upper coordinates of the layers of either the type A or the type B material, η_A is the loss factor of the A layer, E_A is the elastic modulus of A layer, etc. and d_{11} is the component of the compliance matrix $[d]$ obtained from the lamina properties. The results from this equation show reasonable agreement (within 15% error) with experimental values [65].

Damping of Textile Structural Composites

With the advancement of textile structural composites such as 2-dimensional and 3-dimensional fabric based, the damping properties of these materials are likely to become important design considerations in aerospace and automotive applications. Models developed for predicting the elastic constants of textile structural composites include Mosaic, Crimp, and Bridging models for 2-dimensional fabric composites (both mono-fiber type and co-woven hybrids) and Fiber Inclination and Fabric Geometry models for 3-dimensional composites. Jang and Yim [35,36] have extended these models for predicting damping in various textile structural composites.

As an example, consider the Mosaic Model which involves simulating the fabric configuration with an equivalent cross-ply laminate and using the various equations derived in the classical lamination theory for the extensional stiffness [A], coupling stiffness [B], and bending stiffness [D]. Predictive equations for damping are obtained by first expressing [A], [B], and [D] in terms of the $[Q]_k$ stiffness matrix and layer location h_k of each laminate. The stiffness matrices are expressed using the four basic engineering constants E_L, E_T, G_{LT}, and ν_{LT}, which can be replaced with corresponding complex elastic constants by using the elastic-viscoelastic correspondence principle. In this approach, the basic damping factors η_L, η_T, η_{LT}, η_v for each equivalent lamina are obtained from the modified Hashin theory [39] and rule-of-mixtures approach (e.g., Eq 4, 9, and 10). By substituting these constants into the expressions for $\overline{A}, \overline{B}$, and $\overline{D}$ in the Mosaic model (Chapter 5) for both upper and lower bound solutions, one can obtain the damping coefficients $A\eta_{ij} = A_{ij}'' / A_{ij}'$, $B\eta_{ij} = B_{ij}'' / B_{ij}'$, and $D\eta_{ij} = D_{ij}'' / D_{ij}'$. Damping properties of hybrid fabric composites can be similarly obtained by extending the elastic equations obtained by Ishikawa and Chou [41] to the corresponding viscoelastic ones.

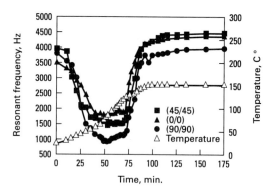

Fig. 6 Cure monitoring data obtained from the frequency response function for carbon fiber/epoxy composites (3 stacking sequences)

Applications

Composites for noise and vibration controls, dynamic behavior of composites in machinery, and vibration of space structures have been recently reviewed [2]. This section will emphasize the application of vibration based techniques for nondestructive evaluation of materials. In particular, the subjects of cure monitoring of both thermosetting resins and composites and damage assessment of composites will be discussed.

Cure Monitoring

A new nondestructive technique, known as Mechanical Impedance Analysis (MIA), has been proposed and evaluated by Jang et al. [66,67] for cure cycle design and cure monitoring of thermosetting resins and composites. The basic technique involves measuring the vibrational response of a composite material under a prespecified excitation. Various forms of excitation can be exerted on the material with the frequency response functions measured and analyzed by a front-end data acquisition and FFT spectrum analyzer. The frequency response spectra of composites being cured in a molding press can be measured and analyzed. Information on the rheological characteristics such as the moment when the matrix starts to flow, the time for cross-linking to start, and the region of minimum viscosity can be obtained.

This MIA technique can be performed in real-time and there is no restriction on sample size and shape. In contrast, most other NDE techniques have limitations which prevent them from being completely effective when applied to integrated composite structures (exceptions include ultrasonic, fiber optic, microdielectric and fluorescence techniques). For a continuous structure, a number of peaks will occur on the frequency plot. For impedance data, each peak corresponds to a natural frequency of vibration. The storage modulus of a material can be obtained from the resonance data. The damping coefficient can also be identified from the shape of the peak by using either the half-power-point method or a

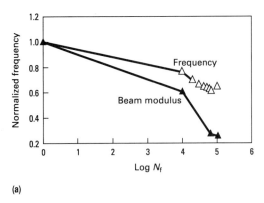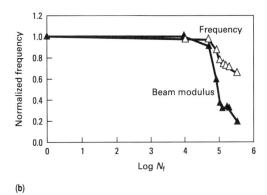

Fig. 7 Variations of resonant frequency and modulus of carbon fiber/epoxy laminates subjected to fatigue loading. (a) Unidirectional laminate, fatigued at 64% UTS. (b) Cross-plied laminate, fatigued at 40% UTS

curve fitting technique. Software packages for modal analysis are available.

Figure 6 shows the resonant frequency values from frequency response functions for three graphite/epoxy laminates heated from 23 to 150 °C (73 to 302 °F) and then isothermally cured at 150 °C (302 °F) for two hours. At room temperature (23 °C, or 73 °F) the uncured prepreg laminate showed a damped viscoelastic behavior. As the temperature increased but before resin gelation occurred, the epoxy resin became more and more liquid-like and the composite sample behaved like a viscoelastic liquid. A corresponding minimum in the resonant frequency curve is observed in Fig. 6. No apparent peak in the frequency response function representing resonance was observed in this cure state (at 40 to 80 °C, or 104 to 176 °F). Chain extension and possibly cross-linking started to become pronounced at about 55 minutes and continued to proceed when the sample was heated and held at 150 °C (302 °F). As the cross-linking reaction proceeded, the composite sample exhibited more and more elastic behavior, indicating the transformation of the material from a liquid to a rubbery and eventually a rigid glassy state. The point at which E' (resonant frequency) starts to increase drastically represents the onset of an active cross-linking reaction, which causes the epoxy resin to change from a viscous liquid to a viscoelastic solid. These transition points would exist at different cure times for different curing temperatures. The steady state value for E' at the latter stage of curing indicates the near completion of the cross-linking reaction. Cure time as defined in ASTM Standard D 4475 can be obtained as a function of cure temperature. An optimum cure region for the production process can then be estimated. This information can be saved as part of a database for the computer aided cure cycle design, so that an optimum composite curing condition can be defined. The feasibility of applying the MIA technique to a molding press with a real composite structure in place for curing also has been demonstrated [66]. Epoxy cure monitoring using the transient excitation technique has also been explored [68].

Damage Assessment

Many different vibration-based techniques have been used for nondestructive evaluation of various types of flaws, defects, and degradation [1,2,3,67]. These include global methods, such as natural frequency measurements and damping measurements, and local measurements, such as vibrothermography, local amplitude measurement, coin-tap test, (local) impedance method, membrane resonance methods, and acoustic spectral methods [1,2]. Resonant frequencies and damping coefficients are sensitive to the variations in the damage state of composites in response to either fatigue or impact loading [67,69].

The equations of motion were developed by Allen *et al.* [70] for laminated composite beams with load-induced matrix cracking. The damage was accounted for by utilizing internal state variables. The net consequence of these variables on the field equations was the introduction of both enhanced damping and degraded stiffness. An enhanced Zener damping effect was shown to be produced by the presence of microstructural damage. The equations derived [70] are particularly useful for predicting the variations in damping caused by matrix cracking, during the early stage of fatigue loading, where the effect of delamination is minimal. Equations to account for the effect of delamination induced by fatigue or impact loading have yet to be developed.

Figure 7 shows the variations of beam natural frequency, and hence modulus, of a unidirectional and a cross-plied carbon fiber/epoxy composite, respectively, in response to stress-controlled fatigue loading. The variations of the same properties of a cross-plied laminate from the same batch of specimens subjected to repeated impacts are shown in Fig. 8. These data demonstrate the sensitivity of vibration-based techniques for condition monitoring of composite structures. These techniques are global methods to indicate the presence of defects or damage in structure, even though in theory the location and severity of the flaws can be identified with modal analysis, for instance. In conjunction with other techniques such as ultrasound or radiography, these tech-

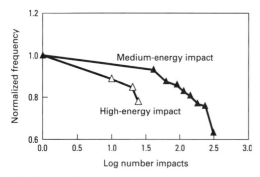

Fig. 8 Variations of resonant frequency and modulus of cross-plied carbon fiber/epoxy laminates subjected to impact fatigue

niques can be very powerful tools for damage assessment in composites.

REFERENCES

1. P. Cawley and R.D. Adams, Vibration Techniques, in *Nondestructive Testing of Fiber Reinforced Plastics Composites*, Vol 1, J. Summerscales, Ed., Elsevier Applied Science, 1987, p 151-200
2. R.F. Gibson, Dynamic Mechanical Properties of Advanced Composite Materials and Structures: A Review, *Shock Vib. Digest*, Vol 19 (No. 7), 1987, p 13-22
3. B.Z. Jang, Vibration Testing and Analysis of Composites, in *International Encyclopedia of Composites*, Vol 6, S.M. Lee, Ed., VCH Pub., 1990
4. C.T. Sun, J.K. Wu, and R.F. Gibson, Prediction of Material Damping of Laminated Polymer Matrix Composites, in *Damping 1986 Proceedings*, AT-WAL-TR-86-3059, Vol 2, Paper FE-1
5. R.F. Gibson, S.K. Chaturvedi, and C.T. Sun, Complex Moduli of Aligned Discontinuous Fiber Reinforced Polymer Composites, *J. Mater. Sci.*, Vol 17, 1982, p 3499
6. S.A. Suarez, R.F. Gibson, C.T. Sun, and S.K. Chaturvedi, "The Influence of Fiber Length and Fiber Orientation on Damping and Stiffness of Polymer Composite Materials," Paper FF-1, presented at the SESA Spring 1985 Conference, 9-13 June 1985, Las Vegas, NV, and published in the Conference Proceedings
7. R.D. Adams and D.G.C. Bacon, Effect of Fiber Orientation and Laminate Geometries on the Dynamic Properties of CFRP, *J. Compos. Mater.*, Vol 7, Oct 1973, p 402-428
8. P. Cawley and R.D. Adams, The Predicted and Experimental Natural Modes of Free-Free CFRP Plates, *J. Compos. Mater.*, Vol 12, Oct 1978, p 336-347
9. N. Alam and N.T. Asnani, Vibration and Damping Analysis of Fiber Reinforced Composite Plates, *J. Compos. Mater.*, Vol 20, Jan 1986, p 2-18
10. N. Alam and N.T. Asnani, Vibration and Damping Analysis of Multilayered Rectangular Plates with Constrained Viscoelastic Layers, *J. Sound Vib.*, Vol 97 (No. 4), Dec 1984, p 597-614
11. C.W. Bert and F. Gordaninejad, "Forced Vibration of Timoshenko Beams Made of Multimodular Materials," ASME Paper No. 83-DET-28, 1983
12. C.W. Bert and C.A. Rebello, Vibration of Sandwich Beams with Bimodular Facings, *1982 Advances in Aero-*
space Structures and Materials, R.M. Laurenson and Y. Yuceoglu, Ed., ASME, AD-03, 1982, p 101-105
13. R.G. Ni and R.D. Adams, The Damping and Dynamic Moduli of Symmetric Laminated Composite Beams—Theoretical and Experimental Results, *J. Compos. Mater.*, Vol 18, Mar 1984, p 104-121
14. L.A. Taber and D.C. Viano, Comparison of Analytical and Experimental Results for Free Vibration of Non-Uniform Composite Beams, *J. Sound Vib.*, Vol 83 (No. 2), July 1982, p 219-228
15. I. Sheinman and S. Greif, Dynamic Analysis of Laminated Shells of Revolution, *J. Compos. Mater.*, Vol 18, May 1984, p 200-215
16. K.P. Soldatos, A Comparison of Some Shell Theories Used for the Dynamic Analysis of Cross-Ply Laminated Circular Cylindrical Panels, *J. Sound Vib.*, Vol 97 (No. 2), Nov 1984, p 305-320
17. K.H. Huang and S.B. Dong, Propagating Waves and Edge Vibrations in Anisotropic Composite Cylinders, *J. Sound Vib.*, Vol 96 (No. 3), Oct 1984, p 363-380
18. A.P. Bhatacharya, Large Amplitude Vibrations of Imperfect Cross-Ply Laminated Cylindrical Shell Panels with Elastically Restrained Edges and Resting on Elastic Foundation, *Fibre Sci. Technol.*, Vol 21 (No. 3), 1984, p 205-222
19. N. Alam and N.T. Asnani, Vibration and Damping Analysis of a Multilayered Cylindrical Shell; Part II: Numerical Results, *AIAA J.*, Vol 22 (No. 7), July 1984, p 975-981
20. P.A.A. Laura, G. Sanchez Sarmiento, P. Verniere de Irassar, and J. Mazumdar, Fundamental Frequency of Composite Circular Membranes with Boundary Disturbances, *Fibre Sci. Technol.*, Vol 21 (No. 4), 1984, p 267-276
21. C.W. Pryor, Jr. and R.M. Barker, A Finite Element Analysis Including Transverse Shear Effects for Applications to Laminated Plates, *AIAA J.*, Vol 9, 1971, p 912-917
22. S.T. Mau, P. Tong, and T.H.H. Pina, Finite Element Solutions for Laminated Thick Plates, *J. Compos. Mater.*, Vol 6, 1972, p 304-311
23. S.T. Mau, T.H.H. Pina, and P. Tong, Vibration Analysis of Laminated Plates and Shells by a Hybrid Stress Element, *AIAA J.*, Vol 11, 1973, p 1450-1452
24. A.K. Noor, Free Vibrations of Multilayered Composite Plates, *AIAA J.*, Vol 11, 1973, p 1038-1039
25. A.S. Mawenya and J.D. Davies, Finite Element Bending Analysis of Multilayered Plates, *Intl. J. Numerical Methods Eng.*, Vol 8, 1974, p 215-225
26. J.N. Reddy, A Penalty Plate-Bending Element for the Analysis of Laminated Anisotropic Composite Plates, *Int. J. Numerical Methods Eng.*, Vol 15, 1980, p 1187-1206
27. J.N. Reddy, A Review of the Literature on Finite Element Modeling of Laminated Composite Plates, *Shock Vib. Digest*, Vol 17 (No. 4), 1985, p 3-8
28. C.T. Sun, S.K. Chaturvedi, and R.F. Gibson, Internal Damping of Short Fiber Reinforced Polymer Matrix Composites, *Computers and Structures*, 1985, p 285
29. C.T. Sun, R.F. Gibson, and S.K. Chaturvedi, Internal Material Damping of Polymer Matrix Composites Under Off-Axis Loading, *J. Mater. Sci.*, Vol 20, 1985, p 2575-2585
30. C.T. Sun, J.K. Wu, and R.F. Gibson, Prediction of Material Damping in Randomly Oriented Short Fiber Polymer Matrix Composites, *J. Reinforced Plast. Compos.*, Vol 4, 1985, p 262-272
31. S.A. Suarez, R.F. Gibson, C.T. Sun, and S.K. Chaturvedi, "The Influence of Fiber Length and Fiber Orientation on Damping and Stiffness of Polymer Composite Materi-

als," presented at the SESA Spring 1985 Conference, 9-13 June 1985, Las Vegas, NV, and published in the Conference Proceedings

32. S.A. Suarez, R.F. Gibson, C.T. Sun, and S.K. Chaturvedi, The Influence of Fiber Length and Fiber Orientation on Damping and Stiffness of Polymer Composite Materials, *Exp. Mech.*, Vol 26 (No. 2), June 1986, p 175-184

33. S.A. Suarez, "Optimization of Internal Damping in Fiber Reinforced Composite Materials," Ph.D. Dissertation, University of Idaho, 1984

34. R.F. Gibson, Development of Damping Composite Materials, AD-06, in *1983 Advances in Aerospace Struct. Mater. and Dynamics*, ASME, 1983, p 89-95

35. B.Z. Jang and J.H. Yim, Damping in Textile Structural Composites, to be submitted to *J. Compos. Mater.*

36. J.H. Yim, "Damping in Laminated, Partially Delaminated and Textile Structural Composites," Ph.D. Dissertation, Auburn University, AL, Dec 1993

37. C.T. Sun and J.K. Wu, Stress Distribution of Aligned Short-Fiber Composites under Axial Load, *J. Reinf. Plast. Compos.*, Vol 3, April 1984, p 130-144

38. R.G. White and T.A. Palmer, Control of the Properties of Carbon Fiber Reinforced Plastics, *AIAA J.*, Vol 22 (No. 11), 1984, p 1662-1669

39. Z. Hashin, *Intl. J. Solids and Struct.*, Vol 6, 1970, p 797-807

40. T.A. Willway and R.G. White, *Compos. Sci. Technol.*, Vol 36, 1989, p 77-94

41. T. Ishikawa and T.W. Chou, *J. Compos. Mater.*, Vol 16, 1982, p 2-19

42. S. Chang and C.W. Bert, Analysis of Damping for Filamentary Composite Materials, *Composite Materials in Engineering Design*, ASM, Metals Park, OH, 1973, p 51-62

43. R.F. Gibson and R. Plunkett, Dynamic Mechanical Behavior of Fiber Reinforced Composite: Measurement and Analysis, *J. Compos. Mater.*, Vol 10, 1976, p 325-341

44. S.J. Hwang and R.F. Gibson, Micromechanical Modeling of Damping in Discontinuous Fiber Composites Using a Strain Energy/Finite Element Approach, *J. Eng. Mater. and Technol.*, Vol 109 (No. 1), Jan 1987, p 47-52

45. S.J. Hwang, "Finite Element Modeling of Damping in Discontinuous Fiber Reinforced Composites," M.S. Thesis, University of Idaho, April 1985

46. C.W. Bert, Composite Materials: A Survey of the Damping Capacity of Fiber Reinforced Composites, in *Damping Applications for Vibration Control*, P.J. Torvik, Ed., ASME, New York, 1980, p 53-63

47. F.H. Chu and B.P. Wang, Experimental Determination of Damping In Materials and Structure, in *Damping Applications for Vibration Control*, P.J. Torvik, Ed., ASME, New York, 1980, p 113-122

48. P. Caravani and W.T. Thomson, Identification of Damping Coefficients in Multidimensional Linear System, *J. Appl. Mech. (Trans. ASME)*, June 1974, p 379-382

49. J.G. Beliveare, Identification of Viscus Damping in Structures from Modal Information, *J. Appl. Mech.*, June 1976, p 335-382

50. F.H. Chu and B.P. Wang, Experimental Determination of Damping in Materials and Structure, in *Damping Applications for Vibration Control*, P.J. Torvik, Ed., ASME, New York, 1980, p 113-122

51. K. Zaveri, *Modal Analysis of Large Structures-Multiple Exciter Systems*, Bruel and Kjaer Co., Nov 1984

52. R.E.D. Bishop and G.M.L. Gladwell, An Investigation into the Theory of Resonance Testing, *Trans. R. Soc. London A*, Vol 255, 1963, p 241-280

53. B.Z. Jang and G.H. Zhu, Monitoring the Dynamic Mechanical Behavior of Polymers and Composites Using Mechanical Independence Analysis (MIA), *J. Appl. Polym. Sci.*, Vol 31, 1986, p 2627-2646

54. C.C. Kennedy and C.D.P. Pancu, Use of Vectors in Vibration Measurement and Analysis, *J. Aeronaut. Sci.*, Vol 14, 1947, p 603-626

55. G.W. Asher, A Method of Normal Mode Excitation Utilizing Admittance Measurements, *Proc. National Specialists Meeting on Dynamics and Aeroelasticity*, Ft. Worth, Institute of Aeronautical Science, 1958, p 69-76

56. J.D. Ferry, *Viscoelastic Properties of Polymers*, 3rd ed., John Wiley and Sons, New York, 1982

57. N.G. McCrum, B.E. Read, and G. Williams, *Anelastic and Dielectric Effects in Polymeric Solids*, John Wiley and Sons, New York, 1967

58. I.M. Ward, *Mechanical Properties of Solid Polymers*, 2nd ed., John Wiley and Sons, New York, 1983

59. T. Murayama, *Dynamic Mechanical Analysis of Polymeric Materials*, Dekker, New York, 1979

60. C.W. Bert, Research on Dynamic Behavior of Composite Plates and Sandwich Plates, *Shock Vib. Digest*, Vol 17, 1985, p 3-15

61. B.E. Read and G.D. Dean, *The Determination of Dynamic Properties of Polymers and Composites*, John Wiley and Sons, New York, 1978

62. Y.H. Yim, J.S. Burmeister, R.L. Kaminski, and J.W. Gillespie, Jr., "Experimental Characterization of Material Damping in Laminated Polymer Matrix Composites," CCM Report 88-39, University of Delaware, 1988

63. L.J. Pulgrano and L.H. Miner, *28th Natl. SAMPE Symp.*, Apr 1983, p 56-64

64. J. Chottiner, Z.N. Sanjana, and R.L. Kolek, in *41st Ann. Conf., Reinforced Plastics/Composites Institute*, SPI, 23 Jan 1986, Sec. 25-C/1-6

65. S.V. Hoa and P. Ouellette, *Polym. Compos.*, Vol 5, 1984, p 334-338

66. B.Z. Jang, G.H. Zhu, and Y.S. Chang, Cure Monitoring of Composite Structures Using Mechanical Impedance Analysis (MIA), in *Nondestructive Characterization of Materials II*, J.F. Bussiere, J.P. Monchalin, and R.E. Green, Jr., Ed., Plenum Press, 1987, p 39-48

67. B.Z. Jang, M.D. Shelby, H.B. Hsieh, and T.L. Lin, Vibration-based NDE Techniques for Cure Monitoring and Damage Assessment in Composites, *19th Intl. SAMPE Tech. Conf.*, Oct 1987, p 265-276

68. B.Z. Jang, H.B. Hsieh, M.D. Shelby, and Y.J. Paig, Real-Time Monitoring and Control of Composite Curing Using Dynamic Mechanical Methods, *14th Conf. on Production Res. and Tech.*, SME, Oct 1987, p 427-33

69. M.D. Shelby, H.J. Tai, B.Z. Jang, Vibration-Based Non-Destructive Evaluation of Polymer Composites, *Polym. Eng. Sci.*, Vol 31, 1991, p 47-55

70. D.H. Allen, C.E. Harris, and A.L. Highsmith, in *The Role of Damping in Vibration and Noise Control*, L. Rogers and J.C. Simonis, Ed., ASME, New York, 1987, p 253-264

Chapter 11

Composites as Intelligent Materials and Structures

Recently, the concept of functionally smart or intelligent materials has received increasing attention in the materials research communities. A smart system can be defined as one "that has built-in or intrinsic sensors, processors, control mechanisms, or actuators making it capable of sensing a stimulus, processing the information, and then responding in a predetermined manner and extent in an appropriate time and reverting to its original state as soon as the stimulus is removed" [1].

In a very strict sense, a smart or intelligent material is defined as a material that possesses intrinsic sensing, controlling, and actuating capabilities. A smart or intelligent structure is a combined material system or device that is, intrinsically or extrinsically, capable of sensing, controlling, and actuating.

At a recent workshop [2], Takagi proposed that "intelligent materials may be defined as the materials which respond to environmental changes at the most optimum conditions and manifest their own functions according to the changes." Takagi [3] explained the concept in terms of three elements: intelligence from the human standpoint, intelligence inherent in materials, and primitive functions of intelligence. The inherent primitive functions of each material originate from its electronic, atomic and molecular structure.

To achieve material systems or structures with high intelligence levels and with the capability to recognize, discriminate and adapt, more advanced functional materials must become available. During the last two decades, new techniques for the design and fabrication of materials have enabled the synthesis of many novel classes of functional materials, including semiconducting, optical, magnetic and superconducting materials. The hybridization of such materials could lead to material systems with intrinsic mechanisms for sensing, control, and response [1]. However, efforts at development of these functional materials have been rather sporadic. In most cases, these efforts were directed towards the development of a spe-

cific functional material or device, such as piezoelectric and pyroelectric polyvinylidene fluoride (PVDF). Little concerted interdisciplinary effort was made to systematically investigate the atomic, molecular and microstructural aspects of polyfunctional materials. An urgent need exists to understand the fundamental dependence of various functional (tensor) properties upon the atomic, molecular and microstructural factors. These tensor properties include thermoelectricity, piezoelectricity, pyroelectricity, electro-optical effect (third-rank tensors) and elastic compliances, elasto-optical coefficients, piezo-optical coefficients, and electrostriction (fourth-rank tensors), and so on.

The complexity of the issues involved in the study of the structure-property relationships in advanced functionally responsive materials requires a comprehensive approach. Any research efforts aiming at the development of new intelligent materials must address the following issues:

- Establishment of the relationships between various functionalities (tensor properties) and the chemical and crystal structure of materials. A combined experimental, analytical and numerical approach must be used to achieve this goal.

- Exploration of novel functional materials with a good level of intelligence. With a clear understanding of the fundamental atomic, molecular and microstructural aspects that dictate the various tensor properties of materials, new functional materials can be developed.

- The material factors and stimulus parameters that affect the functional stability (e.g., dimensional stability) of a material under the influence of chemical, thermal, mechanical, electrical, magnetic, and optical forces must be determined.

Intelligent and Functional Materials: A Brief Review

Intelligent Structures or Systems

The growing use of composite materials in aerospace applications and the trend toward higher performance with sophisticated flight-control systems necessitate the use of a complex network of sensors that can monitor several parameters simultaneously over the entire lifetime of the structure [4]. Fiber-optic sensors can be easily embedded within materials with little effect on the structural shape or external aerodynamic shape of the vehicle. These sensors can be used for

- Monitoring the manufacturing process [5]
- Augmenting nondestructive evaluation techniques
- Enabling vehicle health monitoring and damage assessment systems [6-9]
- Supporting control systems [5]

In particular, fiber-optic sensors are used in the detection of the occurrence of impact damage and the measurement of strains [10,11], for continuous self diagnosis [12], and as part of an aircraft's "smart skin" [13].

Significant progress has been made towards regulating structural radiated noise by active/adaptive means applied directly to the structure. By incorporating electrorheological (ER) fluids in composites, a new class of materials can be produced whose mechanical properties can be changed *in situ* [14-17]. By adjusting the rheological properties of the ER fluid through an electric field, both stiffness and damping capabilities can be altered [17-21]. When the electric field is on, the dynamic stiffness of the composite is high (viscosity of ER fluid raised considerably), which reduces the resonant response. At higher frequencies, the field is turned off and the dynamic stiffness is lowered, thereby effectively damping out the higher frequency vibration [17,21].

Active vibration control can also be achieved by incorporating a shape memory alloy (SMA, for example, Nitinol) in a fiber-reinforced composite as the embedded distributed actuators. In principle, the SMA-embedded laminates have the capability to change their material properties, modify the stress and strain state of the structure, and possibly alter the structure's configuration, all in a controlled manner [22,23]. In this application, fiber-optic sensors can be used to evaluate the dynamic response of the structure [22]. The use of smart materials and structures in aircraft was discussed by Brown [24]. The main functions of smart systems in aircraft are on-line monitoring, nondestructive evaluation, vehicle health monitoring, and flight control. The advantages and limitations of ER fluids, piezoelectric ceramics, and SMAs as the actuators for smart structures were described. Fiber-optic glass with polyimide coatings was evaluated as a potential sensor for the smart structure [24].

Smart Materials

As pointed out by Ahmad [1], the development of synthetic intelligent materials is still at the conceptual stage, although a few intrinsically intelligent materials can be identified in naturally occurring biosystems. Most of the current research work related to smart materials is on the mechanisms of biosystems and biomolecules. These mechanistic studies [2,25] are being conducted in order to develop new prosthetics, drug delivery systems, substitutes, etc. They are a very important source of new ideas and approaches to developing novel materials with various degrees of intelligence.

Montal [2] described the function of channel proteins, an example of intelligent biomaterial, which possess the actual polar pathways that permit the selective passage of ions across the apolar lipid bilayer, and a sensing structure that detects the stimulus and couples it with the gating of the channel. Examples of intelligent biomaterials, including a receptor molecular assembly embedded in a biomembrane, were given by Aizawa at the same conference [2]. Successful synthesis of a polypyrrole/glucose oxidase molecular assembly as a model of intelligent materials was also reported by Aizawa.

Osada [2] reported that hydrogel, a water-swollen synthetic polymer, underwent shape changes by applying DC current and the velocity of shape change was proportional to the charge density of the gel. Based on the gel behavior, Osada [2] developed a model of an electrically activated artificial muscle working in an aerobic and aqueous medium system. The muscle contracts and dilates reversibly by an electrical stimulus under isothermal conditions. Thermosensitive polymers, such as a copolymer of n-isopropyl acrylamide with butyl methacrylate, can be used as on/off switches in a drug delivery system. New self-assembled Langmuir-Blodgett (LB) films were developed for use as part of a visible laser system [2].

Takahashi suggested that intelligent materials can be designed at the atomic level using a process called "genetic control in materials science." To accomplish this successfully, specific rules will have to be developed [1]. These rules could be derived from various established concepts such as (1) atoms with even atomic number make insulators, while those with an odd number form conductors; and (2) the solid state energy band is determined by the periodic potential, and so on [1]. By using the superlattice approach and properly considering the quantum effects, one can create minibands of energy levels and freely control the forbidden bands or energy gaps in the material. Once design and creation of materials can be controlled at the atomic level, functionally intelligent materials can be produced [1].

Select Functional Materials

Sensors and transducers are terms that are interchangeable in many cases. A sensor is a device that detects a change in a physical stimulus and turns it into a signal which can be measured or recorded. A transducer is a device that transfers power from one system to

Table 1 Physical effects and associated sensors

Type	Sensor	Physical Effect
Self-generators	Radiant-electrical	Photovoltaic; radiation-current
	Mechanical-electrical	Electrodynamic; voltage-velocity piezo-electric; strain-charge
	Thermal-electric	Thermal-electric; temperature-voltage
	Magnetic-electrical	Electromagnetic; flux change-volts
Modulators	Electrical-(radiant)-electrical	Photoconductive; radiation-resistance change
	Electrical-(thermal)-electrical	Thermoresistive; temperature-resistance change
	Electrical-(magnetic)-electrical	Magnetoresistive; magnetic field-resistance change
		Hall effect; emf due to current in magnetic field
	Radiant-(mechanical)-radiant	Radiation change due to motion
	Electrical-(mechanical)-electrical	Piezoresistive; strain-resistance change
Modifiers	Radiant-radiant	Temperature change due to collected radiation
	Mechanical-mechanical	Displacement due to pressure; force or pressure change due to flow
	Thermal-thermal	Temperature change due to heat flow
	Electrical-electrical	Change in electrical form

Source: Ref 26

another in either the same or different form. In this chapter, a transducer is either a sensor or an actuator, while the materials that can be used in a transducer are called functional materials, which are parts of, or precursors to, a smart material, or intelligent material.

Sensors are commonly grouped into three types: self-generators, modulators and modifiers. Table 1 outlines the physical processes used in each of these sensors. In some cases, transducers are classified as active or passive. An active transducer requires an external power source to convert or amplify the input into a useful output. A passive transducer directly converts or amplifies the input without a power source [27]. Polymers play an important role in several types of sensors. Some applications require many of the advantageous properties that polymers have to offer. A good example is the chemical inertness of certain polymers for biomedical applications.

Conducting polymers could be used in molecular electronics. Wires could be replaced by polymer chains, silicon replaced by organic molecules. Polyacetylene is the most widely discussed conductive polymer being investigated today [28]. Besides metal- or carbon-filled polymer composites, electron conductive polymers now known can be classified as π-conjugate nonintrinsic conductors (become conductive when doped), redox (oxidation reduction) copolymers with hopping conduction (a material with self-exchange electron transport), or ion-exchange polymers [27,28]. High ionic conduction in polymers also is observed in polymer solid systems (e.g., polyethylene oxide-alkali metal salts) which do not contain deliberately added low molecular-weight solvents [27].

Functional polymer monolayers prepared by the Langmuir-Blodgett (LB) technique are also being evaluated for their potential as sensors. One example is a gas detector that uses conducting polymer monolayers and infrared detectors with LB pyro field-effect transistors. A waveguide consisting of LB molecules doped with Nd^{3+} ions shows non-linear optical properties [2].

Optical fibers provide an inert, fast, reliable method of sensing changes or transmitting sensor information back to a micro-processor. Most optical fibers used today are fused silica (SiO_2). Polymer and halide glasses are promising for the future but are still mainly in the research stage. Plastic fibers have a large aperture and are flexible, but they cannot tolerate heat, have high attenuation, and can only be used in the visible light range. The tensor properties of polymeric and fuse silica fibers in response to various stimuli have not been studied to any significant extent.

Polyvinylidene fluoride (PVDF) is a well-known piezoelectric material. PVDF films are produced by extrusion followed by orientation-stretching below the melting point to cause packing of the unit cells into parallel planes (β-phase). The β-phase films are then exposed to a high electric field to align the dipoles relative to the field direction. The current uses of PVDF film include contact sensors, impact sensors, force sensors, biomedical flow meters, thermal radiation sensors, and ultrasonic transducers. Although the piezoelectric effects of PVDF have been studied extensively, very little has been done on the thermomechanical, elasto-optical, electro-optic (to some extent), thermo-optical, and electrothermal responses of this and other functional materials. Functional composites containing electroceramics such as piezoelectric materials have been investigated [29-44]. However, current models are inadequate to permit reliable predictions of the above functional coefficients of multiphase composite systems. Recent developments of sensor materials, both polymeric and inorganic, are summarized in Table 2 [27].

Design Approaches for Intelligent Materials

Increased sophistication in the use of advanced functional materials in a smart system requires that the struc-

Table 2 Functional sensor materials

Physical effect	Material	
	Polymer	Inorganic
Passive transducers		
Piezoelectricity	Polyvinylidene fluoride Polyvinylidene trifluoroethylene Polyhydroxibutyrate Liquid crystalline polymers (flexoelectricity)	Piezoelectric zirconate titanate (PZT) Quartz
Pyroelectricity	Polyvinylidene fluoride Langmuir-Blodgett ferroelectric superlattices	Triglycine sulfate Lead-based lanthanum-doped zirconate titanate Lithium tantalate
Thermoelectricity (Seebeck effect)	Nitrile-based polymers Polyphthalocyanines	$Cu_{100}/Cu_{57}Ni_{43}$ Lead telluride Bismuth selenide
Photovoltaic	Polyacetylene/n-zinc sulfide Poly(N-vinyl carbazone) + merocyanine dyes	Silicon Gallium arsenide Indium antimonide
Electrokinetic	Polyelectrolyte gel ionic polymers	Sintered ionic glasses
Magnetostriction	Molecular ferromagnets	Nickel Nickel-iron alloys
Active transducers		
Piezoresistivity	Polyacetylene Pyrolized polyacrylonitrile Polyacequinones	Metals Semiconductors
Thermoresistivity	Poly(p-phenylene vinylene)	Metals, oxides Titanate ceramics Semiconductors
Magnetoresistivity	Polyacetylene Pyrolized polyvinylacetate	Nickel-iron alloys Nickel-cobalt alloys
Chemioresistivity	Polypyrrole Polythiophene Ionic conductiing polymers Charge transfer complexes	Palladium Metal oxides Titanates Zirconia
Photoconductivity	Copper phthalocyanines Polythiophene complexes	Intrinsic and extrinsic (doped) semiconductors

Source: Ref 27

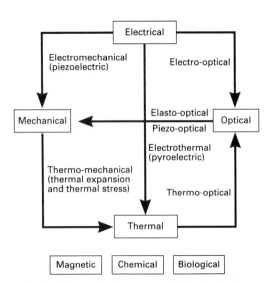

Fig. 1 Various stimulus-response relationships in functional materials

ture of such materials be carefully designed. The great demand on the design and the actual tailoring of such functional materials requires detailed knowledge of structure-property relationships. Not only do we need to know how to design material structures, but also what structures are required to give the desired combination of properties.

In general, combined computational/modeling, theoretical, and experimental techniques are required to study the structure-property relationships in functional materials. Of particular interest are the physical properties (tensors or coefficients) relating a stimulus to a response. As illustrated in Fig. 1, both the stimulus and the response can be thermal, mechanical, electrical, or optical in nature (magnetic, chemical, and so on, are not listed). For instance, if the stimulus is electrical and the response mechanical, piezoelectricity is the tensor or coefficient property.

In general, although piezoelectricity has been studied to some extent, other functionalities (electro-optical, thermo-optical, electrothermal, elasto-optical, and thermomechanical properties) of the same materials are largely overlooked. Further, correlations among various tensor properties have yet to be examined. These properties must be further investigated in relation to the materials structure. Then materials-by-design tasks can be carried out to design a material that meets a specific set of property requirements.

Modeling and Analytical Approaches

Structure-Property Relations and Property Predictions of Functional Polymers. The design of a functionally responsive material requires an estimation of the physical and chemical properties before a new synthesis is attempted. The prediction of certain chemical and physical properties (e.g., moduli, solubility parameter) of a polymer, given its structure, is relatively straightforward compared to the estimation of other properties (for example, piezoelectric and elasto-optical coefficients). Even more difficult is the reverse process: designing a polymer which meets a specific set of property requirements. Now that a wide variety of chemicals are being used as monomers in various polymerization processes, synthesizing polymers with desired characteristics is becoming a practical reality.

The Group Additivity Approach. A number of approaches have been developed in an attempt to predict the properties of polymers based on their composition and structure, including theoretical calculations from first principles, and semi-empirical correlations [45-47]. The most successful method for predicting polymer properties is based on group additivity calculations. In a group additivity calculation the repeat unit is broken up into sub-units. The contribution that each of these sub-units makes to a particular polymer property is known, and by simply adding up the group contributions the property value of the polymer can be estimated. In a group additivity calculation, the property of the polymer repeat unit is assumed to be the same as that of the entire molecule. If this assumption is valid, one need only to calculate the property value of the repeat unit to obtain the property value of the complete polymer. This implies that the molecular weight of the polymer is sufficiently large that molecular weight effects can be ignored. However, when using group additivity, no provision exists for taking into account processing conditions, interchain interactions, or cross-linking.

In many instances, the physical properties which can be derived from group additivity are extensive in nature, and the materials scientist is interested in their intensive counterparts. For instance, using group additivity the polarizability of a polymer can be calculated, but it is the dielectric constant which is of interest to an electrical engineer in the semiconductor industry. In this case, a simple relationship exists between the extensive and intensive properties given by the Claussius-Mossotti equation.

For those properties for which group additivity is valid, calculating a property value for a given chemical repeat unit is simple. This procedure of going from a structure to a property may be of less commercial importance than its reverse, but it represents an important step in the efforts to establish structure-property relationships in materials.

This approach has been found to be extremely accurate in predicting many polymer properties, such as polarizability, molar refractivity, density, molar volume, heat capacity, thermal conductivity, cohesive energy, and solubility. However, this approach is slightly less accurate when used in predicting the Rao function or molar sound velocity function (from which bulk modulus, Poisson's ratio, shear modulus, tensile modulus and ultimate tensile strength can be calculated), coefficient of thermal expansion, glass transition temperature, and crystalline melt temperature. Very little effort has been made to determine if this approach can be used for predicting higher-order tensor properties (piezo-electrical, electrothermal, and electro-optical, among others).

Approaches of Atomic/Molecular Calculations. Both *ab-initio* theory and semi-empirical models can be used to calculate energetics and response functions. For materials that possess lattice periodicity, the conventional calculations based on band structures can be carried out. Molecular simulations can also be undertaken for systems that do not have a well defined lattice structure.

The first-principle calculations are based mainly on the density-functional theory [48]. Although the Hartree-Fock approximation is good for the ground states of atoms and small molecules, it is not accurate for condensed matters. On the other hand, the local density-functional approximation (LDA) [51] has proven to be very accurate for the structural properties of solids. This theory has been extended to deal with excitations, in the Green function self-energy correction (GW) approximation [49].

The most demanding task in the application of the density functional theory is the calculation of the electronic structure associated with the effective one-electron potential. Chen *et al.* [50] have just completed a new LDA code using the multiple-scattering Green's function method. The method deals with the full potential without any shape approximation [51]. This method can also be applied to molecules. It also provides a natural basis to be incorporated in the coherent potential approximation (CPA) [52] to treat disordered systems [53]. While the norm-conserved pseudo-potentials [54] are more efficient and will be used for most systems, this method can also handle full potentials, which will be used whenever necessary.

While *ab-initio* calculations are less biased than semi-empirical models, the former are computationally more demanding and sometimes are more difficult to give physical insight. Therefore, a good strategy will be to employ semi-empirical models along with the *ab-initio* calculations. Chen's semi-empirical theory is based on the tight-binding model (TBM) [55]. This is the equivalent solid-state version of the schemes such as MND and CNDO (software) and the Huckel-Hubbard model used in quantum chemistry [56]. Chen and co-workers have applied this approach very successfully to a variety of properties, e.g., optical [57] and transport properties [58], alloy mixing enthalpies [59] and phase diagrams [60], cleavage energies [61], elastic constants [55], etc., in semiconductors and alloys. This approach is expected to work for polymer materials and ceramics, with perhaps an extension to include the local correlation effects which are important to some of these materials [50].

It may be noted that the TBM, when used in a local molecular-bond picture, is Haarison's bond orbital model (BOM) [62]. The BOM contains the most important energetic contributions from the covalent bonds. When the interactions among the bonds are corrected perturbatively, the extended BOM results [63]. This approach often gives explicit expressions connecting the atomic quantities to materials properties. The fast and easy computations of this approach allow a comprehensive comparison across a broad spectra of systems. This capability will be extremely valuable when one seeks to develop materials to meet the requirements for a smart device. Furthermore, when enough information is available to determine the basic parameters, TBM can be used in full scale calculations [55] for all the response functions. Such an approach has proven to be very successful in molecular dynamic simulations [64].

Physical Properties of Hybrid or Composite Systems. Many functionally responsive material systems are expected to be hybrids (blends, co-polymers, or alloys) or composites composed of two or more components or phases. Although the elastic constants (including the fourth-rank compliance matrix tensor and the second-rank thermal expansion coefficient tensor) of composite systems have been studied thoroughly, other tensor properties have been studied to a much lesser extent or largely overlooked. The compliance or stiffness matrix properties of composites were usually predicted using some kind of rule-of-mixture law. The micro-mechanics approaches employed include strength-of-materials, elasticity, variational and strain energy principles. Within the framework of macromechanics, the elastic properties of a laminate can be predicted by the classical lamination theory.

Piezoelectric constants of semi-crystalline polymers were studied by separately considering the properties of the amorphous and crystalline phases which were then combined together using rule-of-mixture laws [65-67]. Composite systems composed of polymers as the matrix and piezoelectric ceramics such as lead zirconate-titanate (PZT) as the reinforcement exhibit piezoelectricity. These composites are useful because of the inherent good ductility imparted to the composite by the normally ductile polymer matrix. Such composites can be readily molded into various useful shapes, such as a thin film or a wire. Keyboards consisting of a printed circuit board and a piezoelectric composite film have been developed [35]. Piezoelectric constants, along with their associated dielectric and elastic constants, of binary systems have been studied largely based on various geometric models and rule-of-mixture concept [36-43].

The differences among various tensor coefficients must be understood when trying to obtain a composite property. For instance, the electrical conductivity of a multiphase system depends more on the formation of contiguity (percolation) of the more conductive phase rather than just the volume fraction of this phase, although thermal conductivity is less sensitive to contiguity. Optical activities, such as the higher-order χ^3 of composites are also expected to be more complex than the prediction from a simple rule-of-mixture law. The functional property-structure property relationships in functional composites have yet to be systematically studied and represent a fruitful area of research. The following are a list of the design guidelines of functional composites compiled by Newnham and co-workers [44,68]:

- **Sum properties** of a composite are normally obtained by averaging similar properties of the constituent phases, with a mixing law bounded by the series and parallel models. For simple properties such as elastic and dielectric constants, the composite property lies between those of the individual phases. This is not true for *combination properties* based on two or more properties. For instance, the mixing rules for elastic modulus and density are often different, the acoustic velocity (dictated by modulus and density) of a composite could be smaller than those of the constituent phases.

- **Product properties** are even more complex, because different properties in the two constituents combine to produce a third property in the composite. For example, in a magnetoelectric composite, the piezoelectric effect in barium titanate acts on the magnetostrictive effect of cobalt ferrite to generate a composite magnetoelectric effect.

- **Connectivity patterns** are an important feature of composite electroceramics. The self-connectiveness of the phases determines whether series or parallel models apply and thereby minimize or maximize the properties of the composite. The 3-D nature of the connectivity patterns make it possible to minimize some tensor components while maximizing others.

- **Concentrated field and force patterns** are possible with carefully chosen connectivities. For instance, using internal electrodes, electrostrictive ceramics are capable of producing strains comparable to the best piezoelectrics.

- **Periodicity and scale** are key features when considering use of composites for high frequencies or where resonance and interference effects occur. When the wavelengths are on the same scale as the component dimensions, the composite can behave like a non-uniform, possibly non-linear, solid.

- **Symmetry** governs the physical properties of composites, just as in single crystals. The Curie principle of symmetry superposition and Newmann's law can be generalized to cover fine-scale composites, thereby elucidating the nature of their tensor properties.

- **Interfacial effects** can yield novel barrier phenomena in composites.

- **Polychromatic percolation** provides another route of producing novel functional properties. For instance, the SiC-BeO composites are excellent thermal conductors, yet poor electrical conductors at the same time. A thin layer of BeO-rich carbides separate the SiC grains, insulating them from one another electrically, but providing a good acoustic impedance match, ensuring phonon conduction.

- **Coupled phase transformation** in multiphase solids could introduce novel functional properties when the environmental conditions, (temperature or pressure), are altered.

- **Porosity and inner surfaces** in many electroceramic composites can provide special sensing capabilities.

Applications of Theoretical / Modeling Approaches

Elasto-Optical Effects in Single-Phase Materials

Of the six tensor properties specified in Fig. 1, only a small part of the elasto-optical effect has been studied using the group additivity method. The coefficient of proportionality C between the mechanical stress σ and birefringence Δn resulting from the stress is given by [69,70]:

$$\Delta n = C \, \sigma \tag{Eq 1}$$

An empirical relation between C and the parameters of the chemical structure of a repeat unit of a polymer is:

$$C = \Sigma C_i / (N \, \Sigma \, \Delta v) + \pi \tag{Eq 2}$$

where C_i is the increment specifying contribution of the ith atom and ith type of intermolecular interaction into the C coefficient while $\Sigma \, \Delta v$ is the van der Waals volume of the repeat unit, and N is Avogadro's number. The values C_i for various atoms can be found from [69,70]. Other elasto-optical effects (e.g., piezo-optical coefficients π_{ijkl} as defined by $\Delta B_{ij} = \pi_{ijkl}\sigma_{kl}$) can be computed from the properties of the constituent chemical groups. These values can then be compared with those obtained from the direct calculations of atomic/molecular force/potential. These properties of many important functional polymers such as polydiacetylene, PVDF, nitrile-based polymers and liquid crystalline polymers must be measured and compared with the theoretical values. These efforts will construct the foundation for building structure-property relationships in functional polymers.

Piezoelectric Effects

Certain polymers (PVDF, trifluoroethylene, and liquid crystalline polymers, among others) with a proper degree of orientation polarize under applied stress, giving a linear relation between polarization P_i and σ_{jk}: $P_i = d_{ijk}\sigma_{jk}$. Broadhurst et al. [66] proposed a model for calculating the piezoelectric and pyroelectric properties of PVDF. The model consists of an array of crystal lamellae with a net moment from aligned dipoles in the crystals and compensating space charge on the crystal surfaces. Other methods are discussed in [65,67].

A theoretical equation for calculating piezoelectric constants of polymer crystals has been derived by Tashiro et al. [71] based on a point charge model. This model goes beyond the traditional group additivity approach. In this model, one assumes that the ith atom of the mth asymmetric unit in the kth unit cell has an effective point charge Q_i. One then sums up over all the unit cells in the whole system, all the asymmetric units in the unit cell, and all the atoms contained in the asymmetric unit, respectively to obtain the polarization of the whole crystal. The changes of this polarization in response to an applied stress are then calculated from the internal strains of the individual atoms and the compliance tensor matrix. This approach can be followed to calculate the piezoelectric coefficients d_{ijk} for polymers such as PVDF, polyacrylonitrile series and liquid crystalline polymers. The potential energy and force constant calculations may be performed in accordance with the methods discussed earlier in the section "Modeling and Analytical Approaches." The validity of this approach may be tested by comparing the experimental values with calculated results.

Electro-Optic Effects

A material must be noncentrosymmetric in structure in order to have a finite electro-optic coefficient. For organic polymers, noncentrosymmetry is often introduced via high field poling which organizes the electroactive parts preferentially along the field direction. The magnitude of the electro-optic coefficient is proportional to the difference of the dipole moments of the ground and the first excited state of the molecule. Recently, very large electro-optic coefficients (more than fifty times larger than that of the well known inorganic electro-optic material $LiNbO_3$) have been obtained in specific organic crystals in which the molecules have large ground state dipole moments. Polymeric materials with strongly dipolar side groups can also have a large electro-optic coefficient, if the side groups are appropriately aligned. The electro-optic coefficients of polymers such as PVDF, polyhydroxibutyrate, poly(n-vinyl carbazole), and polyacetylene have yet to be measured and compared with calculated results.

Combined Electrical and Mechanical Effects on a Single-Phase Material

The refractive index of a crystal can be specified by the indicatrix, an ellipsoid whose coefficients are the components of the relative dielectric impermeability tensor B_{ij} at optical frequencies:

$$B_{ij}X_iX_j = 1 \tag{Eq 3}$$

where X_is are the principal axes of the dielectric constant (or the permittivity) tensor and $B_{ij} = K_0 \, \delta E_i/\delta D_j$ [72]. Thus, in general, the small change of refractive index produced by the electric field and the mechanical stress is a small change in the shape, size and orientation of the indicatrix. The corresponding changes ΔB_{ij} in the coefficients, under an applied field E_k and an applied stress σ_{kl}, are given by

$$\Delta B_{ij} = Z_{ijk} E_k + \pi_{ijkl} \sigma_{kl} \qquad (\text{Eq 4})$$

where the third-rank tensor Z_{ijk} gives the electro-optical effect and the fourth-rank tensor π_{ijkl} gives the photoelastic effect. This equation is similar to the equation giving the strains of a crystal produced by a field (converse piezoelectric) and by a stress (elasticity):

$$\varepsilon_{ij} = d_{kij} E_k + S_{ijkl}\sigma_{kl} \qquad (\text{Eq 5})$$

where the d_{kij}s are the piezoelectric coefficients and the S_{ijkl}s are the compliances. If the second-order effects, such as electrostriction ($\delta^2\varepsilon/\delta E^2$), are considered, then additional terms like $\gamma_{klij}E_iE_j$ must be added to Eq 5, where electrostriction γ_{klij} is a fourth-rank tensor. Alternatively, Eq 4 can be rewritten as

$$\Delta B_{ij} = Z_{ijk} E_k + P_{ijrs} \varepsilon_{rs} \qquad (\text{Eq 6})$$

where $P_{ijrs} = \pi_{ijkl}C_{klrs}$ represents the elasto-optical coefficients, $\pi_{ijkl} = P_{ijrs}S_{rskl}$ and $\sigma_{kl} = C_{klrs}\varepsilon_{rs}$. For inorganic crystals, typical orders of magnitude for these coefficients are $Z_{ijk} \approx 10^{-12}$m/V, $\pi_{ijkl} \approx 10^{-12}$m^2/N, $d_{kij} \approx 3 \times 10^{-12}$/V, $S_{ijkl} \approx 10^{-11}$ m^2/N and $P_{ijrs} \approx 10^{-1}$. For most organic polymers, the magnitudes of these coefficients (except for S_{ijkl}) are yet to be determined. Even for well-studied functional polymers such as PVDF, coefficients such as Z_{ijk} and π_{ijkl} and their correlations have not been well established. These coefficients have yet to be measured as a function of crystal symmetry and chemical structure of polymers and compared with the theoretical results.

Thermoelectric Effects (Example 5)

When heat and electricity conduction processes occur together they interfere with one another. These thermoelectrical effects include the thermoelectric emf (the Seebeck effect), the Peltier heat, and the Thomson heat [72]. An emf is set up in the circuit made up of two different metals, if the junctions are maintained at different temperatures (Seebeck effect). When current is allowed to flow across a junction between two different metals, it is found that heat must be continuously added or subtracted at the junction in order to maintain its temperature constant, and the heat is proportional to the current flowing (Peltier effect). When a current flows in a wire, of homogeneous material and of constant cross-section but with a non-uniform temperature, heat must be supplied to keep the temperature distribution steady. The heat that must be supplied in unit time to an element of the wire in which the temperature rise is dT is $dQ \approx \tau JdT$, where τ is the Thomson coefficient. The effects were studied for metals and, to a much lesser extent, for other inorganics. Very little work has been done on polymers (except nitrile-based polymers and polyphthalocyanines) [27]. Future research should include a study of these effects in relation to atomic and crystal structure of various functional polymers.

Materials By Design

Going from a set of desired properties to a single polymer or a few polymers is a useful approach for designing intelligent materials. The ability to perform this reverse calculation is non-trivial, since there may not be a unique solution. More than one material may satisfy a given set of property requirements, or property requirements may be too restrictive and no repeat unit (single polymer) can satisfy all the property requirements. In this case, material combinations in the form of co-polymers, blends, hybrids or composites can be attempted.

The structure-property relationships obtained, along with the current state of knowledge on the lower-rank tensor properties [e.g., in 45-47,70,73], may be used to form the basis of an expert system. Simple rules that would govern the decision-making processes have to be developed first. Efforts then can be made to design a few functionally responsive materials for special applications. Much more research is needed to develop effective methodologies and proper guidelines for designing advanced functional materials.

Experimental Verification of Design Approaches

Future experiments are needed to measure the various tensor properties of functional polymers with known structures and compositions. Efforts should be made to vary the chemical structure of a series of functional polymers systematically to facilitate a better understanding of the relationships between various high-rank tensor properties and the chemical structure of polymers. The experimental values obtained should be compared to corresponding values computed from the group additivity approach and atomic force calculations. This comparison will provide a basis for judging the applicability of these approaches in predicting higher-order tensor properties of polymers. When deemed necessary, modifications to these methods or development of other approaches should be attempted to further establish the required property-chemical structure relationships in functional polymers.

Another task will be to specify a set of property requirements to design functionally responsive materials. Attempts may be made to synthesize such conceptually-designed polymers. The chemical composition and molecular structure of these polymers can be determined by FTIR, NMR, and other spectroscopic and chemical analysis techniques. The crystal structure and morphology of each polymer synthesized can be examined by diffraction and analytical electron microscopy techniques. The various tensor properties of these polymers may then be measured and again compared with the theoretical predictions. Techniques for measuring the pie-

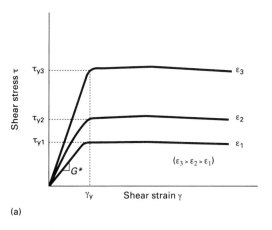
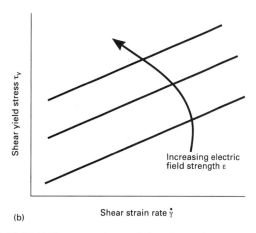

(a)

(b)

Fig. 2 Typical rheological behavior of an electrorheological (ER) fluid. (a) Shear stress-shear strain behavior as a function of varying electric field strength. (b) Viscosity remains constant as the electric field is increased

zoelectric and pyroelectric coefficients are discussed in [65-67,71].

Advanced Functionally Responsive Materials: Case Studies of Model Systems

Dimensionally Stable Structure in Response to Temperature Variations. One way to minimize the dimensional variations of a structure in response to a temperature change would be to counteract the temperature effect by a combination of mechanical, electrical, optical and possibly magnetic means. By tentatively leaving out the optical and magnetic terms without losing the generality, one can write the variation in strain as

$$d\varepsilon = (\delta\varepsilon/\delta\sigma)d\sigma + (\delta\varepsilon/\delta E)dE$$
$$+ (\delta\varepsilon/\delta T)dT + \dots \qquad \text{(Eq 7)}$$

where the coefficient of the first term, $(\delta\varepsilon/\delta\sigma)$, represents the compliance s, that of the second term the converse piezoelectric effect d, and that of the third term the thermal expansion coefficient α. For $d\varepsilon$ to vanish, one must have $(\delta\varepsilon/\delta T)dt = -[(\delta\varepsilon/\delta\sigma)d\sigma + (\delta\varepsilon/\delta E)dE + \dots]$. The thermal strain can be counterbalanced by electric, optical, and magnetic stresses, in addition to the traditional way of mechanical stressing. For traditional non-metallic and non-ferroelectric crystals, the values of s and α are on the order of magnitude (in MKS units) 10^{-11} m^2/N and 10^{-5}, respectively. Suppose that the temperature variation is ΔT = 10 °C, then $\Delta\varepsilon \approx 10^{-4}$, leading to a residual stress of $10^{-4}/10^{-11} \approx 10^7$ Pa, requiring a stress of this magnitude to eliminate the thermal strain. With the added degrees of freedom from the electrical, optical, and magnetic stresses, one can fine tune or proportionally control the thermal strain without having to involve a large mechanical stress.

Control of Thermal Properties of Polymers. The expansion of a given material depends on internal forces. Thermal stresses result from the differential thermal expansion caused by temperature gradients. Therefore, to achieve thermally stable structures one has to offset normal thermal expansion by stress, magnetic or electric-induced contraction. The thermal expansion coefficient constitutes a second rank tensor; therefore, for anisotropic materials up to six values have to be studied. Weak internal forces result in high thermal expansion, so most polymeric materials exhibit high linear coefficients of thermal expansion. The thermal behavior of materials can also be regulated by the volume of the open space and the cooperative rotational effects. In polymeric materials the thermal expansion coefficient is proportional to the van der Waals volume. An introduction of an electric field offers an additional way of controlling the thermal behavior of polymeric materials.

Intelligent Composites Containing Electrorheological Fluids

The examples presented below serve to illustrate the point that many functionally responsive material structures or devices can be produced by properly combining different functional and non-functional but supportive materials together to form a composite. The opportunities of designing and producing functionally smart composites are unlimited.

Electrorheological Materials

Electrorheological (ER) materials are suspensions of very fine dielectric particles (typically hydrophilic) dispersed in insulating media (hydrophobic) which exhibit unique rheological behavior in the presence of an applied electric field [74-84]. Figure 2 shows the typical shear stress-shear strain behavior of an ER material as a function of varying electric field strength. Typically, a higher electric field strength results in a higher shear yield stress τ_y of the Bingham plastic-like suspensions. This variable flow stress characteristic has proven useful in electroactive devices including valves and clutches [18]. The ac-

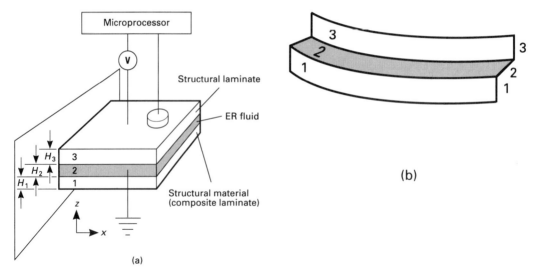

Fig. 3 Use of a viscoelastic material layer to control structural damping. (a) A composite beam containing an ER fluid for active vibration control. (b) The beam after deformation

tual viscosity of the ER material remains relatively constant as the applied electric field is varied.

The shear yield stress of an ER material increases with increasing applied electric field while the yield strain remains relatively constant (normally at approximately 1%). Consequently, the complex shear modulus increases with increasing electric field strength. The magnitude of the complex shear modulus can vary several orders of magnitude when the electric field strength varies between 0 and 4,000 V/mm. This property can be utilized to obtain desirable structural behavior of components containing ER materials.

A full understanding of the mechanisms that govern the observed rheological response of ER fluids has yet to be obtained, but is under active investigation [17,85]. Choi *et al.* [17] demonstrated that the suspension changed from a random structure with no external field to an anisotropic structure with an applied voltage of 2 kV/mm. The particles in the suspension oriented themselves in relatively regular chain-like patterns to form a mixture with globally anisotropic mechanical properties. These field-induced columnar structures are thought to increase the energy-dissipation characteristics of the suspension and to increase the stiffness of the overall suspension-electrode structure. The particles return to a state of random orientation when the electric field is removed.

Active Damping Control of Composite Structures

The optimization strategies with traditional composites lead to an optimal design which is passive in nature and cannot respond to unstructured environments [11]. The vibrational characteristics of a helicopter rotor fabricated from composites are known to depend on factors such as the rotational speed, aerodynamic loading, payload and the ambient hygrothermal environment. A tradi-

tional composite rotor, even when optimally tailored, cannot actively respond to changes in the rotor speed and aerodynamic loading. Choi, Thompson, and Ghandi [17,82-84] therefore proposed the design of intelligent composites containing ER fluids whose response characteristics can be continuously varied in order to achieve the optimal performance under varying service conditions.

When subjected to variations in the electric field strength the ER fluids dramatically alter their rheological properties. The global mass, stiffness and dissipative characteristics of ER fluid-containing composite structures may be varied in a programmable manner. Since such a response is practically instantaneous, the electro-elastodynamic properties of composites can be actively regulated in real time. The modern sensors, microprocessors, actuators and dynamically tunable ER fluids can be integrated to form an intelligent system. A simplified arrangement is shown in Fig. 3, where the sensors monitor the elastodynamic behavior of the composite structure. The signals from the sensors are fed to the appropriate microprocessor which evaluates the signals prior to determining a proper control strategy in order to synthesize the desired elastodynamic response characteristics. This can be accomplished by manipulating the properties of the ER fluid domains in the finite element segment associated with the particular sensor to achieve the desired vibrational response of the structure [17].

Vibrational Response of a Model Three-Layer Structure

Viscoelastic material layers such as those containing ER fluid can be used as components of a structure to obtain desired levels of structural damping. The viscoelastic material can be either placed in a constrained position subjected to shear loading or in a free position subjected to extensional deformation. In such arrangements, the damping behavior of the overall structure can be related, using vibration theories [88-89], to the viscoelastic properties of the constrained or free layers. Coulter *et al.* [90,91] demonstrated the feasibility of applying the well known Ross-Kerwin-Ungar model (RKU) to structures containing ER fluid as viscoelastic damping layers.

Figure 3(a) shows a possible shear damping configuration where a viscoelastic layer (material 2) is sandwiched between two elastic material layers (materials 1 and 3). Flexural deformation of the structure results in not only bending and extension but also shear, the latter occurring primarily in the viscoelastic layer (Fig. 3b). The strain energy associated with this shear tends to dominate the damping behavior of this multilayer structure [91]. Based on the RKU model [89], Coulter *et al.* [91] derived the following equations to describe the loss factor (η) and flexural rigidity per unit length of the structure:

$$\eta = (\eta_2 Y X) / [1 + (2 + Y)X \\ + (1 + Y)(1 + \eta_2^2)X^2] \qquad \text{(Eq 8)}$$

$$B = (B_1 + B_3)[1 + XY \\ + X^2 Y(1 + \eta_2^2)] / [1 + 2X + X^2(1 + \eta_2^2] \qquad \text{(Eq 9)}$$

where η_2 is the loss factor in shear of the viscoelastic layer ($\eta_2 = G''/G'$), Y is a stiffness parameter, and X is a shear parameter. These and other parameters are defined as follows:

$$Y = \left\{ [(E_1 H_1^3 + E_3 H_3^3) / (12 H_{31}^2) \\ \cdot [1 / E_1 H_1) + (1/E_1 H_1)] \right\}^{-1} \qquad \text{(Eq 10)}$$

where E_i and H_i are the tensile modulus and thickness of the *i*th layer, respectively, and $H_{31} = H_2 + (H_1 + H_3)/2$ is the separation between the mid-planes of the elastic constraining layers. When $E_1 = E_3$ and $H_1 = H_3$ (same material and thickness), Y becomes simply a geometric parameter, i.e., $Y = 3 (H_{31}/H_1)^2$.

$$X = (G_2/P_2 H_2) \cdot [(1/E_1 H_1) + (1/E_1 H_1)] \qquad \text{(Eq 11)}$$

where P is the wave number, which for a simply supported beam is given by $P = n\pi/L$ (L = beam length, n = the mode of flexure vibration).

The effective flexural rigidity per unit depth of the *i*th layer of the structure is given by $B_i = (E_i H_i^3)/12$. The change in effective stiffness of composite panels due to the presence of a viscoelastic layer results in a corresponding change in the resultant frequency of flexural vibration, f_n. For simple beams, this can be understood through the relation

$$f_n = (\rho^2/2\pi)\sqrt{B}/\mu \qquad \text{(Eq 12)}$$

where μ is the mass per unit area of the beam, which, for three-layer laminates, can be expressed as

$$\mu = \Sigma e_i H_i \qquad \text{(Eq 13)}$$

Here, ρ_i is the mass density of the material of layer *i*.

The above equations suggest that the damping or loss factor of a composite can be regulated by incorporating a viscoelastic material in an otherwise elastic composite panel. Similarly, the frequency of vibration of a composite panel encompassing a viscoelastic layer varies with corresponding variations in the complex shear modulus of the added viscoelastic material. If this viscoelastic material is an ER fluid, then the dynamic mechanical properties of a composite can be tuned in real time in accordance with the changing stressing conditions.

For composite structures containing a more complex distribution of ER fluid, the damping factor and vibration frequency may become difficult to obtain analytically. Fortunately, other techniques such as modal analysis and numerical techniques may be used in these cases.

Intelligent Composites Containing Shape Memory Alloys (SMAs)

Shape Memory Alloys

A shape memory alloy (SMA) is an alloy which, when plastically deformed with the external stresses removed and then heated, will regain its original (memory) shape. The process of regaining the original shape is related to a reverse transformation of the deformed martensitic phase to the higher temperature austenite phase.

Materials that are known to exhibit shape memory effect include copper alloy systems of Cu-Zn, Cu-Zn-Al, Cu-Zn-Ga, Cu-Zn-Sn, Cu-Zn-Si, Cu-Al-Ni, Cu-Au-Zn, Cu-Sn, and the alloys of Au-Cd, Ni-Al, Fe-Pt, etc. The most well known SMA is the nickel-titanium alloy named "Nitinol" [92-97]. Plastic strains of 6 to 8% typically in Nitinol may be completely recovered by heating the material above the characteristic transition temperature. To restrain the material from regaining its original shape could result in an internal stress of up to 690 MPa (100 ksi). Such a large internal stress within the material or structure may not be always desirable. However, SMAs have the unique ability of changing properties reversibly, and this characteristic can be exploited without embedding SMA fibers nor creating internal stresses in the structure [22,98,99]. One dimensional thermomechanical constitutive relations for SMAs were developed by Liang and Rogers [24].

Composites Containing SMAs

SMA fibers or films may be incorporated in a composite laminate in such a way that the composite can be stiffened or otherwise controlled by the addition of heat, e.g. by applying a current through the fibers. One possible configuration will consist of embedding SMA wires in a material off the neutral axis on both sides of the beam in agonist-antagonist pairs [99]. The SMA wires are plastically deformed before embedding and are constrained from contracting to their original length upon curing the composite. When the wires are heated, they attempt to contract and thereby generate a uniformly distributed shear stress along the entire length of the wires. The shear stress offset from the neutral axis of the structure causes the structure to bend in a predetermined manner. Other ways of incorporating SMAs in a structure for creating both concentrated and distributed loads within the structure are also possible.

These special characteristics can be used to suppress or damp structural vibration by generating internal forces to the structure in such a way as to dissipate the energy within the structure. In addition to such a transient vibration control, a steady state control can also be effected to change the modal characteristics of the structure. This can be accomplished by using the SMA mechanisms properly to increase the stiffness of the composite in different directions by predetermined amounts. As proposed by Rogers et al. [99], SMAs can also be utilized to impart large distributed loads throughout the material to alter the stored strain energy within the composite structure and thereby modify the modal response of the structure. This concept is referred to as "active strain energy tuning" [99].

Piezoelectric Composites—Analysis Approaches

Maxwell and Wagner [100] made probably the first attempt to predict theoretically the dependencies of the electrical properties of composites on the ceramic volume fraction; a spherical model was the basis for the proposed derivation. Theoretical equations for the dielectric constants were derived by Buesson [101] also from parallel and series models. The development of piezoelectric composites containing ceramic particles and polymers was pioneered in Japan by Kitayama et al. [35] and in the United States by Pauer [36]. Pauer investigated the dielectric properties of the piezoelectric composite composed of PZT ceramic particles dispersed in polyurethane rubber using series, parallel, spherical and cube models. Furukawa et al. [37] developed a theoretical equation for the piezoelectric constant d_{31} based on a spherical model. Newnham et al. [38] proposed a concept of connectivity to classify various composites. By using series and parallel models, they derived theoretical equations for the piezoelectric constants of composites. Banno [43] reviewed the developments of piezoelectric composite materials, their applications in Japan, and theoretical considerations for the dielectric and piezoelectric constants of various types of composites. Recently, Newnham and Ruschau [66] presented a review on smart electroceramics, including piezoelectric composites.

Theories

Banno [43] has reviewed various theories developed for predicting piezoelectric coefficients and related elastic and dielectric constants of composites. Normally these theories were developed based on one of the many two-phase models including series, parallel, spherical, ellip-

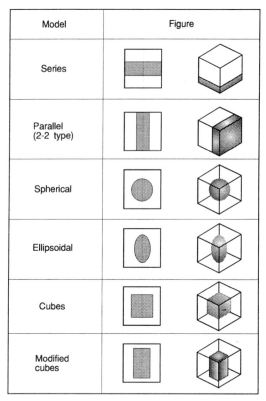

Model	Figure
Series	
Parallel (2-2 type)	
Spherical	
Ellipsoidal	
Cubes	
Modified cubes	

Fig. 4 Models for a two-phase composite for physical property predictions. Source: Ref 43

soidal, cubes and modified cubes models (Fig. 4). The shaded area in each diagram represents the inclusion dispersed in a continuous matrix. When the dispersed phase is spherical, cubic, or ellipsoidal, the model is named spherical, cubic, or ellipsoidal model. On the basis of these models, different equations were developed for predicting the dielectric constants of two-phase composites (Table 3). These equations can be easily extended to include the presence of additional phases. The notation used here (Table 4) is that suggested by Banno [43].

In general, parallel and series models lead to upper and lower limits of the interested physical constant of a binary composite. In addition, Beusson [101] provided an empirical relation for the intermediate state of the system:

$$(\bar{\varepsilon}_{33})^r = {}^1v \cdot ({}^1\varepsilon_{33})^r + {}^2v \cdot ({}^2\varepsilon_{33})^r \qquad \text{(Eq 14)}$$

where $-1 < r < +1$. When r is nearly zero, $(\bar{\varepsilon}_{33})^r$ becomes approximately $(1 + r \cdot \log \bar{\varepsilon}_{33})$. Therefore, Eq 14 would become a logarithmic rule-of-mixture.

$$\log \bar{\varepsilon}_{33} = {}^1v \log {}^1\varepsilon_{33} + {}^2v \log {}^2\varepsilon_{33} \qquad \text{(Eq 15)}$$

This empirical quantity r, which may be obtained by curve fitting, does not carry any physical meaning [43]. Pauer [36] measured the dielectric constants of compos-

Table 3 Theoretical equations for the dielectric constant of a two-phase system

Model	Dielectric constant of composite(a)
Series	$1/\varepsilon_{33} = {}^1v/{}^1\varepsilon_{33} + {}^2v/{}^2\varepsilon_{33}$
Parallel	$\varepsilon_{33} = {}^1v \cdot {}^1\varepsilon_{33} + {}^2v \cdot {}^2\varepsilon_{33}$
Spherical	$\varepsilon_{33} = {}^2\varepsilon_{33}\dfrac{\lvert 2 \cdot {}^2\varepsilon_{33} + {}^1\varepsilon_{33} - 2 \cdot {}^1v\,({}^2\varepsilon_{33} - {}^1\varepsilon_{33})\rvert}{\lvert 2 \cdot {}^2\varepsilon_{33} + {}^1\varepsilon_{33} + {}^1v({}^2\varepsilon_{33} - {}^1\varepsilon_{33})\rvert}$
Ellipsoidal(b)	$\varepsilon_{33} = {}^2\varepsilon_{33}\left\{ 1 + \dfrac{K \cdot {}^1v({}^1\varepsilon_{33} - {}^2\varepsilon_{33})}{K \cdot {}^2\varepsilon_{33} + ({}^1\varepsilon_{33} - {}^2\varepsilon_{33})(1 - {}^1v)} \right\}$
Cubes	$\varepsilon_{33} = \dfrac{{}^1\varepsilon_{33} \cdot {}^2\varepsilon_{33}}{({}^2\varepsilon_{33} - {}^1\varepsilon_{33}) \cdot {}^1v^{-1/3} + {}^1\varepsilon_{33} \cdot {}^1v^{-2/3} + {}^2\varepsilon_{33} \cdot (1 - v^{2/3})}$
Modified cubes(c)	$\varepsilon_{33} = \dfrac{a^2 \cdot [a + (1-a)n]^2 \cdot {}^1\varepsilon_{33} \cdot {}^2\varepsilon_{33}}{a \cdot {}^2\varepsilon_{33} + (1-a)n \cdot {}^1\varepsilon_{33}} + \lvert 1 - a^2 \cdot [a + (1-a)n]\rvert \cdot {}^2\varepsilon_{33}$

(a) ${}^1v = a^3$. (b) When the particles are spherical, $K = 3$. (c) When $n = 0.1$, and $(1/a^2 - a)/(1 - a)$, it becomes the parallel, cubes, and series model, respectively

Table 4 Notation used for constants of two-phase composites

	Phase 1	Phase 2	Composite
Dielectric constant	${}^1\varepsilon_{33}$	${}^2\varepsilon_{33}$	$\overline{\varepsilon}_{33}$
	${}^1\varepsilon_{22}$	${}^2\varepsilon_{22}$	$\overline{\varepsilon}_{22}$
	${}^1\varepsilon_{11}$	${}^2\varepsilon_{11}$	$\overline{\varepsilon}_{11}$
Piezoelectric constant d	${}^1d_{33}$	${}^2d_{33}$	D_{33}
	${}^1d_{32}$	${}^2d_{32}$	D_{32}
	${}^1d_{31}$	${}^2d_{31}$	D_{31}
	${}^1d_h$	${}^2d_h$	D_h
Elastic constant s	${}^1s_{33}$	${}^2s_{33}$	S_{33}
	${}^1s_{22}$	${}^2s_{22}$	S_{22}
	${}^1s_{11}$	${}^2s_{11}$	S_{11}
Mechanical quality factor	${}^1Q_{m33}$	${}^2Q_{m33}$	Q_{m33}
	${}^1Q_{m11}$	${}^2Q_{m11}$	Q_{m11}
	$1Q_{mp}$	$2Q_{mp}$	Q_{mp}
Mechanical loss tangent	${}^1\tan\delta_{m33}$	${}^2\tan\delta_{m33}$	$\tan\delta_{m33}$
	${}^1\tan\delta_{m11}$	${}^2\tan\delta_{m11}$	$\tan\delta_{m11}$
Frequency constant	${}^1f_r \cdot d$	${}^2f_r \cdot d$	$F_R \cdot d$
Radial coupling factor	${}^1K_p$	${}^2K_p$	K_p
Volume fraction	${}^1v$	${}^2v$	$V = {}^1v + {}^2v = 1$

ites composed of PZT ceramic particles dispersed in a urethane rubber matrix. However, the experimental values did not agree well with the calculated values based on the parallel, series, cubes, spherical models and the logarithmic rule-of-mixture (Fig. 5). This prompted Banno and Saito [39] to introduce a modified cubes model, which generalizes the parallel, series, and cube models (Fig. 6). In this model, l, m, and n are parameters related

to the shape of the unit cell deformed from the cubes model. In many cases, l is equal to m and the modified cubes model is determined by the values of n and a, while the volume fraction of the dispersed phase is ${}^1v = a^3$. In the special cases where $n = 0,1$ or $({}^1v^{-2/3} - {}^1v^{1/3})/(1 - {}^1v^{1/3})$ (with $1 = m = 0$), the modified cubes model becomes the parallel, cubes, or series model, respectively. This model appears to give a better fit to the experimental

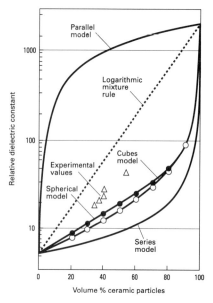

Fig. 5 Theoretical and experimental values for dielectric constants of composites of piezoelectric particles dispersed in a polyurethane rubber. Particles are PZT-5H ceramic. Source: Ref 43

data obtained by Banno [40] on PbTiO$_3$-rubber composites.

The theoretical equations for predicting the piezoelectric constants of composites were first developed from the spherical model by Furukawa *et al.* [37]. Based on the ellipsoidal model, Yamada *et al.* [33] derived more generalized equations for composite piezoelectric constants. These are listed in Table 5, where K is a parameter attributed to the shape of the ellipsoidal particles. The equations fit the experimental data obtained on PZT-PVDF systems if the shape parameter K was taken to be 8.5, requiring that ellipsoidal axes ratio be 2.8 and the long ellipsoids of PZT particles be perpendicular to the composite film surface [34]. These specifications are difficult to accept because, as pointed out by Banno [43], it is more probable that the PZT particles were randomly distributed in the PVDF polymer matrix and that the shape of the PZT particles were approximately spherical.

Table 6 summarizes the theoretical equations derived by Newnham *et al.* [39] based on the parallel and series models. The parallel model equations are derived for the conventional 2-2 type parallel model. For the parallel model of 1-3 type (Fig. 7), the corresponding equations are given in Table 7 [43]. For a 0-3 type and 1-3 type composite, Banno *et al.* [40,41,43] have derived alterna-

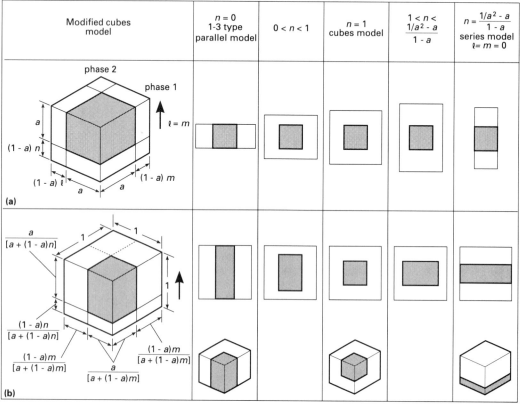

$[a + (1 - a)n] [a + (1 - a)m]^2 = 1, 0 < a \le 1.$

Fig. 6 Modified-cubes model as a function of the n value. (a) Phase 1 is cubic. (b) The unit cell is cubic. Source: Ref 43

tive equations by following the modified cubes model. These equations are listed in Table 8. Tashiro *et al.* [42] investigated select functional properties of sandwich-type PZT ceramic structures. Comparison was made between the experimental and calculated values of piezoelectric constants D_{33} according to the series model-based equations. Coupling factor, frequency constant, and mechanical quality factor of sandwich-type PZT ceramic structures were further discussed by Banno [43], as well as porosity dependencies of dielectric and piezoelectric constants of polycrystalline ceramics. Banno derived several useful equations for predicting porosity dependencies of these constants, which were found to agree well with the experimental data measured by Okazaki, Nagata, and Igarashi [102-106].

Optical Composites

Active research is in progress in the area of optical composites which are composed of two or more phases and exhibit unique optical properties [107-116]. Sol-gel chemical processing has been the primary method for fabricating these optical composites, which usually contain an optically active organic phase dispersed in a sol-gel derived oxide matrix. The chemical stability of the oxide network imparts to the composites substantial advantages in processing, reliability, and performance over the organics by themselves [116]. Liquid organics at ambient temperature are of limited use in an optical system. By encapsulating or embedding the organic in an inorganic matrix these restrictions can be lifted. This subject has recently been reviewed by Hench [116]. Recent developments on nonlinear optical materials were summarized by Eaton [117].

Hench [116] described two methods for preparation of optical composites. In the sol-processing method, the metal organic sol-gel precursor is mixed with the optical organic molecules. Hydrolysis and polycondensation reactions result in formation of a highly porous, interconnected three-dimensional oxide network (the gel) containing the optical organic species within the framework of the gel. In the absorption or impregnation processing, an optically transparent inorganic Type VI gel-silica matrix is first prepared and stabilized thermally. Then an optically active polymer is impregnated into the interconnected three-dimensional pore network. Comparisons were made on processing merits, mechanical properties

Table 5   Theoretical equations for the piezoelectric constant D_{31} of a two-phase system(a)

Model	Piezoelectric constant D_{31}
Spherical	$D_{31} = {}^1d_{31} \cdot \dfrac{5 \cdot {}^1v}{2 + 3 \cdot {}^1v} \cdot \dfrac{3 \cdot {}^2\varepsilon_{33}}{2 \cdot {}^2\varepsilon_{33} + {}^1\varepsilon_{33} + {}^1v({}^2\varepsilon_{33} - {}^1\varepsilon_{33})}$
Ellipsoidal(b)	$D_{31} = \dfrac{\alpha \cdot {}^1v \cdot K\{{}^1\varepsilon_{33}/{}^2\varepsilon_{33} - 1 + K + (K-1)({}^1\varepsilon_{33}/{}^2\varepsilon_{33} - 1) \cdot {}^1v\} \cdot {}^1d_{31}}{({}^1\varepsilon_{33}/{}^2\varepsilon_3 - 1 + K)^2 + ({}^1\varepsilon_{33}/{}^2\varepsilon_{33} - 1)\{(K-1)^2 - {}^1\varepsilon_{33}/{}^2\varepsilon_{33}\} \cdot {}^1v}$

(a) ${}^2d_{31} = 0$, ${}^2S_{11} \gg {}^1S_{11}$, and ${}^2S_{33} \gg {}^1S_{33}$. (b) α is the polarizing factor. When the particles are spherical, $K = 3$

Table 6   Piezoelectric constants D_{33} and D_{31} and elastic constants S_{33} and S_{31} of a two-phase system (composite) derived from series and parallel models

Constant	Model	
	Series	**Parallel**
Piezoelectric D_{33}	$D_{33} = \dfrac{{}^1v \cdot {}^1d_{33} \cdot {}^2\varepsilon_{33} + {}^2v \cdot {}^2d_{33} \cdot {}^1\varepsilon_{33}}{{}^1v \cdot {}^2\varepsilon_{33} + {}^2v \cdot {}^1\varepsilon_{33}}$	$D_{33} = \dfrac{{}^1v \cdot {}^1d_{33} \cdot {}^2s_{33} + {}^2v \cdot {}^2d_{33} \cdot {}^1s_{33}}{{}^1v \cdot {}^2s_{33} + {}^2v \cdot {}^1s_{33}}$
Elastic S_{33}	$S_{33} = {}^1v \cdot {}^1s_{33} + {}^2v \cdot {}^2s_{33}$	$1/S_{33} = {}^1v/{}^1s_{33} + {}^2v/{}^2s_{33}$
Piezoelectric D_{31}	$D_{31} = \dfrac{{}^1v \cdot {}^1d_{31} \cdot {}^2\varepsilon_{33} \cdot {}^2s_{11} + {}^2v \cdot {}^2d_{31} \cdot {}^1\varepsilon_{33} \cdot {}^1s_{11}}{({}^1v \cdot {}^2\varepsilon_{33} + {}^2v \cdot {}^1\varepsilon_{33})({}^1v \cdot {}^2s_{11} + {}^2v \cdot {}^1s_{11})}$	$d_{31} = {}^1v \cdot {}^1d_{31} + {}^2v \cdot {}^2d_{31}$
Elastic s_{11}	$1/s_{11} = {}^1v/{}^1s_{11} + {}^2v/{}^2s_{11}$	$s_{11} = {}^1v \cdot {}^1s_{11} + {}^2v \cdot {}^2s_{11}$

Table 7   Dielectric, elastic, and piezoelectric constants of a two-phase system

Constant	1-3 type parallel model
Dielectric $\bar{\varepsilon}_{33}$ and $\bar{\varepsilon}_{11}$	$\bar{\varepsilon}_{33} = {}^{1}v \cdot {}^{1}\varepsilon_{33} + {}^{2}v \cdot {}^{2}\varepsilon_{33}$
	$\bar{\varepsilon}_{11} = \dfrac{\sqrt{{}^{1}v}}{\dfrac{\sqrt{{}^{1}v}}{{}^{1}\varepsilon_{11}} + \dfrac{1 - \sqrt{{}^{1}v}}{{}^{2}\varepsilon_{11}}} + (1 - \sqrt{{}^{1}v}) \cdot {}^{2}\varepsilon_{11}$
Elastic $\bar{s}_{33}$ and $\bar{s}_{11}$	$\dfrac{1}{\bar{s}_{33}} = \dfrac{{}^{1}v}{{}^{1}s_{33}} + \dfrac{{}^{2}v}{{}^{2}s_{33}}$
	$\dfrac{1}{\bar{s}_{11}} = \dfrac{\sqrt{{}^{1}v}}{\sqrt{{}^{1}v}\,{}^{1}s_{11} + (1 - \sqrt{{}^{1}v})\,{}^{2}s_{11}} + \dfrac{1 - \sqrt{{}^{1}v}}{{}^{2}s_{11}}$
Piezoelectric $\bar{d}_{33}$ and $\bar{d}_{31}$	$\bar{d}_{33} = \dfrac{{}^{1}v \cdot {}^{1}d_{33} \cdot {}^{2}s_{33} + {}^{2}v \cdot {}^{2}d_{33} \cdot {}^{1}s_{33}}{{}^{1}v \cdot {}^{2}s_{33} + {}^{2}v \cdot {}^{1}s_{33}}$
	$\bar{d}_{31} = \sqrt{{}^{1}v} \cdot \dfrac{{}^{2}s_{11}\sqrt{{}^{1}v} \cdot {}^{1}d_{31} + (1 - \sqrt{{}^{1}v}) \cdot {}^{2}d_{31} \cdot {}^{1}s_{11}}{\sqrt{{}^{1}v} \cdot {}^{1}s_{11} + (1 - \sqrt{{}^{1}v})\,{}^{2}s_{11}} + (1 - \sqrt{{}^{1}v})\dfrac{{}^{2}d_{31}}{{}^{2}s_{11}}\,{}^{1}s_{11}$

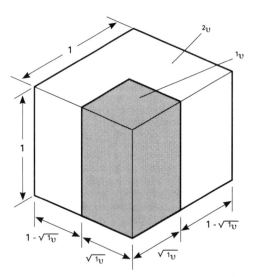

Fig. 7   A parallel model for 1-3 type composites. Source: Ref 43

and optical applications with emphasis on solid state tunable dye lasers [116]. Predictive models for various functional properties (including optical) of this type of composites have yet to be developed.

An optically transparent yet electrically conductive composite containing preferentially arranged conductive paths has been developed recently by Jin *et al.* [118]. The composite contains many vertically aligned but laterally isolated chains of ferromagnetic spherical particles (each coated with a thin layer of highly conductive metal) dispersed in a sheet of transparent polymer such as silicon rubber. The sheet material transmits more than 90% of the incident light and is highly conductive only in the thickness direction. When suitably modified, the sheet exhibits on-off electrical switchability at a certain threshold pressure. The pressure exerted by writing on the surface with a stylus, for instance, causes conduction. This composite medium is potentially useful for a variety of device applications such as write pads, touch-sensitive screens, sensors, and alarm devices.

Conclusions

Functionally responsive materials constitute the framework of a smart material system. Design of a smart material system or device (a sensor, transducer, actuator, etc.) requires the availability of specific functional materials. Increased sophistication of the utilization of smart materials depends on a clear understanding of the functional properties in relation to the chemical and crystal structure of materials.

Fundamental structure-property relations must be established for functionally responsive materials. Research efforts should emphasize the atomic, molecular and microstructural design of such materials. The guidelines or rules for fabricating desired functional materials must be identified. Research efforts must also be aimed at the establishment of computational/modelling, theoretical and experimental methodologies for designing and tailor-making functionally responsive materials.

Table 8   Piezoelectric constants of 0-3 type and 1-3 type composites using the modified cubes method

Composite type	Piezoelectric constants D_{33} and D_{31}
0-3(a)	$$D_{33} = {}^1d_{33} \cdot \frac{a^3 \cdot [a + (1-a)n]}{a + (1-a)n \cdot ({}^1\varepsilon_{33} / {}^2\varepsilon_{33})} \cdot \frac{1}{(1-a)n / [a + (1-a)n] + a^3}$$ $$D_{31} = {}^1d_{31} \cdot \frac{a^2 \cdot [a + (1-a)n]^2}{a + (1-a)\,n\,({}^1\varepsilon_{33} / {}^2\varepsilon_{33})};\, a^3 = {}^1v$$
1-3	$$D_{31} = {}^1d_{31} \cdot \frac{a^2 \cdot [a + (a + (1-a))n]}{a + (1-a)n\,({}^1\varepsilon_{33} / {}^2\varepsilon_{33})} \cdot \frac{a}{1 - a[a + (1-a)\,n^{1/2} + a^3}$$

(a) Assuming ${}^2s_{11} > {}^1s_{11}$ and ${}^2s_{33} > {}^1s_{33}$

For a given material, several important tensor properties must be investigated. These include second-, third-, and fourth-rank tensors, specifying the thermomechanical, thermo-optical, thermo-electrical, elasto-optical, electromechanical, and electro-optical effects. Correlations among these coefficients and their dependence on the material structure must be studied. Reliable prediction of various tensor properties of a multi-phase composites is still difficult to achieve. Systematic efforts are needed on the development of constitutive equations relating various stimuli to responses in hybrid and composite systems. These efforts will contribute greatly to the establishment of a scientific basis from which many intelligent material systems can be designed and fabricated.

REFERENCES

1. Iqbal Ahmad, U.S.—Japan Workshop on Smart/Intelligent Materials and Systems, *ONR Asia Sci. Info. Bull.*, Vol 15, No. 4, 1990, p 67-75
2. *Proc. U.S.—Japan Workshop on Smart/Intelligent Materials and Systems*, Tokyo, Japan, March 1990
3. Toshinori Takagi, The Concepts of Intelligent Materials and the Guidelines on R and D Promotion, *A Report to the Japan Council for Aeronautics, Electronics, and Other Advanced Technologies*, Science and Technology Agency (STA), Tokyo, Japan, 1989
4. A.M. Vengsarkar and K.A. Murphy, Fiber Sensors in Aerospace Applications: The Smart Structures Concept, *Photonics Spectra*, Vol 24, No. 4, 1990, p 119-122
5. E. Udd, Fiber Sensors for Smart Structures, in *Optical Fiber Sensors: Proc. 6th Intl. Conf. OFS' 89*, Springer-Verlag, Berlin, Germany, 1989, p 392-399
6. R. Davidson, D.H. Bowen, and S.S.J. Roberts, Composite Materials Monitoring Through Embedded Fiber Optics, *Intl. J. Optoelectronics*, Vol 5, No. 5, 1990, p 397-404
7. R.M. Measures, N.D.W. Glossop, J. Lymer, M. Leblanc, J. West, S. Dubois, W. Tsaw, and R.C. Tennyson, Structurally Integrated Fiber Optic Damage Assessment System for Composite Materials, *Proc. SPIE—Intl. Soc. Optical Eng.*, 986, 1989, p 120-129
8. R.M. Measures, Fiber Optics In Composite Materials—Materials with Nerves of Glass, in *Proc. SPIE Conf. on Fiber Optic Sensors (IV)*, Intl. Soc. for Optical Eng., The Hague, Netherlands, March 1990, p 241-256
9. E. Udd, Fiber Optic Skin and Structural Sensors, *Industrial Metrology*, Vol 11, No. 1, 1990, p 3-18
10. R.D. Turner, T. Valis, W.D. Hogg, and R.M. Measures, Fiber-Optic Strain Sensors for Smart Structures, *J. Intelligent Mater. Syst. Struct.*, Vol 1, 1990, p 26-49
11. H. Smith, Jr., A. Garrett, and C.R. Staff, Smart Structures Concept Study, in *Proc. SPIE Conf. Fiber Optic Smart Structures and Skins II*, Boston, MA, Sept 1989, Intl. Soc. for Optical Eng., 1990, p 224-229
12. P. Sansonetti, *et al.*, Intelligent Composites Containing Measuring Fiber Optic Networks for Continuous Seft Diagnosis, in *Proc. SPIE Conf. Fiber Optic Smart Structures and Skins II*, Boston, MA, Sept 1989, Intl. Soc. for Optical Eng., 1990, p 211-223
13. M.H. Thursby, B.G. Grossman, T. Alavie, and K.S. Yoo, Smart Structures Incorporating Artificial Neural Networks, Fiber Optic Sensors, and Solid State Actuators, in *Proc. SPIE Conf. Fiber Optic Smart Structures and Skins II*, Boston, MA, Sept 1989, Intl. Soc. for Optical Eng., 1990, p 316-325
14. A.S. Brown, Materials Get Smarter, *Aerospace America*, Vol 28, March 1990, p 30-34,36
15. P. Hagenmuller, Selection of Intelligent Materials Use In Practical Applications, *Chemtronics*, Vol 4, No. 4, 1989, p 254-258
16. H. Aldersey-Williams, A Solid Future For Smart Fluids, *New Scientist*, Vol 125, 1990, p 37-40
17. S.B. Choi, M.V. Gandhi, and B.S. Thompson, An Active Vibration Tuning Methodology for Smart Flexible Structures Incorporating Electrorheological Fluids: A Proof-of-Concept Investigation, *Proc. 1989 Am. Control Conf.*, Vol 1, June 21-23, 1989, Pittsburgh, PA, p 694-699. Also in *Proc. 12th Biennial ASME Conf. on Mechanical Vibration and Noise*, Montreal, Canada, Sept 1989
18. T.G. Duclos, "Design of Devices Using Electrorheological Fluids," SAE Technical Paper 881134, SAE, Warrendale, PA, 1988
19. P.M. Adriani and A.P. Gast, A Microscopic Model of Electrorheology, *Phys. Fluids*, Vol 31, No. 10, 1988, p 2757-2768
20. H. Corand, A.F. Sprecher, and J.D. Carlson, Eds., *1st Intl. Symp. Electrorheological Fluids*, NC State University,

Aug 1989, Technomic Publishing Company, Lancaster, PA, 1990

21. Y. Choi, A.F. Sprecher, and H. Conrad, Vibration Composite Beam Containing an Electrorheological Fluid, *J. Intelligent Mater. Syst. Struct.*, Vol 1, 1990, p 91-104

22. C.A. Rogers, D.K. Barker, K.D. Bennett, and R.H. Wynn, Jr., Demonstration of a Smart Material With Embedded Actuators and Sensors For Active Control, *Proc. SPIE—Intl. Soc. Optical Eng.*, 1989, p 90-105

23. C.R. Fuller, C.A. Rogers, and H.H. Robertshaw, Active Structural Control with Smart Structures, in *Proc. SPIE Conf. Fiber Optic Smart Structures and Skins II*, Boston, MA, Sept 1989, Intl. Soc. for Optical Eng., 1990, p 338-358

24. C. Liang and C.A. Rogers, *J. Intelligent Mater. Sys. Structure*, Vol 1, 1990, p 207-234

25. *Proc. 1st Intl. Workshop on Intelligent Materials*, Soc. of Non-Traditional Technology of Japan, Tsukuba, Japan, April 1989

26. T. Usher, *Sensors and Transducers*, MacMillan Publishing, New York, 1985

27. D. De Rossi, Need Better Sensors? Try Using Polymers, *Research and Development*, May 1989, p 67-70

28. J.C.W. Chien, *Polyacetylene: Chemistry, Physics, and Materials Science*, Academic Press, 1984

29. T.R. Shrout, L.J. Bowen, and W.A. Schulze, *Mater. Res. Bull.*, Vol 15, 1980, p 1371

30. R.E. Newnham, D.P. Skinner, K.A. Klicker, A.S. Bhalla, B. Hardiman, and T.R. Gururaja, *Ferroelectrics*, Vol 27, 1980, p 49

31. H. Yamazaki and T. Kitayama, *Ferroelectrics*, Vol 33, 1981, p 147

32. T. Furukawa, K. Ishida, and E. Fukada, *J. Appl. Phys.*, Vol 50, 1979, p 4904

33. L.J. Bowen and T.R. Gururaja, *J. Appl. Phys.*, Vol 51, 1980, p 5661

34. T. Yamada, T. Ueda, and T. Kitayama, *J. Appl. Phys.*, Vol 53, No. 6, 1982, p 4328-4332

35. I. Namiki, K. Sukiyama, T. Kitayama, and T. Ueda, *IEEE CHMT-4*, 1981, p 304

36. T. Kitayama and S. Sugawara, Report of Tech. Group, Institute of Electronic Communication Eng., Japan, CPM 72-17, 1972

37. L.A. Pauer, *IEEE Intl. Conv. Rec.*, 1973, p 1-5

38. T. Furukawa, K. Fujino, and E. Fukada, *Japan J. Appl. Phys.*, Vol 15, 1976, p 2119-2129

39. R.E. Newnham, D.P. Skinner, and L.E. Cross, *Mater. Res. Bull.*, Vol 13, 1978, p 525-536

40. H. Banno and S. Saito, *Japan J. Appl. Phys.*, Vol 22, Supplement, 1983, p 67-79

41. H. Banno, Proc. 6th Intl. Meeting on Ferroelectricity (IMF-6, Kobe, Japan, 1985); *Japan J. Appl. Phys.*, Vol 24, Supplement, 1985, p 243-245

42. S. Tashiro, N. Arai, H. Igarashi, and K. Okazaki, *Ferroelectrics*, Vol 37, 1981, p 595-598

43 H. Banno, Recent Developments of Piezoelectric Composites in Japan, in *Advanced Ceramics*, S. Saito, Ed., Oxford University Press, Oxford, p 8-26

44. R.E. Newnham, *J. Mater. Edu.*, Vol 7, 1985, p 605-651

45. D.W. van Krevelen and P.J. Hoftyzer, *Properties of Polymers, Their Estimation and Correlation with Chemical Structure*, Elsevier Scientific Publishing Co., Amsterdam, Netherlands, 1976

46. A. Bondi, *Physical Properties of Molecular Crystals, Liquids and Glasses*, Wiley, New York, 1968

47. A.A. Askadskii, Prediction of Physical Properties of Polymers, in *Polymer Year Book*, Vol 4, 1989, p 93-147

48. P. Hohnberg and W. Kohn, *Phys. Rev. B*, Vol 136, 1964, p 864; W. Kohn and L.J. Sham, *Phys. Rev. A*, Vol 140, 1965, p 1133

49. M.S. Hyberston and S.G. Louie, *Phys. Rev. Lett.*, Vol 58, 1987, p 155

50. J. Korringa, *Physica*, Vol 13, 1947, p 392; W. Kohn and N. Rostoker, *Phys. Rev.*, Vol 94, 1954, p 1111; Zhang and A. Gonis, *Phys. Rev. B*, Vol 39, 1989, p 10373

51. C.-Y. Yeh, A.-B. Chen, D.M. Nicholson, and W.H. Butler, *Phys. Rev. B*, Vol 42, 1990, p 10976

52. P. Soven, *Phys. Rev.*, Vol 156, 1967, p 809; H. Ehreneich and L. Schwartz, *Solid State Phys.*, Vol 31, 1976, p 149

53. A.-B. Chen, *Phys. Rev. B*, Vol 7, 1973, p2230; A.-B. Chen and A. Sher, *Phys. Rev. Lett.*, Vol 40, 1978, p 900; *Phys. Rev. B*, Vol 18, 1978, p 5326; S. Krishnamurthy, A. Sher and A.-B. Chen, *Phys. Rev. B*, Vol 33, 1986, p 1026

54. D.R. Hamann, M.S. Schluter, and C. Chiang, *Phys. Rev. Lett.*, Vol 43, 1979, p 1494; G.B. Bachelet, D.R. Hamann, and M.S. Schluter, *Phys. Rev. B*, Vol 26, 1982, p 4199

55. C.-Y. Yeh, A.-B. Chen, and A. Sher, *Phys. Rev. B*, Vol 43, 1991, p 9138

56. R. Silbey in *Polydiacetylenes: NATO ASI Series E No. 102*, D. Bloor and R.R. Chance, Ed., Martinus Nijhoff Publishers, 1985, p 105

57. A.-B. Chen, S. Phokachaipatana, and A. Sher, *Phys. Rev. B*, Vol 28, 1983, p 1121

58. S. Krishnamurthy, A. Sher, and A.-B. Chen, *Appl. Phys. Lett.*, Vol 47, 1985, p 160, and Vol 55, No. 10, 1989, p 1002; *J. Appl. Phys.*, Vol 61, 1987, p 1475; *Appl. Phys. Lett.*, Vol 52, No. 6, 1988, p 468

59. A.-B. Chen and A. Sher, *Phys. Rev. B*, Vol 32, 1985, p 3695

60. R. Patrick, A.-B. Chen, and A. Sher, *Phys. Rev. B*, Vol 36, 1987, p 6585

61. M.A. Berding, S. Krishnamurthy, A. Sher, and A.-B. Chen. *J. Appl. Phys.*, Vol 67, No. 10, 1990, p 6175

62. W.A. Harrison, *Electronic Structure and the Properties of Solids*, W.H. Freeman and Co., 1980

63. W.A. Harrison, *Phys. Rev. B*, Vol 27, 1983, p 3592

64. C.Z. Wang, C.T. Chan, and K.M. Ho, *Phys. Rev. B*, Vol 39, 1989, p 8592, and Vol 40, 1989, p 3390, and private communication

65. T. Furukawa, J. Aiba, and E. Fukada, Piezoelectric Relaxation in PVDF, *J. Appl. Phys.*, Vol 50, 1979, p 3615-3621

66. M.G. Broadhurst, G.T. Davis, J.E. McKinney, and R.E. Collins, Piezoelectricity and Pyroelectricity in PVDF—A Model, *J. Appl. Phys.*, Vol 49, 1978, p 4992-4997

67. R. Hayakawa and Y. Wada, Piezoelectricity and Related Properties of Polymer Films, *Adv. Polymer Sci.*, Vol 11, 1973, p 1-56

68. R.E. Newnham, *J. Mater. Edu.*, Vol 7, 1985, p 605-651

69. A.A. Askadskii, Prediction of Physical Properties of Polymers, in *Polymer Year Book*, Vol 4, 1989, p 93-147

70. A.A. Askadskii, S.N. Prozorova, and G.L. Slonymsky, *Visokomolek Soedin, A*, Vol 18, No. 3, 1976, p 633

71. K. Tashiro, M. Kobayashi, H. Tadokoro, and E. Fukada, Calculation of Elastic and Piezoelectric Constants of Polymers by a Point Charge Model, *Macromolecules*, Vol 13, 1980, p 691-698

72. J.F. Nye, *Physical Properties of Crystals*, Clarendon Press, Oxford, 1957

73. D.H. Kaelble, *Computer Aided Design of Polymers and Composites*, Marcel Dekker, Inc., New York, 1985

74. W. Konig, *Ann. Physics*, Vol 25, 1985, p 618-624

75. W.M. Winslow, *J. Appl. Phys.*, Vol 20, 1949, p 1137-1140

76. J.E. Stangroom, *J. Phys. Technol.*, Vol 14, 1983, p 290-296

77. V.N. Makatun, K.N. Lapko, A.D. Matsepuro, and V.F. Tikeoyi, *Inzhenerno-Fizicheskii Zhurnal*, Vol 45, 1983, p 597-602

78. W.A. Bullough and D.J. Peel, *Japanese Hydraulic and Pneumatic Soc.*, Vol 17, 1986, p 510-526

79. Z.P. Shulman, R.G. Gorodkin, E.V. Korobko, and V.K. Gleb, *J. Non-Newtonian Fluid Mech.*, Vol 8, 1981, p 29-41

80. H. Vejima, *Jpn. J. Appl. Phys.*, Vol 11, No. 3, 1972, p 319-326

81. R.G. Gorodkin, Y.V. Korobko, G.M. Blokh, V.K. Gleb, G.I. Sidorova, and M.M. Ragotnot, *Fluid Mech.—Soviet Res.*, Vol 8, No. 4, 1979, p 48-61

82. M.V. Gandhi, B.S. Thompson, S.B. Choi, and S. Shakir, in *Advances in Design Automation, Vol 2, Robotics, Mechanisms and Machine Systems*, S.S. Rao, Ed., 1987, p 1-10

83. M.V. Gandhi, B.S. Thompson, and S.B. Choi, *Proc. 4th Japan—U.S. Conf. Comp. Mater.*, Washington, D.C., June 1988

84. M.V. Gandhi and B.S. Thompson, *Proc. ARO Workshop on Smart Materials, Structures and Mathematical Issues*, Blacksburg, VA, Sept 1988

85. P.M. Adriani and A.P. Gast, *Phys. Fluids*, Vol 31, No. 10, 1988, p 2757-2768

86. P. Lienard, *La Recherche Aeronautique*, Vol 20, 1951, p 11-22

87. H. Oberst, *Acastica*, Vol 2, No. 4, 1952, p 181-194

88. E.M. Kerwin, Jr., *J. Acoustic Soc. Am.*, Vol 31, July 1959, p 1-12

89. D. Ross, E.E. Ungar, and E.M. Kerwin, in *Structural Damping*, J.E. Ruzicka, Ed., ASME, New York, 1959, Sec. 3

90. T.G. Duclos, J.P. Coulter, and L.R. Miller, Applications for Smart Materials in the Field of Vibration Control, in *Proc. ARO Workshop on Smart Materials Structures and Mathematical Issues*, Blacksburg, VA, Sept 1988, Technomic Publishing Company, Lancaster, PA, 1989, p 132-146

91. J.P. Coulter, T.G. Duclos, and D.N. Acker, "The Wage of Electrorheological Materials in Viscoelastic Layer Damping Applications," in "Damping '92," Paper CAA-1

92. W.J. Buehler and R.C. Wiley, U.S. Patent 3,174,851, March 23, 1965

93. C.M. Wayman and K. Shimizu, *Metal Sci. J.*, Vol 6, 1972, p 175

94. J. Perkins, Ed., *Shape Memory Effects in Alloys*, Plenum Press, New York, 1975

95. L. Schetky, *Sci. Am.*, Vol 241, 1979, p 74

96. R.V. Delaey, H. Tas Krishnan, and H. Warlimont, *J. Mater. Sci.*, Vol 9, 1974, p 1521-1545

97. T. Saburi and C.M. Wayman, *Acta Metall.*, Vol 27, 1979, p 979

98. C.A. Rogers and H.H. Robertshaw, Eng. Sci. Preprints 25, ESP25.88027, Soc. of Eng. Sci., Inc., June 20-22, 1988

99. C.A. Rogers, C. Liang, and D.K. Barker, in *Proc. ARO Workshop on Smart Materials, Structures and Mathematical Issues*, Blacksburg, VA, Sept 1988, Technomic Publishing Company, Lancaster, PA, p 39-62

100. K.W. Wagner, *Arch. Electrotech.*, Vol 2, 1914, p 371-387

101. W.R. Buesson, in *The Physics and Chemistry of Ceramics*, C. Klinsberg, Ed., Gordon and Breach Science Publishing, Inc., 1963, p 22-29

102. K. Okazaki and N. Nagata, Trans. IECE Japan, 53-C, 1970, p 815-822

103. K. Okazaki and K. Nagata, *Proc. 1971 Intl. Conf. Mechanical Behavior of Materials IV*, 1972, p 404-412

104. K. Okazaki and H. Igarashi, *Proc. 6th Intl. Mater. Symp. Ceramic Microstructure*, 1976, p 564-583

105. H. Igarashi, *Ceramics Japan*, Vol 10, 1975, p 799-806

106. H. Igarashi, *Ceramics Japan*, Vol 12, 1977, p 126-131

107. D. Avnir, D. Levy, and R. Reisfeld, *J. Phys. Chem.*, Vol 88, 1984, p 5956

108. D. Avnir, V.R. Kaufman, and R. Reisfeld, *J. Non-Cryst. Solids*, Vol 74, 1985, p 395

109. R. Reisfeld, *J. Non-Cryst. Solids*, Vol 121, 1990, p 254-266

110. E.J.A. Pope and J.D. Mackenzie, *MRS Bull.*, March/May, 1987, p 29-31

111. B. Dunn, J.D. Mackenzie, J.I. Zink and O.M. Stafsudd, *Proc. Conf. SPIE Sol-Gel Optics*, San Diego, July 1990

112. T.A. King, *Proc. Conf. SPIE High Power Solid State Lasers*, The Hague, 1990, p 1277

113. J.L. Nogues, S. Majewski, J.K. Walker, M. Bowen, R. Wajcik, and W.V. Morsehead, *J. Am. Ceram. Soc.*, Vol 71, 1988, p 1159-1163

114. L.L. Hench, J.K. West, B.F. Zhu, and R. Ochoa, *Proc. Conf. SPIE Sol-Gel Optics*, San Diego, July 1990

115. S.M. Melpolder, M.R. Roberts, B.K. Coltrain, and R.W. Wake, in *Advanced Composite Materials: Am. Ceram. Soc. Comp. Conf.*, 1991, p 287

116. L.L. Hench, Optical Composites, in *Advanced Composite Materials: Am. Ceram. Soc. Comp. Conf.*, 1991, p 265-286

117. D.F. Eaton, Nonlinear Optical Materials, *Science*, Vol 253, July 1991, p 281-287

118. S. Jin, T.H. Tiefel, R. Wolfe, R.C. Sherwood, and J.J. Morttiner, Jr, Optically Transparent, Electrically Conductive Composite Medium, *Science*, Vol 255, 1992, p 446-448

Chapter 12

Polymer Composites for Automotive Applications

Advances in structural materials and processing technology are of key importance to transportation as well as defense and space industries. In transportation, there has been a continued drive to reduce weight, number of components, and process waste through improved materials and processes. These three attributes, along with the recyclability of automotive parts, underlie the overall goal of optimizing vehicle lifetime efficiency in the automotive sector of the industry.

Automobiles account for approximately two-thirds of the gasoline consumption and one-third of the total energy use in the United States [1]. Each year, the U.S. automotive industry uses more than 20 million tons of material to produce more than 10 million vehicles. Approximately 75% of a vehicle's fuel consumption is directly related to weight-related factors, the other 25% related to air drag. A small weight reduction of current vehicles (for example, by 25%) would result in tremendous gasoline savings (e.g., 750,000 barrels/day, or about 13% of U.S. gasoline consumption), as well as a reduction of a hundred million tons/year in CO_2 emissions [1].

Automobile companies have made a commendable effort to reduce the weight of vehicles over the past two decades. The average weight of a vehicle has been reduced nearly 25%, doubling fuel economy without sacrificing other desirable performance features. However, further improvements are possible and would likely come from the following efforts [1]:

- To enhance processing of traditional materials to allow further weight reduction
- To develop new processes and material forms that make the current light-weight materials more cost effective
- To develop cost-effective recycling technology for in-plant and scrapped vehicle waste

- To develop capabilities to model processes, materials, and components to optimize the design, performance, and material usage of vehicle systems
- To accelerate the development and introduction of new materials to automotive applications

With a low density, high strength and stiffness, and good chemical resistance, reinforced plastics and composites are prime candidate materials for future transportation designs. In this chapter, polymers and polymer composites will be discussed that are used or being considered for use in automotive industries. The processing techniques for automotive polymer composites are then reviewed. The performance and design issues with these materials are then considered. The environmental impacts and recycling of automotive polymer composites are also discussed in terms of lifetime efficiency or cost. This chapter concludes by proposing fruitful areas of future research and development on this subject.

Automotive Polymers and Polymer Composites

Polymeric materials are expected to replace steel in many automotive applications [2-8]. Steel is heavy, expensive to tool and subject to corrosion. Plastics are lighter, offer better dent resistance, are more flexible, and statistically are highly desirable for dent-prone vertical surfaces [8]. However, unreinforced plastics are typically much weaker and more expensive than traditional materials [5]. High energy levels are required to break current automotive plastics; but fracture modes are brittle in nature when failure occurs.

The plastic materials that are currently employed to manufacture a typical family passenger car are exemplified in Table 1 [2]. Plastics are only a small fraction of car weight. Plastics can substitute for metals in bumpers,

Table 1 Use of plastics in the automotive industry

Material	1980		1990 (estimated)	
	million lb.	million kg	million lb	million kg
Polypropylene	495	225	529	240
Polyethylene	121	55	214	97
Polyurethane	484	220	499	226
Vinyls	400	181	166	75
ABS	193	88	186	84
Unsaturated polyester	290	132	340	154
Polyamide (nylon)	55	25	114	52
Phenolics	44	20	55	25
Acrylic	57	26	57	26
Others	59	27	640	290
Applications				
Use per car	200	90.8	300	136
Weight of an average car	3500	1587	2500	1134
% Plastics		5.7		12

Source: Ref 2

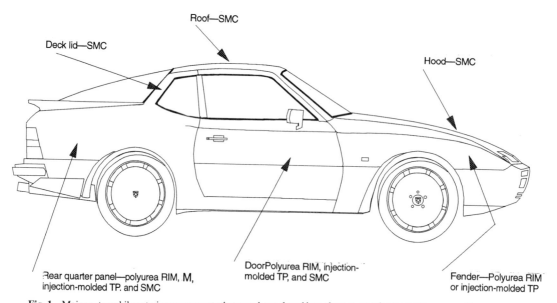

Fig. 1 Major automobile exterior components that may be replaced by polymers or polymer matrix composites

front and rear ends, radiator end tanks, and door handles. The metals replaced are respectively rolled steel, pressed copper alloy, and a variety of cast or wrought metals [9,10].

Major Plastic Parts in a Modern Car

Figure 1 and Table 2 illustrate the major penetration areas for plastic or polymer composite body panels. Fenders appear to be the leading area where polymer-based materials can replace metals. Polyurea resins processed by reaction injection molding (RIM) are the current front runners in fender applications. These resins have

been used on the Pontiac Fiero and GM APV [10] and Chevrolet Camaro four-seat sport coupe (debuting mid-1993). In this later application, front fenders and both fascias are RIM-polyurea reinforced with mica while the rear fascia is reinforced polyurethane. Thermoplastics (TP) being used include on-line paintable nylon/polypropylene (PP) alloys and glass fiber-reinforced/impact-modified polyethylene terephthalate (PET) and polybutylene terephthalate (PBT). For off-line painting, prime candidates are PBT/Polyphenylene oxide (PPO) alloy, modified PPs, and polycarbonate/acrylonitrile-butadiene-styrene (PC/ABS) alloys [11].

Table 2 Plastic automotive parts and the polymers used

Part	Polymer
Interior Parts	
Trim	ABS, ABS alloys, modified PBT, acetal
Instrument panel	RIM polyurethane (PU), RIM polyurea
Instrument panel skin	ABS/PVC alloy, PPO
Console	ABS
Functional Interior Parts	
Radiator header tanks	Nylon
Brake reservoirs	Nylon, acetal
Electrical distributor caps	PBT, PPO blends
Fuel tanks	Acetal, nylon, modified PET, PBT
Gears	Acetal, nylon
Seat backs	PPO
Seat cushions	PU
Engine fans and shrouds	Nylon
Air cleaner housing	Nylon
Exterior Parts	
Bumper fascia	RIM-PU, RIM polyurea, SRIM-PU, PPs, toughened PPs, fiber-reinforced PPs, and other thermoplastic polyolefins (TPOs), PC/PBT alloys, thermoplastic elastomers
Body panels	SMC, BMC, PET/PBT blends, TPOs, nylon, nylon/PPO blends
Exterior trim	PPO/HIPS blends, PC, PC/PET blends, PC/PBT blends
Hood	SMC
Functional Exterior Parts	
Headlamp assemblies	Nylon, PC, modified polyesters
Wheel hubs, covers	Nylon, PC, modified polyesters
Mirror housing	Nylon

Sheet molding compound (SMC) is the leading candidate material for hoods. SMC is a composite of thermoset polyester modified with fillers like calcium carbonate and reinforced with chopped glass fibers. The roof, doors, hatch, and spoiler assembly of Camaro four-seat sport coupe and those of the fourth-generation Pontiac Firebird are made of chopped glass/polyester SMC. SMC's material cost (typically less than $1/lb) is very attractive to auto makers.

SMC, injection-molded thermoplastics (TPs), and RIM polyurea are being used in car doors. Examples are SMC door skins on the GM APV and RIM polyurea doors on the Fiero; both are space frame construction. SMCs have also found applications in specialty roofs for light-duty vehicles such as the Blazer and Jeep Wranglers. SMC lift gates are utilized on Jeep Cherokees and Ford Broncos. SMC is being incorporated in side panels on the box of GM support trucks. Rear quarter panels can be made of RIM polyurea, injection molded TPs, or SMC. Deck lid application of SMC is still only a small

volume. Car interior uses of plastics include the instrument panels and the driver/passenger seats.

Unreinforced Polymers

Plastics have found increasing use in automotive body panels largely due to the following advantages [1,6,12-40]:

- Plastic parts are corrosion resistant
- Plastics can be used to reduce part weight by 30 to 40% for improved fuel economy
- New models can be made by simply changing the design of the outer panels leading to cost savings for lower volume productions
- The design flexibility of plastics makes it less difficult to fabricate parts with curved surfaces
- The cost of changing molds for new plastic panel designs is only about one-third that of dies for steel sheet forming

However, using plastics for external panels is not without its problems. Unreinforced thermoplastics typically have much higher coefficients of thermal expansion (CTEs) than steel, which causes large expansion and contraction when plastics are subjected to paint oven temperatures. A new way of coloring automotive components obviates the need for paint ovens by using a dry-paint laminate [8]. This new procedure also eliminates volatile organic compounds which reduce work place hazards and environmental pollution. Wider uses of plastics in body panels are expected.

For components that contact methanol-containing fuels, plastics are capable of satisfying both the corrosive effects of methanol and the cost criteria. Thermoplastics are also used under the hoods for air out-take ducts in pick-up trucks and for clean-air intake ducts for turbocharged passenger cars. These ducts are blow-molded and cost less than the wire-reinforced parts they replace. Other under the hood applications for thermoplastics include air-filter housings, radiator end tanks, fuse covers, air ducts, distributor caps and cooling fans [9,39].

The important performance features, applications, and processing of various unreinforced plastics for automotive applications are summarized in the following paragraphs.

ABS polymers offer a unique combination of toughness, rigidity, chemical resistance, quality surface appearance, and low cost. ABS can be well processed into complex parts by almost any thermoplastic processing technique. Injection-molding extrusion and various thermoforming techniques (vacuum, pressure, and mechanical) are most widely used for fabricating ABS parts. Their automotive applications include instrument panels and large grille parts that have demonstrated excellent performance through extreme temperature (up to about 90 °C, or 194 °F) and weather conditions.

Acetals are highly crystalline linear polymers that exhibit predictable mechanical, chemical, and electrical properties over a broad temperature range for long durations of time. The high crystallinity imparts to acetals excellent creep resistance and fatigue endurance. Acetals are readily processed in standard injection-molding machines, extruders, and rotational-molding machines. Automotive applications include electrical switches, body hardware, seat belt components, fuel system components, gears, window support brackets, steering columns, and cranks. Acetal resins exhibit a better strength-retaining ability in a methanol-containing fuel environment than typical polyester resins like PBT.

Nylons, being melt processable by a wide range of thermoplastic processing techniques, offer a combination of properties including high strength (especially at elevated temperatures), toughness at low temperatures, stiffness, wear and abrasion resistance, low coefficient of friction, and good chemical resistance. Included in automotive applications are speedometer and windshield wiper gears, wire harness clips and fasteners, connectors, emission canisters, fluid reservoirs, engine fans and shrouds, air cleaner housings, fuel system components, painted exterior body parts, lamp assemblies, mirror housings, wheel hubs, and door and window hardware [40]. Light-weight air-intake manifolds molded of nylon are now widely used in many European cars.

Polybutylene Terephthalate (PBT). The rapid crystallization rate of PBT makes injection molding the preferred method of processing. PBT's chemical resistance to automotive fluids is an attractive attribute. Its thermal resistance and strength properties are necessary for underhood applications including distributor caps, connectors, and other electrical parts. High strength is needed for door and window hardware and grille opening panels. Impact-modified PBT blends are also used in bumpers.

Polycarbonate (PC). The excellent balance of toughness, clarity, and high heat-deflection temperature qualifies PC as an engineering thermoplastic. PC and PC-based blends can be melt-processed using practically all of the traditional methods for thermoplastics. PC is used for tail and side marker lights, headlamps, and supports. PC blends are being used in bumpers and instrument panels due to their high-impact resistance. A potential large market for PC blends exists in such applications as exterior body parts, such as body panels and wheel covers. PC resins are also used in traffic light housings and signal lenses.

Polypropylene (PP) compositions can be tailored in accordance with the requirements of many fabricated methods and applications. Excellent chemical resistance, low density, high melting point (highest among the high-volume thermoplastics), and moderate cost contribute to the value and versatility of PP. In the automotive industry, PP copolymers and filled grades are used under the hood and dash. Elastomer-modified PP copolymers exhibit very high impact resistance and are well suited for exterior parts like bumper covers (fascia), side molding, and rocker panels. Chrysler will replace RIM PU with impact-modified PP with molded-in accent color for fascias of most of the company's low-line vehicles for the mid-1990s.

Polyphenylene oxide (PPO) blends (typically containing polystyrene) are lower-cost alternatives to PC, nylons, polyesters, and die-cast metals in the automotive industry. They are used in major internal parts such as instrument panels and seat backs, and exterior parts such as rear spoilers, wheel covers, and mirror housing. Some connectors and fuse blocks are also made of PPO blends [41].

Polyurea. Urea resin is a common example of thermosetting amino resins. The RIM process involves only a low pressure (less than 1380 kPa, or 200 psi) and therefore can be used to mold larger panels with polyureas than with thermoplastics. One RIM machine can produce over 1000 polyurea fenders per day with a typical cycle time of about 90 s [15]. With glass reinforcement, polyurea is commonly used in rear quarter panels, tailgate covers, fascias and truck fenders. Polyurea resins provide good dimensional stability, thermal and moisture resistance combined with very short cycle times.

Polyurethane (PU). The most common method to form PU is by the reaction of a polyisocyanate with an organic polyhydroxyl (polyol). A third category of monomers commonly used in the PU reaction are chain extenders. They are typically low molecular weight diols or diamines that react with isocyanates to form the rigid phase of the polymer. Commonly used chain extenders include ethylene glycol, butanediols, and glycerol. The most widely used aromatic isocyanates are toluene diisocyanate (TDI) and methylene diphenylene diisocyanate (MDI). The polyol or the "softblock resin" is the precursor to the elastomeric or flexible portion of the polymer structure. The fraction of this ingredient in the polymer composition determines the degree of softness (or hardness) and flexibility (or rigidity) of the resulting polymer. High density PU elastomers for automobile exterior parts typically contain 10 to 70% soft domains in the structure. The most widely used polyols are aliphatic polyethers and polyesters.

The chemical and structural diversity of PUs provide an unusually wide spectrum of properties that can be tailored to meet various engineering applications. PUs are resistant to impact, chipping (for paints), weather, chemicals and solvents [11]. MDI-based flexible molded foams have recently gained a large share of European auto seating, headrests, and armrests [16]. PUs were traditionally blown using chlorinated fluorocarbon (CFC) blowing agents. Because of the perceived damaging effects of CFCs on the ozone layers, CFC-free agents are now being actively investigated. Interior components molded of CFC-free PU systems include steering wheels, instrument panels, headliners, and door panels. Steering wheels for some heavy trucks now can be made of water-blown PUs.

An alternative to thermoset PUs will be thermoplastic elastomers (TPEs) that can be processed like thermoplastics but exhibit the strength and elasticity of thermoset rubbers. TPEs may be designed for applications requiring soft touch and feel, such as steering wheels, radio knobs, armrests, and pads.

Reinforced Polymers

Manufacturers are developing processes for producing parts from fiber-reinforced plastics to reduce both the cost and the weight in vehicles [7]. Developing composite-intensive auto bodies is now a global trend in the automotive industry. This trend will lead to development of new materials and improvement of existing materials. Advanced composites provide the potential to outperform steel in strength and stiffness at a lower total weight. Polymer composites have higher damping capacities (reduced noise), can be fabricated in integrated parts, corrode less, need less tooling and equipment, render Class A surfaces, can be engineered to more varied specifications and provide more design freedom [1].

Advanced polymer composites can be disassembled and shredded into short fiber-reinforced polymers [28]. Gloss and paint adhesion to composite substrates are two performance requirements for future automotive composites. Total system costs, reflecting saving from part consolidation and reduced overhead, may eventually make composites cost-effective for the automotive industry.

Sheet Molding Compounds (SMC). Automotive body panels are required to exhibit good surface finish, heat sag resistance, low CTEs, design flexibility, impact resistance and stiffness [1]. SMCs offer most of these features. SMCs provide as good or better Class A finish than steel and have shown the ability to withstand E-coat bake temperatures. Weight savings of approximately 20 to 30% can be achieved with SMC over steel [3]. SMCs have an added advantage that they can be molded as a single part without any joints [22]. The lower cooling costs of SMCs also give them the edge over steel in shorter runs; but steel retains a per part advantage in cost at high volumes.

Many European automakers, including Audi, BMW, Mercedes, Peugeot, and Volkswagen, now utilize SMC bumpers. The use of high-performance SMC is expected to extend to other automotive structural applications beyond exterior body panels. SMC parts containing 30% glass fibers are lighter than steel stampings, less expensive than aluminum or magnesium die castings, dampen sound more effectively than metals and are more heat resistant than thermoplastic moldings [32]. SMC can be used to replace cast aluminum for engine valve covers, offering advantages in reduced valve-train and other engine noise, reductions in machining by parts consolidation and improved seal integrity due to better dimensional stability [39]. SMC is considered a good substitute for aluminum in the oil pan for the engine; cost and weight savings of 30% are achievable [39].

Reinforced Thermoplastics (R-TPs). A variety of R-TPs are currently either in use or as candidates for use by the automotive industry. Freightliner is using a glass fiber-reinforced polyester hood for the company's medium-duty trucks. This R-TP part has a surface finish that exceeds minimum requirements for a Class A surface. The weight of this one-piece hood is 15% lighter than a similar SMC part. As compared to SMC, this thermoplastic part has higher strength and lower tooling and fabrication costs [33].

Reinforced nylon materials are being designed into a number of engine-related applications. These include timing chain tensioners, bearing cages, transmission components, sensors, and valve covers [39]. Sound damping valve covers made of glass fiber-reinforced nylon can cost up to 25% less than comparable aluminum parts [32]. Air-intake manifolds made of glass-nylon using fusible core injection molding are displacing their cast aluminum counterparts. Such a substitution leads to a 50% weight reduction, better engine performance due to lower friction, and reduced total cost due to lower tooling costs [32].

Front and rear bumper backup bars made of 40% fiberglass-reinforced polypropylene were introduced on 1987 Ford and General Motors passenger car models [20]. Composite bumpers are simpler, perform as designed, and attachment parts are reduced [22,35].

Continuous Fiber-Reinforced Composites. Advanced polymer composites (APCs) are finding new applications in automotive parts that require higher performance. New or potential applications include transmission, universal joints, and other power-train components [41]. Composite drive shafts are of lower weight, with good design flexibility, and relatively easy to fabricate. GM has used a carbon fiber composite (in one piece) to replace a heavier two-piece steel drive shaft for trucks [30]. The axle and transmission housings are potential areas where APCs can substitute for steel.

The advantages of using APCs for automotive applications became more apparent when a one-piece composite leaf spring weighing 3.6 kg (8 lb) replaced a 7-piece steel spring weighing 18.6 kg (41 lb) [5]. Filament wound leaf springs are now being used on high-volume car and van programs [1]. Composite brake drums showed 30% better performance than conventional drums with added benefits of increased resistance to cracking and heat cracking [36]. Glass/epoxy composites are also being considered for use in wheels and strut while carbon/carbon composite brakes have been developed for large trucks and sports cars. Hollow glass microspheres in a three-dimensional honeycomb formation offer exceptional energy absorption for improved car crashworthiness [37]. Many composites, in various fiber placement styles, have demonstrated superior energy-absorbing capabilities [29]. Carbon fiber composites have been used to fabricate demonstration parts such as engine blocks, engine brackets, suspension components, bumper beams, bumper components, drive shafts and bearings, and leaf springs [27].

Sandwich structure composites are widely used in the defense and aerospace industry due to their superior specific stiffness (stiffness/weight ratio). These composites are also being considered for use in automotive production. Ford has developed a one-piece integrated sandwich structure consisting of skins of preformed glass mat around a foam-plastic core as part of a prototype program to develop a composite tailgate for the Sierra pick-up truck [7]. Buick and Oldsmobile have developed

Table 3 Effects of filler or reinforcement on selected properties of polypropylene

Material	Flexural modulus, GPa (ksi)	Notched Izod impact strength, J/m (lbf · ft/in.)	Heat deflection temperature, °C (°F)
PP homopolymer	1.7 (247)	45 (0.84)	65 (149)
PP + 40 wt% calcium carbonate	2.8 (406)	53 (1.00)	75 (167)
PP + 40% wt% talc	4.2 (609)	35 (0.66)	97 (207)
PP + 30% wt% glass fiber	6.5 (943)	100 (1.87)	148 (298)

Source: Ref 42

a two-piece sandwich tailgate (13.4 kg, or 29.7 lb) to replace a seven-piece structure (19.5 kg, or 42.9 lb). These composite sandwich structures meet all design load and deflection criteria. Foam cores have been used in the design of composite front-end structures. A modular automobile front energy absorber and beam/support replaced an assembly of 20 plastic and stamped-steel parts with a weight saving of 12.9 kg (28.5 lb) [7]. This front-end structure successfully passed various static and dynamic crash tests [7,29]. Honeycomb panels are widely used in European and Japanese railway cars. Advantages of using composite structures in mass-transit vehicles include weight savings, increased passenger comfort (lower noise) and lower assembly cost (lighter parts can be installed quickly and easily) [38]. Weight savings result in faster acceleration and deceleration, reduced wear on bogeys and track, and enhanced energy economy [38].

Processing of Automotive Polymer Composites

Injection and Blow Molding of Thermoplastic Composites

Injection molding entails injecting a molten polymer into a closed mold, where the polymer solidifies to give the product. The molded part is recovered by opening the mold and ejecting the part. The mold is then closed and the cycle is repeated. Virtually every thermoplastic can be injection-molded. Injection molding is particularly suitable for large production runs of small- or medium-sized parts. But large parts like automotive bumpers can also be injection molded without much difficulty. Many filled or reinforced thermoplastics are routinely injection molded for a variety of applications. The benefits of filling or reinforcing a thermoplastic are shown in Table 3. Several desirable properties of PP can be improved by incorporating 30 to 40% talc, calcium carbonate, or glass fibers.

Under-hood applications demand that the materials be of high temperature resistance, chemical resistance, and mechanical integrity. Glass fiber-reinforced nylon 66 is now widely accepted for use in the radiator tanks, where the material has to withstand temperatures of 130 °C (266 °F) and the effect of many aggressive chemicals (antifreeze, hydrocarbons, cleaning products, etc.). Glass fiber-reinforced nylon is also used for fans, fan brackets,

and gear box covers. For non-critical applications such as fan shrouds, horn shells and heater housings, glass fiber reinforced PP is chosen because of its low cost. Reinforced PPs are also specified for use in bumpers. Front fenders and rear quarter panels on the Saturn cars are injection molded of PPO/nylon alloys, door outer panels molded of a PC/ABS alloy, and fascias molded of modified PPs [18]; all three are unreinforced thermoplastics.

A slightly modified version of injection molding is lost-core molding, which is gaining wide acceptance by the automotive industry as a highly efficient process for producing lightweight and complex hollow parts [43]. This process involves the injection-molding of a thermoplastic part around a metallic core (a low melting-point alloy). After part ejection, the core is melted out, recast, and reused as the core. What would drive the automotive industry to adopt this process is the need to enhance fuel economy and to reduce production costs. This process is being used to mold air intake manifolds, water pump impellers and housing. Glass fiber-reinforced, heat-stabilized nylon 66 resins are the materials of choice for most air-intake manifolds. This material can withstand a temperature range of –40 to 215 °C (–40 to 419 °F) and a pressure range of 10 to 500 kPa (0.1 to 5 bar, or 1.45 to 73 psi). With a low thermal conductivity, the reinforced nylons heat up far less than aluminum when used in manifolds. The materials are also resistant to oils, greases, gasoline, and coolants.

A variety of vehicle parts are processed presently by blow molding, including windshield and cooling system overflow bottles, strut/chock dust covers, fuel pump housings, and HVAC duct work. This process is expected to become an important means for fabricating large structural parts requiring such features as high strength-to-weight ratio, chemical and weight resistance, and weatherability. Auto body panels, doors, bumpers, and spoilers are suitable candidates [19].

Liquid Composite Molding (RTM AND SRIM)

Liquid composite molding (LCM) refers to either resin transfer molding (RTM) or structural reactive injection molding (SRIM). LCM is widely used for fabricating structural components in a variety of automotive applications. The development of snap-cure SRIM resin systems, low-viscosity RTM systems and the associated preform technology has made it economically more feasible for LCM to penetrate into the automotive market.

Resin transfer molding (RTM) involves placing a reinforcement preform in a mold, closing the mold and then injecting resin into the mold. The injected resin flows through the reinforcement preform, expelling the air in the cavity and impregnating the reinforcement. When excess resin starts to flow from the vent areas of the mold, the resin flow is stopped and curing begins to occur. Completion of curing can take from several minutes to hours.

The structural reaction injection molding (SRIM) is a very similar process. However, mold release and reinforcement sizings are varied to optimize their chemical characteristics for the SRIM chemistry. Once the mold has been closed, the resin is rapidly injected into the mold and reacts quickly to cure fully within a few seconds. This chemical reaction proceeds as the resin penetrates through the preform and, therefore, SRIM requires fast fiber wet-out and air displacement. Thorough impregnation of reinforcement is quickly followed by complete cure of resin. The resin viscosity rapidly becomes too thick to permit resin flash through vents.

Normally RTM resins have two components mixed at a ratio of approximately 100:1 at low pressure. The reaction rates are much slower to permit the resin with a viscosity range of 100 to 1000 cp to fully impregnate a large preform. SRIM resins are highly reactive and require very fast, high pressure impingement mixing to achieve thorough mixing before injection. While RTM resins are being pushed to achieve cure as fast as 20 s, some SRIM resins are being slowed down to allow filling of large molds and impregnation of higher-content reinforcements. This trend is also leading to a convergence in resin dispensing technology; some SRIM metering systems are being adopted for fast RTM dispensing.

The process of LCM to produce automotive structural composites has gained considerable attention over the last decade. RTM and SRIM prototype parts for automotive uses include pick-up truck boxes, bumper beams, auto instrument-panel retainers and floor pans, and horizontal auto panels cross members. The LCM process is characterized by low temperatures and low pressures, and therefore, lower equipment costs. Lower tooling costs are economically attractive for limited production runs. The reinforcements can be placed exactly where they are needed. Complex sub-assemblies can be consolidated into one-shot molding, thus requiring lower assembly labor.

Preforming Technology for LCM

One hurdle to overcome before the LCM process becomes more widely accepted has been the lack of adequate expertise in the cost-effective production of reinforcement preforms. Two basic input forms of fiberglass are currently available to the LCM molder for producing a stiff 3-D preform. These are a thermoformable continuous strand mat and a multi-end roving [44]. In addition, either of the above forms of reinforcement can be combined with unidirectional or bi-directional forms of reinforcement in selected areas during preform fabrication. The specialty reinforcements are incorporated to provide additional strength to areas where the composite may experience higher stresses.

In the automotive industry, two basic routes of fabricating the LCM preforms from the two basic input forms of reinforcement (mats and rovings) are directed fiber spray-up and stamping of thermoformed mats. The directed fiber spray-up process utilizes an air-assisted chopper/binder guns which carry glass and binder to a perforated metal screen shaped identical to the part to be molded. The chopped fibers are held in place on the screen by a large blower drawing air through the screen. Once the desired thickness of reinforcement has been achieved, the chopping system is turned off and the preform is formed by curing the binder. Once stabilized, the preform is cooled and removed from the screen [44].

The thermoformed mat process requires an oven to heat the mat, a frame to hold it while being stretched into shape, and a tool to form the mat into a preform [45]. In a typical process, several plies of mat would be cut to the approximate desired shape of the molded part, allowing extra material to be held in a frame. The frame containing the material is then placed in an oven to be heated (up to 170 °C, or 338 °F) and then quickly transferred to the forming tool. The tool is closed, forming and cooling the mat for a short period of time. After removing the frame and trimming the waste fibers clamped in the frame, the preform is ready for molding. Both thermoplastic and thermoset binder systems are available to retain the formed shape [44].

Many recent efforts have been directed to improving the processes involved in LCM [45-70]. Automation of preforming techniques is essential to the success of producing high-volume, high-quality, near-net shaped preforms at a predictable low cost [62]. The simultaneous engineering needed for effective preforming in higher volume composites was discussed by Gembinsky [63], who also gave a brief review on topics such as resin specification and preformable reinforcement selection, surface treatments, costs of preformed composites, and design criteria. Taguchi-designed experiment was followed to investigate the mechanical properties of LCM composites as a function of strand size, filament diameter, chop length, binder type, and binder amount in fiber-directed preform reinforcement [64]. Fielder et al. [65] discussed technology developed for consolidation of the preforming step with the molding operation in LCM. This consolidation eliminates the need for a separate preform tool, reduces the overall cycle time, and possibly improves the quality of the molded parts through precise control of the preform trim and placement [65]. Chavka and Johnson [66] presented a good review on critical preforming issues for large automotive structures. The advances that have taken place in thermoformable glass mat preforming were reviewed by Mazzoni and Riley [67]. Preforms of large dimensions can be progressively built into complex geometries by combining numerous component preform sections with preforming technology such as "CompForm" [68]. Central to this technology is the use of a new energy directing methodology involving ultraviolet radiation curing of proprietary binder resins to

Table 4 Mechanical properties of some SMCs and BMCs

Property	SMC-R25	SMC-R65	BMC with 10 to 25 wt% E glass
Specific gravity	1.83	1.82	1.8 to 2.0
Tensile strength, MPa (ksi)	82.4 (11.9)	227 (32.9)	27 to 54 (3.92 to 7.83)
Tensile modulus, GPa (ksi)	13.2 (1914)	14.8 (2146)	3.4 to 10.3 (493 to 1494)
Poisson's ratio	0.25	0.26	...
Strain at failure, %	1.34	1.63	...
Flexural strength, MPa (ksi)	220 (31.9)	403 (58.4)	69 to 172 (10 to 24.9)
Flexural modulus, GPa (ksi)	...	14.9 (2161)	5.5 to 8.3 (798 to 1204)
Coefficient of thermal expansion, $10^{-6}/°C$ ($10^{-6}/°F$)	...	...	14.4 to 27 (26 to 49)

Source: Ref 72

rigidize and hold basic preform shapes, which can then be assembled into final preform structures. Textile structural preforms were studied by a number of investigators [e.g., 69,70].

Major advantages of LCM include [44]:

- Very large and complex shapes can be made efficiently and inexpensively
- Production times are much shorter than lay-up
- Inserts and special reinforcements can be added easily

It is possible for separate preforms to be produced at different stations and then joined together to form the required geometry. In this manner, higher reinforcement content can be placed in selected areas of the preform. Denser fiber areas would require a low-viscosity resin to achieve complete impregnation. Current LCM technology still suffers from the following major shortcomings:

- The mold design is critical and requires great skill
- Properties of LCM parts are not generally as good as with those produced by vacuum bagging, filament winding or pultrusion
- Control of resin uniformity and elimination of reinforcement movement during resin injection are still quite difficult
- Preform technology for carbon fibers is not as mature as that for glass fibers

Molding of SMC and BMC

SMC and BMC utilize the same type of compression molding press, but differ in the form of the material that is placed in the mold to form the part. Commonly used materials in these techniques are fiberglass-reinforced resins (polyester, vinyl ester, phenolic, and epoxy), plus a sizable amount of fillers such as clay, alumina, or calcium carbonate. A typical SMC composition would contain 30 to 50% fibers (generally shorter than 75 mm, or 2.95 in., long), 25% resin, plus the balance amount of fillers, viscosity modifier, and low-profile agents. For automotive uses, polyester and vinyl ester resins are expected to remain the resin systems of choice while epoxy and

phenolic resins may gain more acceptance for underhood applications [71]. GM has a strong interest in SMCs for roofs, rear decks, and door outer panels for Camaro and Firebird models [72].

The SMC process consists of chopping glass fibers onto a sheet of polyethylene film on which a mixture of resin-catalyst-fillers has been doctored. A predetermined amount of this mixture paste, placed on top of another film, is then conveyed forward to receive the dropping chopped fibers. The sandwich of resin mixture and chopped glass fibers is passed between compaction rolls to wet the fibers and thoroughly mix the ingredients. The mixture is then cured slightly (called aging, maturing, or B-staging) to produce a leather-like texture, and rolled up for shipment. At the matched-die molding facilities, SMC is cut off the roll, the plastic film backing removed, and loaded into the molding presses. Molding temperatures and pressures can be as high as 160 °C and 30 MPa (320 °F and 4350 psi), respectively, depending on the resin chemistry and filler content.

BMC usually has a composition similar to SMC, but the fiberglass content is slightly lower and lengths are shorter. BMC generally contains 15 to 20% of 6 to 12-mm (0.24 to 0.47-in.) long E-glass fibers in polyester resin. BMC can be prepared by compounding chopped glass fibers with the resin formulation in an intensive mixer and then extruding the mixture into a continuous log. Such a premix can also be conveniently prepared at the molding facilities. The extruded log is cut into desired lengths for molding. Parts produced from BMC are compression-molded, injection-molded, or transfer-molded. Tensile strength, modulus, and impact energy of BMCs are lower than those of SMCs, largely due to shorter fiber length and lower fiber content in BMCs (Table 4). The processing technology, material compositions, and mechanical properties of SMC, BMC, and other types of molding compounds have been discussed in detail [71,72].

Performance and Design Issues

In a high-volume production industry like automotive, the rate of composite manufacture is a major concern when choosing between composite and metal parts. Manufacturing processes capable of meeting the production output requirement are the primary consideration, and design of a structural component must be within the constraints of the selected fabrication process [73]. Presently, the uses of polymer composites in high-volume vehicles are primarily for decorative and semi-structural applications.

Structural Properties of SMC and BMC

The mechanical properties of SMCs have been reviewed by Burns [74] and Mallick [72]. SMC composites generally do not exhibit any yielding and their tensile failure strains are low (<2%). Both tensile and flexural strength increase significantly with increasing fiber content while the modulus is affected only moderately. Fillers do not affect tensile properties of SMCs to any great degree, but they do affect flexural modulus [74]. Watanabe and Yasuda [75] developed empirical equations for the tensile strength and tensile modulus of SMC-R composites in relation to fiber volume fraction:

$$\sigma_u = 0.33\ \sigma_f V_f + 0.31\ \sigma_m (1 - V_f) \qquad \text{(Eq 1a)}$$

$$E = 0.59\ E_f V_f + 0.71\ E_m (1 - V_f) \qquad \text{(Eq 1b)}$$

where σ_u is the ultimate tensile strength of the SMC-R composites; σ_f and σ_m are the ultimate tensile strength of the fiber and that of the matrix, respectively; E is the initial tensile modulus of the SMC-R composite, E_f is the fiber modulus, E_m is the matrix modulus, and V_f is fiber volume fraction. In these equations, the matrix refers to the resin-filler combination, and not just the resin. In general, flexural strengths of SMCs are higher than their tensile strengths, but the flexural modulus and tensile modulus values are similar.

Commercial grades of SMC contain glass fibers 12 to 50 mm (0.47 to 1.97 in.) long. Both tensile strengths and flexural strength of SMCs increase as the chopped fiber length is increased. The modulus is relatively insensitive to the fiber length variation within this range. A large strand length could lead to poor flow characteristics, particularly in complex moldings where rib, bosses, and other restricted areas of the mold might be bridged by long fibers [74].

The fiber type has a significant effect on the fiber-dominated properties, such as the strength and modulus, of the SMC composites. The resin type may affect the tensile properties of SMCs, but to a relatively small extent. Resin and filler type and concentrations do significantly affect toughness of these composites. These factors are now being carefully designed so that SMC composites can meet competition from glass fiber-reinforced thermoplastics and SRIM polyurethanes in light of energy-absorbing capabilities and impact resistance. Resin modifications can dramatically improve impact strength. Impact properties of SMC composites, including damage caused by low energy impact and post-impact residual

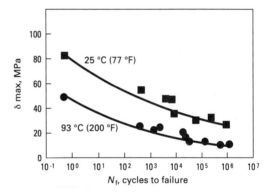

Fig. 2 Typical *S-N* curve for a SMC automotive part

strength, have been studied by several researchers [75-79].

In general, *S-N* diagrams of SMC composites subjected to tension-tension cyclic loading do not exhibit a fatigue limit (see Fig. 2). The number of cycles to failure increases with decreasing stress level. Typically, the fatigue strength of SMC-R composites increases with increasing fiber content [72]. SMC composites also tend to exhibit a decrease in apparent modulus after prolonged cyclic loading; a 50% drop in modulus has been observed just prior to failure [74].

Fatigue damage in SMC composites typically involves matrix cracking and fiber-matrix interfacial debonding. Matrix cracking occurs preferentially at poorly dispersed clusters of fillers and at the chopped fiber ends. If the fibers are inclined at an angle with loading direction, fatigue damage tends to occur in the form of interfacial debonding. The accumulated fatigue damage in SMC composites often leads to reduction in residual tensile strength [72].

Wang *et al.* [80] reported that the fatigue damage behavior in SMC-R composites can be described by a power law equation:

$$(dD/dN) = AN^{-B} \qquad \text{(Eq 2)}$$

where the damage parameter is defined as

$$D = 1 - (E/E_0) \qquad \text{(Eq 3)}$$

Here, E is the tensile modulus after N cycles, E_0 is the initial tensile modulus, B is a constant, and A is a function of the damage state which depends on the loading history, stress amplitude, mode of loading, temperature and other variables.

Shirrell and Onachuk [81] reported that the Whitney-Nuismer point stress failure criterion can be used to predict the notched tensile strength of SMC composites. For isotropic composites, this criterion can be expressed as:

$$(\sigma_R / \sigma_0) = 2 / [2 + f(R)^2 + 3f(R)^4] \qquad \text{(Eq 4)}$$

Table 5 Major weight savings in a carbon fiber-reinforced polymer (CFRP) vehicle compared with a steel vehicle

Component	Steel		CFRP		Reduction	
	kg	lb	kg	lb	kg	lb
Body-in-white	192.3	423	72.7	160	115	253
Front end	43.2	95	13.6	30	29.5	65
Frame	128.6	283	93.6	206	35	77
Wheels (5)	41.7	91.7	22.3	49	19.4	42.7
Hood	22.3	49	7.8	17.2	14.7	32.3
Decklid	19.5	42.8	6.5	14.3	13.1	28.9
Doors (4)	64.1	141	25.2	55.5	38.9	85.5
Bumpers (2)	55.9	123	20	44	35.9	79
Driveshaft	9.6	21.1	6.8	14.9	2.8	6.2
Total vehicle	**1705**	**3750**	**1138**	**2504**	**566**	**1246**

Source: Ref 73

where σ_R = the tensile strength of a specimen containing a central hole of radius R, σ_0 = the unnotched tensile strength, and $f(R)$ in its simplest form may be given by $f(R) = R/(R + R_0)$. Here, R_0 is a characteristic distance from the hole edge at which the tensile stress due to the applied load is equal to the unnotched tensile strength; R_0 depends on the fiber content in SMC-R composites [82].

Higher-Performance Composites

Carbon fiber-reinforced epoxy resins are widely used in the aerospace and defense industries because of the superior combination of stiffness, strength, and fatigue resistance that these materials offer. The automotive industry has yet to take full advantage of this class of high performance structural materials due to the high costs of carbon fibers. Ford built a prototype carbon fiber-based LTD sedan to compare directly with a production steel vehicle [83,84]. The weight savings for the various components in this case study are given in Table 5 [73]. The study concluded that, from a ride quality and vehicle dynamics viewpoint, a vehicle with a structure made entirely of fiber composites was at least equivalent to a steel vehicle.

Judging on the basis of optimal combination of cost and performance, Beardmore [73] concluded that E-glass fibers hold the greatest potential for automotive structural applications. Most of the structural parts that involve significant loads will likely utilize a combination of both chopped and continuous glass fibers with the particular proportions depending on the intended application. The associated processes must be capable of handling mixtures of chopped and continuous fibers [73]. If the raw material price of carbon fibers can be lowered to a more affordable level (< $5/lb), these materials will be the reinforcement of choice. The next logical step will be to bring the carbon fiber preform technology for the liquid composite molding to a mature stage. There is no reason to expect a major difference between the preform technology for carbon and carbon-glass hybrid fibers and the more advanced preform technology for glass fibers.

In addition to ride quality (vehicle dynamics), glass fiber composites must also exhibit adequate fatigue resistance and energy absorption capability. These composites typically exhibit a well-defined fatigue limit of approximately 35 to 40% of their ultimate strength. In contrast, the fatigue limit of the chopped glass composite is typically about 25% of its ultimate strength. By properly adjusting the proportion of continuous fiber with respect to chopped fibers, one can design glass fiber composites that meet rigorous fatigue resistance requirements.

Thornton and coworkers [85-87] have shown that polymer composites are capable of absorbing high impact energy and are suitable for crashworthy structural applications. These materials (in a proper form like tubes) are designed to progressively crush under impact such that the kinetic energy of the vehicle is dissipated and the passenger compartment is brought to rest without the driver/passengers having to experience excessive deceleration. Krishna [88] has demonstrated that progressive crushing can be achieved in knitted fabric-reinforced composite tubes. With relatively low fiber content (30%), these knitted fabric composite tubes display good energy absorption, which is comparable to conventional woven fabric reinforced composite tubes with higher fiber content (40 to 50%).

Crush strength and failure modes of composite structures, including plate-type columns, were studied by Mahmood and co-workers [89-91]. The main factors that dictate the crushing mode of plate-type composite columns are the material strength and the buckling strengths of column. Other important factors include fiber orientation, thickness-to-width ratio, and the edge constraints of column walls. An equation was developed to predict the maximum crush strength, σ_{max} [89]:

$$(\sigma_{max}/\sigma_{UTS}) = [\beta \, \sigma_{cr}/\sigma_{UTS}]^n \qquad \text{(Eq 5)}$$

where σ_{UTS} is the material strength, n and β are empirical material constants, and σ_{cr} is the buckling strength of a column which can be calculated by:

Table 6   Mechanical properties of a representative mini-van fender

Property	Guidelines	Bayflex 110-80	Dow Spectrum	du Pont Bixloy
Specific gravity	...	1.18	1.25	1.40
Flexural modulus, GPa (ksi)	1.72 (250)	1.38 (200)	1.93 (280)	2.61 (378)
Elongation, %	20	20	35	15
Tensile strength, MPa (ksi)	27.6 (4)	27.6 (4)	31.0 (4.5)	30.3 (4.4)
Heat sag, 1 h at 190 °C (375 °F), mm (in.)	...	Complete	12.7 (0.5)	12.7 (0.5)
Warp test, mm (in.) 45 min. at 190 °C (375 °F)	...	...	2.5 (0.1)	2.5 (0.1)
Gardner impact strength, J (in. · lbf)				
RT	9.0 (80)	2.8 (25)	10.8 (96)	5.8 (51)
−29 °C (−20 °F)	5.6 (50)	0.90 (8)	3.4 (30)	2.4 (21)
Notched Izod strength, J (in. · lbf)	0.45 (4.0)	0.22 (2.0)	0.32 (2.8)	0.28 (2.5)
Coefficient of thermal expansion, $10^{-6}/°C$ ($10^{-6}/°F$)	11.1 (20)	16.7 (30)	13.9 (25)	14.4 (26)

Source: Ref 93

$$\sigma_{cr} = \phi \, E^* \, \pi^2 \, (t/b)^2 \qquad \text{(Eq 6)}$$

Here, t and b are the thickness and width of column wall, respectively, E^* is the equivalent Young's modulus of a laminated plate, and ϕ is the buckling coefficient [90].

The potential utilization strategy of composites in automotive structural applications can be divided into two categories [73]: one-on-one component substitution, and integration of multiple steel components into one composite component. Examples for one-on-one part replacement include driveshafts and leafsprings. However, one-on-one substitution of composite for steel is not considered to be a viable approach for most of the automotive parts due to economic reasons and complex attachment considerations. Multiple-part integration to reduce assembly costs will be a better alternative [73].

High-speed resin transfer molding (HSRTM) and SMC were used by Beardmore [73] to demonstrate the part integration concept. HSRTM appears to offer potential benefits of combining high production rates, precise fiber control and high degree of part integration. In the case of body-side assembly, 300 major steel parts will be required, as opposed to 10 to 20 SMC parts and 2 to 10 HSRTM parts. This high degree of part integration with HSRTM reflects the great versatility of this process. A widespread use of composites in critical load-bearing, integrated structures will require an expanded database of design parameters for glass fiber-reinforced polymer composites and major innovations in fabrication technology [73].

An ever increasing amount and diversity of polymers and polymer composites are being utilized in the automotive applications. Fuel economy, light-weight, high specific stiffness and strength, toughness, and versatility are the words that can be used to describe modern engineering polymers and polymer composites, which make them very attractive to automotive engineers. The fuel crisis required weight reduction in modern cars, where polymers and polymer composites replaced some of the steel parts. Polymers and polymer composite materials for exterior body panels can reduce overall system cost,

improve customer perception of automotive products, and make model changes flexible in response to a highly competitive marketplace. The demand for an occupant-friendly car interior led to the development of soft skin materials and other materials with aesthetic value for the trim parts. The increasingly stringent federal car safety requirements resulted in the application of foam materials in automotive interior components. To reduce cost, polymer composites are used to replace metals as engine components. New car models, characterized by smaller size, aerodynamic shape, and improved safety features, require the development of high performance materials that possess high impact strength (body panels), high heat stability (interior), and high energy absorption (safety and comfort).

Exterior Body Panels

Materials and Processing Requirements. Requirements of the exterior panel materials include [92-94]:

- Smooth surface and good paint adhesion
- Thermal and dimensional stability
- Sufficient flexural modulus or stiffness
- Good impact resistance
- Capable of functioning in existing assembly plant environment while maintaining required field performance
- Being economically competitive with existing steel panels on a total cost basis

To meet the above-cited requirements, Chrysler and its suppliers have investigated two types of exterior body panel materials. The first type is glass fiber reinforced polyester thermoplastics. The polyester resins have high melting points, which help to withstand the baking process at 205 °C (400 °F) of the car steel body part with E-coating (an anti-corrosion coating). Further, the resin itself has a high CTE which is about 5 times that of steel. Glass fibers were added to cut material shrinkage, and improve both dimensional and thermal stability.

Table 7   Pendulum impact response of Type-I bumper beam

	Hybrid resin A	Hybrid resin B	Unsaturated polyester	Vinyl ester
Pendulum weight, kg (lb.)	1066 (2345)	1306 (2873)	1066 (2345)	1066 (2345)
Caused load, kg (ton)	3336 (3.67)	4273 (4.70)	…	…
Dynamic deflection, mm (in.)	11.5 (0.45)	15.2 (0.60)	…	12.0 (0.47)
Appearance	pass	pass	cracked at first stroke	hair crack

Impact velocity 5.1 mph. Beam tested with fascia and foam. Results are after fifth impact at the center of the bumper. Source: Ref 96

The parts were produced by injection molding, which is highly automated and thus a cost-effective process. Like other injection-molded reinforced thermoplastics, surface quality and part performance of these polyester products were affected by gate design, glass level and type, material composition, attachment, and flange design and location.

The second type is polyurea RIM elastomer systems which can be produced with a fast cycle. In this system, no catalyst is needed and internal mold release ingredients are incorporated in the material, which allow for fast and easy release of parts from molds without spraying of an external release agent. A wide processing window, superior properties, and high productivity of the system make polyurea RIM a preferred candidate over polyurethane RIM and injection molded thermoplastics. Polyurea RIM elastomers are very tough and are more corrosion resistant as compared with the conventional stamped steel body panels [94]. For exterior body panel applications, this material also has very high thermal stability which enables it to survive E-coat bake temperatures [93].

Performance. The mechanical properties and physical properties of thermoplastic polyester and polyurea RIM, for use as mini-van fenders, are presented in Table 6 [93]. BAYFLEX110-80 refers to the commercial polyester thermoplastics, and the other two materials are polyurea RIM from different sources. The data indicates that the polyurea RIM has higher flexural modulus, higher tensile strength, smaller linear CTE, and higher impact strength than those of polyester. Their properties can be tailored to meet various special requirements.

Design with Polymers/Composites. The performance data of polymers and polymer composites are very important in the conventional design of engineering parts. A new concept of design was introduced to automotive industry which was called Metal Alternative Design (MAD) [95]. The development of this concept was driven by the recognition that the effective replacement of metal components and assemblies requires more than just the substitution into old designs. Polymers and polymer composites are different from steels. These differences are recognized by MAD, which addresses metal replacement as an overall process that integrates material selection, process design, and part design. This same approach can also be applied in the replacement of metal engine parts and components.

Cost Issues. Injection molding and reaction injection molding are two cost-effective processes that are well-suited to large scale production. Automobile parts produced in these ways are competitive in cost to the parts made of steel.

Bumper Beams

Materials and Processing Requirements. A typical bumper consists of a soft fascia of polyurethane or polypropylene, a shock-absorbing molded foam, and a bumper beam which is generally made of steel. The configuration of the bumper beam requires that a bumper beam material be of high tensile strength, high compressive strength, high flexural modulus, and generally high impact resistance.

The primary advantage of replacing steel by polymer composites is weight reduction. One type of bumper beam mounted on Mazda 323, Mazda 626, Mazda 929, and Mazda RX-7 is fabricated from a random glass fiber-reinforced hybrid resin composite. The hybrid resin was composed of unsaturated polyester polyol and polyisocyanate. This new PU-ester resin produces a cured system that is much tougher than the conventional unsaturated polyesters [96].

This H-SMC (High-strength sheet molding compound) can be processed with the existing compression molding machines for conventional polyester SMCs. During compounding the urethane-forming reaction occurs between the unsaturated polyester polyol and the polyisocyanate impregnated in the glass fibers. This H-SMC is made to have a viscosity suitable for handling and press molding. Upon heat compression molding, the H-SMC cures through radical copolymerization to obtain a high strength and high toughness material.

Performance. Tables 7 and 8 give pendulum impact test data of two H-SMCs with different hybrid resins. Table 9 lists the physical properties of H-SMC moldings. The impact response of the materials for the bumper beam is determined not only by material property but also by the type of bumper involved. For example, hybrid resin A did not pass the impact test when it was in the type II bumper beam, but it still passed the impact test for Bumper type I. This is reasonable because the difference in design requires different material performance. Despite these, both hybrid resin A and B are superior to unsaturated polyester and vinyl ester (Table 9).

Table 8 Pendulum impact response of Type-II bumper beam

Impact velocity, mph	Hybrid resin A		Hybrid resin B		Unsaturated polyester		Vinyl ester	
	load	deflection	load	deflection	load	deflection	load	deflection
3.1	4.78	23	3.63	22	4.10	24	4.65	29
4.0	6.28	30	5.12	30	4.05 (fail)	72 (fail)	5.05	31
5.1	5.61 (fail)	46 (fail)	6.61	37	...	...	6.88	37
5.3	...	...	6.88	38	...	...	6.61 (fail)	39 (fail)
6.0	...	...	6.93 (pass)	56 (pass)	...	...	...	...

Bumpers tested without fascia and foam. Load is caused load at impact in tons; to convert to kg, multiply by 909. Deflection is dynamic deflection at the center of the bumper beam in mm; to convert to in., multiply by 0.04. Source: Ref 96

Table 9 Properties of H-SMC moldings

	Hybrid resin A	Hybrid resin B	Unsaturated polyester	Vinyl ester
Flexural strength, MPa (ksi)	522 (75.7)	483 (70.0)	432 (62.6)	440 (63.8)
Flexural modulus, GPa (ksi)	19.6 (2843)	18.2 (2640)	17.5 (2538)	17.7 (2567)
Tensile strength, MPa (ksi)	309 (44.8)	288 (41.8)	256 (37.1)	279 (40.5)
Tensile modulus, GPa (ksi)	20.1 (2915)	20.5 (2973)	18.2 (2640)	18.4 (2669)
Tensile elongation, %	2.0	2.2	1.8	2.1
Notched Izod impact strength (edgewise), J/m (lbf · ft/in.)	1667 (31.2)	1550 (29.0)	1295 (24.3)	1069 (20.0)
Poisson's ratio	0.32	0.43	0.37	0.35

Glass fiber content: 65%. Source: Ref 96

Table 10 Impact response of cured resins

	Hybrid resin A	Hybrid resin B	Unsaturated polyester	Vinyl ester
Maximum stress, MPa (ksi)	186 (27.0)	284 (41.2)	127 (18.4)	206 (29.8)
Impact modulus, GPa (ksi)	2.91 (422)	3.03 (439)	3.07 (445)	2.97 (431)
Initiation energy, J (ft · lbf)	0.44 (0.32)	2.08 (1.53)	0.25 (0.18)	0.52 (0.38)
Propagation energy, J (ft · lbf)	0.03 (0.022)	0.10 (0.074)	0.02 (0.015)	0.04 (0.029)
Deflection at F_{max}, mm (in.)	1.1 (0.04)	2.7 (0.11)	0.8 (0.03)	1.2 (0.05)

The maximum stress (σ_{max}) and impact modulus (M) are determined from the following equations:

$$\sigma_{max} = (F_{max} / t^2)(1 + v)\left[0.485 \ \text{in} \ (a / t) + 0.52\right]$$
$$M = 0.217(F_{max} / \delta)(a^2 / t^3)$$

where F_{max} is Maximum load in N; δ is deflection in m; t is specimen thickness in m; a is radius of specimen clamp in m; and v is Poisson's ratio. The circumference of the thin test specimen is fixed, and the concentrated load is given at the center of it. Source: Ref 96

Design with Polymers/Composites. Among the H-SMC resins, only those hybrid resins with larger crack initiation energy (E_i) in impact are suitable for the bumper beam. Table 10 gives a comparison of the crack initiation energy values of several cured resins. The results were obtained from the instrumented falling dart impact test. Hybrid resin B appears to have twice as much crack initiation energy as that of hybrid resin A. Resin B may be more suitable for the bumper beam although it has a lower flexural strength, lower flexural modulus, lower tensile strength, and lower tensile modulus than those of resin A. In specifying the polymeric bumper beam, crack initiation energy is an important factor to take into ac-

count. Geometry of the bumper beam is also a factor that influences the initiation energy of the beam; the MAD method is also required to address this issue in the design process.

Cost Issues. The capital equipment costs for SMC are relatively high; therefore, only large production quantities of the parts can make this process cost-effective.

Automobile Interior Trims

Materials and Processing Requirements. The current trend in automotive body designs and in interior aesthetics have lead to the interior auto compartment downsize and the increase of glass ratio for these com-

Table 11   Properties of interior trim resins as injection-molded

	Standard heat ABS	Medium heat stable ABS	High heat stable ABS	AMS medium heat ABS	AMS high heat ABS	MA high heat ABS	High-crystalline heat PP
Tensile yield strength, MPa (ksi)	39.3 (5.7)	42.1 (6.1)	37.2 (5.4)	43.4 (6.3)	47.6 (6.9)	35.9 (5.2)	41.4 (6)
Flexural modulus, MPa (ksi)	2183 (310)	2414 (350)	2255 (327)	2276 (330)	2414 (350)	2138 (310)	2069 (300)
Notched Izod impact strength, J/m (lbf · ft/in.)	176 (3.3)	245 (4.6)	223 (4.2)	170 (3.2)	154 (2.9)	239 (4.5)	37 (0.7)
Peak energy, J (ft · lbf)	29 (21.4)	28 (20.7)	26 (19.2)	26 (19.2)	23 (16.7)	19 (14.0)	...
Heat distortion temperature, °C (°F)	104 (219)	112 (234)	121 (250)	113 (235)	120 (248)	128 (262)	132 (270)
Intrinsic melt viscosity, Pa · s (226 °C or 439 °F, and 250 kPa or 36.3 psi)	90	280	300	680	1170	1200	...
Aesthetics (gloss)	low	low	low	high	high	medium	medium

partments. These new changes create new challenges to the interior trim materials. Studies have indicated that, in a car under an ambient temperature of 40 °C (105 °F), the temperature of certain parts of the trim can be as high as 110 to 115 °C (230 to 240 °F) [97]. Such temperatures have already exceeded the critical performance temperature limits of standard ABS resins. The new interior trim materials are expected to have better thermal stability, higher impact resistance, higher UV resistance, and lower gloss aesthetics than conventional ABS. Materials for interior trim resins include various kinds of ABS and PP. Injection molding is commonly used to produce trim compartments.

Performance. Typical properties of ABS and PP are listed in Table 11. In general, the impact resistance of high-heat ABS resins used for interior trim has been adequate [97]. However, recent trends of reducing part weight and part assembly costs require thinner part walls and part consolidation. The low temperature(–29 °C, or –20 °F) instrumented impact resistance of these resins is approximately 2 to 7 J (20 to 60 in.-lb), a marginal value for the intended application. Furthermore, due to a high intrinsic viscosity (approximately 680 to 1200 Pa · s), these resins have been rather difficult to process. Because of these drawbacks, conventional high-heat ABS resins may not continue to be adequate. Table 11 also shows that the low-temperature impact resistance, 16 to 17 J (140 to 150 in.-lb), of the new high-heat ABS resins approaches that of standard ABS resins. The new resins provide an additional level of flexibility in engineering performance for design of thinner walls and for part consolidation. The high-heat ABS interior components are usually protected from UV degradation by expensive post-molding painting. In contrast, the new thermally stable low-gloss ABS resins can be internally stabilized against UV degradation and therefore will not require protective and costly painting [97].

Design and Cost. When selecting materials for the interior trim one should also consider cost. In the area where maximum temperature is very high, such as the instrument panel, new high-heat stable ABS may be adopted. But, in the areas where high impact resistance is required, the less expensive conventional ABS materials may be adequate in meeting the requirements. Where UV

degradation to conventional ABS trim materials is a concern, new high-heat stable ABS may be specified to avoid using expensive post-molding protective painting. For the door trim, there is a trend to use injection-molded polypropylene instead of ABS due to lower cost and better acoustic barrier characteristics of PP versus ABS [98]. Generally, all the interior trim materials can be processed by injection molding, which is a cost-effective process in large scale production. As shown in Table 11, many trim materials are of low or medium gloss which are consistent with the current aesthetic designing trend. Amorphous polymers tend to experience less shrinkage than crystalline polymers and, therefore, exhibit better surface quality.

Trim Skin Materials

Materials and Processing Requirements. Trim skin materials are required to exhibit softer touch, be compatible with human skin, feel smooth, and show low gloss. PVC dry powder compounds may be processed by the dry powder slush molding process [98]. The soft touch skin material produced in this method has been used in BMW 700 and 500 series, Audi 100, Peugeot 605 and Austin Rover 800. Vacuum formed ABS/PVC alloy are used in most American-made cars. These materials are used as instrument panel skins. PVC dry powder has also been used for door panel skins. The usage of PVC for skin material in seating has declined appreciably, since fabric cloth has a more luxurious appearance. Seat cushion can be made from cold-cured polyurethane foam, a single density material. One way to make dual density cushion is via bonding. Another way is to use the insert method [98]. Table 12 provides a comparison of the cost and performance of various skin materials produced via different processes.

Automotive Interior Components with Improved Safety Features

Materials and Processing Requirements. The following criteria are important to the protection of a vehicle occupant in an impact environment [99]: efficient energy absorption, insensitivity to impact angle, and ability to withstand automotive interior environment. Various types of foamed polymers may be used for automotive

Table 12 Comparison of technical and economic characteristics of instrument panel skin materials and processes

	DP-PVC (single slush)	DP-PVC (double slush)	Wet slush	Negative thermoform	Positive thermoform	Thermoform	Spray
Skin material	PVC	PVC	PVC	PVC/ABS alloy	PVC/ABS alloy	EPVC	PU
Skin thickness, mm (in.)	1.5 (0.06)	1.5 to 2 (0.06 to 0.08)	1.8 (0.07)	1.1 (0.04)	1.1 (0.04)	1.1 (0.04)	1.1 (0.04)
Grain reproduction	excellent	excellent	fair	fair	excellent	fair	fair
Foam/process	PUF/injected	PVC/integral	PUF/injected	PUF/injected	PUF/injected	PVC/integral	PUF/injected
Mold cost, 000$	200	20	20	50	50	50	40
Relative cost index	200	…	260	100	120	100	157
Cycle time, sec	315	320	420	45	45	45	150
Softness	excellent	excellent	fair	fair	excellent	fair	fair

Source: Ref 97

occupant impact protection. Expanded bead foams are low-cost, light-weight, efficient energy absorbing media. The materials are manufactured from small, solid beads containing a blowing agent, such as pentane. The solid plastic is transformed to a low density foam by the application of heat, which softens the polymer and causes the blowing agent to vaporize and expand. The expansion of the blowing agent leads to the formation of many small, closed cells in each bead. The cell size and the foam density can be varied by controlling the heat input and exposure time. The beads are then introduced into a mold where pressurized steam comes into contact with the particles. This second heating operation causes the beads to expand further, fill in the interstices between them and fuse the foamed beads together [99].

Performance. Most foams are relatively isotropic and hence insensitive to impact angle. Expanded polyolefin materials have good energy absorbing properties where multiple impacts are anticipated, such as in bumper applications. They are less effective in managing energy than a foam which deforms permanently on impact; their resiliency could cause occupant rebound. Expanded PS will permanently deform, but it does not meet the service temperature criterion for long term automotive interior use. Expanded high-heat copolymers SMA (styrene maleic anhydride) appear to meet all the requirements [92]. The SMA foams are being used in a full-length knee bolster assembly in the Eagle Premier. The expanded SMA foam knee bolsters are currently also used in a variety of vehicles, such as General Motors' Grand Am/Skylark/Cutlass, Bonneville, Corsica/Beretta, Park Avenue/LeSabre, Grand Prix, Olds 98/Delta 88, and Chrysler's LeBaron/Daytona [99].

Design and Cost. When compared to other types of energy management systems, foamed plastics are of relatively low cost. The design of interior safety components should emphasize optimizing the energy absorption of the interior compartments.

Engine Components

The major driving force behind replacing metal parts in engine components with polymer/composite parts is to reduce the cost. The consolidation of complex-shaped engine parts via integrated polymer/composite molding made it possible to reduce the cost through reducing machine and assembly operations. Engine components such as the transmission torque converter, brake piston, commutator, pulley and other components call for stiffness, strength, dimensional stability, chemical resistance, fatigue resistance, and creep resistance under high loads at elevated temperatures.

Certain high-temperature thermoplastics, such as polyether ether ketone, polyetherimide, polyamide-imide and polyether sulfone, can meet the required performance. The engine components can be made by injection molding any of these resins, which are still relatively expensive. Fiber-reinforced phenolics have found commercial applications in the engine parts mentioned above. A typical composition contains roughly one-third resin and two-thirds fibers, fillers and other additives. The properties of the material can be tailored to meet specific needs by varying the type and amount of filler or reinforcement. The properties of this reinforced resin are also sensitive to processing conditions. Part designs, gate location and molding conditions all affect the development of fiber orientation during molding, which in turn influences local part stiffness, strength and thermal expansion. Molding and postcuring procedures dictate the degree of cross-linking, which controls the component's final dimensions, creep resistance and retention of stiffness and strength in the 150 to 205 °C (300 to 400 °F) range [95].

Performance. Table 13 lists typical values for the static strength at room temperature of four injection-molded short glass-reinforced phenolic compounds [95]. The properties of the reinforced phenolics are closely related to fiber orientations. Near mold surface, shear stress tend to align the fibers in the direction of flow. Fibers midway between mold surfaces tend to orient transverse to the flow direction, leading to very different properties between the part interior and the part surface. Another important property of the phenolic engine parts is the thread strength, which refers to the pull-out force and strip torque that have been measured on tapped

Table 13   Static strength values for phenolic molding compounds

Property	RX630	RX660	RX865	XB-22
Tensile strength, MPa (ksi)	82.7 (12)	55.2 (8)	68.9 (10)	75.8 (11)
Flexural strength, MPa (ksi)	193 (28)	137.9 (20)	186 (27)	151.7 (22)
Compressive strength, MPa (ksi)	275.8 (40)	227.5 (33)	241.3 (35)	234.4 (34)
Shear strength, MPa (ksi)	89.6 (13)	68.9 (10)	68.9 (10)	68.9 (10)
Tensile modulus, GPa (Msi)	13.8 (2.0)	11.7 (1.7)	17.2 (2.5)	7.6 (1.1)
Tensile fracture strain, %	0.6	0.5	0.4	1.0
Coefficient of thermal expansion, $10^{-6}/°C$ ($10^{-6}/°F$)	4.4 (8)	5.6 (10)	3.9 (7)	3.3 (6)

Source: Ref 95

threads in the phenolic material for a range of bolt diameters and engagement lengths.

Engine and transmission components must be designed to withstand various dynamic stresses, including impact and fatigue loads. The highly cross-linked molecular structure of a thermoset resin gives rise to excellent stiffness and creep resistance. But the same 3-D cross-linked network also makes the resin inherently brittle. Typical notched Izod impact strengths of the phenolic resins discussed range from 26.7 to 80 J/m (0.5 to 1.5 lbf · ft/in), unnotched Izod impact from 80 to 133.4 J/m (1.5 to 2.5 lbf · ft/in.).

Design with Phenolic Composites. Since short glass fiber-reinforced phenolic materials are not resistant to crack propagation, components made from these materials must be designed to prevent crack initiation. Again, the effective replacement of metal assemblies by polymers or composites requires more than just the substitution of phenolic material into old designs. These two classes of materials have their own strengths, weaknesses, and processing methods. A design concept such as MAD may be used to address the design issues.

Recycling of Automotive Polymers and Composites

Post-Consumer Plastic Recycling

The plastic portion of post consumer waste contributes less than 10% by weight, but 18 to 38% by volume, to the solid waste problem [100]. The American public perceives plastics as non-degradable, producing toxic fumes on burning, non-recyclable, and creating toxic waste [101]. The primary reason for this public perception is that there are still few aggressive recycling efforts for plastics. This is largely due to (1) little economic incentive to recycle plastics; (2) difficulty in differentiating and identifying one plastic from another; and (3) difficulty in separating products manufactured with different plastics [102].

Substitution of the conventional plastics by degradable plastics may not be a viable approach to solving the solid waste problem. Properly built landfills are dry, cold, and oxygen-deficient and, as such, degradation of plastics simply is too slow to make any difference in volume

reduction of solid waste [101]. The burning of packaging plastics hot with sufficient air could eliminate virtually all potentially problematic pyrolysis products [101]. An independent study by Franklin Associates concluded that, in a life cycle, soft drink bottles generated much fewer pounds of air and water pollution than glass and aluminum at current recycling rates [101].

In principle, all plastics can be recycled; but some plastics cannot be remolded or reused in the same type of application as the original product [102]. Thermoplastics are recyclable as is demonstrated daily by regrind levels, which are regularly 10 to 25% [101]. Those polymers that cannot be remelted can always be granulated and used as fillers or property modifiers in other polymers, asphalt, concrete, or other materials. The problems of plastic recycling are mainly collection and sorting by generic type. PET, the leading packaging polymer, has been successfully included in curbside recycling programs. The programs of curbside collection of other plastics are just now underway. Sorting technology is being developed to build on the SPI codes that are shown on the bottle of rigid containers. Recycling of plastics will gradually become economically attractive.

Recycling of Automotive Polymers and Composites

Although the weight of plastics used is still small in comparison to total vehicle weight, the number of plastic applications is significant. In a modern vehicle there may be, on the extreme, up to 700 different grades of 40 different families of resins [103]. Polymers are typically incompatible with one another and different polymers may require to be recycled in a different manner. To properly identify the various polymers used in a vehicle, the Society of Automotive Engineers (SAE), in collaboration with automotive manufacturers, suppliers, and recyclers, has arrived at a labeling protocol (SAE J1344) that permits the resin family to be marked into every plastic part. This protocol will be extended to include filler types and quantity. Labeling is crucial to recycling because separated materials provide better control over the properties, resulting in a better recycling value, than do mixed materials [103].

Removing the plastic components from the vehicles is another issue of recycling. Today's vehicles are designed to be assembled quickly without regard to the difficulty of disassembly. Design for future cars should take disassembly into consideration. Re-usable fastening systems such as bolts, plastic fastening systems and some molded-in fastening could be utilized. Design for ease of recyclability would also imply specification of the most recyclable materials in the first place. Re-use of recycled polymers in the automotive applications is also an avenue to pursue in the design process [103]. In the pursuit of reduced vehicle weights and improved fuel economy, more light-weight high-performance polymers and polymer composites must be used. In principle, this can be accomplished economically without undue proliferation of resin types.

The total volume of polymers (excluding tires) used in the automotive industry was approximately 907 million kg (2 billion lb) in 1990 and was expected to increase to at least 1360 million kg (3 billion lb) by the year 2000 [100]. Each car manufactured in the United States in 1990 had on average almost 136 kg (300 lb) of plastics that cover a wide variety of resin families [102]. After removal of the metallic parts via magnetic separation in a shredder shop, the automotive shredder residue (ASR) or "fluff" is usually landfilled and rarely incinerated [104].

Since sorting of the ASR into individual families of resins is difficult to accomplish, many efforts have been made to evaluate the ASR as a plastic raw material stream in its own right. Potential applications of this material include use as a high-loading filler in a polyolefin matrix or as a concrete additive [105]. The poor compatibility between different polymers often leads to low impact resistance and flexural strength of the resulting blends. To date, no universal compatibilizing agents are available that would help in fabrication of "commingled" plastics with adequate mechanical properties. A glass-filled structural foam core product of recycled plastics (plastics lumbers) was claimed to possess adequate mechanical properties for long-term use [106]. Some of the mechanical properties of post consumer plastics were improved by the addition of recycled polystyrene [107]. Since the supply of automotive fluff is potentially huge, it is worthwhile to spend efforts to determine whether usable products can be prepared from this mixture. At present, recycling of automotive polymers and composites has yet to overcome two problems: (1) unsteady supply of adequate quantities of uncontaminated plastics parts and (2) high labor cost of hand-dismantling of plastics parts from the junked cars.

Recycling Technologies for Automotive Polymers and Composites

Thermoplastics. Many reprocessing technologies have been developed for recycling of post-consumer plastics [107]. For instance, most PET is reclaimed by a system that washes the PET and HDPE and separates them by the density differences (flotation or hydrocyclone Process). PET recycling companies are also interested in depolymerizing PET polymers back to their monomers or oligomers in order to repolymerize them into a good-grade recycled PET [107]. PET is a conden-

sation polymer derived from terephthalic acid (TPA) or dimethyl terephthalate (DMT) and ethylene glycol (EG). Polymerization is effected by heating these systems, typically with an antimony catalyst, while continuously removing the by-product (water or methanol) from the reaction systems. The reaction is reversible and, therefore, depolymerization is technically feasible. This can be accomplished in several ways. One is to treat PET with water in excess at an elevated temperature of 150 to 250 °C (300 to 480 °F) in the presence of a catalyst (e.g., sodium acetate) to produce TPA and EG. Alternatively, PET can be treated with an excess of methanol to produce DMT and EG (methanolysis). If the reverse reaction is carried out with glycols (glycolysis) instead of water or methanol, aromatic polyols can be produced. These polyols can be further treated with isocyanates or unsaturated dibasic acids to produce polyurethanes or unsaturated polyesters, respectively. PBT, which is made via the condensation of DMT and 1,4-butanedial, can be depolymerized by methanolysis to recover DMT [107].

ABS polymers can be remelted and reused several times if properly stabilized with various antioxidants and thermal stabilizers. Contamination of ABS parts by other plastics can lead to significant property degradation. Nylons can be remelted and reused several times, but the hygroscopic nature of nylon requires the reground product be dried to a moisture level of about 0.2% prior to remelting. The technical feasibility of depolymerizing nylon-based materials has also been demonstrated [107].

Acetal polymers are immiscible with most other polymers and, therefore, are not found in polymer alloys or blends. Consequently, recycling of acetals with the presence of other polymers will present a serious technical challenge. Pyrolysis technology of polyacetals to produce formaldehyde or trioxane monomer has yet to be established. Polycarbonate (PC) and PPO are melt-reprocessable, so are polyolefin-based polymers such as PPs. Engineering thermoplastics include polyarylates, polyphenylene sulfides (PPS), fluorocarbon polymers, polysulfones, polyether ether ketone, polyamide imide and liquid crystalline polymers. The difficulty of collection and sorting of these polymers, coupled with their high processing temperatures, has made it difficult to attempt post-consumer recycling [107].

Thermoset Resins and Composites. Both in-plant scrap and post-use thermoset materials are being recycled. The recycling techniques for thermoset materials include [108]:

- Incorporation of regrind in thermoplastic or thermoset systems
- Recovery of raw materials via hydrolysis or glycolysis
- Recovery of chemical values via pyrolysis
- Recovery of heat or energy through incineration

The thermoset materials to be reprocessed must go through processes of size reduction. Size reduction typically entails feeding scrap into a series of pulverizers to

Table 14 Properties of polyurethane RIM fascia containing regrind

Property	No regrind	Regrind, unfilled, unpainted	Regrind, unfilled, painted	Regrind, filled, painted
Specific gravity	1.04	1.04	1.04	1.07
Flexural modulus, MPa (ksi)	350 (50.8)	329 (47.7)	355 (51.5)	378 (54.8)
Tensile strength, MPa (ksi)	24.9 (3.6)	24.9 (3.6)	26.0 (3.8)	25.1 (3.64)
Tensile elongation, %	240	265	245	250
Gardner impact strength, J (ft · lbf)	>36.2 (>26.7)	>36.2 (>26.7)	>36.2 (>26.7)	>36.2 (>26.7)
Tear strength, kN/m (10^3 lbf/ft)	100 (1459)	99 (1445)	102 (1488)	100 (1459)

Source: Ref 32

Table 15 Properties of BMC containing recycled SMC

	BMC	BMC + coarse recycled SMC		BMC + fine recycled SMC	
Ground SMC, %	0	10	20	10	20
Tensile stress, MPa (ksi)	27.9 (4.05)	16.1 (2.34)	14.5 (2.10)	17.3 (2.51)	16.1 (2.34)
Tensile modulus, GPa (Msi)	13.1 (1.9)	9.9 (1.4)	9.2 (1.3)	12.7 (1.8)	10.2 (1.5)
Elongation, %	0.44	0.68	0.34	0.22	0.14
Flexural modulus, GPa (Msi)	10.5 (15)	9.8 (1.4)	9.4 (1.36)	10.2 (1.5)	9.2 (1.3)
Notched Izod impact strength, J/m (lbf · ft/in.)	270 (5.06)	270 (5.06)	158 (2.96)	209 (3.91)	221 (4.14)
Unnotched Izod impact strength, J/m (lbf · ft/in.)	361 (6.76)	278 (5.21)	183 (3.43)	267 (5.00)	301 (5.64)

Source: Ref 41

achieve a 20 mesh size, which is further reduced to 200 mesh by a series of hammer mills. The material is then discharged through sorting screens to ensure the proper particle size is obtained. The equipment requirement and economics for the grinding of thermoset scrap for use as filler in both thermoplastic and thermosetting resins has recently been discussed by Bauer [109].

PU Flexible Foams. Production of cushioning materials generates approximately 10% scrap. The scrap is used primarily in the production of rebonded carpet underlay [110-112] and re-used in the foam formulation [113]. In the first application, the uncontaminated foam is reduced to suitable size, and the foam pieces are then coated with binder (urethane-based: TDI or MDI with a polyether polyol). The mixtures of foam binder and catalysts are subsequently compression-molded into the carpet underlay products. The re-use of foam scrap requires intensive grinding at cryogenic temperature to a fine powder. The physical and mechanical properties of the foam with regrind are comparable to control materials [113].

PU RIM. The scrap of PU RIM or post-use PU RIM can be recycled by (1) putting the regrind into RIM parts [114-118], (2) compression molding of regrind into other parts [117], and (3) extrusion of scrap into profiles and tubing [118]. With proper size reduction of the regrind, the RIM PU elastomer containing regrind exhibit mechanical properties similar to the controls (Table 14). Further, the surface finish of the painted parts is excellent and comparable to the control materials [114].

Phenolics and Epoxy Resins. Several problems have been identified with epoxy and phenolic materials con-

taining their regrind versions: (1) increased processing difficulty due to high viscosity of the mix; (2) reduced tensile strength; and (3) poorer impact resistance [119-123]. Glass fiber reinforced epoxy may be recycled by calcining to remove the epoxy matrix, leaving fibers that can be reused as composite reinforcement [108]. Hand lay-up of aerospace-grade carbon fiber-epoxy prepreg materials typically generates 10 to 40% scrap, which is a large source of high-value, high-performance materials. At present, no effective program is in place to recycle these materials for lower-performance applications.

SMC and BMC. Several attempts have been made to reuse ground SMC or BMC scrap as filler in SMC/BMC compounds and in thermoplastics [124-126]. Graham and Jutte [124,125] investigated the effect of adding SMC regrind on the properties of SMC. Two material sizes were used: a coarse fraction which passed a 9.5-mm screen, and a fine fraction which passed a 4.8 mm screen. The results, summarized in Table 15, indicate some loss in tensile strength and modulus with the addition of the coarse regrind and somewhat less reduction with the fine regrind. Significant loss in impact resistance was also observed in both cases. Parallel studies on PE and PP containing recycled SMC indicated that the regrind SMC powder acted as a filler and not as a reinforcement [124,125].

The mixed plastic scrap from junked automobiles can be ground and bound together with an isocyanate-based binder to produce useful composite panels [127]. Automotive plastic waste can be treated with a separation process to recover less heterogeneous materials [128]. Flexible PU foam successfully separated from ASR may

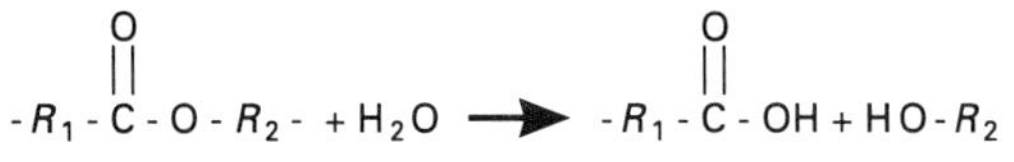

$$-R_1 - \overset{\overset{\textstyle O}{\|}}{C} - O - R_2 - \ + H_2O \ \longrightarrow \ -R_1 - \overset{\overset{\textstyle O}{\|}}{C} - OH + HO - R_2$$

Fig. 3 Hydrolysis of polyester

$$-R_3 - NH - \overset{\overset{\textstyle O}{\|}}{C} - O - R_4 \ + H_2O \ \longrightarrow \ -R_3 - NH_2 + HO - R_4 - + CO_2$$

Fig. 4 Hydrolysis of polyurethane. Source: Ref 129-135

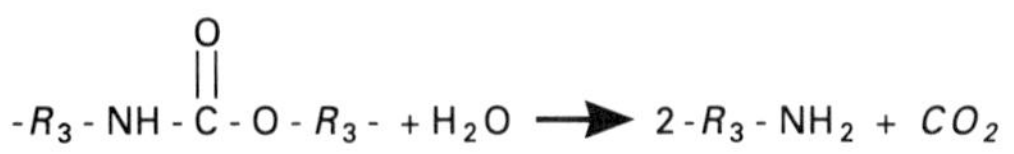

$$-R_3 - NH - \overset{\overset{\textstyle O}{\|}}{C} - O - R_3 - \ + H_2O \ \longrightarrow \ 2 - R_3 - NH_2 + CO_2$$

Fig. 5 Hydrolysis of urea

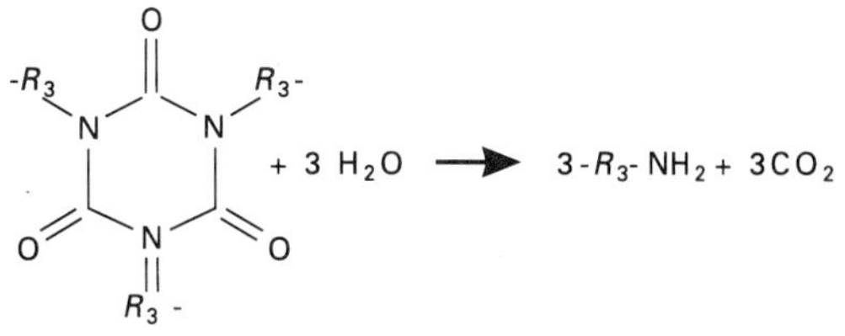

Fig. 6 Hydrolysis of isocyanurate

$$-R_3 - NH - C - O - R_4 - \ + \ HO - R_5 - OH \ \longrightarrow$$

$$-R_3 - NH - C - O - R_5 - OH \ + \ HO - R_4 -$$

Fig. 7 Glycolysis of polyurethane

be suitable for rebond or other recycle techniques such as pyrolysis [128].

Chemical Recycling Techniques. As indicated earlier, polymers containing carbonyl groups such as PET, PBT, PU, and nylon can be reversed back into monomers by hydrolysis or glycolysis. Hydrolysis of PET follows a reversed condensation reaction to recover the original monomers (Fig. 3). Hydrolysis of PU produces the diamine and the polyol (Fig. 4). The urea functional groups in some flexible foams and the isocyanurate groups in some rigid forms can also be hydrolyzed back to amine and carbon dioxide (Fig. 5, 6). High-pressure steam [135,136], a vertical reactor [137,138], a twin-screw extruder [139] and other processes [140-142] have been employed to hydrolyze PU foam materials. These techniques are simple in that they convert everything to amines and polyols which, however, must be separated before either is used [108].

Glycolysis of PU resins produces a mixture of poly-hydroxy compounds which can be used directly without further separation (Fig. 7). However, when urea or isocyanurate functions are present amine functional moieties and hydroxycarbonates are produced [131]. Scrap from a variety of sources (rigid foam, flexible foam, RIM PU, and microcellular PU elastomers) has been treated with glycols by several processes [143-163] to recover reusable products.

Pyrolysis and Incineration. Alternatively, thermal energy can be used to break down polymers to recover chemical values [164-171]. Pyrolysis of the plastic mixture from automotive waste resulted in a product composition of about 45% char, 35% liquid, and 20% gas at a final temperature of 550 to 600 °C (1020 to 1110 °F) [170]. Pyrolysis of PU materials has been examined by many investigators [171-179]. Wooley *et al.* [176-178] reported that pyrolysis of PU resins based on polypropylene glycol and TDI in an inert atmosphere at 200 to 250 °C (390 to 480 °F) occurred by random scission of polymer bonds to isocyanate and hydroxyl. At 200 to 300 °C (390 to 570 °F), pyrolysis of rigid PU foams produces isocyanate and polyol [178].

Large-scale pyrolysis trials of SMC scrap by SMC Auto Alliance and General Motors [180] showed that SMC pyrolysis could generate a fuel gas in quantities sufficient to sustain the pyrolysis. Other recovered products include a heavy oil pyrolysate of potential heating value and solid products such as carbon, calcium carbonate and glass fibers. The solid products, upon grinding, could be substituted for calcium carbonate in SMC formulations [181].

Studies have shown that incineration of PU foams [182], mixed plastics containing PU and PVC [183], and ASR [184], can be combusted under various conditions with environmentally acceptable emissions. Energy recovery from PU RIM scrap was reported to have an energy content equivalent to coal [185]. Combustion technologies for PU RIM include fluidized bed, rotary oven, and mass-burn combustors. Each technique is capable of undergoing clean combustion with stock gas emissions well within allowable limits [185].

Outlook for Automotive Composites

Polymer composites are expected to be more widely accepted in the automobile industry, provided that the total manufacturing and life-cycle costs can be further reduced. The notion of reducing the manufacturing cost embraces the development of cost-effective processing technologies for mass production of automotive parts with improved part quality and reliability. In view of reliability improvement and cost reduction, current polymer composite manufacturing technology issues include:

- Manufacturing process modeling, monitoring and control
- NDE and quality control
- Alternative and innovative processing technologies
- Cost-effective recycling technologies for both thermoplastic and thermoset composites

Improved and guaranteed quality of a composite product can only be achieved with the development of process monitoring sensors and process models. The ability to adjust processing conditions in real time to accommodate material chemistry, storage, and composite manufacturing variations will be essential as costly parts are attempted. Since composites are difficult to repair, the concept of process-controlled built-in quality to eliminate manufacturing defects will be cost effective in eliminating rejection of large parts. Select sensors developed for process control could be conceivably left on the fabricated structure for "condition monitoring". Whether these sensors can serve as real-time smart displays of actual structural strength, stiffness, or post-damage performance, remains to be demonstrated.

Selection or development of a proper process technology is probably the single most important factor influencing total cost reduction. Cost effectiveness of a composite manufacturing process is dictated by the capital equipment investment, tooling cost, cycle time, and part reliability and consistency. Injection molding, a highly automated technique, is restricted to short fiber composites. The process also tends to further degrade the fibers. Thermoplastic molding compounds containing longer fibers (10 to 15 mm, or 0.39 to 0.59 in., before molding) can be obtained by combined pultrusion and pelletizing operations. Injection molding of thermoset compounds typically suffers from poor control of fiber orientation distribution and the difficulty of processing high fiber concentrations. While SMC has found a broad range of applications, compression molding of SMC involves relatively high molding pressures, resulting in large machine costs and limiting part size. Also, the SMC process requires considerable resin flow to achieve optimum mechanical and surface properties. This flow leads to complex and difficult-to-control fiber orientations and concentrations.

Resin transfer molding (RTM) and structural reaction injection molding (SRIM) are capable of achieving precise control in the placement of fibers in high concentration. The ability to produce large integrated sections and complex part geometries with reasonably short processing cycles would make these LCM techniques ideal composites fabrication methods for automotive applications. However, further development of preform technology is needed to optimize cycle time, fiber wetout, mechanical properties and surface appearance. More efforts are needed to develop preform technology for carbon fibers and carbon/glass hybrid systems. In SRIM of high reactivity systems, fast fill, complete penetration of the preform, good wetout, reduction of void content, minimal movement of the reinforcement fibers, and avoidance of resin-rich areas are some of the challenges that future research must address. The following paragraphs outline a few fruitful research subjects related to polymer composites for automotive applications.

Development of Process Sensor Techniques. The efforts for developing process sensor techniques have been sporadic. A concerted effort is needed to evaluate critically the advantages and limitations of existing and developing sensors in terms of feasibility for use in a real RTM manufacturing environment. Techniques that hold greatest potential include microdielectrometry, acoustic emission-ultrasonic, fluorescence, fiber optics, and vibration-based mechanical impedance analysis. Critical process parameters will be better understood with the use of process sensors. Conversely, real-time data acquired by the sensors may be inverted to extract critical process parameters. Advancement in process sensor technology would result in the understanding of initial process parameters, devices for *in situ* monitoring, and lower-noise reliability robust signal acquisition.

Real-Time Manufacturing Built-In-Quality Sensor System. The composites manufacturing industry needs to be able to adjust processing conditions in real time to accommodate material chemistry, storage and composites manufacturing variations. Research and development efforts are needed to survey and select sensor systems and utilize, modify, or develop sensor models which correlate to polymer viscoelasticity or other properties, and which interface to control system software. The main goals of these efforts are to produce sensor selection guidelines based on potential information and cost effectiveness, and to establish sensor models to convert signal output to meaningful cure information during RTM. The ultimate objective will be to improve manufacturing confidence through a convincing benefit analysis.

Life monitoring measuring science for composites should be investigated. This effort should include:

- Survey of feedback sensor systems
- Evaluation of "sensible" composites with an internal nervous system concept
- Demonstration of life-time monitoring potential for stress-strain, impact, and fatigue real time measurement
- Development of scientific quantitative materials characterization
- Demonstration of sensors in prototype full-scale applications

At present, acousto-ultrasonic, fiber optic, and vibration-based MIA techniques appear to be viable as condition monitors. This investigation could eventually result in improved selection of feedback sensors and smart materials with "life remaining" gages, instantaneous damage warning, and readout of quantifiable structural conditions and performance.

Composite Processing Expert Systems. A small number of organizations have attempted to develop expert systems for composite processing, particularly for aerospace applications. Similar programs are needed to ensure that the developed expert systems be more adaptable (applicable to a wider range of material systems) and more user-friendly in a real automotive manufacturing environment. A team of participants from related industries should be assembled to study and recommend standardization of the expert systems. This effort should also incorporate the AI systems for the starting materials, and the capability to interact with the process sensors. Areas of particular significance include the development of user-friendly rule-based systems for thermoplastic composite process models.

Emphasis probably should be placed on alternative RTM materials and processing technology. RTM is now widely recognized as a viable process for fabricating automotive components, provided part production time can be shortened to minutes or less, and variable costs can be kept at acceptable levels. Major research and development needs in RTM include the further development of preform technology (including preform binder), textile structural preform technology, and modified resins (both thermoplastic and thermoset) with larger processing windows.

Preform Binder. Additional work is needed on preform rigidization by developing new binder resins and innovative ways of applying the binder. Unsaturated polyester has been the primary binder resin of choice. However, polyurethane-based fast-curing resins appear to hold great promise in acting to hold the fibers together to facilitate resin impregnation and wetout of fibers without undesirable fiber movement. Preforms that contain hybrid constructions by design (different styles of mats or fabrics or different fiber types) should be investigated. Preform technology for carbon fibers and hybrid fibers lags far behind that for glass fibers. Systematic studies should be carried out to establish the relationships between preform characteristics and the mechanical properties of the resulting composites.

Textile Structural Preforms. The introduction of the thickness-direction fibers through three dimensional braiding, stitching, or weaving is very effective in restricting the growth of damage in composites. For instance, a braided three dimensional composite exhibits a much smaller degree of damage than a quasi-isotropic laminate of carbon/epoxy, even though the damage threshold energy values are very similar. Stitched multidimensional (3-D) composites were found to possess greater damage resistance than conventional two dimensional composite materials. The stitched third-direction Kevlar fibers have proven to be effective in arresting delamination propagation. A three dimensional composite had a smaller damage area (with little or no visible delamination) than a two dimensional composite. These textile structural preforms generally possess much higher fiber densities than do the chopped fiber- or mat-based preforms, making it more difficult for the resin to impregnate and wetout the fibers. Process modeling work should be carried out with RTM composites based on these specialty preforms.

Modified Thermoset and New Thermoplastic Resins for RTM. The demand for a reduced RTM cycle time with high reactivity systems places stringent requirements on the process variables: fast fill, complete penetration of the preform, good fiber wetout, low void content, and few resin-rich areas. These requirements cannot be met without a clear understanding of the simultaneous resin flow and chemical reaction phenomena. Limited efforts have been directed toward process modeling of RTM in terms of mass and energy transport. These modeling efforts were typically based on very simple part geometry and highly simplifying assumptions. Before large integrated parts of complex geometry are attempted, the level of modeling sophistication must be raised considerably. This effort is critical to the goal of eliminating non-uniformity in resin penetration and in fiber orientation and concentration, particularly in the hard-to-reach zones such as around sharp corners. Additional process modeling efforts, in conjunction with process monitoring sensor development, must be made to establish a science-based RTM process control system.

A need exists for developing modified thermosetting resins that possess larger processing windows. These include the resins of which the compositions can be readily formulated or modified by the RTM users so that the resin chemistry would be compatible with the preform types and dimensions and the RTM equipment capabilities. In addition, RTM-processable thermoplastics are important materials to develop. These resins are required to be of low viscosity during the resin filling stage, possibly at the oligomer, 'cyclics', or other intermediate state, and then transformed into a high-molecular-weight state upon completion of resin impregnation and fiber wetout. Thermoplastic matrix resins have high damage resistance and recyclability. However, the viscosity of conventional thermoplastic resins is normally too high to use in a RTM situation.

When designing with polymer-based parts, one must also consider competition against and compatibility with steel parts. Mixing steel with plastic on a vehicle in the same assembly process may present a difficult problem for the automotive engineers. The plastic parts, such as glass fiber-reinforced polyester fenders, may have to go through the E-coat ovens at 177 to 204 °C (350 to 400°F) on-line with the rest of the steel body. The resins must withstand the heat without undergoing degradation. Thermal stresses caused by differences in thermal expansion must be given special attention.

The total cost of the polymer body panels, including the costs of tooling and secondary operations, is significantly lower than for steel. However, the piece cost is still appreciably higher. The steel industry is spending great efforts to reduce tooling and fixture costs for both high- and low-volume production. Fewer die-stamping stages are desired. Efforts to reduce tooling costs for low-volume steel production include using alternative processes such as hydroforming, and reducing machining requirements by incorporating cast elements. Polymer-based materials are getting stiffer competition from steel, even in the medium-volume market (50,000 to 100,000 vehicles) where polymers have established cost advantages [92].

REFERENCES

1. "Materials for Lightweight Vehicles," Office of Transportation Materials, U.S. Department of Energy, Washington, D.C., July 1992

2. "Automotive Crash Avoidance Research," SAE International Congress and Exposition, Detroit, 1987
3. M. O'Malley, Checking the Status of Plastic Materials for Auto Body Panels, *J. Metals*, Jan 1990, p 17-19
4. S.A. Wood, For Large Exterior Car Parts, the Boom Begins, *Modern Plast.*, Oct 1990, p 50-55
5. R.J. Cleereman, Effective Design with Plastics, *Adv. Mater. Proc.*, Feb 1990, p 33-38
6. T. Stevens, RIM Polyureas, Ready to Dent the Automotive Body Panel Market, *Mater. Eng.*, May 1990, p 16-19
7. H. Baker, Liquid Molding Technology Rolls Ahead, *Adv. Mater. Proc.*, Feb 1991, p 39-34
8. S.A. Wood, The All-Plastic Car Comes Several Laps Closer, *Modern Plast.*, Oct 1991, p 50-54
9. M.A. Dorgham, The Economic Case for Plastics, in *Designing with Plastics and Advanced Plastic Composites*, Interscience Enterprises, Ltd., Switzerland, 1986, p 10-13
10. Factors Affecting Material Selection in Automotive Body Panels, *Mater. Eng.*, Vol 107, May 1990, p 38-44
11. Plastics in Body Panels: Slow Penetration in Early 1990s, *Plast. Technol.* Vol 36, Sept 1990, p 131
12. Automotive Materials for Changing Times, *Mater. Eng.*, Vol 107, May 1990, p 20-23
13. PP : Not Just a Cheap Plastic, *Mater. Eng.*, Vol 103, Jan 1990, p 39-43
14. Automotive Plastics, *Plast. Eng.*, Vol 46, Sept 1990, p 19-24
15. Polymer Auto Parts, *Manufact. Eng.*, Vol 105, July 1990, p 46
16. RIM Polyureas, Ready to Dent the Automotive Body Panel Market, *Mater. Eng.*, Vol 107, May 1990, p 16-19
17. Materials and Machinery Advances Will Shine at Polyurethane Conference, *Plast. Technol.*, Vol 36, Sept 1990, p 17-23
18. What's Down the Road for Captive Automotive Molders, *Plast. Technol.*, Vol 36, Sept 1990, p 72-81
19. For Large Exterior Car Parts, the Boom Begins, *Modern Plast.*, Vol 75, Oct 1990, p 50-57
20. PPO Appearance Part is Blow Molded, *Modern Plast.*, Vol 67, Feb 1990, p 19-20
21. R.G. Dubensky, D.E. Jay, and R.K. Salansky, Process Driven Design of a Plastic Bumper Beam, *Modern Plast.*, Vol 67, Feb 1990, p 29-32
22. S.A. Wood, In Subtle Ways, Plastics Tighten an Alliance with the Automotive Market, *Modern Plast.*, April 1991, p 46-49
23. Toray Research Center, Tokyo, Auto-Plastics Application Increasing World-Wide, *Adv. Mater. Proc.*, May 1990, p 16-21
24. T. Stevens, New Engineering with Old Resins, *Mater. Eng.*, Nov 1990, p 29-32
25. Truck Grille Wins Award, *Adv. Mater. Proc.*, July 1990, p 24
26. Composites Get It in Gear, *Mater. Eng.*, Oct 1990, p 8
27. L.K. Enigh, The Next Generation(s) of Nylon, *Mater. Eng.*, Feb 1989, p 47-51
28. D. Stover, Waiting for Detroit: Fuel Efficiency to Pave the Way for Composites in Automobiles, *Adv. Compos.*, June 1991, p 47-53
29. J. Koster, The Composite-Intensive Vehicle—The 3rd Generation Automobile!—A Bio-Cybernetical Engineering Approach, *Adv. Compos.*, June 1991, p 129-137
30. R.L. Frutiger, S. Baskar, K.H. Lo, and R. Farris, Design Synthesis and Assessment of Energy Management in a Composite Front End Vehicle Structure, *Adv. Compos.*, June 1991, p 33-43
31. P. Beardmore, Automotive Components: Fabrication, *Engineering Materials Handbook*, J. Weeton, Ed., ASM International, Materials Park, OH, 1987, p 24-31
32. L. Dodyk, High-Performance Composites for the Automotive Industry, *Adv. Comp.*, June 1991, p 21-27
33. D.F. Baxter, Jr., Green Light to Plastic Engine Parts, *Adv. Mater. Proc.*, May 1991, p 26-31
34. Plastic Hood has Ultrasmooth Surface, *Adv. Mater. Proc.*, Nov 1991, p 9-10
35. T.P. Schroeter and R.K. Leavitt, "1990 Corvette Rear Underbody-The Case for Preform," p 275-281
36. D.A. Kleymeer and J.R. Stimpson, "Production of an Automotive Bumper Using LCM," p 283-290
37. Composite Brake Drum Won't Break, *Adv. Mater. Proc.*, May 1990, p 24
38. Composite Frame on Fast Track, *Adv. Mater. Proc.*, Sept 1990, p 73
39. D. Stover, Composites Move into Mass Transit, *Adv. Compos.*, Sept/Oct 1990, p 29-44
40. D.F. Baxter, Plastic Beat the Heat in Underhood Components, *Adv. Mater. Proc.*, May 1990, p 36-41
41. *Modern Plastics Encyclopedia*, R. Juran, Ed., McGraw-Hill, Inc., New York, 1991, p 19-154
42. B. Fisa, Injection Molding of Thermoplastic Composites, in *Composite Materials Technology*, P.K. Mallick and S. Newman, Ed., Hanser Publishing, Munich, 1990, p 267-320
43. E. Sattler, Lost-Core Molding Offers a Route to Hollow Parts with Complex Geometry, *Modern Plast.*, Jan 1993, p 50-53
44. A.B. Strong, *Fundamentals of Composites Manufacturing*, SME, Dearborn, MI, 1989
45. D. Weyrauch and W. Michaeli, "A New Preforming Technology for RTM and SRIM," ANTEC SPE, 1992, p 761-766
46. P.K. Mallick and S. Newman, Ed., *Composite Materials Technology*, Hanser Publishing, Munich, 1990
47. H.S. Kliger and B.A. Wilson, Resin Transfer Molding, in *Composite Materials Technology*, P.K. Mallick and S. Newman, Ed., Hanser Publishing, Munich, 1990, p 149-178
48. E.P. Carley, J.F. Dockum, and P.L. Schell, in *Proc. 5th Ann. Conf.* Advanced Composites, ASM International, Sept 1989, p 259-273
49. C.F. Johnson and N.G. Chavka, in *Proc. 4th Ann. Conf.* Advanced Composites, ASM International, Sept 1988
50. B.C. Mazzoni, in *Proc. 43rd Ann. Conf. Reinforced Plastics*, Composites Institute, SPI, Feb 1988, Sec. 11-F
51. S.G. Dunbar, in *Proc. 43rd Ann. Conf. Reinforced Plastics*, Composites Institute, SPI, Feb 1988, Sec. 4-D
52. M.J. Cloud, in *Proc. 4th Ann. Conf. Advanced Composites*, ASM International, Sept. 1988
53. E.P. Carley, J.F. Duckum, and P.L. Schell, in *Proc. 44th Ann. Conf. Reinforced Plastics*, Composites Institute, SPI, Feb. 1989, Sec. 10-B
54. B. Miller, *Plast. World*, Jan 1988, p 23
55. M. Kallaur, in *Proc. 43rd Ann. Conf. Reinforced Plastics*, Composites Institute, SPI, Feb 1988, Sec. 22-A
56. J. Scirvo, in *Proc. 43rd Ann. Conf. Reinforced Plastics*, Composites Institute, SPI, Feb 1988, Sec. 22-B
57. S.L. Voeks *et al.*, in *Proc. 43rd Ann. Conf. Reinforced Plastics*, Composites Institute, SPI, Feb 1988, Sec. 4-E
58. C.D. Shirrel, *AutoCom'88 Conf.*, SME, May 1988, Paper No. EM88-228
59. P. Emrich, in *Proc. 3rd Ann. Conf.* Advanced Composites, ASM International, Sept 1987
60. J.K. Rogers, *Plast. Technol.*, March 1989, p 50-58
61. D.A. Kleymeer, in *Proc. 4th Ann. Conf. Advanced Composites*, ASM International, Sept 1988, p 289-295

62. T. Drummond, Automated Near Net Shape Preform Manufacturing for High Speed RTM, *in Structural Composites: Design and Processing Technologies, Proc. 6th ASM/ESD Adv. Compos. Conf.*, Detroit, Oct 1990, p 373-376

63. J.C. Gembinsky, The Role of Preforms in Higher Volume Composites, in *Structural Composites: Design and Processing Technologies, Proc. 6th ASM/ESD Adv. Compos. Conf.*, Detroit, Oct 1990, p 377-392

64. J.F. Dockum, Jr. and P.L. Shell, Fiber Directed Preform Reinforcement, in *Structural Composites: Design and Processing Technologies, Proc. 6th ASM/ESD Adv. Compos. Conf.*, Detroit, Oct 1990, p 393-406

65. J.K. Fielder, D.A. Barbington, and D.P. Waszeciak, Single-Step SRIM: Consolidation of the Preform and Molding Steps, in *Structural Composites: Design and Processing Technologies, Proc. 6th ASM/ESD Adv. Compos. Conf.*, Detroit, Oct 1990, p 407-411

66. N.G. Chavka and C.F. Johnson, Critical Preforming Issues for Large Automotive Structures, in *Structural Composites: Design and Processing Technologies, Proc. 6th ASM/ESD Adv. Compos. Conf.*, Detroit, Oct 1990, p 413-422

67. B. Mazzoni and J. Riley, Update for Glass Mat Preformers, in *Structural Composites: Design and Processing Technologies, Proc. 6th ASM/ESD Adv. Compos. Conf.*, Detroit, Oct 1990, p 423-428

68. S.W. Horn, Complex Structural Shape Preforming Process Technology, in *Structural Composites: Design and Processing Technologies, Proc. 6th ASM/ESD Adv. Compos. Conf.*, Detroit, Oct 1990, p 429-432

69. S. Hamilton, Multiaxial Stitched Preformed Reinforcement, in *Structural Composites: Design and Processing Technologies, Proc. 6th ASM/ESD Adv. Compos. Conf.*, Detroit, Oct 1990, p 433-434

70. H.B. Soebroto, C.M. Pastore, and F.K. Ko, Engineering Design of Braided Structural Fiberglass Composites, in *Structural Composites: Design and Processing Technologies, Proc. 6th ASM/ESD Adv. Compos. Conf.*, Detroit, Oct 1990, p 435-440

71. R.W. Meyer, *Handbook of Polyester Molding Compounds and Molding Technology*, Chapman and Hall, New York, 1987

72. P.K. Mallick, Sheet Molding Compounds, in *Composite Materials Technology*, P.K. Mallick and S. Newman, Ed., Hanser Publishing, Munich, 1990, p 25-65

73. P. Beardmore, Automotive Components: Fabrication, in *Engineering Materials,* J. Weeton, Ed., ASM International, Materials Park, OH, 1987, p 24-31

74. R. Burns, *Polyester Molding Compounds*, Marcel Dekker, Inc., New York, 1982, p 181-228

75. T. Watanabe and M. Yasuda, *Composites*, Vol 13, 1982, p 54

76. A. Golovoy, M.F. Cheung, and H. Van Oene, *Proc. 39th Ann. SPI Technical Conf.*, 1984

77. S.K. Chaturvedi and R.L. Sierakowski, *Compos. Structure,* Vol 1, 1983, p 137

78. R.P. Kheatan and D.C. Chang, *J. Compos. Mater.*, Vol 17, 1983, p 137

79. S.K. Chaturvedi and R.L. Sierakowski, *J. Compos. Mater.*, Vol 19, 1985, p 100

80. S.S. Wang, E.S.M. Chim, and N.M. Zalan, *J. Compos. Mater.*, Vol 17, 1983, p 114 and 250

81. C.D. Shirrel and M.G. Onachuk, in *Composite Materials: Fatigue and Fracture*, STP 907, H.T. Hahn, Ed., ASTM, Philadelphia, PA, 1986, p 32

82. P.K. Mallick, *Composites*, Vol 19, 1988, p 283

83. P. Beardmore, J.J. Harwood, and E.J. Horton, Design and Manufacture of a GrFRP Concept Automobile, *Proc. Intl. Conf. Compos. Materials,* Elsevier, Paris, Aug 1980, p 47-60

84. H.T. Kulkarni and P. Beardmore, Design Methodology for Automotive Components Using Continuous Fiber Reinforced Materials, *Composites*, Vol 12, 1980, p 225-235

85. P.H. Thornton, Energy Absorption in Composite Structures, *J. Compos. Mater.*, Vol 13, 1979, p 262-274

86. P.H. Thornton and P.J. Edwards, Energy Absorption in Composite Tubes, *J. Compos. Mater.*, Vol 16, 1982, p 521-545

87. P.H. Thornton, J.J. Harwood, and P. Beardmore, Fiber Reinforced Plastic Composites for Energy Absorption Purposes, *Compos. Sci. Technol.*, Vol 24, 1985, p 275-298

88. S.R. Krishna, "Knitted Fabric Reinforced Polymer Composites," Ph.D. Dissertation, Dept. of Materials Science and Metallurgy, Univ. of Cambridge, Aug 1992

89. H.F. Mahmood and A. Paluszny, Design of Thin Walled Columns for Crash Energy Management, *Proc. 4th Intl. Conf.* Vehicle Structural Mechanics, Detroit, Nov 1981, Society of Automotive Engineers, Paper No. 811302

90. H.F. Mahmood, J.H. Zhou, and M.S. Lee, Axial Strength and Mode of Collapse of Composite Components, *Proc. 7th Intl. Conf.* Vehicle Structural Mechanics, Detroit, April 1988, Society of Automotive Engineers, Paper No. 880891

91. H.F. Mahmood, R.A. Jeryan, and J.H. Zhou, Crush Strength Characteristics of Composite Structures, in *Structural Composites: Design and Processing Technologies, Proc. 6th Ann. ASM/ESD Adv. Compos. Conf.*, Detroit, Oct 1990, p 1-10

92. V. Wigotsky, Plastics in Automotives, *Plast. Eng.*, April 1993, p 34-40

93. D. Vesey and S. Abouzadbr, E-Coat-Capable Plastics—A New Generation of Materials for Automotive Exterior Body Panels, in *Automotive Exterior Body Panels*, Society of Automotive Engineers, Detroit, 1988, p 1-6

94. D.J. Primeaux, II *et al.* Polyurea RIM in Exterior Body Panel Applications, *Automotive Exterior Body Panels*, Society of Automotive Engineers, Detroit, 1988, p 1-14

95. J. Arimond and B.B. Fitts, Design Data for Phenolic Engine Components, *Polymer Composite for Automotive Applications: International Congr. Expos.*, Detroit, Feb 29-March 4, 1988, Society of Automotive Engineers, p 79-83

96. J.W. Berg *et al.*, H-SMC Bumper Beam Endured 5 mph Impact, *Plastics in Automobiles: Bumper Systems, Interior Trim, Instrument Panels, and Exterior Panels*, Society of Automotive Engineers, Detroit, 1989, p 47-63

97. J.J. Plomer and T. Traugott, New High Heat Stable, Low Gloss, Automotive Interior Trim Resins Having Excellent Processability, *Plastics in Automobiles: Bumper Systems, Interior Trim, Instrument Panels, and Exterior Panels*, Society of Automotive Engineers, Detroit, 1989, p 103-113

98. M. O'Malley and R. Eller, New Developments in Materials and Fabrication Process for Automotive Interior Trim Skin Materials, *Plastics in Automobiles: Bumper Systems, Interior Trim, Instrument Panels, and Exterior Panels*, Society of Automotive Engineers, Detroit, 1989, p 73-82

99. K.M. Bartosiak and M.A. Huber, The Use of Expanded Bead Foam Materials for Improved Safety in Automotive Interior Components, *Plastics in Automobiles: Bumper Systems, Interior Trim, Instrument Panels, and Exterior Panels*, Society of Automotive Engineers, Detroit, 1989, p 83-92

100. *Mod. Plast.*, Vol 68, No. 1, Jan 1991, p 120

101. D. Cornell, Post-Consumer Plastic Recycling: Wading Through the Myths, Understanding the Cost of Collection, in *Structural Composites: Design and Processing Technologies, Proc. 6th Ann. ASM/ESD Adv. Compos. Conf.*, Detroit, Oct 1990, p 265-270

102. D. Manganaj, *et al.* "Recycling and Disposal of Waste Plastics from Automotive and Allied Industries," Report by Battelle Institute, Columbus, OH, 1989, p 1-22-1-33

103. S.L.D. Day, Plastics Recycling and the Automotive Industry, in *ANTEC*, SPE, 1992, p 1542-1544

104. T.R. McClellan, *Mod. Plast.*, Vol 60, No. 2, Feb 1983, p 50-52

105. M.J. Cury, "Research Report—Phase I: Secondary Reclamation of Plastics," Plastics Institute of America, 1987, p 3-5

106. W. Mack, "Turning Plastic Waste into Engineering Products Through Advanced Technology," presented at Recyclingplas V, Plastics Institute of America, Washington, DC, 1990, p 81

107. R.H. Burnett and G.A. Baun, Engineering Thermoplastics, in *Plastics Recycling*, R.J. Ehrig, Ed., Hanser Publishing, New York, 1993, p 153-168

108. W.J. Farrissey, Thermosets, in *Plastics Recycling*, R.J. Ehrig, Ed. Hanser Publishing, New York, 1993, p 233-265

109. S.H. Bauer, *Plast. Eng.*, Vol 33, No. 3, 1977, p 44; *ANTEC*, SPE, 1976, p 650

110. Hull *et al.*, "End Use Market Survey of the Polyurethane Industry in the United States and Canada," Prepared for the Society of the Plastics Industry, Inc., Polyurethane Division, 29 May 1990

111. *Polyurethane Handbook*, G. Oertel, Ed., Hanser Publishing, Munich, 1985, p 176

112. G. Woods, *The ICI PURs Book*, J. Wiley, New York, 1987, p 203

113. B.D. Baumann, P.E. Burdick, M.L. Bye, and E.A. Galla, *SPI Intl. Tech. Conf.*, 1983, p 139

114. R.E. Morgan and J.D. Weaver, SAE Intl., 25-28 Feb 1991, Paper No. 910580

115. J.A. Vanderhider, *International Polyurethane Forum*, Nagoya, Japan, 1990

116. M.C. Cornell, *Polyurethanes 1990: 33rd SPI Ann. Tech./Marketing Conf.*, Oct 1990, p 440

117. R.P. Taylor, R. Eiben, W. Rasshofer, and U. Liman, *SAE Intl.*, 25-28 Feb 1991, Paper No. 910581

118. S.N. Singh, E. Piccolino, J.B.. Bergenholz, and R.C. Smith, *SAE Intl.*, 25-28 Feb 1991, Paper No. 910582

119. D.M. Kaylon, M. Hallouch, and N. Fares, *ANTEC*, SPE, 1984, p 640

120. D.M. Kaylon and N. Fares, *Plast. Rubber Proc. Appl.*, Vol 5, 1985, p 369

121. R.D. Deanin and B.V. Ashar, *Organic Coatings Plast.*, Vol 41, 1979, p 495

122. R.A. Kruppa, *Plast. Technol.*, Vol 23, No. 5, 1977, p 63

123. C.V. Tran-Bruni, R.D. Deanin, *ANTEC*, SPE, 1987, p 446

124. W.D. Graham and R.B. Jutte, *SPI Composites Institute: Press Molders Conf.*, 2 May 1990

125. R.B. Jutte and W.D. Graham, *SPI Composites Institute: 46th Ann. Conf.*, 18-21 Feb 1991

126. K. Butler, K. *SPI Composites Institute: 46th Ann. Conf.*, 18-21 Feb 1991

127. W. Carroll and T.R. McClellan, U.S. Patent 4 382 108, 1983

128. B.J. Jody, E.J. Daniels, and P.V. Bonsignore, *SAE Int.*, 25-28 Feb 1991, Paper No. 910854

129. T.M. Chapman, *J. Polymer Sci., A Polymer Chem.*, Vol 27, 1989, p 1993

130. E. Grigat, *Kunststoffe*, Vol 68, 1978, p 281

131. H. Ulrich, A. Odinak, B Tucker, and A.A.S. Sayigh, *Polymer Eng. Sci.*, Vol 18, 1978, p 844

132. M.B. Sheratte, *SPI Tech. Conf.*, 1977, p 59

133. J.F. Kinstle, L.D. Forshey, R. Valle, and R.R. Campbell, *Polymer Reprints*, Vol 24 (No. 2), 1983, p 446

134. H. Ulrich, H., *Adv. Urethane Sci. Technol.*, Vol 5, 1978, p 49

135. G.A. Campbell and W.C. Meluch, *Environ. Sci. Technol.*, Vol 10, No. 2, 1976, p 182

136. W.C. Meluch and G.A. Campbell, U.S. Patent 3 978 128, 1976

137. G.A. Campbell and W.C. Meluch, *J. Appl. Polymer Sci.*, Vol 21, 1977, p 581

138. R.J. Salloum, *ANTEC*, SPE, 1981, p 491

139. E. Grigat and H. Hetzel, U.S. Patent 4 051 212, 1977

140. L.R. Mahoney, S.A. Weiner, and F.C. Ferris, *Environ. Sci. Technol.*, Vol 8, No. 2, 1974, p 135

141. J. Gerlock, J. Braslaw, and M. Zimbo, *Indus. Eng. Chem. Proc. Des. Dev.*, Vol 23, 1984, p 545

142. Japan Patent 54-068883, 1979

143. H. Ulrich, B. Tucker, A. Odinak, A.R. Gamache, *Elast. Plast.*, Vol 11, 1979, p 208

144. F.F. Frulla, A. Odinak, and A.A.R. Sayigh, U.S. Patent 3 709 440, 1973

145. F.F. Frulla, A. Odinak, and A.A.R. Sayigh, A.A.R. U.S. Patent 3 738 946, 1973

146. B. Tucker and H. Ulrich, U.S. Patent 3 983 087, 1976

147. F. Simioni, S. Bisello, and M. Cambini, *Macplas*, Vol 8, No. 47, 1983, p 52

148. F. Simioni, S. Bisello, and M. Tavan, *Cellular Polym.*, Vol 2, No. 4, 1983, p 281

149. F. Simioni, M. Modesti, and G. Navazio, *Macplas*, Vol 12, No. 88, 1987, p 157

150. F. Simioni, M. Modesti, and C.A. Brambilla, *Cellular Polym.*, Vol 8, 1989, p 387

151. B. Prajsnar, *Recycling*, 1979, p 11213

152. T.R. Ten Broeck and D.W. Peabody, U.S. Patent 2 937 151, 1960

153. Bridgestone Tire Co., French Patent 1 484 107, 1967

154. W. Schutz, U.S. Patent 4 339 358, 1982

155. H.O. Wolf, U.S. Patent 3 225 094, 1965

156. L.C. Pizzini and J.T. Patton, U.S. Patent 3 441 616, 1969

157. T. Yukutam, T. Ishiwaka, K. Usui, and M. Akoh, Great Britain Patent 2 062 660, 1980

158. K. Iwasaki, H. Kawakami, and T. Hirose, Japan Patent 77-95849, 1977

159. *Chem. Market Rept.*, Vol 233, No. 1, 1988, p 5

160. *Urethane Technol.*, Apr/May 1989, p 19

161. *APME Newsletter*, No. 29, 1988

162. G. Baurer, *Recycle 1990*, Davos, Switzerland, 29-31 May 1990

163. G. Baurer, Germany Patent 2 738 572; Europe Patent 948

164. G. Tesoro, *Polym. News*, Vol 12, 1987, p 265

165. M. Roy, A.L. Rollin, and H.P. Schreiber, *Polym. Eng. Sci.*, Vol 18, 1978, p 721

166. J. Bhatia, *Polymer Reprints*, Vol 24, 1983, p 436

167. J. Bhatia and R.A. Rosi, *Chem. Eng.*, Vol 10, 1982, p 58

168. W. Kaminsky and H. Sinn, *Kunststoffe*, Vol 68, 1978, p 284

169. L. Hammerling, W. Kaminsky, and R. Reichardt, *Recycle Intl.*, Vol 4, 1984, p 603

170. J. Braslaw, R.L. Gealer, and R.C. Wingfield, *Polymer Reprints*, Vol 24, 1983, p 434

171. S.L. Madorsky and S.J. Straws, *Polym. Sci.*, Vol 36, 1959, p 183

172. J.D. Ingham and N.S. Rapp, *J. Polymer Sci., A2*, Vol 689, 1964, p 4941

173. J.N. Tilley, H.G. Nadeau, H.E. Reymore, P.H. Waszeciak, and A.A.R. Sayigh, *J. Cellular Plast.*, Vol 4, No. 2, 1968, p 56

174. B.C. Levin, NBSIR 85-3267, 1986

175. P. Maya and B.C. Levin, *Fire Mater.*, Vol 11, No. 1, 1987, and references cited therein

176. W.D. Wolley and P.J. Fardell, *Fire Res.*, Vol 1, 1977, p 11

177. W.D. Wolley, *Br. Polym. J.*, Vol 4, 1972, p 27

178. W.D. Wolley, P.J. Fardell, and I.G. Buckland, Fire Research Note No. 1039, Aug 1975

179. J. Chambers, J. Jiricny, and C.B. Reese, *Fire and Mater.*, Vol 5, 1981, p 133

180. D. Norris, *SPI Composites Institute, 45th Ann. Conf.*, 12-15 Feb 1990

181. SMC Auto Alliance, *SPI Composites Institute, 46th Ann. Conf.*, 18-21 Feb 1991

182. N.C. Hilyard, A.I. Kinder, and G.L. Axelby, *Cellular Polym.*, Vol 4, 1985, p 367

183. J.J. Binder, *Proc. 11th Natl. Waste Process Conf.*, 1984, p 458

184. P. Freimann, *Voest-Alpine Report on High Temperature Gasification of Shredder Residue*, 1989

185. J.I. Myers and W.J. Farrissey, *SAE Intl.*, 25-28 Feb 1991, Paper No. 910583

186. B. Fisa, Injection Molding of Thermoplastic Composites, in *Composite Materials Technology*, P.K. Mallick and S. Newman, Ed., Hanser Publishing, Munich, 1990, p 267-320

Appendix

Glossary of Abbreviations

ABS	acrylobutadiene styrene
AC	alternating current
ACCC	automated composites cure control
ACK	Ackerman, Cooper, and Kelly
AE	acoustic emission
AES	Auger electron spectroscopy
AI	artificial intelligence
APC	advanced polymer composites
ASR	automotive shredder residue
ASTM	American Society for Testing and Materials
ATR	attenuated total reflectance
BET	Brunauer-Emmett-Teller
BHE	Budiansky, Hutchinson, and Evans
BMC	bulk molding compound
BMI	bismaleimide
BOM	bond orbital model
BTDA	benzophenone tetracarboxylic dianhydride
CAI	compression after impact
CAT-SCAN	computed tomography
CDS	characteristic damage state
CFC	chlorinated fluorocarbons
CFRP	carbon fiber-reinforced plastic
CIB	controlled interlaminar bonding
CIP	controlled interlaminar phase
CLS	cracked lap shear
CMC	ceramic matrix composite
CNDO	complete-neglect-differential overlap
CPA	coherent potential approximation
CSL	coincident site lattice
CT	computed tomography
CTBN	carboxyl terminated butadiene acrylonitrile
CTE	coefficient of thermal expansion
CVD	chemical vapor deposition
DC	direct current
DCB	double cantilever beam
DDA	dynamic dielectric analysis
DETA	dielectric thermal analysis; diethylene triamine

DMA	dynamic mechanical analysis
DMAC	dimethylacetamide
DMF	dimethylformamide
DMSO	dimethyl sulfoxide
DMT	dimethyl terephthalate
DSC	differential scanning calorimetry
DSZ	diffuse shear zone
DTA	differential thermal analysis
EDT	edge delamination tension
EDX	energy-dispersive X-ray spectroscopy
EG	ethylene glycol
EMI	electromagnetic interference
ENF	end-notched flexure
ER	electrorheological
ESCA	electron spectroscopy for chemical analysis
FED	free-edge delamination
FEM	finite element model
FFT	fast Fourier transform
FGM	fabric geometry model
FPF	first-ply failure
FRF	frequency response function
FTIR	Fourier transform infrared analysis
GC/MS	gas chromatography/mass spectroscopy
GFRP	glass fiber-reinforced plastic
GPC	gel permeation chromatography
GW	green function / coulomb potential
HEC	highest-elongation constituent
HIPS	high-impact polystyrene
HPLC	high-performance liquid chromatography
H-SMC	high-strength sheet molding compound
HSRTM	high-speed resin transfer molding
HTT	heat treatment temperature
HVAC	heating, ventilation, and air conditioning
ILSS	interlaminar shear strength
IPN	interpenetrating network
IR	infrared
LB	Langmuir-Blodgett
LCM	liquid composite molding
LCP	liquid crystalline polymers

LDA	local density-functional approximation
LEC	low-elongation constituent
LRS	laser Raman spectroscopy
LXT	cross-linked thermosetting
MA	molar cross-sectional area
MAD	metal alternative design
MDA	methyl dianiline
MDI	methylene diphenylene diisocyanate
MEC	medium-elongation constituent
MIA	mechanical impedance analysis
MMC	metal matrix composite
MND	modified-neglect-differential overlap
MW	molecular weight
NASA	National Aeronautics and Space Administration
NASP	National Aerospace Plane
NDE	nondestructive evaluation
NG	nucleation and growth
NMP	n-methylpyrolidone
NMR	nuclear magnetic resonance
ON	orthogonal nonwoven
ORNL	Oakridge National Laboratories
PAI	polyamide imide
PAN	polyacrylonitrile
PAS	positron annihilation spectroscopy
PBO	polybenzobisoxazoles
PBT	polybutylene terephthalate; poly(p-phenylene benzobisthiazole)
PC	polycarbonate
PE	polyethylene
PEE	photoelectron emission
PEEK	poly(ether ether ketone)
PEI	poly(ether imide)
PEK	poly(ether ketone)
PES	poly(ether sulfone)
PET	polyethylene terephthalate
PI	polyimide
PMC	polymer matrix composite
PMR	polymerization of monomeric reactants
PPD-T	poly(p-phenylene terephthalate)
PP	polypropylene
PPO	polyphenylene oxide
PPS	polyphenylene sulfide
PU	polyurethane
PVDF	polyvinylidene fluoride
PZT	lead zirconate-titanate
RDC	ringdown count
RF	radio frequency
RHEED	reflection high-energy electron diffraction

RIM	reaction injection molding
RKU	Ross-Kerwin-Ungar
RTM	resin transfer molding
R-TP	reinforced thermoplastics
SAE	Society of Automotive Engineers
SAMPE	Society of Advanced Manufacturing and Processing Engineers
SB	shear band
SCB	split cantilever beam
SD	spinodal decomposition
SEA	surface energy analysis
SEM	scanning electron spectroscopy
SHPB	split Hopkinson pressure bar
SIMS	secondary ion mass spectroscopy
SMA	shape memory alloy
SMC	sheet molding compound
SMC-R	sheet molding compound reinforced
SME	Society of Manufacturing Engineers
SPD	surface potential difference
SPE	Society of Plastics Engineers
SPI	Society of the Plastics Industry
SRF	strength reduction factor
SRIM	structural reaction injection molding
SRP	surface remission photometry
STM	scanning tunneling microscopy
SWF	stress wave factor
TBA	torsional braid analysis
TBM	tight-binding model
TDI	toluene diisocyanate
TEM	transmission electron microscopy
TGA	thermal gravimetric analysis
TMA	thermal mechanical analysis
TPA	terephthalic acid
TPE	thermoplastic elastomer
TPI	thermoplastic polyimide
TTT	time-temperature-transformation
UD	unidirectional
UHMWPE	ultrahigh molecular weight polyethylene
UV	ultraviolet
UV-VIS	ultraviolet-visual
VMD	one type of short carbon fibers by Amoco
VME	one type of short carbon fibers by Amoco
WBL	weak boundary layer
WF	woven fabric-based
XPS	X-ray photoelectron spectroscopy

Glossary of Selected Symbols

α coefficient of shear or thermal expansion; dimensionless shape parameter; shear-lag parameter

α_c degree of cure

β_0 Weibull scale parameter

γ surface tension

Γ fracture surface energy

δ logarithmic decrement

$\Delta\delta$ cyclic stress amplitude

Δn birefringence

Δv van der Waals' volume

ΔW total energy dissipation

ε strain; dimensionless damping ratio

η loss factor; loss tangent

θ contact angle

μ coefficient of friction

ν Poisson's ratio

π_{ijkl} piezo-optical coefficients

ρ density

σ stress; tensile strength

τ shear strength; stress; Thomson coefficient

ϕ specific damping capacity

ω_n resonant frequency

A   stiffness constant

$[A]$ extensional stiffness matrix

A_{ij} extensional stiffness coefficients

B bulk modulus; stiffness constant

$[B]$ coupling stiffness matrix

B_{ij} relative dielectric impermeability tensor

C coefficient of proportionality; compliance

$[C]$ stiffness matrix

D damage parameter; stiffness constant

$[D]$ bending stiffness matrix

E energy; modulus of elasticity

f_n frequency of flexural vibration

F strength coefficient

G shear modulus; strain energy release rate; toughness

H_R heat of reaction

I moment of inertia

I_x iteration factor

K curvature; stress intensity factor

L lower bound

m coefficient of mutual influence

M bending or twisting moment

N Avogadro's number; normal force; shear force; total number

N_f cycles to failure

P load

P_i polarization

Q quality factor

Q_{ij} reduced stiffness coefficients

r surface tension

R radical; radius; strength ratio

S stiffness; strength

$S\text{-}N$ stress vs. number of cycles

T_g glass transition temperature

T_m melting temperature

T_R vibration reverberation time

U energy; Rao function; upper bound

U_i stiffness

V volume; volume fraction

Y_g group additivity

Z point impedance

Index

Index

F